Feature Papers in Mathematical and Computational Applications 2023

Feature Papers in Mathematical and Computational Applications 2023

Collection Editors

Gianluigi Rozza
Oliver Schütze
Nicholas Fantuzzi

Basel • Beijing • Wuhan • Barcelona • Belgrade • Novi Sad • Cluj • Manchester

Collection Editors

Gianluigi Rozza
International School for
Advanced Studies
Trieste
Italy

Oliver Schütze
Cinvestav
Mexico City
Mexico

Nicholas Fantuzzi
University of Bologna
Bologna
Italy

Editorial Office
MDPI AG
Grosspeteranlage 5
4052 Basel, Switzerland

This is a reprint of the Topical Collection, published open access by the journal *Mathematical and Computational Applications* (ISSN 2297-8747), freely accessible at: https://www.mdpi.com/journal/mca/topical_collections/ZGRQX7CNGE.

For citation purposes, cite each article independently as indicated on the article page online and as indicated below:

Lastname, A.A.; Lastname, B.B. Article Title. *Journal Name* **Year**, *Volume Number*, Page Range.

ISBN 978-3-7258-2631-5 (Hbk)
ISBN 978-3-7258-2632-2 (PDF)
https://doi.org/10.3390/books978-3-7258-2632-2

Contents

About the Editors

Gianluigi Rozza

Gianluigi Rozza is a Full Professor of Numerical Analysis and Scientific Computing at SISSA MathLab—International School for Advanced Studies, Trieste, Italy. He received his Master of Science in Aerospace Engineering (2002) at Politecnico di Milano and his Ph.D. in Applied Mathematics (2005) at Ecole Polytechnique Fédérale de Lausanne, Switzerland. He was a Research Assistant (2002–2006) and Researcher and Lecturer (2008–2012) at École Polytechnique Fédérale de Lausanne; a Post-Doctoral Associate Researcher (2006–2008) at Massachusetts Institute of Technology, Boston, MA, USA; and a Researcher (2012–2014) and Associate Tenured Professor (2014–2017) at the International School for Advanced Studies, Trieste, Italy. His research interests include Numerical Analysis; Numerical Simulation; Scientific Computing; Reduced Order Modeling and Methods; Efficient Reduced-Basis Methods for Parametrized PDEs and Posteriori Error Estimation; Computational Fluid Dynamics with applications in Aero-Naval-Mechanical Engineering and Environmental Fluid Dynamics; Fluid–Structure Interaction Problems; Parametrized Navier–Stokes Equations for Bifurcations and Stability of Flow; Optimal Control and Flow Control based on PDEs. He has authored 180 scientific papers, receiving more than 4500 citations (H-index 32). He has received the Bill Morton CFD Prize (2004) from the Institute of Computational Fluid Dynamics, University of Oxford (UK), the ECCOMAS Ph.D Award (2005) from the European Community on Computational Methods in Applied Sciences, the Springer Computational Science and Engineering Prize (2009), the ECCOMAS Jacques Louis Lions Award in Computational Mathematics (2014), the ERC consolidator grant 'Advanced Reduced Order Methods with Applications in Computational Fluid Dynamics' (AROMA-CFD, 2016–2021), and the ERC Proof of Concept Grant 'Advanced Reduced Groupware Online Simulation'(ARGOS, 2022).

Oliver Schütze

Dr. Oliver Schütze is a Full Professor at the Cinvestav in Mexico City, Mexico. His main re-search interests are numerical and evolutionary optimization. He is the co-author of more than 150 publications, including two monographs, five school textbooks, and ten edited books. Two of his papers have received the IEEE Transactions on Evolutionary Computation Outstanding Paper Award (in 2010 and 2012). He is the founder of the Numerical and Evolutionary Optimization (NEO) workshop series. He is Editor-in-Chief of the journal *Mathematical and Computational Applications* and is a member of the Editorial Board of the journals *Engineering Optimization*, *Computational Optimization and Applications*, *IEEE Transactions on Evolutionary Computation*, *Research in Control and Optimization*, and *Applied Soft Computing*. He is a member of the Mexican Academy of Sciences and the National System of Researchers (SNI III).

Nicholas Fantuzzi

Nicholas Fantuzzi has been an Associate Professor at the University of Bologna since 2021, where he teaches Advanced Structural Mechanics and Modeling of Offshore Structures at the Ravenna Campus. Here, he performs research on numerical modeling of advanced composites with innovative numerical methods. He has co-organized 14 international conferences within the composite materials field and has been invited to deliver seven keynote lectures at international events. He has received three international awards and has been a visiting professor in Croatia, China, and Hong Kong. He is a reviewer for more than 110 international journals and the author of more than 130 publications in international journals, nine books, and four book chapters, with more than 100 abstracts at national and international conferences.

Mathematical and Computational Applications

MDPI

Editorial

Feature Paper Collection of *Mathematical and Computational Applications*—2023

Gianluigi Rozza [1], Oliver Schütze [2,*] and Nicholas Fantuzzi [3]

[1] SISSA MathLab, International School for Advanced Studies, Office A-435, Via Bonomea 265, 34136 Trieste, Italy

[2] Computer Science Department, Cinvestav-IPN, Mexico City 07360, Mexico

[3] Department of Civil, Chemical, Environmental, and Materials Engineering, University of Bologna, Viale del Risorgimento 2, 40136 Bologna, Italy

* Correspondence: schuetze@cs.cinvestav.mx

Citation: Rozza, G.; Schütze, O.; Fantuzzi, N. Feature Paper Collection of *Mathematical and Computational Applications*—2023. *Math. Comput. Appl.* **2024**, *29*, 99. https://doi.org/10.3390/mca29060099

Received: 24 October 2024

Accepted: 24 October 2024

Published: 1 November 2024

This Special Issue comprises the second collection of papers submitted by both the Editorial Board Members (EBMs) of the journal *Mathematical and Computational Applications* (*MCA*) and the outstanding scholars working in the core research fields of *MCA*. Therefore, this collection typifies the most insightful and influential original articles that discuss key topics in these fields. More precisely, this issue contains 13 research articles published in *MCA* between May and December 2023. All papers are briefly outlined below, organized chronologically by their publication times.

In [1], Romo-González and Mezura-Montes use deep learning techniques to discriminate between healthy individuals and patients with breast cancer, based on the banding patterns obtained from the Western Blot strip images of the autoantibody response to antigens of the T47D tumor line. The authors propose that neuroevolving convolutional neural networks (CNNs) can be used to find the optimal architecture to achieve competitive ranking, taking Western Blot images as the input. The CNN obtained reached 90.67% accuracy, 90.71% recall, 95.34% specificity, and 90.69% precision in classifying three different classes (healthy, benign breast pathology, and breast cancer).

Bacterial Vaginosis (BV) is a common disease and recurring public health problem for which all possible combinations of the pathogens of a possible case of infection are not known, complicating diagnosis at the onset of the disease. Salvador-González et al. contribute to this line of research in [2]. The experimental results obtained by the authors allowed a reduced subset of biologically meaningful association rules to be selected for the numerical treatment of the considered objective function.

In [3], Vázquez-Santiago et al. propose an open-set recognition (OSR) strategy with an extension for new class discovery aimed at vehicle make-and- model recognition (VMMR). The results show that the presented strategy can effectively address this problem as an OSR problem, and furthermore, it is able to simultaneously recognize the new classes hidden within the rejected objects. The proposed VMMR method is a benchmark for future domain-specific OSR.

In [4], Eivazi et al. provide a clear description of the algorithmic FE^2 structure together with a particular integration of deep neural networks. This allows for a suitable training strategy, where particular knowledge of the material behavior is considered to reduce the required amount of training data. The resulting method yields a significant speed-up of the FE^2 computations, and an efficient implementation of the trained neural network in a finite element code is provided. Moreover, the deep neural network surrogate model is able to overcome the load-step size limitations of the representative volume element (RVE) computations in step-size controlled computations.

In [5], Kakuli et al. use Lie symmetry to analyze the Hunter–Saxton equation, an equation relevant to the theoretical analysis of nematic liquid crystals. The proposed method has two main advantages over the classical double-reduction method: firstly, it is more efficient as it can reduce the number of variables and order of the equation in a single step. Secondly, by incorporating conservation laws, physically meaningful solutions that satisfy important physical constraints can be obtained.

Páez-Rueda et al.[6] study the approximation of one-dimensional smooth closed-form functions with compact support using a mixed Fourier series. Their method improves the signal processing performance in a wide range of scenarios. Moreover, this paper provides comprehensive examples of one-dimensional problems to showcase the advantages of this approach.

In [7], González Flores and Barrera Sánchez review some grid quality metrics and define some new quality measures for quadrilateral elements. Furthermore, they define new discrete functionals, which are implemented as objective functions in an optimization-based method for quadrilateral grid generation and improvement. These functionals are linearly combined with a discrete functional whose domain has an infinite barrier at the boundary of the set of unfolded grids to preserve convex grid cells in each step of the optimization process.

Middle East respiratory syndrome coronavirus (MERS-CoV) is a highly infectious respiratory illness that poses a significant threat to public health. In [8], Fatima et al. develop a precise mathematical model to capture the transmission dynamics of MERS-CoV. Stability theory is employed to analyze the local and global properties of the model, providing insights into the system's equilibrium states and their stability. Sensitivity analysis is conducted to identify the most critical parameter affecting the transmission dynamics. The model can serve as a valuable tool for public health authorities when designing effective control and prevention strategies, ultimately reducing the burden of MERS-CoV on global health.

In [9], Deb and Ehrgott analyze and outline the properties of generalized dominance structures for multi-objective optimization which help provide insights into the resulting optimal solutions. The theoretical and deductive results of this study can be utilized to create more meaningful dominance structures for practical problems, understand and identify resulting optimal solutions, and help develop better test problems and algorithms for multi-objective optimization.

In [10], Wang et al. present an innovative cascade predictor to forecast the state of recurrent neural networks (RNNs) with delayed output. The new predictor is more useful than the conventional single observer in predicting neural network states when the output delay is arbitrarily large but known. In contrast to examining the stability of error systems solely employing the Lyapunov–Krasovskii functional (LKF), several new global asymptotic stability standards are obtained by combining the application of the Linear Parameter Varying (LPV) approach, LKF, and convex principle. The latter is verified by several numerical simulations.

In [11], Sánchez-García et al. analyze the determination of interplanetary trajectories from Earth to Mars to evaluate the cost of the required impulse magnitudes for an areostationary orbiter mission design. The results show that, for the dates of the minimum-energy Earth–Mars transfer trajectory, a low value for the maneuvers to achieve an areostationary orbit is obtained for an arrival hyperbola with the minimum possible inclination, in addition to a capture into an elliptical trajectory with a low periapsis radius and an apoapsis in the stationary orbit.

The preventive measures taken to curb the spread of COVID-19 have emphasized the importance of wearing face masks to prevent potential infection with serious diseases during daily activities or for medical professionals working in hospitals. In [12], Melin et al. investigate various existing methods employing artificial intelligence and deep learning to detect whether individuals are wearing face masks. The results demonstrate that the bat algorithm obtained better results than the other metaheuristics analyzed in this study.

Finally, in [13], Annunziato and Borzí present a method for the analysis of super-resolution microscopy images. The method is based on the analysis of stochastic trajectories of particles moving on the membrane of a cell with the assumption that this motion is determined by the properties of this membrane. The results demonstrate the ability of the proposed method to reconstruct the potential of a cell membrane by using synthetic data similar those captured by super-resolution microscopy of luminescent activated proteins.

Conflicts of Interest: The authors declare no conflicts of interest.

References

1. Llaguno-Roque, J.-L.; Barrientos-Martínez, R.-E.; Acosta-Mesa, H.-G.; Romo-González, T.; Mezura-Montes, E. Neuroevolution of Convolutional Neural Networks for Breast Cancer Diagnosis Using Western Blot Strips. *Math. Comput. Appl.* **2023**, *28*, 72. [CrossRef]
2. Salvador-González, M.C.; Canul-Reich, J.; Rivera-López, R.; Mezura-Montes, E.; de la Cruz-Hernandez, E. Evolutionary Selection of a Set of Association Rules Considering Biological Constraints Describing the Prevalent Elements in Bacterial Vaginosis. *Math. Comput. Appl.* **2023**, *28*, 75. [CrossRef]
3. Vázquez-Santiago, D.-I.; Acosta-Mesa, H.-G.; Mezura-Montes, E. Vehicle Make and Model Recognition as an Open-Set Recognition Problem and New Class Discovery. *Math. Comput. Appl.* **2023**, *28*, 80. [CrossRef]
4. Eivazi, H.; Tröger, J.-A.; Wittek, S.; Hartmann, S.; Rausch, A. FE^2 Computations with Deep Neural Networks: Algorithmic Structure, Data Generation, and Implementation. *Math. Comput. Appl.* **2023**, *28*, 91. [CrossRef]
5. Kakuli, M.C.; Sinkala, W.; Masemola, P. Conservation Laws and Symmetry Reductions of the Hunter–Saxton Equation via the Double Reduction Method. *Math. Comput. Appl.* **2023**, *28*, 92. [CrossRef]
6. Páez-Rueda, C.-I.; Fajardo, A.; Pérez, M.; Yamhure, G.; Perilla, G. Exploring the Potential of Mixed Fourier Series in Signal Processing Applications Using One-Dimensional Smooth Closed-Form Functions with Compact Support: A Comprehensive Tutorial. *Math. Comput. Appl.* **2023**, *28*, 93. [CrossRef]
7. González Flores, G.F.; Barrera Sánchez, P. New Quality Measures for Quadrilaterals and New Discrete Functionals for Grid Generation. *Math. Comput. Appl.* **2023**, *28*, 95. [CrossRef]
8. Fatima, B.; Yavuz, M.; ur Rahman, M.; Althobaiti, A.; Althobaiti, A. Predictive Modeling and Control Strategies for the Transmission of Middle East Respiratory Syndrome Coronavirus. *Math. Comput. Appl.* **2023**, *28*, 98. [CrossRef]
9. Deb, K.; Ehrgott, M. On Generalized Dominance Structures for Multi-Objective Optimization. *Math. Comput. Appl.* **2023**, *28*, 100. [CrossRef]
10. Wang, W.; Chen, J.; Huang, Z. Observer-Based State Estimation for Recurrent Neural Networks: An Output-Predicting and LPV-Based Approach. *Math. Comput. Appl.* **2023**, *28*, 104. [CrossRef]
11. Sánchez-García, M.M.; Barderas, G.; Romero, P. Modelization of Low-Cost Maneuvers for an Areostationary Preliminary Mission Design. *Math. Comput. Appl.* **2023**, *28*, 105. [CrossRef]
12. Melin, P.; Sánchez, D.; Pulido, M.; Castillo, O. Comparative Study of Metaheuristic Optimization of Convolutional Neural Networks Applied to Face Mask Classification. *Math. Comput. Appl.* **2023**, *28*, 107. [CrossRef]
13. Annunziato, M.; Borzì, A. Fokker–Planck Analysis of Superresolution Microscopy Images. *Math. Comput. Appl.* **2023**, *28*, 113. [CrossRef]

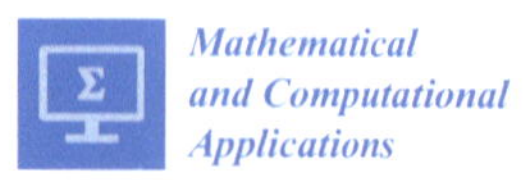
Mathematical and Computational Applications

Article

Neuroevolution of Convolutional Neural Networks for Breast Cancer Diagnosis Using Western Blot Strips

José-Luis Llaguno-Roque [1], Rocio-Erandi Barrientos-Martínez [2], Héctor-Gabriel Acosta-Mesa [2,*], Tania Romo-González [1] and Efrén Mezura-Montes [2]

[1] Instituto de Investigaciones Biológicas, Universidad Veracruzana, Dr. Luis Castelazo Ayala S/N, Industrial Animas, Xalapa C.P. 91190, Veracruz, Mexico; lllaguno@uv.mx (J.-L.L.-R.); tromogonzalez@uv.mx (T.R.-G.)

[2] Instituto de Investigaciones en Inteligencia Artificial, Universidad Veracruzana, Campus Sur, Calle Paseo Lote II, Sección Segunda N° 112, Nuevo Xalapa, Xalapa C.P. 91097, Veracruz, Mexico

[*] Correspondence: heacosta@uv.mx

Abstract: Breast cancer has become a global health problem, ranking first in incidences and fifth in mortality in women around the world. In Mexico, the first cause of death in women is breast cancer. This work uses deep learning techniques to discriminate between healthy and breast cancer patients based on the banding patterns obtained from the Western Blot strip images of the autoantibody response to antigens of the T47D tumor line. The reaction of antibodies to tumor antigens occurs early in the process of tumorigenesis, years before clinical symptoms. One of the main challenges in deep learning is the design of the architecture of the convolutional neural network. Neuroevolution has been used to support this and has produced highly competitive results. It is proposed that neuroevolve convolutional neural networks (CNN) find an optimal architecture to achieve competitive ranking, taking Western Blot images as input. The CNN obtained reached 90.67% accuracy, 90.71% recall, 95.34% specificity, and 90.69% precision in classifying three different classes (healthy, benign breast pathology, and breast cancer).

Keywords: Western blot; breast cancer; neuroevolution; convolutional neural networks

Citation: Llaguno-Roque, J.-L.; Barrientos-Martínez, R.-E.; Acosta-Mesa, H.-G.; Romo-González, T.; Mezura-Montes, E. Neuroevolution of Convolutional Neural Networks for Breast Cancer Diagnosis Using Western Blot Strips. *Math. Comput. Appl.* **2023**, *28*, 72. https://doi.org/10.3390/mca28030072

Academic Editor: Suchuan Dong

Received: 16 February 2023
Revised: 9 May 2023
Accepted: 22 May 2023
Published: 24 May 2023

1. Introduction

Breast cancer has become a global health problem as it ranks first in the world in terms of incidence and fifth in terms of cancer-related mortality [1]. In Mexico, breast cancer is the first cause of death in women between 30 and 50 years of age, and since 2006, it has replaced cervical cancer as a public health concern, and it is a major challenge for the health system [2].

Breast cancer is identified by an accelerated and uncontrolled proliferation of mammary epithelial cells. These are healthy cells that have an increased reproductive capacity; they multiply and increase until they form tumors that, depending on their characteristics, can be malignant or benign [3].

There are several complementary approaches to the diagnosis of breast cancer. The tests traditionally used for diagnosis are breast examination, ultrasound, mammography, and biopsy. During a breast exam, the doctor checks the lymph nodes in both breasts and armpits for lumps or other abnormalities. This test identifies lumps of at least 3 mm, and detection of this size has been clinically shown to be beneficial for patient survival. The diagnostic percentage of this test is 40% to 69% [4].

Mammography is a diagnostic test, where an image is obtained and then analyzed and interpreted by a specialist. It is an expensive, painful procedure, generally performed on patients over 40 years of age. The percentage of diagnosis is from 63% to 87%, depending on the age of the patient, as well as the density of the mammary tissue. Ultrasound or sonography is a diagnostic procedure that uses sound waves to detect cysts or malformations in the breasts. It is used complementarily to mammography and allows guiding the taking of the biopsy. The diagnostic percentage is 68% to 98% [5]. A biopsy is a diagnostic test that determines the presence or absence of cancer cells in a patient's breast tissue. If the type of biopsy is surgical, it can be a painful and invasive procedure [6].

The aforementioned diagnostic tests could be expensive, invasive, subjective, and painful. In addition, they can be ineffective in the early detection of cancer since these tests identify the disease when it is present in the patient and, most of the time, in an advanced state. The detection of breast cancer in Mexico usually occurs in late stages because Mexican women feel embarrassed when being examined by doctors, which decreases the possibility of providing an effective and successful treatment. In addition, in México, not having sufficient infrastructure to perform the procedure and not having enough trained and certified radiologists to interpret the tests [7,8] is a limitation, which is why the number of tests recommended by international organizations (19.9 mammograms per million inhabitants) is not met. Thus, in Mexico, life expectancy is very low in relation to developed countries [4]. Therefore, it is necessary to have tests that diagnose breast cancer early before it manifests as tumors in patients.

Since breast cancer is a heterogeneous disease in which tumors express a variety of aberrant proteins (antigens), which creates an immune response by the production of autoantibodies against such tumor-associated antigens, it is possible to use this antitumor reaction as an oncogenic signal before tumor formation manifests itself in the body. Therefore, methods are being developed that identify autoantibodies that recognize tumor proteins that are present up to four years before the disease is detected using the traditional test [9]. Desmetz et al. [10], by evaluating autoantibody responses to some tumor-associated antigens, have been able to accurately distinguish healthy patients from those with early stage breast cancer, particularly carcinoma in situ. Thus, developing these methods could help in the early detection of breast cancer, supporting mammographic screening, especially in women under 50 years of age. However, it is necessary to probe its efficacy since this kind of test changes with the genetical and phenotypical background of patients.

To that respect, Romo et al. [11] developed a method specific to Mexican women, which confirms the presence of autoantibodies reacting to tumor cells in the T47D cell line (ductal carcinoma of the breast), which are capable of discriminating between women with and without breast disease. This was achieved by analyzing the bands expressed in the one-dimensional Western Blot images of the autoantibody response to antigens of the T47D tumor line. Although the results obtained are promising, the analysis of the images is complex, subjective, and slow since it takes a month to create a binary base (1 present and 0 absent proteins), from which the data are obtained for discrimination between healthy patients and those with breast disease. On the other hand, an expert, with the help of commercial software, is required to align the strip bands for each patient, but the identification and final position of the bands depend exclusively and subjectively on the expert. Consequently, more precise and automated tools are needed to identify these banding patterns.

In recent years, artificial intelligence (AI) has used machine learning and computer vision techniques to support processes such as the prevention and diagnosis of breast cancer. Contributions have been made, for example, in image processing, to identify patterns that make it possible to distinguish women with breast disease from those who do not have the disease [12]. The images usually used to diagnose breast cancer are obtained from mammary tissue by means of mammography, ultrasound, thermography, histopathology (Whole Slide Image—WSI) [13,14], or they are images obtained from the reaction of the immune system from a blood sample and processed with the Western Blot technique (proteomic images) [15].

In addition, afterward, Sánchez-Silva et al. [15] proposed a semi-automated system to avoid subjectivity and shorten image analysis time in Western blot images by analyzing protein bands from the classification of patterns represented as time series [11]. These time series were obtained from the change in tone in the pixels of the bands. Because the time series are of different lengths, they were manually standardized to a predefined length using a geometric scaling transformation. The K-Nearest Neighbor (KNN) algorithm was used to classify the time series, using the Euclidean, Mahalanobis, and correlation similarity distances, achieving a classification percentage of 65.40% with three classes (healthy, benign breast pathology, and breast cancer), and an 86.06% classification percentage with two classes (healthy and breast cancer). The classification percentages achieved are similar to those of the expert of reference [11]. However, the method is considered semi-automatic since, to obtain the time series, an area is subjectively selected in each strip, which causes the variation in the lengths of the time series and needs to be standardized. To improve the work previously described in [16], it was proposed to analyze the bands of the Western Blot images of antibodies that are reactive to antigens (tumor line T47D—ductal carcinoma) using convolutional neural networks (CNN), and dispense by obtaining the time series of a subjectively chosen area to perform the classification. A classification percentage of 68.24% for three classes (healthy, benign breast pathology, and breast cancer) is obtained. The classification percentage was statistically equivalent to that seen in [15], obtaining for two classes (healthy and breast cancer) 86.00%. It is important to remark that the architecture of the CNN used was handcrafted, so the architecture used does not ensure that the best performance, in terms of accuracy, will be reached.

In the work developed in [17], they propose to automate the detection of breast cancer, analyzing the regions of invasive ductal carcinoma (IDC) tissues in 162 whole slide images (WSI), from which 277,524 patches were obtained in digital format, RGB with a size of 50×50 pixels. Patches were labeled with the value of 1 for IDC positive and 0 for IDC negative. Three CNN's architectures obtained through experimentation were used, achieving a classification accuracy of 87%. In [18], detecting breast cancer using thermographic images is proposed. Thermographic images capture the heat map of the breasts and their surroundings. The analysis of this type of images is based on the assumption that in a breast cancer process, blood vessels are formed and inflamed, producing an increase in temperature in that area. They used 3895 thermographic images of breasts in JPEG format with a dimension of 640×480 pixels, obtaining the information to generate 140 patients, of which 98 were healthy patients and 42 were cancer patients. For the classification, a CNN, whose parameters were optimized by means of the Bayes optimization algorithm, was used, obtaining an accuracy of 98.95%. In the work presented in [19], the objective was to differentiate malignant from benign breast cancer tumors, classifying histopathology images using convolutional neural networks. They use the BreakHis database, formed with histopathological images of mammary tissues with breast cancer from 82 patients. This database consists of 7909 images of microscopic biopsies, of which 2480 are benign and 5229 are malignant, each image has four magnification levels ($40\times$, $100\times$, $200\times$, and $400\times$). The CNN architecture was obtained from the importation of previously trained layers from CNN AlexNet [20], achieving a classification accuracy of 89.66%. In [21], it was proposed to predict HER2 expression (a protein that is used as a marker of breast cancer) by analyzing ultrasound images of preoperative breast cancer patients, using a deep learning model based on DenseNet. The model was trained with 108 patients and validated with 36 patients, obtaining an accuracy of 80.56%. In [22], a framework for the classification of breast cancer from mammographic images is proposed. A pre-trained network (EfficientNet-b0) is used to classify two databases of mammography images. The first database is CBIS-DDSM, achieving a classification accuracy of 95.4%, and the second database is INbreast, achieving a classification accuracy of 99.7%.

Although CNNs are very competitive, their main disadvantage is the necessity to design their components (architecture), which in most cases is performed manually and by trial and error, consuming a lot of time in finding a suitable architecture that adapts to the requirements. Given that most network architectures have many convolution layers, filters of different sizes, and some hyperparameters at the moment of being executed, they demand excessive computational costs, both in time and in memory [23].

Several solutions have been proposed to deal with this matter; one of the most used in recent years is neuroevolution, a technique inspired by the biological process of the evolution of the human brain, through the use of evolutionary computing, which has made good progress toward optimizing the design of CNN architectures [24].

One of the most important parts of neuroevolution for the design of CNNs is neural coding, which corresponds to the computational representation of an artificial neural network. A suitable coding will allow for the creation of a design with a competitive performance and more efficient and less complex structures.

In this work, the DeepGA neuroevolution algorithm proposed by Vargas-Hakim et al. [25] is used as a framework for neuroevolution. It is based on the fundamentals of genetic algorithms, exploitation (by crossing) and exploration (by mutation), and has three fundamental characteristics: (1) A hybrid coding, which combines blockchains and binary codings; (2) The use of evolutionary operators to handle this type of encoding; (3) A linear aggregation fitness function to evaluate individuals based on their classification accuracy and the number of parameters. The goal of this work, which uses neuroevolution, is to automatically obtain a convolutional neural network architecture suitable for our problem and to classify the bands of the Western Blot images of antibodies reactive to antigens (tumor line T47D—carcinoma ductal). According to studies [26–28], the reaction of antibodies to tumor antigens occurs early in the process of tumorigenesis, years before clinical symptoms appear, contrary to mammographic images, WSI (Whole-Slide Images/histopathology), and ultrasound, that detect a tumor process that already exists. On the other hand, the CNN architecture obtained by neuroevolution prevents either configuring a CNN by hand or using a trained CNN, in addition to improving the classification obtained, as described in [16].

2. Materials and Methods

The pipeline process proposed in this work is described in Figure 1.

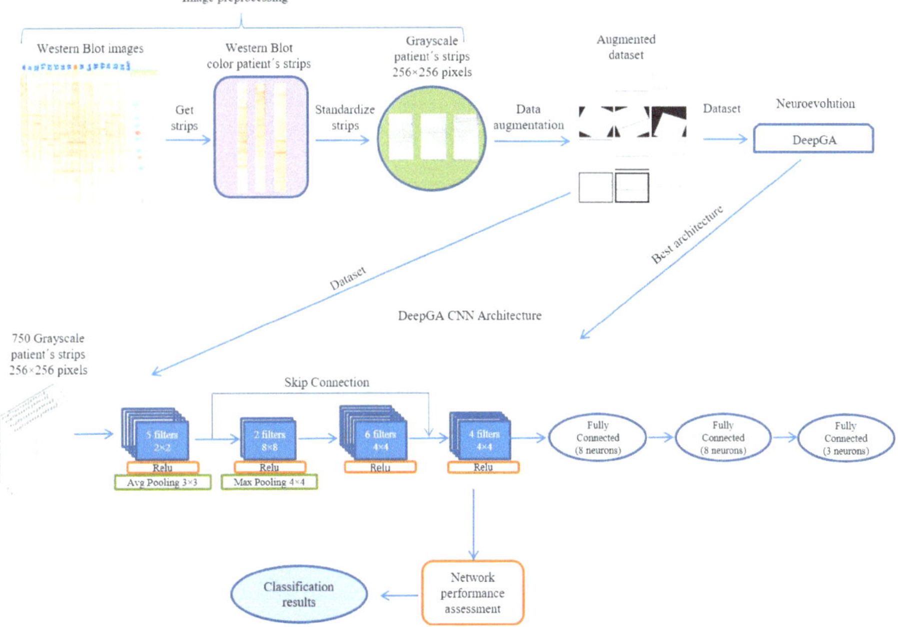

Figure 1. Proposed pipeline process.

2.1. Western Blot Strips Database

For this study, a database containing 150 images corresponding to nitrocellulose membrane strips with the expression of bands obtained with the Western Blot of the reaction of antibodies to specific protein antigens (T47D) has been used. Image acquisition was performed following a protocol in a controlled environment, in addition to using commercial editing software for image enhancement, as described in [11]. A total of 50 of the images correspond to patients with breast cancer, 50 to patients with benign pathology, and 50 to healthy patients. These images have been provided by the Biology and Integral Health area of the Biological Research Institute of the Universidad Veracruzana, following ethical standards and the acquisition of informed consent from the patients who participated. The protocol was reviewed and approved by the Research Ethics Committee of the General Hospital of Mexico "Dr. Eduardo Liceaga" (DI/12/11/03/064). The study conforms to the Code of Ethics of the World Medical Association (Declaration of Helsinki), printed in the *British Medical Journal* (18 July 1964).

2.2. Image Preprocessing

The color images provided by the area of Biology and Integral Health of the Institute of Biological Research of the Universidad Veracruzana are composed of an average of 18 strips in which the bands of patients of the antibody reaction to specific protein antigens are expressed (T47D). In total, 50 strips were obtained from healthy patients, 50 strips from patients with benign breast disease, and 50 strips from patients with breast cancer.

Sánchez-Silva et al. [15] carried out experiments with color and grayscale images and determined that color was not relevant, so they chose to work with grayscale images. Due to the above and for the sake of simplicity in image processing, the color images were converted to grayscale in this study. On the other hand, based on previous experiments carried out with the CNNs, it has been established that the ideal transformation for the size of the strips in this work is 256×256 pixels.

2.3. Data Augmentation

CNNs require a large amount of data for feature extraction, as well as for training and testing, which are used for network architecture evaluation. In the medical area, it is difficult to have many images. To solve this problem, data augmentation is used, which consists of applying affine transformations (such as rotation, scaling, and/or translation) to the images of the original database to generate additional images and increase the diversity of the training set, since CNNs can classify objects in different orientations. It is recommended that the applied transformations are carried out on small scales so as not to alter the nature of the images.

For this study, 200 additional images have been generated for each of the classes, with which a database containing 750 images has been obtained. The affine transformations that were used randomly and with a range of degrees, movement or size, are: (a) Rotation, with a degree range of 10 to 30; (b) Translation with a movement range of 0.1 to 0.3; (c) Scaling with a size range of 0.5 to 1; (d) Gaussian blur, with a kernel size of 7.

2.4. CNN Neuroevolution

Neuroevolution is an approach that harnesses evolutionary algorithms to optimize the artificial neural networks, inspired by the fact that natural brains are the products of an evolutionary process [29].

To find a CNN architecture that achieves a balance between complexity and efficiency for the classification of Western Blot strips, the DeepGA neuroevolution algorithm [25] has been used. The first step was adjust the parameters of the algorithm, which are shown in Table 1. DeepGA is formed by a neuroevolutionary framework based on genetic algorithms. Their goal is to obtain competitive CNNs through flexible hybrid coding combined with binary and blockchain coding. The parameters required by DeepGA are the population size (N = 20), the number of generations (T = 50), the crossover rate (CXPB = 0.7), the mutation rate (CXPB = 0.3), and the size of the tournament (S = 5); these values were manually adjusted experimentally. The adjustment of mutation rate (MUPB) and crossover rate (CXPB) was performed until sufficient diversity was obtained throughout the scan. Both the size of the population and the number of generations were established by virtue of time and available computational resources, for which it was not necessary to use automatic methods for parameter adjustment. The best architecture obtained by DeepGA is shown in Figure 2.

Table 1. List of parameters used in DeepGA.

Parameters	Values
Population Size	20
Number of Generation	50
Crossover Rate	0.7
Mutation Rate	0.5
Tournament Size	4

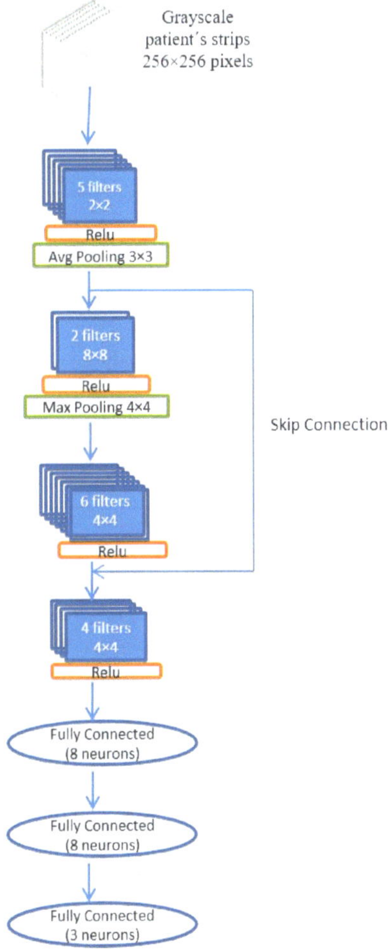

Figure 2. Deep-CNN architecture.

2.5. Evaluation of the Convolutional Neural Network

From the best architecture obtained in DeepGA, we proceeded to evaluate the convolutional neural network. For this, a set of 750 Western Blot strips was used, and through the hold-out technique, 70% of the data were used to train the network and the remaining 30% to test it. From the results obtained, the accuracy, recall, specificity, and precision of the network for the classification of the Western Blot strips were calculated.

The accuracy is calculated from the total number of predictions that the algorithm classified correctly divided by the total number in the data set (Equation (1)).

$$\text{Accuracy} = (\text{correctly classified images})/(\text{total images}) \tag{1}$$

The recall is the number of elements correctly identified as positives out of the total number of true positives (Equation (2)).

$$\text{Recall} = TP/(TP + FN) \tag{2}$$

Specificity is the number of items correctly identified as negative out of the total number of negatives (Equation (3)).

$$\text{Specificity} = TN/(TN + FP) \tag{3}$$

Precision is the number of elements correctly identified as positive out of a total of elements identified as positive (Equation (4)).

$$\text{Precision} = TP/(TP + FP) \tag{4}$$

2.6. Comparison and Statistical Analysis

The result of Western Blot strip classification accuracy obtained in this work was compared by statistical test with the classification accuracy obtained in [15,16], with the aim of obtaining statistical significance between them.

The data were analyzed using one-way analysis of variance (ANOVA) for independent groups, with treatment as the factor, followed by the Tukey post hoc test for multiple mean comparisons. The results are expressed as mean + standard error of the mean, and the significance level was set at $p < 0.05$. The assumptions of normality and homogeneity were verified. The data were analyzed using the MINITAB17 software program.

3. Experimentation and Results

To obtain the classification accuracy of the Western Blot strips with the support of neuroevolution and convolutional neural networks, the following process was carried out:

1. The CNN obtained through the DeepGA neuroevolution algorithm (CNN-DeepGA) was trained, taking as input data the database of 750 Western blot strips; 250 belong to the class of healthy patients, 250 to the class of patients with benign pathology, and 250 to the class of patients with cancer. The parameters with which CNN-DeepGA was executed have been shown in Table 1;
2. Training CNN-DeepGA consisted of only 10 epochs (as suggested by [30]); Adam's optimizer was used with a learning rate of 1×10^{-4}. For training, we used 70% of the data set (525 images out of 750 total), while accuracy/error was calculated using 30% (225 images out of 750 total) of the remaining set for testing;
3. To evaluate the performance of CNN-DeepGA, 10 executions were carried out, obtaining the average and the standard deviation of the accuracy in each of the executions, as shown in Table 2;
4. To handle biases, such as overfitting and underfitting, a data augmentation was performed by increasing the original size of examples for each class five times, going from 50 to 250 images in each class. On the other hand, the images were obtained in a controlled environment and an editing software program was used to improve

them [11]. The hold-out technique was used for the evaluation of the model; 70% of the data were used for training the network and the remaining 30% for testing it;

5. The performance of the Alexnet pretrained CNN [20] was tested with 150 Westen Blot strip images (50 healthy, 50 benign breast pathology, and 50 breast cancer). For the training consisting of 100 epochs, Adam's optimizer was used with a learning range of 1×10^{-4}. For the training set, 70% of the data set was used, while the accuracy/errors were calculated using 30% of the data set;

6. Regarding the ANOVA statistical test that was applied to establish if there was a significant difference between the results obtained in this work and those achieved in [15] and [16], the results are shown in Tables 3–5, respectively.

Table 2. Results of CNN-DeepGA performance evaluation.

	Accuracy
1	95.83
2	94.82
3	94.76
4	83.33
5	87.77
6	87.50
7	83.48
8	87.50
9	91.67
10	100.00
Average	90.67
Standard deviation	5.60

Table 3. Accuracy results.

Executions	KNN Time Series-Geometric Scaling [15]	Handcrafted CNN [16]	Alexnet [20]	CNN-DeepGA
1	71.11	68.89	50.00	95.83
2	66.66	66.67	45.95	94.82
3	62.22	64.44	54.05	94.76
4	64.44	68.89	39.19	83.33
5	64.44	62.22	45.95	87.77
6	60.50	64.44	48.65	87.50
7	71.11	71.11	55.41	83.48
8	60.50	66.67	54.05	87.50
9	68.88	66.67	55.41	91.67
10	64.44	64.44	45.95	100.00
Average	65.43	66.44	49.46	90.67
Stand. Dev.	3.94	2.66	5.34	5.60

Table 4. Analysis of the mean and standard deviation results.

Factor	N	Average	Stand. Dev.	95% CI
KNN	10	65.43	3.94	(62.52, 68.35)
Handcrafted-CNN	10	66.44	2.662	(63.531, 69.357)
Alexnet pretrained CNN	10	49.46	5.34	(46.55, 52.37)
CNN-DeepGA	10	90.67	5.60	(87.75, 93.58)

Table 3 shows the results obtained from the accuracy averages of the time series classification with the KNN classification algorithm (65.43%), the handcrafted CNN (66.44%), Alexnet pretrained CNN (49.46), and CNN-DeepGA (90.67%). The classification accuracy with KNN and the handcrafted CNN are statistically equivalent, and the Alexnet pretrained

CNN accuracy showed the lowest values. However, the accuracy of the CNN-DeepGA classification is better and statistically significant considering the other three compared approaches. It is important to remark that the same data set was used on the different runs executed on all algorithms. Figure 3 shows the confusion matrix obtained. Table 5 shows the different metrics used to evaluate the CNN architecture obtained through DeepGA. As mentioned above, an accuracy of 90.67% was obtained; likewise, a recall of 90.71%, a specificity of 95.34%, and a precision of 90.69% were also obtained.

Table 5. Metrics used to evaluate the CNN architecture obtained through DeepGA.

Metric	%
Accuracy	90.67
Recall	90.71
Specificity	95.34
Precision	90.69

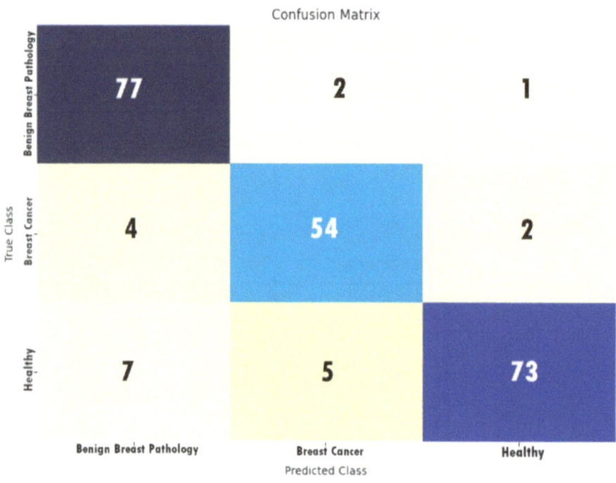

Figure 3. Confusion matrix.

4. Conclusions

Breast cancer is a pathology that has spread throughout the world; it is the leading cause of death in adult women in our country. Commonly used diagnostic tests provide the existence and stage of the disease. However, it is necessary to develop effective detection techniques for this pathology. The response of the immune system to tumor antigens could be the answer to this problem. As mentioned throughout this study, there have been attempts to detect breast cancer early, using the immune response supported by artificial intelligence techniques, such as computer vision and machine learning. Early detection of breast cancer will improve the prognosis, provide adequate treatment, and reduce patient mortality.

It have been reported in some studies that the architecture of the convolutional neural network used has been obtained either manually through experimentation [17], by optimizing the CNN parameters using other algorithms, such as Bayes optimization [18], using a previously trained CNN [19,31], or by taking advantage of the structure of a predefined network [21]. In this work, it was proposed to use neuroevolution to generate a convolutional network architecture that has competitive complexity and efficiency for the classification of Western Blot strips. This was achieved by generating a CNN of four convolutional layers, which allowed a satisfactory execution in terms of time and memory, and a classification accuracy of 90.67%, a recall of 90.71%, a specificity of 95.34%, and precision of 90.69%.

Comparing our results with state-of-the-art research [15,16] and the Alexnet pretrained CNN, which also uses images of the reaction of antibodies to tumor antigens (proteomic images), we observed that the classification percentage was exceeded. Through the ANOVA statistical test, we observe that the best results are statistically significant, as we can see in Table 4.

However, the literature also shows that the diagnosis of breast cancer is carried out using images of breast tissue, coupled with machine learning. It has been mentioned in various works that images of breast tissue are obtained by histopathology (Whole Slide Image—WSI), thermography, ultrasound, and mammography. In [17,19], WSI images were used, reaching an accuracy of 87% and 89.66%, respectively. In [18], thermographic images were analyzed, achieving an accuracy of 98.95%. In [21], ultrasound images are used, and they obtain an accuracy of 80.56%. In [22], the authors used two databases of mammographic images (CBIS-DDSM, INbreast) and obtain an accuracy of 95.4% and 99.7%, respectively.

The architecture of the convolutional network obtained with the DeepGA algorithm allowed us to reach an adequate performance for it and to minimize the time used to find the best configuration of the CNN. On the other hand, the time and subjectivity in the analysis of Western Blot strips continue to be reduced when compared to a proteomics specialist.

While a good rank percentage was achieved with CNN DeepGA, improvement is possible. To achieve this, as future work is proposed to change the DeepGA hyperparameters to obtain a CNN that provides a better classification percentage than the one obtained in addition to exploring the use of another classifier in the last layer of CNN DeeppGA, as well as changing the percentage of data used in training and testing.

This work allowed us to obtain a fast and efficient automatic method for the discrimination of Western Blot images of healthy patients, benign breast pathology patients, and breast cancer patients.

Author Contributions: The authors of this paper contributed to the work as follows: Conceptualization, J.-L.L.-R., R.-E.B.-M., H.-G.A.-M., T.R.-G. and E.M.-M.; methodology, J.-L.L.-R., R.-E.B.-M. and H.-G.A.-M.; software, J.-L.L.-R. and R.-E.B.-M.; validation, J.-L.L.-R., R.-E.B.-M. and H.-G.A.-M.; formal analysis, J.-L.L.-R., R.-E.B.-M. and H.-G.A.-M.; investigation, J.-L.L.-R. and R.-E.B.-M.; data curation, J.-L.L.-R. and R.-E.B.-M.; writing—original draft preparation, J.-L.L.-R., R.-E.B.-M., H.-G.A.-M. and T.R.-G.; writing—review and editing, J.-L.L.-R., R.-E.B.-M., H.-G.A.-M., T.R.-G. and E.M.-M.; supervision, H.-G.A.-M. and T.R.-G. All authors have read and agreed to the published version of the manuscript.

Funding: This research received no external funding.

Acknowledgments: The authors appreciate the help of Gustavo Vargas Hakim with the use and support of the DeepGA algorithm.

References

1. Sung, H.; Ferlay, J.; Siegel, R.L.; Laversanne, M.; Soerjomataram, I.; Jemal, A.; Bray, F. Global Cancer Statistics 2020: GLOBOCAN Estimates of Incidence and Mortality Worldwide for 36 Cancers in 185 Countries. *CA Cancer J. Clin.* **2021**, *71*, 209–249. [CrossRef] [PubMed]
2. Hernández-Nájera, O.; Cahuana-Hurtado, L.; Ávila-Burgos, L. Costos de atención del cáncer de mama en el Instituto de Seguridad y Servicios Sociales de los Trabajadores del Estado, México. *Salud Publica Mex.* **2021**, *63*, 538–546. [CrossRef] [PubMed]
3. Hablemos de El Cáncer de Mama. Available online: https://www.seom.org/seomcms/images/stories/recursos/infopublico/publicaciones/HABLEMOS_CANCER_MAMA.pdf (accessed on 15 February 2023).
4. Torres-Arreola, L.d.P.; Vladislavovna Doubova, S.C.d.M. Detección Oportuna En El Primer Nivel de Atención. *Rev. Méd. Inst. Mex. Seguro Soc.* **2007**, *45*, 157–166. [PubMed]
5. Lara-Tamburrino, M.d.C.; Tapia-Vega, A.X.; Quiróz-Rojas, L.Y. Integración de la imagen en la patología mamaria. *Gac. Mex. Oncol.* **2013**, *12*, 116–123.
6. Cadavid-Fernández, N.; Carretero-Barrio, I.; Moreno-Moreno, E.; Rodríguez-Villena, A.; Palacios, J.; Pérez-Mies, B. The Role of Core Needle Biopsy in Diagnostic Breast Pathology. *Rev. Senol. Patol. Mamar.* **2022**, *35*, S3–S12. [CrossRef]
7. Chávarri-Guerra, Y.; Villarreal-Garza, C.; Liedke, P.E.; Knaul, F.; Mohar, A.; Finkelstein, D.M.; Goss, P.E. Breast Cancer in Mexico: A Growing Challenge to Health and the Health System. *Lancet Oncol.* **2012**, *13*, e335–e343. [CrossRef]
8. Brandan, M.E. Detección del Cáncer de Mama: Estado de la Mamografía en México. *Rev. Inst. Nac. Cancerol.* **2006**, *27*, 16.

9.   Chapman, C.; Murray, A.; Chakrabarti, J.; Thorpe, A.; Woolston, C.; Sahin, U.; Barnes, A.; Robertson, J. Autoantibodies in Breast Cancer: Their Use as an Aid to Early Diagnosis. *Ann. Oncol.* **2007**, *18*, 868–873. [CrossRef]

10. Desmetz, C.; Lacombe, J.; Mange, A.; Maudelonde, T.; Solassol, J. Autoanticorps et diagnostic précoce des cancers. *Med. Sci.* **2011**, *27*, 633–638. [CrossRef]

11. Romo-González, T.; Esquivel-Velázquez, M.; Ostoa-Saloma, P.; Lara, C.; Zentella, A.; León-Díaz, R.; Lamoyi, E.; Larralde, C. The Network of Antigen-Antibody Reactions in Adult Women with Breast Cancer or Benign Breast Pathology or without Breast Pathology. *PLoS ONE* **2015**, *10*, e0119014. [CrossRef]

12. Yue, W.; Wang, Z.; Chen, H.; Payne, A.; Liu, X. Machine Learning with Applications in Breast Cancer Diagnosis and Prognosis. *Designs* **2018**, *2*, 13. [CrossRef]

13. Mahoro, E.; Akhloufi, M.A. Applying Deep Learning for Breast Cancer Detection in Radiology. *Curr. Oncol.* **2022**, *29*, 8767–8793. [CrossRef] [PubMed]

14. Nasser, M.; Yusof, U.K. Deep Learning Based Methods for Breast Cancer Diagnosis: A Systematic Review and Future Direction. *Diagnostics* **2023**, *13*, 161. [CrossRef]

15. Sánchez-Silva, D.M.; Acosta-Mesa, H.G.; Romo-González, T. Semi-Automatic Analysis for Unidimensional Immunoblot Images to Discriminate Breast Cancer Cases Using Time Series Data Mining. *Int. J. Patt. Recogn. Artif. Intell.* **2018**, *32*, 1860004. [CrossRef]

16. Llaguno-Roque, J.-L.; Barrientos-Martínez, R.-E.; Acosta-Mesa, H.-G.; Romo, T. Western Blot Pattern Classification Using Convolutional Neural Networks for Breast Cancer Diagnosis. In Proceedings of the 4th Workshop on New Trends in Computational Intelligence and Applications (CIAPP 2022), Monterrey, Mexico, 24 October 2022.

17. Alanazi, S.A.; Kamruzzaman, M.M.; Islam Sarker, M.N.; Alruwaili, M.; Alhwaiti, Y.; Alshammari, N.; Siddiqi, M.H. Boosting Breast Cancer Detection Using Convolutional Neural Network. *J. Healthc. Eng.* **2021**, *2021*, 5528622. [CrossRef]

18. Ekici, S.; Jawzal, H. Breast Cancer Diagnosis Using Thermography and Convolutional Neural Networks. *Med. Hypotheses* **2020**, *137*, 109542. [CrossRef] [PubMed]

19. Yamlome, P.; Akwaboah, A.D.; Marz, A.; Deo, M. Convolutional Neural Network Based Breast Cancer Histopathology Image Classification. In Proceedings of the 2020 42nd Annual International Conference of the IEEE Engineering in Medicine & Biology Society (EMBC), Montreal, QC, Canada, 20–24 July 2020; IEEE: Montreal, QC, Canada; pp. 1144–1147.

20. Krizhevsky, A.; Sutskever, I.; Hinton, G.E. ImageNet Classification with Deep Convolutional Neural Networks. *Commun. ACM* **2017**, *60*, 84–90. [CrossRef]

21. Xu, Z.; Yang, Q.; Li, M.; Gu, J.; Du, C.; Chen, Y.; Li, B. Predicting HER2 Status in Breast Cancer on Ultrasound Images Using Deep Learning Method. *Front. Oncol.* **2022**, *12*, 829041. [CrossRef]

22. Jabeen, K.; Khan, M.A.; Balili, J.; Alhaisoni, M.; Almujally, N.A.; Alrashidi, H.; Tariq, U.; Cha, J.-H. BC2NetRF: Breast Cancer Classification from Mammogram Images Using Enhanced Deep Learning Features and Equilibrium-Jaya Controlled Regula Falsi-Based Features Selection. *Diagnostics* **2023**, *13*, 1238. [CrossRef]

23. Zhu, Z.; Wang, S.-H.; Zhang, Y.-D. A Survey of Convolutional Neural Network in Breast Cancer. *Comput. Model. Eng. Sci.* **2023**, *136*, 2127–2172. [CrossRef]

24. Baldominos, A.; Saez, Y.; Isasi, P. Evolutionary Convolutional Neural Networks: An Application to Handwriting Recognition. *Neurocomputing* **2018**, *283*, 38–52. [CrossRef]

25. Vargas-Hákim, G.-A.; Mezura-Montes, E.; Acosta-Mesa, H.-G. Hybrid Encodings for Neuroevolution of Convolutional Neural Networks: A Case Study. In Proceedings of the GECCO'21: Genetic and Evolutionary Computation Conference, Lille, France, 10–14 July 2021; Association for Computing Machinery: New York, NY, USA; pp. 1762–1770.

26. Macdonald, I.K.; Parsy-Kowalska, C.B.; Chapman, C.J. Autoantibodies: Opportunities for Early Cancer Detection. *Trends Cancer* **2017**, *3*, 198–213. [CrossRef]

27. Rauf, F.; Anderson, K.S.; LaBaer, J. Autoantibodies in Early Detection of Breast Cancer. *Cancer Epidemiol. Biomark. Prev.* **2020**, *29*, 2475–2485. [CrossRef] [PubMed]

28. Yang, R.; Han, Y.; Yi, W.; Long, Q. Autoantibodies as Biomarkers for Breast Cancer Diagnosis and Prognosis. *Front. Immunol.* **2022**, *13*, 1035402. [CrossRef] [PubMed]

29. Stanley, K.O.; Clune, J.; Lehman, J.; Miikkulainen, R. Designing Neural Networks through Neuroevolution. *Nat. Mach. Intell.* **2019**, *1*, 24–35. [CrossRef]

30. Sun, Y.; Xue, B.; Zhang, M.; Yen, G.G. Automatically Designing CNN Architectures Using Genetic Algorithm for Image Classification. *IEEE Trans. Cybern.* **2020**, *50*, 3840–3854. [CrossRef]

31. Zhou, L.-Q.; Wu, X.-L.; Huang, S.-Y.; Wu, G.-G.; Ye, H.-R.; Wei, Q.; Bao, L.-Y.; Deng, Y.-B.; Li, X.-R.; Cui, X.-W.; et al. Lymph Node Metastasis Prediction from Primary Breast Cancer US Images Using Deep Learning. *Radiology* **2020**, *294*, 19–28. [CrossRef]

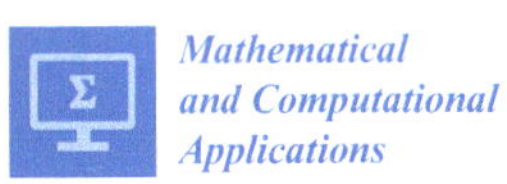
Mathematical and Computational Applications

Article

Evolutionary Selection of a Set of Association Rules Considering Biological Constraints Describing the Prevalent Elements in Bacterial Vaginosis

María Concepción Salvador-González [1], Juana Canul-Reich [1,*], Rafael Rivera-López [2], Efrén Mezura-Montes [3] and Erick de la Cruz-Hernandez [4]

[1] DACyTI, Universidad Juárez Autónoma de Tabasco, Cunduacán 86690, Tabasco, Mexico; mcsalvadorg@gmail.com
[2] DSC, Tecnológico Nacional de México, Instituto Tecnológico de Veracruz, Veracruz 91897, Veracruz, Mexico; rafael.rl@veracruz.tecnm.mx
[3] IIIA, Universidad Veracruzana, Xalapa 91097, Veracruz, Mexico; emezura@uv.mx
[4] DAMC, Universidad Juárez Autónoma de Tabasco, Comalcalco 86658, Tabasco, Mexico; erick.delacruz@ujat.mx
* Correspondence: juana.canul@ujat.mx

Abstract: Bacterial Vaginosis is a common disease and recurring public health problem. Additionally, this infection can trigger other sexually transmitted diseases. In the medical field, not all possible combinations among the pathogens of a possible case of Bacterial Vaginosis are known to allow a diagnosis at the onset of the disease. It is important to contribute to this line of research, so this study uses a dataset with information from sexually active women between 18 and 50 years old, including 17 numerical attributes of microorganisms and bacteria with positive and negative results for BV. These values were semantically categorized for the Apriori algorithm to create the association rules, using support, confidence, and lift as statistical metrics to evaluate the quality of the rules, and incorporate those results in the objective function of the DE algorithm. To guide the evolutionary process we also incorporated the knowledge of a human expert represented as a set of biologically meaningful constraints. Thus, we were able to compare the performance of the rand/1/bin and best/1/bin versions from Differential Evolution to analyze the results of 30 independent executions. Therefore the experimental results allowed a reduced subset of biologically meaningful association rules by their executions, dimension, and DE version to be selected.

Keywords: differential evolution; association rules; bacterial vaginosis

Citation: Salvador-González, M.C.; Canul-Reich, J.; Rivera-López, R.; Mezura-Montes, E.; de la Cruz-Hernandez, E. Evolutionary Selection of a Set of Association Rules Considering Biological Constraints Describing the Prevalent Elements in Bacterial Vaginosis. *Math. Comput. Appl.* **2023**, *28*, 75. https://doi.org/10.3390/mca28030075

Academic Editor: Suchuan Dong

Received: 16 March 2023
Revised: 27 May 2023
Accepted: 12 June 2023
Published: 14 June 2023

1. Introduction

Bacterial Vaginosis (BV) is a common disturbance of the balance of vaginal flora; about 25% of women of childbearing age suffer BV [1]. It is a disease that can be asymptomatic, but symptoms such as discharge, bad vaginal odor, and increased PH can also occur. It can also increase the risk of contracting other infections such as Neisseria gonorrhea, Chlamydia trachomatis, Herpes type 2, and papillomavirus infection, among other sexually transmitted diseases, in addition to being a recurring disease [2,3]. Diseases such as BV and those sexually transmitted can lead to contracting more severe illnesses, including cervical cancer, as has been demonstrated by studies that evaluate the 16S rRNA sequencing to measure the diversity of the vaginal microbiota in women with different BV, human papillomavirus (HPV), and cervical intraepithelial neoplasia (CIN) status [4]. This study compares the microbiota composition of several women to gain insight into a marker of vaginal dysbiosis. The authors use logistic regression to identify risk factors for CIN, such as age, gestational and childbirth history, contraceptive methods, number of sexual partners, BV status, HPV infection status, and condom use. The results show that BV and HR-HPV infection are risk factors for CIN.

Bacterial Vaginosis is a public health problem. The literature mentions that, in healthy vaginal microbiota, lactobacilli are predominating. Otherwise, when lactobacilli are replaced by several bacteria, such as Gardnerella vaginalis and Atopobium vaginae, among others, there exists an imbalance in the vaginal flora that, in most cases, corresponds to a BV. The Nugent score and molecular biology are commonly used for BV diagnosis in the medical area. However, there is no certainty about the causes of this disease. Neither are all the possible pathogen combinations that can cause BV known, because it has a high recurrence rate, and this is essential to identify and treat this disease appropriately [5].

On the other hand, association rules (AR) are one of the four primary data mining tasks [6]. AR algorithms try to find relationships and frequent patterns between data. Quality metrics such as support, confidence, lift, hyper lift, and Fisher's exact test, among others, are used to identify the best patterns [7].

Association Rule Mining (ARM) has been combined with other methods, such as Differential Evolution, in optimization problems with a single objective. In the case of numerical data, they are discretized by grouping into consecutive intervals [8].

Various DE-based methods exist for ARM, such as the DE for ARM using numerical and categorical attributes (ARM-DE), where categorical attributes are discretized into numerical values such as 0 or 1 and encoded in a real-valued parameter vector [9]. Another approach is the Numerical Association Rule Minning (NiaARM), available in its free version with Python libraries [10]. Moreover, other authors have proposed the DE for mining a significant fuzzy association rules (DESigFAR) algorithm that uses fuzzy intervals to discretize the attributes. The authors evaluate each candidate rule using statistical tests and compare their proposal against one genetic algorithm [11]. Although several approaches have been proposed, the use of DE-based algorithms to reduce association rules previously generated by the Apriori algorithm, applied to discover patterns leading to VB, has yet to be studied.

For the reasons mentioned above, the interest of this study is to reduce the number of association rules for contributing to the identification of possible combinations between pathogens of possible bacterial vaginosis with Association Rule Mining and Evolutionary Computation techniques using an adaptation of the Differential Evolution (DE) algorithm to find biologically meaningful association rules from a set of association rules derived from the Apriori algorithm. The use of DE is proposed to decrease the number of rules that were previously generated with Apriori [12]. Another advantage of DE is that it allows the application of biological constraints, thus we claim that the set of rules obtained with Differential Evolution is smaller than that obtained with Apriori alone, and still meets the biological significance required for the diagnosis of diagnostic BV.

2. Materials and Methods

The data used for this study are a dataset with 17 numerical attributes with medical information from 201 sexually active women aged 18 to 50 years who underwent their routine annual gynecological examination at the Metabolic and Infectious Diseases Research Laboratory of the Universidad Juárez Autónoma de Tabasco, and who gave their written consent. The study was designed according to international standards for responsible publication of (COPE) and registered (protocol No. UJAT–20160006) and approved by the Institutional Review Board of the Universidad Juarez Autonoma de Tabasco [13]. We considered 186 records with a positive and negative diagnosis for bacterial vaginosis only. The numerical attributes of integer type used are the density of Lactobacillus crispatus, gasseri, jensenii, and iners. In addition to microorganisms mainly related to BV, Megasphaera type 1, Atopobium vaginae, and Gardnerella vaginalis.

Association rule mining is responsible for discovering interesting patterns within a dataset and is one of the most important knowledge-discovery techniques [14]. An association rule has the form $X \Rightarrow Y$, where X in the rule is called the antecedent, and Y is called consequent [15]. To measure the quality of the association rules, quality metrics are used. The interest of this research is to find association patterns between the pathogens that cause bacterial vaginosis. According to other authors [16–18], metrics such as Confidence and Lift calculate their values according to the relationship between the antecedent and consequent of a rule.

The quality metrics of our interest are described below [19]:

- Support: It is the number of times the element appears.
- Confidence: It is based on the support of frequent itemsets to generate significant rules according to the value of the confidence that one wants to look for.
- Lift: Calculate the number of times the antecedent and consequent occur together.

Other metrics that were evaluated in this work are the following: [19]:

- Fisher Exact Test: Each rule represents a one-sided Fisher's exact statistical test and the correction is used for multiple comparisons.
- Hyperlift: It is a more robust metric than the lift metric. It is used at low counts and its false positives are less frequent.

The Apriori algorithm is one of the most effective methods for discovering valid, novel, and meaningful rules among data and stands out for its simplicity. However, its results exponentially grow when making associations, generating many rules [15].

The Apriori Algorithm consists of three repetitive cycles where k is the length of the pattern generated in the previous step, i are the generations, $Ck + 1$ is the cycle that generates the candidate patterns that join the patterns in Fk, the cycle continues with the pruning and validation of patterns for all the database transactions in T until the set of frequent k-patterns Fk in one iteration is empty. The Algorithm 1 shows the pseudocode of the Apriori Algorithm [20].

Algorithm 1 Apriori pseudocode

Require: n
  1: Generate frequent 1-patterns and 2-patterns using specialized counting methods and denote by $F1$ and $F2$;
  2: $k := 2$;
  3: **while** Fk is not empty do **do**
  4:     Generate $Ck + 1$ by using joins on Fk;
  5:     Prune $Ck + 1$ with $Apriori$ subset pruning trick;
  6:     Generate $Fk + 1$ by counting candidates in $Ck + 1$ with respect to T at support s;
  7:     $k := k + 1$;
  8: **end while**
  9: **return** $\cup_{i=1}^{k} F_i$;

On the other hand, the DE process begins with the random creation of the initial population. The values of each individual in the population must fit within the pre-established limits of the search space. Then, for each individual, three vectors are combined using the mutation and crossover operators to create a new candidate solution. By comparing current with new individuals, one new population is built. The parameters used by the DE algorithm are the population size (NP), crossover rate (CR), mutation factor (F), and also the bounds of the search space [21].

The DE algorithm simulates natural evolution using vectors. Starting from a target vector $\vec{x}_{i,g}$, the search direction is calculated according to the difference of the vectors $\vec{x}_{r1,g}$ and $\vec{x}_{r2,g}$ chosen at random within the population, and its scale factor F is calculated and added to the base vector $\vec{x}_{r0,g}$ and its result is the mutated vector. The mutated vector is recombined by a binomial crossover defined by the parameter CR. Finally, a binomial cross-type is used. The pseudocode of the DE/rand/1/bin version can be found in Algorithm 2. The difference between the DE/rand/1/bin version and DE/best/1/bin is that in the latter the base vector is the best vector of the current population. The pseudocode of the DE/best/1/bin version is depicted in Algorithm 3 [22].

Algorithm 2 DE/rand/1/bin pseudocode

Require: $g = 0$

  1: Create a random initial population $\vec{x}_{i,g} \forall i, i = 1, \ldots, NP$
  2: Evaluate $f(\vec{x}_{i,g}) \forall i, i = 1, \ldots, NP$
  3: **for** $g = 1$ to $MAXG$ **do**
  4:   **for** $i = 1$ to NP **do**
  5:     Select randomly $r_0 \neq r_1 \neq r_2 \neq i$
  6:     $j_{rand} = randint[1, n]$
  7:     **for** j=1 to n **do**
  8:       **if** $rand_j[0,1] < CR \vee j = j_{rand}$ **then**
  9:         $u_{j,i,g+1} = x_{j,r_0,g} + F(x_{j,r_1,g} - x_{j,r_2,g})$
10:       **else**
11:         $u_{j,i,g+1} = x_{j,i,g}$
12:       **end if**
13:     **end for**
14:     **if** $(f(\vec{u}_{i,g+1}) \leq f(\vec{x}_{i,g})$ **then**
15:       $\vec{x}_{i,g+1} = \vec{u}_{i,g+1}$
16:     **else**
17:       $\vec{x}_{i,g+1} = \vec{x}_{i,g}$
18:     **end if**
19:   **end for**
20:   $g = g + 1$
21: **end for**

The main objective of the proposed approach, named the Apriori rules reduction by Differential Evolution (AR2DE) approach, is to apply the DE algorithm to select the most important set of association rules generated by the Apriori algorithm. The individuals in the population encode the original association rules using an integer-valued vector (Figure 1). Several AR metrics are included in the fitness function to identify their biological significance to c.

Algorithm 3 DE/best/1/bin pseudocode

Require: $g = 0$

  1: Create a random initial population $\vec{x}_{i,g} \forall i, i = 1,\ldots, NP$
  2: Evaluate $f(\vec{x}_{i,g}) \forall i, i = 1,\ldots, NP$
  3: **for** $g = 1$ to $MAXGEN$ **do**
  4:   **for** $i = 1$ to NP **do**
  5:     Select randomly $r_0 \neq r_1 \neq r_2 \neq i$
  6:     $j_{rand} = randint[1, n]$
  7:     **for** j=1 to n **do**
  8:       **if** $(rand_j[0,1] < CR$ or $j = j_{rand})$ **then**
  9:         $u_{j,i,g+1} = x_{j,best,g} + F(x_{j,r_1,g} - x_{j,r_2,g})$
10:       **else**
11:         $u_{j,i,g+1} = x_{j,i,g}$
12:       **end if**
13:     **end for**
14:     **if** $(f(\vec{u}_{i,g+1}) \leq f(\vec{x}_{i,g})$ **then**
15:       $\vec{x}_{i,g+1} = \vec{u}_{i,g+1}$
16:     **else**
17:       $\vec{x}_{i,g+1} = \vec{x}_{i,g}$
18:     **end if**
19:   **end for**
20:   $g = g + 1$
21: **end for**

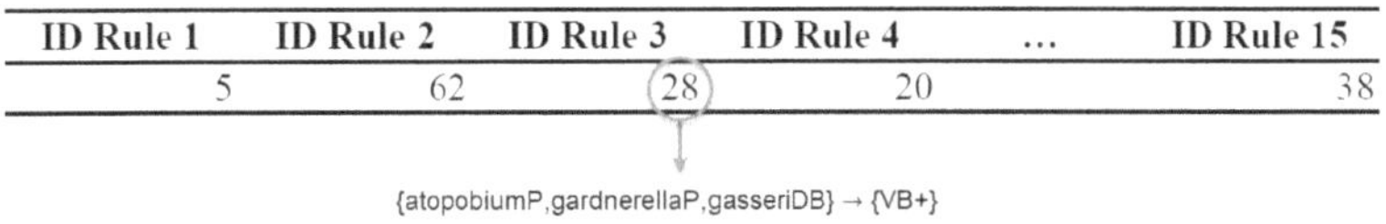

ID Rule 1	ID Rule 2	ID Rule 3	ID Rule 4	...	ID Rule 15
5	62	28	20		38

{atopobiumP,gardnerellaP,gasseriDB} → {VB+}

Figure 1. Encoding scheme from 1 at 15 association rules. The integers represent the ID of each association rule.

3. Results

3.1. Experimental Study

In the first part of the experimental study, of the data set consisting of 184 records of positive and negative BV cases, the attributes were discretized according to their numerical value and cataloged into linguistic concepts according to Table 1 to obtain the transactions used by the Apriori algorithm to generate the association rules, resulting in 5248 association rules.

Table 1. Antecedents.

Antecedent	Range	Classification	Type
Age	1	menoredad	Under 30 years old
	2	mayoredad	Over 30 years old
Cristpatus	1	crispatusDB	Low density
	2	crispatusDA	High density
Gasseri	1	gasseriDB	Low density
	2	gasseriDA	High density
Iners	1	inersDB	Low density
	2	inersDA	High density
Jensenii	1	jenseniDB	Low density
	2	jenseniDA	High density
Megasphaera	1	megasphaeraP	Positive
	2	megasphaeraN	Negative
Atopobium	1	atopobiumP	Positive
	2	atopobiumN	Negative
Gardnerella	1	gardnerellaP	Positive
	2	gardnerellaN	Negative

Antecedents itemset values used in the experimental study.

Below, the cases of interest in this research are the rules that have BV+, after applying the filter 91 rules are evaluated in the DE process to reduce according to their biological significance.

3.2. Analysis of Evaluation Metrics

The next part of the experiment was the analysis of the quality metrics, which evaluate the association rules generated by the Apriori algorithm. Since this study focused on the rules that had as a consequent VB+, which represents one element as a consequent, the metrics Fishers Exact Test (Figure 2), Hyperlift (Figure 3), Lift (Figure 4), and Confidence (Figure 5) were evaluated using scatter plots that allow us to visualize their range of data, maximum, and minimum values. The comparison between the graphs shows that the lowest value range is for the Fishers Exact Test metric (Figure 2), followed by Confidence. In the study that metrics are used in DE the very low value ranges do not favor the evolutionary process.

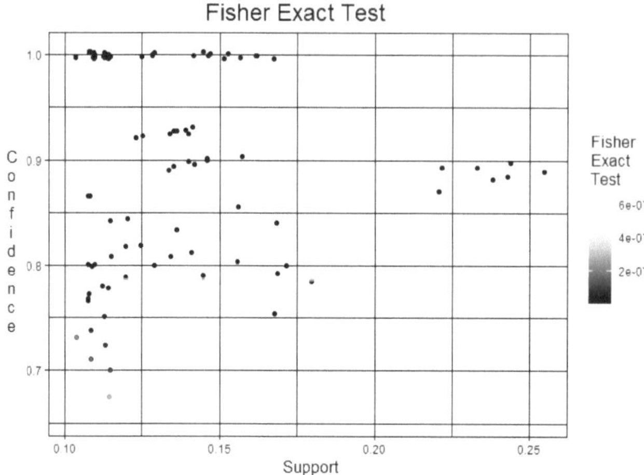

Figure 2. Fisher exact test metric scatter plot.

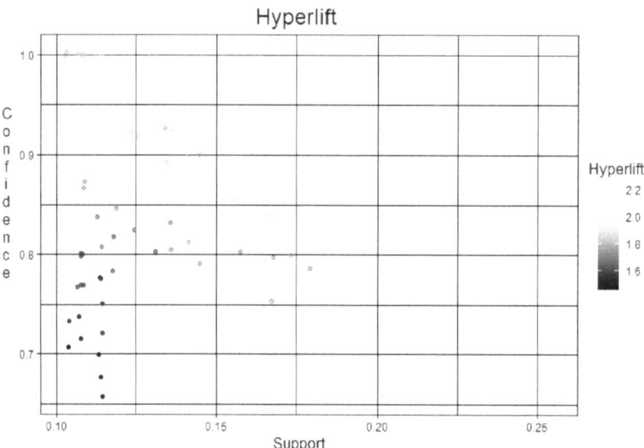

Figure 3. Hyperlift metric scatter plot.

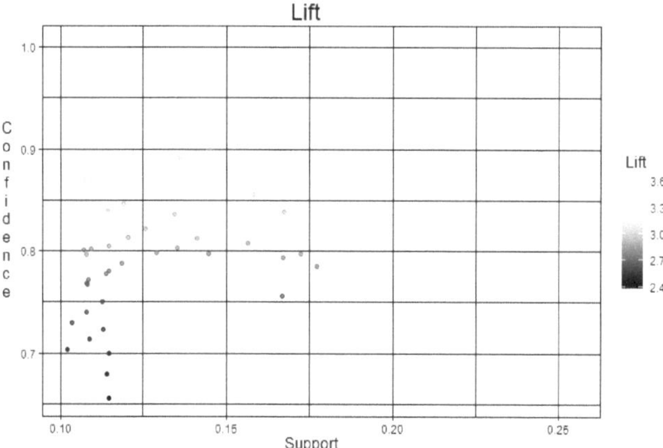

Figure 4. Lift metric scatter plot.

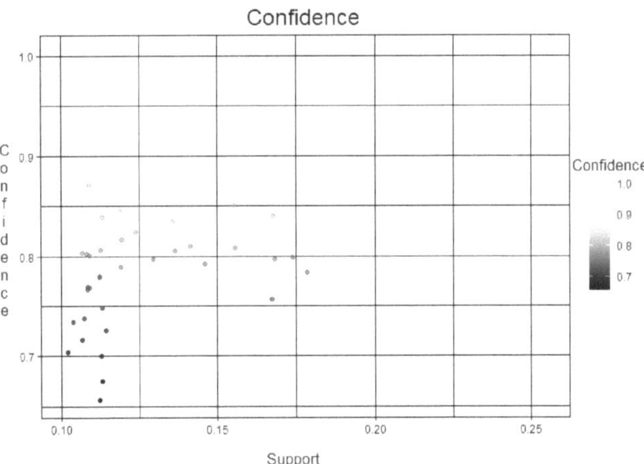

Figure 5. Confidence metric scatter plot.

The scatter plots also show that the highest ranges of values were for the Lift (Figure 4) and Hyperlift metrics (Figure 3).

The metrics that show the best correlation strength between their values according to the scatter plots are Lift (Figure 4) and Confidence (Figure 5). The main metrics according to the literature [16–18] are confidence and lift so lift represented the best option among the four evaluated metrics, and lift, having the highest range of values, favors the evolutionary process in DE.

The result of the analysis of the metrics reveals that lift and confidence are the best-evaluated metrics for this study in addition to evaluating the frequency of positive bacteria and the presence of lactobacillus iners as factors of biological significance in the association rules.

3.3. Implementation of the Differential Evolution (DE) Algorithm

The three main elements in the DE algorithm are the individuals' encoding scheme, the fitness function, and the variation operators.

1. **Encoding scheme:** An individual of the population is a subset of R association rules each identified with an ID number. Figure 1 shows an example of this codification from 1 at 15 rules by ID rule.
    In this work, the value of R is set to 1 to 15 since in [18] authors obtained five rules with a biological significance which were determined by a human expert, so the 15 tests ensure the algorithm will find this minimal set of rules.
2. **Fitness function:** Each j-th individual in the population is evaluated to define the fitness value. In this work, the fitness function $f(x_j)$ is the sum of the S metrics of the association rules encoded on the individual as follows:

$$f(x_j) = \sum_{u=1}^{R} \sum_{w=1}^{S} m_{u,w} \tag{1}$$

where R is the number of association rules, S is the number of metrics involved to define the solution quality, and $m_{u,w}$ is the w-th metric computed for the u-th rule. Since metrics are parameters that allow us to know the quality of attributes quantitatively, support and confidence are normally used [23]. The metrics used in the objective function and described in Section 2 are support, confidence, and lift. In addition, the frequency of positive bacteria in the rules, and the occurrences of high values of lactobacillus iners are included to define the biological significance of the association rules [13] in this sense higher results from the addition of the metrics have a higher significance.

3. **Variation operators:** Differential mutation and crossover operators are defined to create feasible offspring.

- **Mutation:** Three randomly chosen individuals of the current population (x^{r_1}, x^{r_2} and x^{r_3}), are different from each other and also different from the target vector, these individuals are linearly combined to yield a *mutated vector* v^i using a user-specified scale factor F to control the differential variation, as follows:

$$v^i = \lfloor x^{r_1} + F(x^{r_2} - x^{r_3}) \rceil, \tag{2}$$

Equation (2) is related to the DE/rand/1/bin variant defined in [24]. Another commonly used variant is known as DE/best/1/bin, where the best individual in the population x^{best} is combined with two randomly chosen individuals of the current population, as follows:

$$v^i = \lfloor x^{best} + F(x^{r_1} - x^{r_2}) \rceil, \tag{3}$$

- **Crossover:** The mutated vector is recombined with the target vector to build the trial vector u^i. For each $j \in \{1, \ldots, |x^i|\}$, either x^i_j or v^i_j is selected based on a comparison between a uniformly distributed random number $r \in [0,1]$ and the crossover rate CR. The recombination operator also uses a randomly chosen index $l \in \{1, \ldots, |x^i|\}$ to ensure that u^i acquires at least one value from v^i, as follows:

$$u^i_j = \begin{cases} v^i_j & \text{if } r \leq CR \text{ or } j = l, \\ x^i_j & \text{otherwise.} \end{cases} \tag{4}$$

In the Equations (2) and (3), $\lfloor w \rceil$ symbol denotes that the w value is rounded to the nearest integer since the encoding scheme defined for this work indicates that the parameter values are only integers. If a parameter value of a mutated vector is outside its range, it is replaced with a random value between 1 and 91.

3.4. Algorithm Parameters

It is well known that the performance of the Differential Evolution algorithm is affected by the values of its parameters: F (Scale factor), CR (Crossover rate), and NP (Population Size) [25]. The parameter values used in this work are shown in the Table 2 and are based on those commonly used in the existing literature [24]. Since this experimental study is a work in progress, no parameter-tuning process has been carried out.

Table 2. Parameters values.

Parameter	Value
F (Scale factor)	0.7
CR (Crossover rate)	0.5
NP (Population size)	30
MAXGEN (Number of generations)	30
li (lower limit)	1
ls (upper limit)	91

Parameters used in DE/rand/1 and DE/best/1 versions.

4. Discussion

In this work, 30 independent runs were made for the rand/1/bin and best/1/bin versions and 15 tests were made with each version by changing the value of the individual's dimension from 1 to 15.

Table 3 shows the results of 30 independent runs with the two DE variants included in this study (rand/1/bin and best/1/bin). The best results for both versions were when D = 15, the best fitness value in the rand/1/bin version is 96.4201 on test number 14, and for the best/1/bin version is 95.6184 on test 6.

Table 3. Results of 30 independent runs for each DE variant.

Test	Rand/1/Bin	Best/1/Bin	Test	Rand/1/Bin	Best/1/Bin
1	95.5457	94.9080	16	93.2444	92.7573
2	93.4321	94.3232	17	93.1759	93.8795
3	95.0492	95.3183	18	92.96241	94.3069
4	92.6088	93.2915	19	93.1063	95.0708
5	93.1151	94.2030	20	94.2296	93.1564
6	93.0887	**95.6184**	21	95.2363	92.8379
7	92.8634	95.4351	22	93.4080	92.9451
8	94.0896	94.9295	23	96.0235	93.0003
9	95.2359	92.6744	24	<u>94.0122</u>	94.0149
10	92.5891	94.1667	25	92.6781	93.1910
11	94.4798	94.3319	26	94.1964	93.6972
12	93.7070	95.5788	27	92.9309	92.5566
13	94.7591	93.9418	28	94.1645	93.9121
14	**96.4201**	<u>94.1041</u>	29	94.5672	94.0574
15	93.7882	94.4250	30	94.4028	94.2564

The best fitness values are highlighted in bold, and the median value of each variant is underlined.

The statistical comparison for each variant is shown in Table 4, and Figure 6 depicts the convergence plot of the run reaching the median value of the two variants. When comparing the results of the two variants using the Wilcoxon signed-rank exact test by the function wilcox.test from R, V = 163 and the p-value = 0.1579 indicated the data in each group are significant correlated.

Table 4. Statistical values.

Statistical Measure	Rand/1/Bin	Best/1/Bin
Best value	96.4201	95.6184
Mean	93.9703	94.0296
Median	93.9002	94.0808
Standard deviation	1.0428	0.8942
Worst value	92.5891	92.5566
Best test number	14	6
Median test number	24	14

Statistical measure for rand/1/bin and best/1/bin.

According to the statistical results, the best value is obtained with the rand/1/bin variant. However, the results obtained in the independent runs and the behavior of the convergence graph show that the best/1/bin variant had better performance in selecting the association rules.

The best individuals of each variant were taken for the 15 tests and decoded to their corresponding association rule according to their ID. Repeated rules were removed and a count of occurrences in both groups of rules was made to know the most frequent ones as shown in the table. Table 5 shows the rules encoded by the best individuals of each variant. Likewise, most of the rules comply with the biological significance requirement of having at least two bacteria present [13]. The biological significance of the items adds weight to rules that carry bacteria positivity, concurrently with showing low-density levels of lactobacillus iners. This result is concordant with clinical findings observed in women with bacterial vaginosis [26].

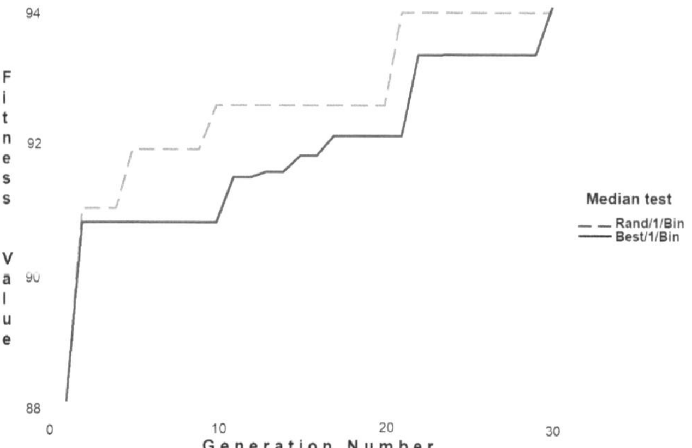

Figure 6. Convergence plot for the median values of the two DE variants.

Table 5. Set of best association rules.

ID	Association Rule
ine 1	{atopobiumP,megasphaeraP} → {VB+}
3	{jenseniiDB,megasphaeraP} → {VB+}
5	{atopobiumP,gardnerellaP} → {VB+}
10	{gardnerellaP,gasseriDB} → {VB+}
14	{atopobiumP,inersDB} → {VB+}
15	{atopobiumP,crispatusDB} → {VB+}
19	{atopobiumP,crispatusDB,megasphaeraP} → {VB+}
20	{atopobiumP,jenseniiDB,megasphaeraP} → {VB+}
21	{atopobiumP,gasseriDB,megasphaeraP} → {VB+}
22	{crispatusDB,jenseniiDB,megasphaeraP} → {VB+}
25	{atopobiumP,crispatusDB,gardnerellaP} → {VB+}
26	{atopobiumP,gardnerellaP,megasphaeraN} → {VB+}
27	{atopobiumP,gardnerellaP,jenseniiDB} → {VB+}
28	{atopobiumP,gardnerellaP,gasseriDB} → {VB+}
37	{atopobiumP,inersDA,jenseniiDB} → {VB+}
42	{atopobiumP,jenseniiDB,mayoredad} → {VB+}
46	{atopobiumP,gasseriDB,inersDB} → {VB+}
53	{atopobiumP,crispatusDB,jenseniiDB,megasphaeraP} → {VB+}
54	{atopobiumP,crispatusDB,gasseriDB,megasphaeraP} → {VB+}
55	{atopobiumP,gasseriDB,jenseniiDB,megasphaeraP} → {VB+}
57	{atopobiumP,crispatusDB,gardnerellaP,megasphaeraN} → {VB+}
58	{atopobiumP,crispatusDB,gardnerellaP,jenseniiDB} → {VB+}
59	{atopobiumP,crispatusDB,gardnerellaP,gasseriDB} → {VB+}
60	{atopobiumP,gardnerellaP,jenseniiDB,megasphaeraN} → {VB+}
61	{atopobiumP,gardnerellaP,gasseriDB,megasphaeraN} → {VB+}
62	{atopobiumP,gardnerellaP,gasseriDB,jenseniiDB} → {VB+}
65	{crispatusDB,gardnerellaP,gasseriDB,jenseniiDB} → {VB+}
72	{atopobiumP,crispatusDB,gasseriDB,mayoredad} → {VB+}
74	{atopobiumP,crispatusDB,inersDB,jenseniiDB} → {VB+}
75	{atopobiumP,crispatusDB,gasseriDB,inersDB} → {VB+}
81	{atopobiumP,crispatusDB,gasseriDB,jenseniiDB,megasphaeraP} → {VB+}
82	{atopobiumP,crispatusDB,gardnerellaP,jenseniiDB,megasphaeraN} → {VB+}
83	{atopobiumP,crispatusDB,gardnerellaP,gasseriDB,megasphaeraN} → {VB+}
84	{atopobiumP,crispatusDB,gardnerellaP,gasseriDB,jenseniiDB} → {VB+}
85	{atopobiumP,gardnerellaP,gasseriDB,jenseniiDB,megasphaeraN} → {VB+}
91	{atopobiumP,crispatusDB,gardnerellaP,gasseriDB,jenseniiDB,megasphaeraN} → {VB+}

Set of the best association rules selected from the two DE variants of the tests with dimensions from 1 to 15.

The 5 most frequent rules of the 15 tests for the rand/1/bin variant by ID are 1, 58, 21, 62, and 19. For the variant, the best/1/bin by ID are 19, 58, 62, and 83. The details for both variants are shown in Figure 7.

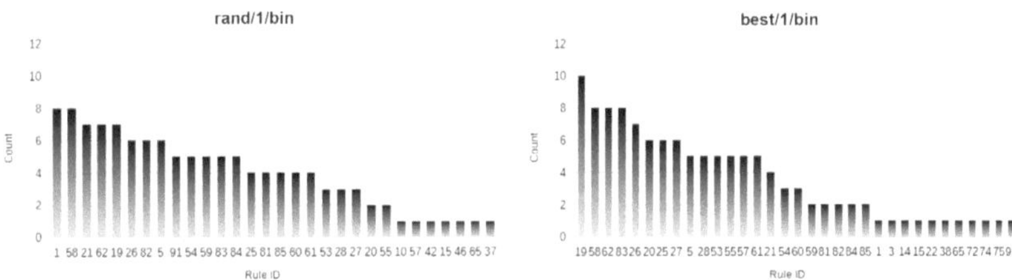

Figure 7. Detail of the frequency of rules per ID for rand/1/bin and best/1/bin variants.

The elements frequently found in the antecedent of the rules of both variants are atopobiumP, crispatusDB, gardnerellaP, jenseniiDB, gasseriDB, megasphaeraP, megasphaeraN, inersDB, inersDA, and mayoredad. The details are shown in Figure 8.

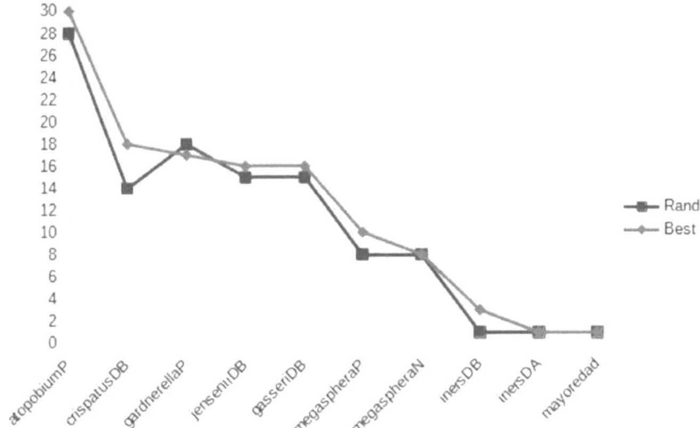

Figure 8. Frequency of antecedent elements for rand/1/bin and best/1/bin variants of the 15 tests.

5. Conclusions

The combination of the Apriori and DE algorithms enables the generation of subsets of rules with biological significance by utilizing a fitness function that incorporates the biological criteria used by experts. The analysis presented in this study demonstrates that the DE/rand/1/bin and DE/best/1/bin algorithms reveal that microorganisms such as Atopobium positive, Gardnerella positive, and L. Crispatus in low density have a greater interaction to present a VB+. The clinical findings coincide with the presence of these microorganisms, which reduce the density of lactobacilli such as L. Crispatus. However, age is not determining factor of a VB+ according to DE algorithms since it is the least frequent antecedent. This study highlights the use of DE algorithms and the integration of biologically significant rules into the objective function.

In that context, the use of DE algorithms and the integration of biological significance rules to the objective function give the expected results, obtaining mostly high-quality association rules. They comply with the requirements of the objective function by having at least two positive bacteria present.

From this perspective, the following three rules were found where there is only one bacterium and one lactobacillus:

- {jenseniiDB,megasphaeraP} → {VB+}

- {gardnerellaP,gasseriDB} → {VB+}
- {atopobiumP,crispatusDB} → {VB+}

In this sense, the validation of the expert indicates that the rules where a bacterium and a lactobacillus are present are those that can be useful for the classification of indeterminate cases, specifically in cases where L. crispatus and iners are not informative. For this reason, they cannot be ruled out and should be validated in other databases and biologically to find out their contribution to the development of the condition.

This approach provides concrete support to experts in identifying relationships that have not been explored or analyzed in the laboratory. The use of computational intelligence approaches in this field of study can be considered highly beneficial for designing new strategies to identify diseases and improve patient health. In future work, it is very important to continue with the validation of the rules by an expert and to carry out tests with a more robust dataset to integrate indeterminate cases, and other rules of biological significance to add penalties to the objective function. It is also proposed to create a new individual coding scheme that allows comparison with other evolutionary computation algorithms for association rule mining and includes parameter adjustment of the DE algorithm.

Author Contributions: Conceptualization, J.C.-R. and R.R.-L.; Formal analysis, E.M.-M.; Investigation, M.C.S.-G.; Resources, E.d.l.C.-H. All authors have read and agreed to the published version of the manuscript.

Funding: This research receives no funding from any agency.

Data Availability Statement: Restrictions apply to the availability of these data. Data was obtained from [13] and are available with the permission of [13].

Acknowledgments: The first author (CVU 769227) acknowledges support from the National Council of Science and Technology (CONACYT) of Mexico through a scholarship to pursue graduate studies at the Universidad Juárez Autónoma de Tabasco.

Conflicts of Interest: The authors declare no conflict of interest.

Abbreviations

The following abbreviations are used in this manuscript:

ARM	Association Rule Mining
DE	Differential Evolution
BV+	Bacterial Vaginosis Positive

References

1. Noormohammadi, M.; Eslamian, G.; Kazemi, S.N.; Rashidkhani, B. Association between dietary patterns and bacterial vaginosis: A case–control study. *Sci. Rep.* **2022**, *12*, 12199. [CrossRef] [PubMed]
2. Coudray, M.S.; Madhivanan, P. Bacterial vaginosis—A brief synopsis of the literature. *Eur. J. Obstet. Gynecol. Reprod. Biol.* **2020**, *245*, 143–148. [CrossRef] [PubMed]
3. Onywera, H.; Anejo-Okopi, J.; Mwapagha, L.M.; Okendo, J.; Williamson, A.L. Predictive functional analysis reveals inferred features unique to cervicovaginal microbiota of African women with bacterial vaginosis and high-risk human papillomavirus infection. *PLoS ONE* **2021**, *16*, e0253218. [CrossRef] [PubMed]
4. Xu, X.; Zhang, Y.; Yu, L.; Shi, X.; Min, M.; Xiong, L.; Pan, J.; Liu, P.; Wu, G.; Gao, G. A cross-sectional analysis about bacterial vaginosis, high-risk human papillomavirus infection, and cervical intraepithelial neoplasia in Chinese women. *Sci. Rep.* **2022**, *12*, 6609. [CrossRef] [PubMed]
5. Abou Chacra, L.; Fenollar, F.; Diop, K. Bacterial vaginosis: What do we currently know? *Front. Cell. Infect. Microbiol.* **2022**, *11*, 1393. [CrossRef] [PubMed]
6. Dhaenens, C.; Jourdan, L. Metaheuristics for data mining: Survey and opportunities for big data. *Ann. Oper. Res.* **2022**, *314*, 117–140. [CrossRef]
7. Telikani, A.; Gandomi, A.H.; Shahbahrami, A. A survey of evolutionary computation for association rule mining. *Inf. Sci.* **2020**, *524*, 318–352. [CrossRef]
8. Varol Altay, E.; Alatas, B. Performance analysis of multi-objective artificial intelligence optimization algorithms in numerical association rule mining. *J. Ambient Intell. Humaniz. Comput.* **2020**, *11*, 3449–3469. [CrossRef]

9. Fister, I.; Iglesias, A.; Galvez, A.; Del Ser, J.; Osaba, E.; Fister, I. Differential evolution for association rule mining using categorical and numerical attributes. In Proceedings of the 19th International Conference on Intelligent Data Engineering and Automated Learning (IDEAL 2018), Madrid, Spain, 21–23 November 2018; Springer: Berlin/Heidelberg, Germany, 2018; pp. 79–88.

10. Stupan, Ž.; Fister, I. NiaARM: A minimalistic framework for Numerical Association Rule Mining. *J. Open Source Softw.* **2022**, *7*, 4448. [CrossRef]

11. Zhang, A.; Shi, W. Mining significant fuzzy association rules with differential evolution algorithm. *Appl. Soft Comput.* **2020**, *97*, 105518. [CrossRef]

12. SuryaNarayana, G.; Kolli, K.; Ansari, M.D.; Gunjan, V.K. A traditional analysis for efficient data mining with integrated association mining into regression techniques. In *Proceedings of the 3rd International Conference on Communications and Cyber Physical Engineering (ICCCE 2020)*; Springer: Berlin/Heidelberg, Germany, 2021; pp. 1393–1404.

13. Sanchez-Garcia, E.K.; Contreras-Paredes, A.; Martinez-Abundis, E.; Garcia-Chan, D.; Lizano, M.; de la Cruz Hernandez, E. Molecular epidemiology of bacterial vaginosis and its association with genital micro-organisms in asymptomatic women. *J. Med. Microbiol.* **2019**, *68*, 1373–1382. [CrossRef] [PubMed]

14. Lin, H.K.; Hsieh, C.H.; Wei, N.C.; Peng, Y.C. Association rules mining in R for product performance management in industry 4.0. *Procedia CIRP* **2019**, *83*, 699–704. [CrossRef]

15. Agrawal, R.; Srikant, R. Fast algorithms for mining association rules. In Proceedings of the 20th International Conference on Very Large Data Bases (VLDB 1994), Santiago, Chile, 12–15 September 1994; Volume 1215, pp. 487–499.

16. Shigetoh, H.; Nishi, Y.; Osumi, M.; Morioka, S. Combined abnormal muscle activity and pain-related factors affect disability in patients with chronic low back pain: An association rule analysis. *PLoS ONE* **2020**, *15*, e0244111. [CrossRef] [PubMed]

17. Olow, A.K.; van't Veer, L.; Wolf, D.M. Toward developing a metastatic breast cancer treatment strategy that incorporates history of response to previous treatments. *BMC Cancer* **2021**, *21*, 212. [CrossRef] [PubMed]

18. de la Cruz Ruiz, F.; Canul-Reich, J. Reglas de asociación para el estudio de la vaginosis bacteriana. *Komputer Sapiens* **2022**, *II*, 26–30.

19. Hahsler, M. A Probabilistic Comparison of Commonly Used Interest Measures for Association Rules. Available online: https://mhahsler.github.io/arules/docs/measures (accessed on 15 March 2023).

20. Aggarwal, C.C.; Bhuiyan, M.A.; Hasan, M.A. *Frequent Pattern Mining Algorithms: A Survey*; Springer: Berlin/Heidelberg, Germany, 2014.

21. Storn, R.; Price, K. Differential Evolution—A Simple and Efficient Heuristic for Global Optimization over Continuous Spaces. *J. Glob. Optim.* **1997**, *11*, 341–359. [CrossRef]

22. Mezura-Montes, E.; Miranda-Varela, M.E.; del Carmen Gómez-Ramón, R. Differential evolution in constrained numerical optimization: An empirical study. *Inf. Sci.* **2010**, *180*, 4223–4262. [CrossRef]

23. Bramer, M. *Principles of Data Mining*; Springer: Berlin/Heidelberg, Germany, 2007; Volume 180, pp. 205–209.

24. Price, K.; Storn, R.M.; Lampinen, J.A. *Differential Evolution: A Practical Approach to Global Optimization*; Springer: Berlin/Heidelberg, Germany, 2006. [CrossRef]

25. Das, S.; Suganthan, P.N. Differential Evolution: A Survey of the State-of-the-Art. *IEEE Trans. Evol. Comput.* **2011**, *15*, 4–31. [CrossRef]

26. Zariffard, M.R.; Saifuddin, M.; Sha, B.E.; Spear, G.T. Detection of bacterial vaginosis-related organisms by real-time PCR for Lactobacilli, Gardnerella vaginalis and Mycoplasma hominis. *FEMS Immunol. Med. Microbiol.* **2002**, *34*, 277–281. [CrossRef] [PubMed]

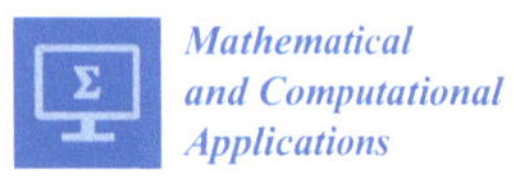

Mathematical and Computational Applications

Article

Vehicle Make and Model Recognition as an Open-Set Recognition Problem and New Class Discovery

Diana-Itzel Vázquez-Santiago, Héctor-Gabriel Acosta-Mesa * and Efrén Mezura-Montes

Artificial Intelligence Research Institute, Universidad Veracruzana, Veracruz 91097, Mexico;
diana.v.s@hotmail.com (D.-I.V.-S.); emezura@uv.mx (E.M. M.)
* Correspondence: heacosta@uv.mx

Abstract: One of the main limitations of traditional neural-network-based classifiers is the assumption that all query data are well represented within their training set. Unfortunately, in real-life scenarios, this is often not the case, and unknown class data may appear during testing, which drastically weakens the robustness of the algorithms. For this type of problem, open-set recognition (OSR) proposes a new approach where it is assumed that the world knowledge of algorithms is incomplete, so they must be prepared to detect and reject objects of unknown classes. However, the goal of this approach does not include the detection of new classes hidden within the rejected instances, which would be beneficial to increase the model's knowledge and classification capability, even after training. This paper proposes an OSR strategy with an extension for new class discovery aimed at vehicle make and model recognition. We use a neuroevolution technique and the contrastive loss function to design a domain-specific CNN that generates a consistent distribution of feature vectors belonging to the same class within the embedded space in terms of cosine similarity, maintaining this behavior in unknown classes, which serves as the main guide for a probabilistic model and a clustering algorithm to simultaneously detect objects of new classes and discover their classes. The results show that the presented strategy works effectively to address the VMMR problem as an OSR problem and furthermore is able to simultaneously recognize the new classes hidden within the rejected objects. OSR is focused on demonstrating its effectiveness with benchmark databases that are not domain-specific. VMMR is focused on improving its classification accuracy; however, since it is a real-world recognition problem, it should have strategies to deal with unknown data, which has not been extensively addressed and, to the best of our knowledge, has never been considered from an OSR perspective, so this work also contributes as a benchmark for future domain-specific OSR.

Keywords: open-set recognition; new class discovery; VMMR; CNN; contrastive loss; clustering; neuroevolution

Citation: Vázquez-Santiago, D.-I.; Acosta-Mesa, H.-G.; Mezura-Montes, E. Vehicle Make and Model Recognition as an Open-Set Recognition Problem and New Class Discovery. *Math. Comput. Appl.* **2023**, *28*, 80. https://doi.org/10.3390/mca28040080

Academic Editor: Leonardo Trujillo

Received: 14 March 2023
Revised: 28 June 2023
Accepted: 29 June 2023
Published: 3 July 2023

1. Introduction

Automatic vehicle make and model recognition (VMMR) aims to offer innovative services to improve the efficiency and safety of transportation networks. These services include intelligent traffic analysis and management, electronic toll collection, emergency vehicle notifications, the automatic enforcement of traffic rules, etc. In recent years, several authors have proposed and implemented different approaches and techniques to present solutions to the various challenges of VMMR such as the similar appearance of different vehicle models [1,2], variations in the images due to weather conditions or resolution [3–5], recognition through different key points or regions [6,7], etc. However, most of these solutions are designed within a *closed-set* approach, where it is assumed that all query data are well represented by the training set, and therefore these solutions lack mechanisms to detect during testing when an input sample does not belong to any of the predefined classes. These unforeseen situations are very likely to happen in real-life scenarios and drastically weaken the robustness of the models.

Every day, we have more and more access to labeled data, which makes data-hungry algorithms such as classification algorithms that employ supervised learning improve their classification accuracy by having more training information. However, it is unrealistic to think that we will be able to train these algorithms to recognize any object that may be presented to them. In the specific case of this application domain, it is estimated that there are currently more than 3300 vehicle makes in the world, for which models have been added and removed from the market, modifying the design in each generation and producing different versions of the same vehicle, which has made it very difficult to have a database containing enough examples of all the existing vehicles in circulation to correctly train a model. This limitation is very common in real-world recognition/classification tasks such as VMMR and, in most cases, results in misclassified vehicles because the algorithms were not prepared to deal with objects of unknown (novel) classes.

To solve this problem, some strategies have been proposed, such as periodically retraining the algorithms, incorporating an incremental update mechanism [8,9], using zero-shot [10,11] or one-shot (few-shot) [12,13] learning, etc. Although these strategies provide models with greater flexibility or the possibility of eventually increasing their classification potential, they do not address the fundamental problem of recognizing a novel class during testing (*open-set* problem). Scheirer et al. were the first to describe a more realistic scenario in which new classes not seen in training appear in testing and require classifiers not only to accurately classify objects of known classes but also to effectively deal with classes not considered in the training set [14]. They formalized this problem as *open-set recognition* (OSR) and proposed a solution called 1-vs-Set machine, where the risk of labeling a sample as known if it is far from the known data (*open space*) is measured, and its objective is to minimize this risk (*open-space risk*) by rejecting queries that lie beyond the reasonable support of the known data.

OSR led to extensive research that mostly focused on more effectively limiting the *open-space risk* [15–18], and little research was developed around efficiently performing *open-set recognition* and simultaneously discovering new classes hidden in the rejected instances. Some of the proposed solutions employed incremental learning [19], transfer learning [20,21], or clustering [22,23]. Although they achieved good results, most of them present limitations such as the determination of the number of new classes in a later or separate event from the recognition of novel instances, or the use of examples of unknown classes during validation, pretraining, or retraining stages as a strategy to fine-tune their representations/parameters; however, in OSR, there is almost never information of unknown classes.

In the specific case of VMMR, few works have been proposed that, although not described within an OSR framework, have mechanisms to deal with new classes. One of these studies was conducted by Nazemi et al. [3] from an anomaly detection approach. Their base system is capable of classifying 50 specific vehicle models, to which they added an anomaly detection based on a confidence threshold to identify vehicles that do not belong to any of these 50 classes. The "anomalies" are further classified based on their dimensions within two new classes: "unknown heavy" and "unknown light". Another approach was proposed by Kezebou et al. [12], with a few-shot learning approach requiring between 1 and 20 images for the generation of new classes.

In this paper, we propose to approach VMMR as an OSR problem extended for new class discovery. Since the known classes are supported by numerous well-labeled examples, we can very effectively train an image classification algorithm that employs supervised learning like convolutional neural networks (CNNs), which are the most widely used tool for this task. While these networks cannot deal with the recognition of new classes, their ability to extract meaningful features can be exploited to design a mechanism that can detect objects of new classes based on the distribution of feature vectors in the embedded space that, when aggregated between feature extraction and classification, would adopt an OSR approach. However, feature vectors are usually of high dimensionality, their distribution is not always clear, and there is no assurance that the behavior will be maintained in instances of unknown classes, which can complicate the representation and interpretation of the space to detect new classes. To tackle these problems, we propose to train a CNN with contrastive learning using the contrastive loss function during the training stage to reorganize the space where the feature vectors are mapped. Instead of separating the images with a hyperplane, the contrastive loss function brings similar images in near space (in terms of, e.g., Euclidean distance, cosine similarity, or some other metric) and moves dissimilar images away, generalizing this behavior on new unseen data.

Although there are CNN architectures such as VGG16, AlexNet, etc., that have achieved state-of-the-art results in the most well-known benchmarks such as ImageNet, CIFAR-100, etc., we propose a new CNN architecture designed from images of the application domain of this work (VMM) and the contrastive loss function using a neuroevolution technique to ensure consistent distribution of feature vectors within the embedded space, which serves as the main guide for a probabilistic model and a clustering algorithm that carry out the detection of objects of new classes and simultaneously discover their classes.

The remainder of this paper is organized as follows: Section 2 presents the related work. Section 3 describes the proposed methodology to approach VMMR as an OSR problem with an extension for new class discovery. This section also presents the proposed global scheme and delves deeper into each stage, detailing how the techniques of neuroevolution, contrastive loss function, the probabilistic model, and clustering are linked so as to achieve the general purpose. Section 4 details the tests performed, including the parameters and justifications for each test and the results obtained at each stage with their respective interpretation. Finally, the conclusions are drawn, and future work is discussed in Section 5.

2. Literature Review

2.1. Open-Set Recognition

OSR [14] introduced a more realistic scenario for real-world recognition/classification tasks, where new classes not seen during training appear at query time during testing. To deal with these unforeseen situations, OSR algorithms have to consider that their knowledge of the world is incomplete and formulate strategies to minimize the risk of considering an unknown instance as known. The authors of [14] formalized this risk as an *open-set risk* (R_O) in a probabilistic formulation (Equation (1)) as the relative measure of positively labeled open space O compared with the overall measure of positively labeled space S_O.

$$R_O(f) = \frac{\int_O f(x)dx}{\int_{S_O} f(x)dx'}$$

(1)

where f denotes a measurable recognition function.

Numerous studies have been conducted to minimize the risk of open sets and more effectively reject objects of unknown classes [15–18], which is the main goal of OSR. However, in a more desirable context, an OSR should go further and discover the unknown classes hidden inside the rejected objects. Within this context, some authors have proposed the use of incremental learning [19], transfer learning [20,21], or clustering [22]. Bendale and Boult [19] extended the *open-set recognition* problem to open-world recognition (OWR) to jointly consider the OSR and incremental learning of new classes. They proposed that an effective OWR system must perform four tasks: detecting unknown objects, choosing which objects to label for addition to the model, labeling these objects, and updating the model. In their paper, they presented the NNO algorithm. However, the tasks they proposed are not automated in the NNO, they require human supervision for labeling, and the determination of the number of classes happens in a later event after the recognition of new instances. In [20], Wang et al. studied the OWR problem in more detail by incorporating transfer learning to transfer knowledge from old classes to new ones. However, they needed to retrain their model with the presence of samples of unknown classes, which is a limitation since, in an OSR context, information from unknown classes is almost never available. A similar knowledge transfer proposal was presented by Han et al. [21]; however, they have the same limitation since their idea was to pretrain their model with images of known and unknown classes. Another interesting proposal was developed in [22] by some authors of [21], where they also took advantage of the knowledge transfer approach but used clustering. The main limitation of this work is that they determined the number of new classes in a separate event from the discovery of new instances, which, as in [19], can lead to suboptimal solutions.

To our knowledge, the most related work to ours, in terms of simultaneously discovering the objects of new classes and these classes themselves, is [23]. They introduced a collective/batch decision-strategy-based OSR framework (CD-OSR) by slightly modifying the hierarchical Dirichlet process (HDP). CD-OSR first involves a co-clustering process in the training phase to obtain the appropriate parameters. In the testing phase, each known class is modeled as a group using a Gaussian mixture model (GMM) with an unknown number of subclasses (one or more subclasses representing the same class can be obtained), and the entire test set (collective/batch) is treated in the same way. Then, all the groups are co-clustered under the HDP framework, and each one is labeled as one of the known classes or as unknown, depending on whether the subclass assigned to it is associated with a known class or not. Other works on OSR such as [24] also took advantage of Gaussian distributions to obtain discriminative representations of the data to detect unknowns and classify knowns.

Another proposal that may be related to our work was presented in [18], where the OSR problem was addressed within a transfer learning approach using contrastive learning to model the data. They also highlighted the importance of developing and testing OSR solutions with domain-specific databases to test their efficiency in dealing with real-world applications. Unfortunately, this solution only rejects objects of new classes and does not include the discovery of their classes.

2.2. Neuroevolution and Contrastive Learning

In the field of evolutionary computation (EC), a technique called neuroevolution (NE) emerged to optimize artificial neural networks (ANNs) at different levels. Its current overall process can be summarized as follows: A random population of individual networks is generated (with a neural coding), real networks are created from them, and the networks are evaluated with a function that measures the quality of the results (the fitness function). The networks with the highest fitness are selected, certain random changes are introduced to generate offspring from them, and a new population (generation) is selected. This process is repeated until a certain level of fitness or number of generations is reached.

NE has achieved excellent results in this optimization task and has rapidly advanced toward the optimization of CNN topologies [25–32]. A crucial point in the performance of NE algorithms is neural encodings, which contain the topology information of an ANN and therefore have a great impact on the complexity of the search space. So, in order to implement this technique in CNNs, NE was faced with the problem of designing neural encodings that could abstract the parameters of CNNs in order to deal with these highly complex architectures. There are two types of neural encodings that are commonly employed: direct and indirect. Among the proposals using an indirect coding framework, we find works such as [25–27], and in the case of works that used indirect encodings, we find proposals such as [28–30]. In recent years, researchers have started to study a "hybrid" neural coding, which combines elements of the encodings mentioned above to eliminate some of their limitations. These "hybrid" representations have proved to be very useful to distribute the CNN representations in different substructures, leading to improvement in the search [31–33]. The advantages and disadvantages of different encoding schemes, as well as important niches of opportunity for future research, were described in detail in [34].

Although NE algorithms have a strategy to determine how well individuals are meeting the criterion (or criteria) being optimized (the fitness function), CNNs have their own strategy to quantify how close their predictions are to the expected output (the loss function), for which cross-entropy or negative log-likelihood are some of the most frequently used functions. However, the study of these functions has continued, and alternatives have been proposed that have achieved superior results. In these advances, the supervised contrastive loss function (SupCon) [35] was developed following the contrastive learning approach but in a supervised environment, which allowed it to maintain the principle of mapping examples into the embedded space of contrastive learning (distance is minimized in terms of Euclidean distance, cosine similarity, etc., between similar objects and maximized for dissimilar objects) but take advantage of labeled data.

Although there are marked differences between various versions of contrastive loss functions, the family of contrastive loss functions, in general, considers the following: For a set of N randomly sampled sample/label pairs (batch), $\{x_k, y_k\}\ k = 1 \ldots N$ is considered; the corresponding batch used for training (multiviewed batch) consists of $2N$ pairs, $\left\{\tilde{x}_\ell, \tilde{y}_\ell\right\}_{\ell=1\ldots 2N}$, where $\tilde{x}_{2k}$ and $\tilde{x}_{2k-1}$ are two random augmentations ("views") of $x_k (k = 1 \ldots N)$ and $\tilde{y}_{2k-1} = \tilde{y}_{2k} = y_k$. Given the above, SupCon is calculated as follows:

$$\mathcal{L}_{in}^{sup} = \sum_{i \in I} \mathcal{L}_{in,i}^{sup} = \sum_{i \in I} -\log\left\{\frac{1}{|P(i)|} \sum_{p \in P(i)} \frac{\exp(z_i \cdot z_p / \tau)}{\sum_{a \in A(i)} \exp(z_i \cdot z_a / \tau)}\right\} \tag{2}$$

where $i \in I \equiv \{1 \ldots 2N\}$ is the index of an augmented sample (anchor), $A(i) \equiv I \setminus \{i\}$ is the set of all the indices of the samples different than i, $P(i) \equiv \left\{p \in A(i) : \tilde{y}_p = \tilde{y}_i\right\}$ is the set of all positive sample indices different than i, $|P(i)|$ is its cardinality, the $\bullet$ symbol denotes the inner (dot) product, and τ is a scalar temperature parameter. SupCon's formulation generalizes the SimCLR loss function [36] to an arbitrary number of positive examples to deal with scenarios in which labels are available so that it is known that more than one sample can belong to the same class.

3. Materials and Methods

This section describes the methodology proposed to approach VMMR as an OSR problem with an extension for new class discovery. Figure 1 shows the overall process of our proposal, and the following subsections describe the process in detail, covering the following objectives:

1. Employ an NE algorithm and contrastive learning to design a domain-specific CNN that generates feature vectors spatially close in terms of cosine distance if the instances belong to the same class and distant if they belong to different classes, preserving this behavior in instances of unknown classes.
2. Implement a mechanism between the feature extraction and classification sections of the CNN capable of detecting objects of unknown classes and simultaneously discovering their classes, taking the mapping of feature vectors, described in the previous objective, as the main guide.
3. Run a series of tests using the test set that includes images of classes with which the CNN was designed and trained (known) and images of new classes (unknown) to test that the algorithm is able to detect objects of unknown classes and simultaneously discover their classes.
4. Classify images of known classes with a classification accuracy above 90%.

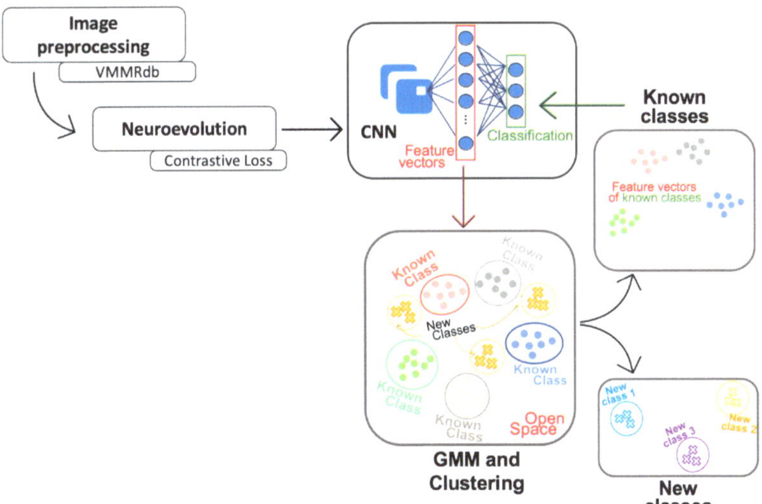

Figure 1. The proposed global process to approach VMMR as an OSR problem with an extension for new class discovery.

3.1. Dataset

The VMMRdb database [37] (available at https://github.com/faezetta/VMMRdb, accessed on 13 March 2023) was used in this work since it is one of the most cited in the specialized literature [12,38,39]. Only eight classes from the VMMRdb database were used and were manually filtered to retain only unduplicated images showing the rear view of the vehicles (i.e., samples of each class were not balanced). The filtered images were transformed to grayscale, resized to 28×28 pixels, and normalized with a mean of 0.456 and a standard deviation of 0.224.

Of the eight classes, five classes were used as *"known classes"*: Chevrolet Silverado 2004, Ford Explorer 2002, Ford Mustang 2000, Honda Civic 2002, and Nissan Altima 2005. A sample of six images of each *"known class"* was used in the NE process to design the domain-specific CNN for VMMR. For the training of the resulting CNN from the NE process, the largest number of examples per class was needed, which had to be the same among different classes. However, due to the number of available samples in the database and the image filtering mentioned above, the final number of functional samples per *"known class"* varied between 75 and 250 images. Among the functional samples, three images of each class were kept for testing the complete OSR framework, and the rest were subjected to a data augmentation process to balance the number of examples per class, resulting in 250 images of each *"known class"*. Furthermore, 200 images were used to train the CNN and model the *"known classes"* with a Gaussian mixture model (GMM), and the remaining 50 images were used to test the CNN classification accuracy and define the threshold of *"known classes"* in the GMM.

From the three remaining classes chosen from the database (Acura RSX 2003, Chevrolet Avalanche 2009, and Ford Escape 2011), three images of each class were chosen to only be used during the testing stage to represent *"unknown classes"* and validate that the proposed approach can detect them and discover their classes.

3.2. Neuroevolution and Contrastive Loss

One of the main objectives of this work is to exploit the ability of a CNN to extract meaningful features for designing a mechanism to detect objects of unknown classes based on the distribution of the feature vectors in the embedded space. To facilitate the interpretation of the embedded space, feature vectors extracted using the CNN are considered to be spatially close in terms of cosine similarity if they belong to the same class and spatially distant if they belong to different classes, maintaining this behavior even if the classes are unknown. According to the state-of-the-art review, adding the contrastive loss function to the CNNs causes the feature vectors to be mapped in near space (in terms of, e.g., Euclidean distance, cosine similarity, or some other metric) if they are similar and far if they are dissimilar.

Although there are CNN architectures such as VGG16, AlexNet, etc., that have achieved state-of-the-art results in the most well-known benchmarks such as ImageNet, CIFAR-100, etc., we propose a new domain-specific architecture that would generate the previously described behavior in feature vectors, using the images mentioned in Section 3.1, an NE algorithm called DeepGA [33] (shown in Algorithm 1), and SupCon [35] expressed in Equation (2).

In [35], the authors made their PyTorch implementation of SupCon generally available (https://t.ly/supcon, accessed on 13 March 2023), and this was used in this work as a loss function in the CNNs generated in the NE process with DeepGA. (Originally, the negative log-likelihood loss was used.) The fitness function of DeepGA (Algorithm 1, line 15) was also modified to measure the desired behavior in feature vectors since optimization was the objective of our study. Thus, as the fitness function, we used the value of SupCon in the last training epoch of each generated CNN. Since the loss function decreases as the desired output is approached, DeepGA was set to work as a minimization problem, i.e., as the generated CNNs approached the desired target, the value of the loss/fitness function decreased.

The hybrid coding employed in DeepGA allows the algorithm to consider the number of fully connected layers and their corresponding number of neurons in its search for the best solution. However, during the NE process, it was detected that leaving the number of fully connected layers to DeepGA only increased the complexity and execution time since with only two fully connected layers, classification accuracies above 90% were achieved. To limit the number of fully connected layers during the evolutionary process, the first level of the mutation operator was modified. At the first level of the mutation operator, if $U_1(0,1) > 0.5$, a new block is added, and if $U_2(0,1) > 0.5$, the added block is a fully connected layer; then, the operator was modified so that if $U_2(0,1) > 0.5$, no block is added. This modification is shown in line 4 of Algorithm 2, which shows the mutation operator of DeepGA. This ensures that, during the whole evolutionary process, the generated networks only have two fully connected layers, allowing the algorithm's search to focus on the blocks of convolutional layers since they would be in charge of generating the feature vectors with the desired behavior.

To access the feature vectors generated using the CNNs, the *CNN class* of DeepGA, which builds the model for training and testing, was modified. As output, this class only generated the probabilities of the images belonging to the different classes. The modification consisted of the addition of the flattened outputs of the convolutional block (feature vectors) to the original output to be able to access them in the next process (i.e., to distinguish objects from new classes and simultaneously discover these classes).

The last modification to the DeepGA algorithm was an improvement in image reading. The PyTorch ImageFolder function was used to be able to read the images of all classes in a single process instead of reading the images of each class individually.

Algorithm 1: DeepGA pseudocode.

1   **Input:** A population P of N individuals. The number of generations T,
2   crossover rate $CXPB$, mutation rate $MUPB$, tournament size $TSIZE$.
3   **Output:**
4   Initialize population (training the networks).
5   $t \leftarrow 1$
6   **while** $t \leq T$ **do**
7   Select $N/2$ parents with probabilistic tournament selection
8   Offs $\leftarrow$ {}
9   **while** $|$ Offs $| < N/2$ **do**
10   Select two random parents $p1$ and $p2$.
11   **if** random$(0,1) \leq CXPB$ **then**
12              $O1, O2 \leftarrow$ Crossover$(p1, p2)$ // Crossover
13   **if** random$(0,1) \leq MUPB$ **then**
14   Mutation$(O1, O2)$ // *Mutation (modified)*
15   fitness$(O1, O2)$ (Equation (1)) // *Evaluation (modified)*
16          $P \leftarrow P \cup$ Offs
17   Select the best N individuals in P as survivals.
18   **end**
19 **end**

Algorithm 2: Mutation process DeepGA.

1   **if** random$(0,1) \leq MUPB$ **then**
2   **if** random$(0,1) \leq U1$ **then** // Adding a new block
3   **if** random$(0,1) \leq U2$ **then**
4   A convolutional block is added // *Removed*
5   **else**
6   A fully connected block is added
7   **else** // Restarting a block
8   **if** random$(0,1) \leq W1$ **then**
9   Restarting a convolutional block
10   **else**
11   Restarting a fully connected block

3.3. Neuroevolved CNN

The CNN architecture with the best fitness generated using DeepGA and SupCon was split to fulfill two purposes. First, the goal was to train the convolutional block with the contrastive loss function and the fully connected block with the cross-entropy loss function, using the full test set described in Section 3.1 (200 images of each of the five *"known classes"*), and to perform a classification accuracy test momentarily assuming a closed-set environment to validate that a good classification accuracy could be obtained since it is an essential point for OSR. Second, we sought to have the feature extraction process and the classification process separate since the detection of new class objects and the discovery of their classes must be accomplished between these events.

3.4. Gaussian Mixture Model (GMM) and Clustering

The main objective of this work is to approach VMMR as an OSR problem with an extension for the discovery of new classes. To achieve this, we divided our strategy into two phases, both relying on the consistent distribution of feature vectors in the embedded space generated using DeepGA and SupCon.

The first phase consisted of extracting the feature vectors from the images used to train the CNN and validating its classification accuracy. The feature vectors were compressed using principal component analysis (PCA) where the second and third components, which contributed 27.81 and 20.97 to the percentage of variance, respectively, were selected to perform a linear regression on the original feature vectors to obtain their projections. As mentioned in Section 3.1, with the same proportion of data with which the CNN was trained and tested (80–20%), the 2D projections of the feature vectors were used to model each *"known class"* with a Gaussian mixture model (GMM) and define a recognition threshold of *"known classes"*. In the test stage, where objects of both known and unknown classes were included, the GMM divided the objects as a group of unknown classes that did not pass the threshold and subsets of known classes whose probabilities matched the *"known class"* models. The above only served as a partial guide in the recognition of new class objects since, in the second phase of the strategy, a multiobjective clustering algorithm with automatic determination of the number of clusters (MOCKs) was employed and optimized with a multiobjective evolutionary algorithm (MOEA), called NSGA-II [40].

In the second phase, the clustering algorithm grouped the feature vectors extracted using the domain-specific CNN without any modification in their dimensionality. Since the GMM can determine the objects of known classes and their respective classes with some confidence, due to the threshold, we compared the subgroups of known classes generated using the GMM with the solutions of MOCK/NSGA-II to select the individual from the population with the highest similarity, where different criteria were used. First, the solutions that grouped the instances that the GMM determined as known and were in the same structures (subgroups) as the GMM had a higher score (one point for each shared structure). Although all the solutions of the clustering algorithm were optimal for the problem, we selected the solution that had the highest score (higher match with the GMM in the known classes) and was closest to the knee point as the "best solution".

Finally, we determined which clusters of the "best solution" contain known objects and separated them from the clusters containing unknown objects in a similar way to how the solutions were scored. Then, since the GMM also detected the objects of unknown classes (the objects that did not pass the threshold) with some confidence, those clusters that only contained objects that the GMM determined as unknown were automatically determined as new classes. After these processes, if there were still undetermined clusters as known or unknown, the number of known and unknown instances within the undetermined clusters were counted (according to the GMM determination), and the clusters were defined in the same category as that containing the majority of instances or as unknown if it contained the same number of examples to try to mitigate the *open-set risk*.

At the end of this strategy, the objects of the clusters that were determined as known were entered into the CNN's fully connected block to be classified, and the clusters that were determined to be unknown were the newly discovered classes of the objects detected as unknown.

The original version of the MOCK algorithm was proposed in 2004 by Handl et al. [41] and employed the MOEA called PESA-II. In 2016, Martinez-Peñaloza et al. [42] managed to improve the results by using the MOEA NSGA-II instead of PESA-II. In the MOCK version improved with NSGA-II, individuals are ranked and sorted according to their non-dominated level, and a crowding distance is used to perform niching. This distance is calculated for each member to be used by the selection operator to maintain a diverse front by ensuring that each member stays a crowding distance apart. Algorithm 3 shows NSGA-II's pseudocode.

Algorithm 3: NSGA-II pseudocode.

1	Initialize Population
2	Generate random population -size M
3	Evaluate Objective values
4	Assign Rank (level) Based on Pareto Dominance -"sort"
5	Generate Child Population
6	Binary Tournament Selection
7	Recombination and Mutation
8	**for** $i = 1$ to *Number of Generations* **do**
9	**for** *each Parent and Child in Population* **do**
10	Assign Rank (level) Based on Pareto –"sort"
11	Generate sets of non-dominated fronts
12	Loop (inside) by adding solutions to next generation starting
13	from the "first" front until M individuals found determine
14	crowding distance between points on each front
15	**end**
16	Select points (elitist) on the lower front (with lower rank) and
17	are outside a crowding distance. Create next generation
18	Binary Tournament Selection
19	Recombination and Mutation
20	**end**

4. Experiments and Results

This section describes the experiments and results obtained from our proposal to approach VMMR as an OSR problem with an extension for new class discovery.

For the neuroevolution process of CNNs performed with DeepGA [33] and Sup-Con [35], as mentioned in Section 3.1, six images of each *"known class"* taken from the VMMRdb [37] database were used.

The parameters described in Table 1 were used to initialize the population. Due to time constraints, it was not possible to use a parameter calibration program. Then, the parameters for the evolutionary process were calibrated manually. The parameters with which the best results were obtained, and which were used to generate the CNNs are shown in Table 2. The different values that each hyperparameter could have during the evolutionary process were the same as those established by the author of DeepGA and are presented in Table 3.

Table 1. Parameters for the population initialization.

Parameter	Values
Min number of convolutional layers	4
Max number of convolutional layers	9
Min number of fully connected layers	1
Max number of fully connected layers	1

Table 2. Parameters for the evolutionary process.

Parameter	Values
Population size	20
No. of generations	100
Tournament size	5
Crossover rate	0.7
Mutation rate	0.7
No. of epochs per individual	20

Table 3. Values that each hyperparameter could have during the evolutionary process.

Hyperparameter	Values
No. of filters *	{2, 4, 8, 16, 32}
Filter size *	{2, 3, 4, 5, 6, 7, 8}
Pooling type *	{Max, Avg}
Pooling size *	{2, 3, 4, 5}
No. of neurons	{4, 8, 16, 32, 64, 128}

Rows marked with * correspond to the convolutional block, while the unmarked row corresponds to the fully connected block.

Seven executions of the NE process with DeepGA and SupCon were performed. Figure 2 shows the convergence curves of the seven executions and a short analysis of the fitness values obtained in each one. As can be seen, all the executions started with a fitness within a range of 3.3 and 3.5, and most reached premature convergence or stalled at local optima. However, one of the executions (marked in red) achieved a more accurate search space that led to a fitness value of 1.96096.

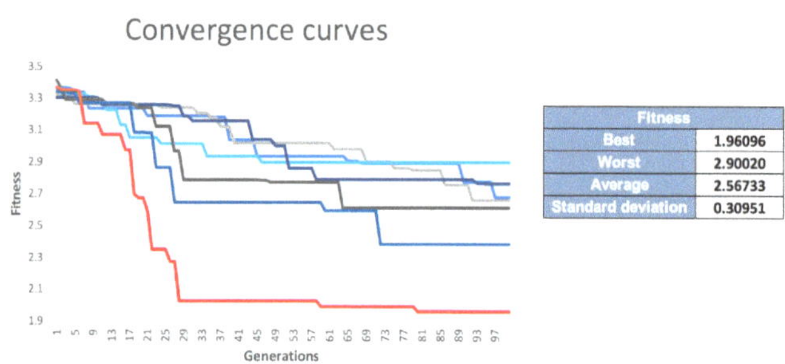

Figure 2. Convergence curves and the most relevant data of the fitness values of the seven runs of the NE process.

The CNN with the best fitness obtained in the NE process (henceforth referred to as the "domain-specific CNN") had a value of 1.96096, which was the value of SupCon in the last training epoch of the CNN (its justification is explained in Section 3.2 in more detail) and took 7 h to execute in the Visual Studio Code editor running on a MacBook Pro with a 2.2 GHz Quad-Core Intel Core i7 processor with 16 GB 1600 MHz DDR3 of memory. Figure 3 illustrates the architecture in terms of its encoding. In the first level, it can be observed that the architecture has 13 convolutional blocks (each one consisting of a single convolutional layer) and 2 fully connected blocks (each one comprising a single layer and a fully connected layer). The last convolutional block/layer generates feature vectors of 288 features. At the second level, the binary string defines the connectivity between convolutional blocks. Each bit represents the connectivity of a previous non-consecutive layer, starting from the third block. For a better understanding, we will explain three examples to understand the connections. The third convolutional block (first bit marked in red) can only have connections with previous blocks that are not its immediately previous consecutive block, so the third block cannot have a connection with the second convolutional block, but it can with the first one, which is why only one bit is assigned to it, and the bit value is 1. This means that there is a connection, which is represented by the red line on the first level. The next two bits (green) are for the fourth block, which can have a connection with the first or second block, and since the bit values are 1, both connections exist (represented by the green lines on the first level). A different case is shown in the next three bits (highlighted in yellow) assigned to the fifth block, which can have connections with the first, second, and third blocks; however, of those three bits, only the second one has a value of 1, which means that the fifth block only connects to the second block.

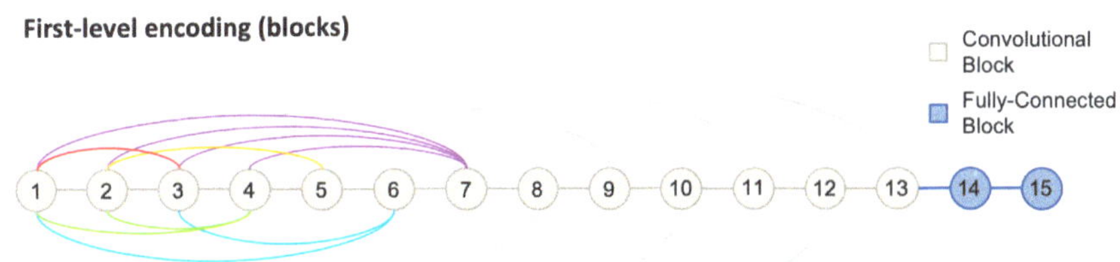

Figure 3. CNN architecture with the best fitness obtained using DeepGA and SupCon. The first level (blocks) represents simple convolutional operations instead of a set of convolutional layers. The second level (binary string) determines the skip connections received from the third block onward. Each bit represents the connectivity from previous layers, from the third layer onward.

To verify that the domain-specific CNN could generate the feature vectors extracted spatially close in terms of cosine similarity if they belong to the same class and far apart if they belong to different classes, a distance matrix using cosine similarity as the metric was generated with the feature vectors obtained in the last training epoch of the domain-specific CNN. On the same feature vectors, the t-SNE [43] technique was used to reduce the dimensionality from 288 to 2 in order to visualize them in a two-dimensional plane. The results of the distance matrix and t-SNE are shown in Figure 4. By means of these two techniques, it could be seen that the desired behavior in the feature vectors was achieved.

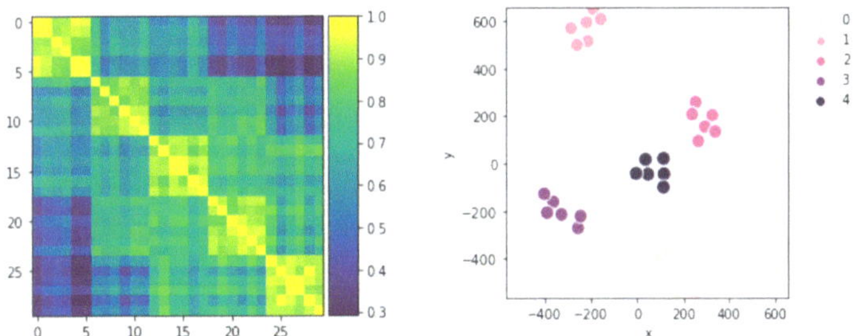

Figure 4. Distance matrix (with cosine similarity) and the projection in a two-dimensional plane of the feature vectors of the last training epoch of the domain-specific CNN.

As mentioned in Section 3.3, the domain-specific CNN was split to train the convolutional block with the contrastive loss function and the fully connected block with the cross-entropy loss function. For the training process, 1000 images of rear views of vehicles of the five *"known classes"* (200 images of each class) were used. A classification accuracy test was performed using 250 images of rear views of vehicles of the five *"known classes"* (50 images of each class) momentarily assuming a closed-set environment to validate that good classification accuracy was being achieved since it is an essential point for OSR. A 90% classification accuracy was reached during this test; more details regarding the data used are presented in Section 3.1.

The next test was to verify that the domain-specific CNN could generate feature vectors spatially close in terms of cosine similarity if they belong to the same class and far apart if they belong to different classes. This behavior was maintained in objects of unknown classes since the detection of objects of new classes and the discovery of their classes depended on this behavior. For this, the testing images, both the nine testing images of *"unknown classes"* shown in Figure 5 on the right and the fifteen images of the five *"known classes"* shown in Figure 5 on the left, were entered into the convolutional block of the domain-specific CNN to extract their feature vectors. To visualize the results, which are shown in Figure 6, a distance matrix using cosine similarity as the metric was generated, and a two-dimensional projection was performed using linear regression with the two components described in Section 3.4. Figure 6 shows that the domain-specific CNN managed to generate feature vectors close in terms of cosine similarity if they belonged to the same class and distant if they belonged to different classes and managed to maintain such behavior even in objects of *"unknown classes"*.

Figure 5. Sample images from the VMMRdb database. Both *"known"* (left) and *"unknown"* (right) classes were used during testing.

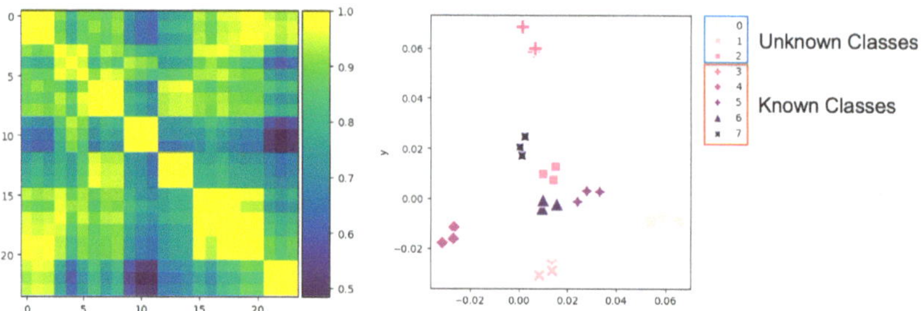

Figure 6. Distance matrix (with cosine similarity) and the projection in a two-dimensional plane of the feature vectors of the samples shown in Figure 5.

Later, the feature vectors of the images used to train the domain-specific CNN and validate its classification accuracy were compressed to two dimensions, and a linear regression was performed on these feature vectors to obtain their projections using the two components described in Section 3.4. With the projections of the 1000 images used to train the domain-specific CNN, we modeled the *"known Classes"* using a GMM, and the distribution of the Gaussians is shown in Figure 7. We then defined a *"known class"* recognition threshold within the GMM with a value of 9.999, using the projections of the 250 images that were used in the domain-specific CNN classification accuracy test.

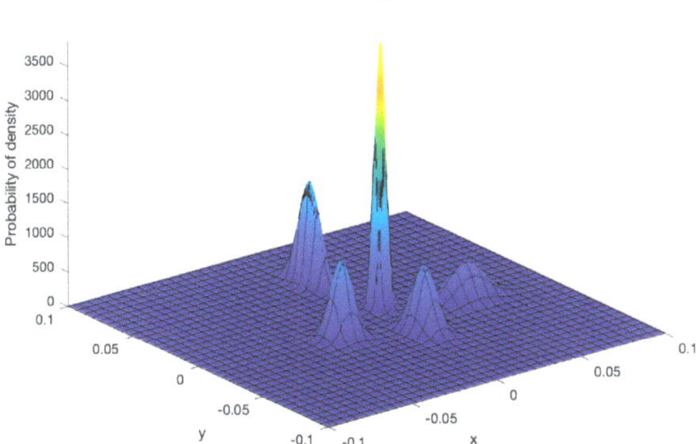

Figure 7. Distribution of the "known classes" using GMM.

Finally, the strategy proposed in Section 3.4 was carried out to detect the objects of new classes and discover their classes. The first step was to enter the two-dimensional projections used in Figure 6, which contain objects of both known and unknown classes (Figure 5), into the GMM to obtain their density probabilities. As its output, the model provided the probability of each object belonging to the known classes, and the threshold allowed us to set a probability limit for known or unknown classes. As can be seen in Table 4, the objects of known classes were correctly identified within their classes, and in the case of the objects of unknown classes, it can be seen that with the limit marked by the threshold, eight of the nine objects were correctly identified as unknown. A clearer representation of the results obtained can be seen in Figure 8, which indicates that the GMM divided the objects as a set of unknown classes that did not pass the threshold as well as the subsets of known classes whose probabilities matched the known class models.

Table 4. Probabilities of the test instances obtained with the GMM. The probabilities that exceeded the threshold (9999) are marked in orange.

	x	y	Real Label		Probability (Class 0)	Probability (Class 1)	Probability (Class 2)	Probability (Class 3)	Probability (Class 4)
0	5.84×10^{-2}	-7.59×10^{-3}	0		1.000×10^{0}	8.152×10^{-27}	7.294×10^{-57}	1.650×10^{-89}	1.483×10^{-73}
1	5.40×10^{-2}	-9.16×10^{-3}	0		1.000×10^{0}	4.920×10^{-22}	6.332×10^{-48}	8.417×10^{-77}	1.826×10^{-67}
2	6.52×10^{-2}	-9.41×10^{-3}	0		1.000×10^{0}	1.412×10^{-33}	3.304×10^{-75}	1.581×10^{-112}	1.321×10^{-88}
3	1.37×10^{-2}	-2.67×10^{-2}	1		4.068×10^{-17}	1.000×10^{0}	6.451×10^{-33}	3.341×10^{-22}	8.823×10^{-29}
4	1.38×10^{-2}	-2.92×10^{-2}	1		6.396×10^{-19}	1.000×10^{0}	1.408×10^{-37}	1.556×10^{-23}	2.039×10^{-31}
5	8.56×10^{-3}	-3.10×10^{-2}	1		1.170×10^{-22}	1.000×10^{0}	3.308×10^{-43}	6.810×10^{-24}	3.034×10^{-27}
6	1.43×10^{-2}	7.13×10^{-3}	2	Known	1.020×10^{-4}	3.865×10^{-6}	9.999×10^{-1}	4.394×10^{-9}	2.252×10^{-11}
7	1.52×10^{-2}	1.24×10^{-2}	2		1.451×10^{-4}	9.982×10^{-9}	9.999×10^{-1}	7.854×10^{-8}	1.672×10^{-11}
8	1.02×10^{-2}	9.51×10^{-3}	2		5.701×10^{-5}	4.896×10^{-6}	9.999×10^{-1}	5.240×10^{-7}	1.290×10^{-8}
9	1.53×10^{-3}	6.82×10^{-2}	3		7.202×10^{-36}	3.243×10^{-35}	5.417×10^{-87}	1.000×10^{0}	4.510×10^{-58}
10	6.30×10^{-3}	5.86×10^{-2}	3		3.311×10^{-28}	6.318×10^{-30}	2.282×10^{-61}	1.000×10^{0}	1.830×10^{-42}
11	6.97×10^{-3}	5.96×10^{-2}	3		2.972×10^{-29}	1.191×10^{-30}	1.055×10^{-63}	1.000×10^{0}	1.461×10^{-43}
12	-2.66×10^{-2}	-1.18×10^{-2}	4		2.727×10^{-28}	2.310×10^{-17}	1.384×10^{-65}	1.521×10^{-56}	1.000×10^{0}
13	-2.68×10^{-2}	-1.64×10^{-2}	4		8.075×10^{-32}	2.380×10^{-17}	1.771×10^{-72}	2.738×10^{-59}	1.000×10^{0}
14	-3.14×10^{-2}	-1.81×10^{-2}	4		3.329×10^{-36}	3.051×10^{-21}	3.378×10^{-87}	2.106×10^{-71}	1.000×10^{0}
15	2.79×10^{-2}	2.76×10^{-3}	5		9.995×10^{-1}	2.116×10^{-6}	5.257×10^{-4}	6.300×10^{-19}	3.457×10^{-21}
16	3.32×10^{-2}	2.46×10^{-3}	5		1.000×10^{0}	2.820×10^{-9}	3.713×10^{-9}	7.638×10^{-27}	2.355×10^{-27}
17	2.41×10^{-2}	-1.54×10^{-3}	5		9.779×10^{-1}	4.606×10^{-3}	1.751×10^{-2}	2.796×10^{-15}	1.285×10^{-18}
18	9.74×10^{-3}	-4.05×10^{-3}	6	Unknown	4.675×10^{-5}	9.986×10^{-1}	1.350×10^{-3}	2.120×10^{-9}	2.322×10^{-8}
19	1.57×10^{-2}	-2.33×10^{-3}	6		1.208×10^{-2}	4.648×10^{-1}	5.231×10^{-1}	9.067×10^{-10}	1.078×10^{-11}
20	1.01×10^{-2}	-7.46×10^{-4}	6		9.939×10^{-4}	5.405×10^{-1}	4.585×10^{-1}	7.914×10^{-8}	2.899×10^{-7}
21	4.78×10^{-4}	2.02×10^{-2}	7		5.100×10^{-2}	1.262×10^{-4}	1.164×10^{-2}	9.322×10^{-1}	5.056×10^{-3}
22	1.33×10^{-3}	1.69×10^{-2}	7		1.022×10^{-1}	1.459×10^{-3}	6.103×10^{-1}	2.326×10^{-1}	5.334×10^{-2}
23	2.57×10^{-3}	2.45×10^{-2}	7		2.779×10^{-4}	8.184×10^{-8}	1.406×10^{-5}	9.997×10^{-1}	4.287×10^{-7}

Known 5 Groups

[0,1,2,16] [3,4,5] [6,7,8] [9,10,11] [12,13,14]

Unknown 1 Group

[15,17,18,19,20,21,22,23]

Figure 8. Instances grouped according to GMM probabilities and the threshold.

As previously mentioned, the GMM results were the first phase of the strategy and only served as a partial guide in the recognition of objects of unknown classes. In the second phase of the strategy, a clustering algorithm called MOCK was used, which was enhanced with NSGA-II. For the second phase, the 24 feature vectors without projection (288 features) were entered into the clustering algorithm. For execution, the algorithm was run with the parameters shown in Table 5.

Table 5. Parameters for the clustering algorithm MOCK/NSGA-II.

Parameter	Values
Population size (M)	9
Nearest neighbors (L)	2
Number of generations	10

The nine final individuals of the clustering process are shown in Figure 9 in terms of their fitness values, and Figure 10 shows how the vectors were grouped in different structures.

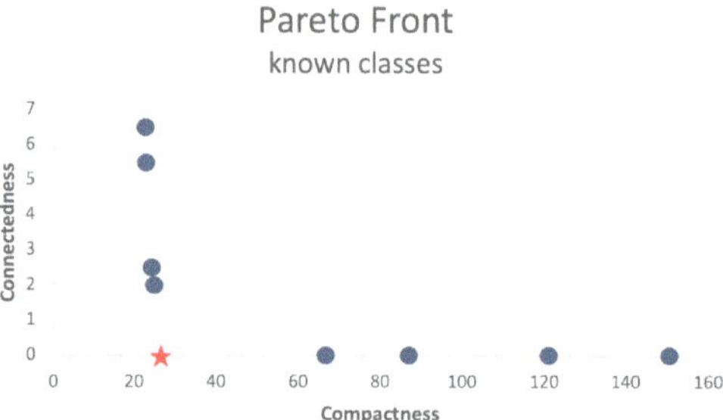

Figure 9. The Pareto front generated using the MOCK clustering algorithm improved with NSGA-II. The individual marked with a red star is the knee point.

```
[[0, 2, 1], [10, 9, 11], [12, 13, 14], [16], [17, 15], [18, 20], [19], [21, 22, 23], [3, 4, 5], [6, 8, 7]]
[[0, 2, 1], [10, 9, 11], [12, 13, 14], [16, 15], [17], [18, 20], [19], [21, 22, 23], [3, 4, 5], [8, 6, 7]]
[[0, 2, 1], [10, 9, 11], [12, 13, 14], [16, 15], [17], [18, 19, 20], [21, 22, 23], [3, 4, 5], [7, 8, 6]]
[[0, 2, 1], [10, 9, 11], [12], [13, 14], [16, 17, 15], [18, 19, 20], [21, 22, 23], [3, 4, 5], [8, 6, 7]]
[[0, 1, 2], [10, 9, 11], [12, 13, 14], [16, 17, 15], [18, 19, 20], [21, 22, 23], [3, 4, 5], [8, 6, 7]]
[[0, 2, 1], [10, 23, 21, 9, 11, 22], [12, 13, 14], [16, 17, 15, 19, 18, 3, 20, 4, 5], [8, 6, 7]]
[[0, 2, 1, 17, 16, 15], [10, 23, 21, 9, 11, 22, 12, 13, 14], [18, 19, 20], [3, 4, 5], [7, 8, 6]]
[[0, 2, 1, 17, 16, 19, 15, 18, 3, 20, 4, 5, 8, 6, 7, 21, 22, 23, 10, 9, 11], [12, 13, 14]]
[[0, 2, 1, 17, 16, 19, 15, 18, 3, 20, 4, 5, 8, 6, 7, 21, 22, 23, 12, 10, 13, 9, 11, 14]]
```

9 Final Individuals

Figure 10. The final nine individuals generated using the MOCK clustering algorithm improved with NSGA-II. The individual marked in red is the knee point.

Subsequently, the comparison described in Section 3.4 was performed to select the "best solution". The individuals generated using the MOCK/NSGA-II algorithm (Figure 10) and the known class subgroups generated using the GMM (Figure 8) were compared. The results of this comparison are shown in Table 6. It can be seen that Solutions 1, 2, 3, and 5 have four structures shared with the subgroups of known classes generated with the GMM. However, since the solution closest to the knee point was selected as the "best solution",

Solution 5 was chosen (marked with *), which in this case, was found to be the knee point, marked in red in Figure 10.

Table 6. Scores obtained from the comparison of the individuals generated using MOCK/NSGA-II and the known class subgroups generated using GMM. The solution marked with * was selected as the "best solution".

	Pareto Frontier Position	**Score**
Solution 1	1	4
Solution 2	2	4
Solution 3	3	4
Solution 4	4	3
Solution 5 *	**5**	**4**
Solution 6	6	2
Solution 7	7	2
Solution 8	8	1
Solution 9	9	0

Given the "best solution", the four clusters with shared structures with the known class subgroups generated with the GMM were determined as *"known classes"*. Then, the clusters containing only objects that the GMM determined as unknown, shown in Figure 8, were determined as *"new classes"*. The result of these processes can be seen in Figure 11.

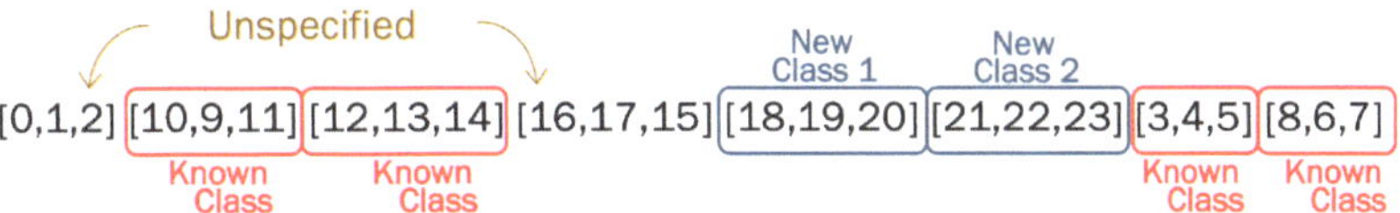

Figure 11. Selection of *"known classes"* and *"new classes"*.

Since there were still unspecified clusters as *known* or *new*, we counted the number of known and unknown instances (as determined using the GMM) in the indeterminate clusters and defined the clusters in the same category as that comprising the majority of instances. Thus, we obtained five groups of *"known classes"* and three *"new classes"*, as shown in Figure 12. Given the data in Table 4, we can confirm that indeed the vectors of the *"new classes"* corresponded to the instances of unknown objects and that they were grouped in the same structure as their "unknown class", thus confirming that both the "new classes" of objects of "unknown classes" can indeed be discovered.

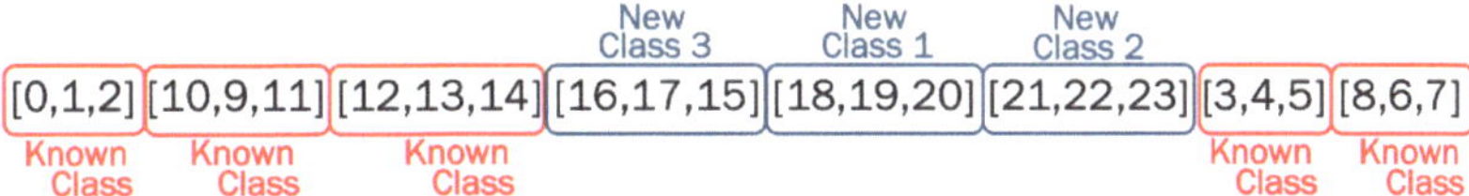

Figure 12. Final selection of *"known classes"* and *"new classes"*.

Finally, the objects of known classes were entered into the classification section of the domain-specific CNN where a classification accuracy of 100% was obtained. Given the classification results obtained, we calculated the critical values of true positive (TP), false positive (FP), and false negative (FN) of both known and unknown classes. Subsequently, we calculated the micro-F1 score since it is one of the most commonly used metrics in OSR algorithms. The results obtained are shown in Table 7.

Table 7. Calculation of the micro-F1 score.

	Label	True Positive (TP)	False Positive (FP)	False Negative (FN)	Micro-F1
Known Classes	Chevrolet Silverado 2004	3	0	0	Precision = 1.0
	Ford Explorer 2002	3	0	0	
	Ford Mustang 2000	3	0	0	
	Honda Civic 2002	3	0	0	
	Nissan Altima 2005	3	0	0	Recall = 1.0
Unknown Classes	Unknown Class 1	3	0	0	
	Unknown Class 2	3	0	0	
	Unknown Class 3	3	0	0	Micro-F1 Score = 1.0
	Total	24	0	0	

5. Discussion and Conclusions

The main contribution of this work is to present a strategy to approach the VMMR as an OSR problem that is extended to the discovery of new classes, taking the distribution of feature vectors generated using a domain-specific CNN as the main guideline. This work seeks to highlight the importance of generating domain-specific OSR strategies and the need to apply them to real-world classification/recognition problems such as VMMR in order to obtain classifiers that are not only more accurate but also more robust, as they are prepared to face real-life scenarios. Although we focused on VMMR, the proposed methodology can be used as a benchmark for future domain-specific OSR problems and can be applied to other domains like handwritten digit recognition, chest X-ray classification, etc.

For the development of this work, we considered four main objectives to fulfill the purpose of approaching VMMR as an OSR problem extended for new class discovery. The fulfillment of our first objective could be validated with the results shown in Figure 6, where it can be seen that the CNN designed through the NE process with contrastive loss managed to map within the embedded space the feature vectors close in terms of cosine distance if they belonged to the same classes and far away if they belonged to different classes, maintaining this behavior for both known and unknown classes.

The second objective was described in detail in Section 3.4, which is the theoretical part of the third objective. In the Section 4, the proposed methodology was described step by step, and the experiments carried out validated that the proposed mechanism is able to detect objects of unknown classes and simultaneously discover their classes. One point to highlight is that our strategy is not restricted by training data, as it can be adjusted as these data change. More precisely, by using contrastive learning to train the feature extraction of the domain-specific CNN, the distribution of feature vectors is not only guided by "known classes" but is able to perform a consistent mapping even for objects of "unknown classes", which allows us to effectively detect objects of known classes and discover their classes simultaneously.

From the outset, we decided to employ a CNN not only to exploit the powerful ability of CNNs to extract meaningful features but also because these networks are known to be powerful classifiers. Therefore, since our domain-specific CNN was trained with numerous well-labeled examples, we could rely on its accuracy in classifying instances of known classes. Therefore, the last objective was met by achieving 100% classification accuracy of the images of the known classes in the test set.

Overall, the entire algorithm achieved a micro-F1 score of 1.00 by accurately classifying instances of known classes and effectively discovering the classes of instances whose classes were not included in the training. In a closed-set context, which is where most classification algorithms are developed, all instances of unknown classes would have been classified into some known class, so the model would not have been able to achieve a classification accuracy higher than 62.5% with the test set used in this work since 9 of the 24 test images belonged to unknown classes. The poor classification accuracy in this specific context, which simulates a real-life scenario, would be due to the incomplete knowledge of the world and not due to the classification potential that the classifier could achieve. Therefore, in this work, we proposed to add a mechanism to one of the most used image classifiers such as CNNs in order to detect objects of unknown classes and identify these classes. This highlights the possibility to expand the classification potential of CNNs and increase their robustness to work more effectively in real-life scenarios, thus enabling these classifiers not only to react to queries but also to continue learning even after being trained.

One of the limitations of this work was that due to time constraints, the neuroevolution algorithm was executed only seven times with the specified parameters, and it is left as future work to create a statistically more representative sample of executions and use a parameter calibration algorithm to possibly have better and more efficient results. It is also left as future work to increase the number of "known classes" to be able to classify more models with the domain-specific CNN and apply other OSR strategies to the VMMR problem for a more representative comparison.

Author Contributions: Conceptualization, D.-I.V.-S., H.-G.A.-M. and E.M.-M.; methodology, D.-I.V.-S., H.-G.A.-M. and E.M.-M.; software, D.-I.V.-S.; validation, H.-G.A.-M. and E.M.-M.; formal analysis, D.-I.V.-S.; investigation, D.-I.V.-S.; resources, D.-I.V.-S., H.-G.A.-M. and E.M.-M.; data curation, D.-I.V.-S.; writing—original draft preparation, D.-I.V.-S.; writing—review and editing, D.-I.V.-S., H.-G.A.-M. and E.M.-M.; visualization, D.-I.V.-S.; supervision, H.-G.A.-M. and E.M.-M. All authors have read and agreed to the published version of the manuscript.

Funding: This research received no external funding.

Acknowledgments: The first author would like to thank the Consejo Nacional de Ciencia y Tecnología (CONACYT), an institution of the Government of Mexico, for the financial support provided through the "Beca Nacional" with CVU 1141251 as part of the Programa de Becas para Estudios de Posgrado.

Conflicts of Interest: The authors declare no conflict of interest.

References

1. Naseer, S.; Shah, S.M.A.; Aziz, S.; Khan, M.U.; Iqtidar, K. Vehicle Make and Model Recognition using Deep Transfer Learning and Support Vector Machines. In Proceedings of the IEEE 23rd International Multitopic Conference (INMIC), Bahawalpur, Pakistan, 5–7 November 2020; pp. 1–6. [CrossRef]
2. Agarwal, A.; Shinde, S.; Mohite, S.; Jadhav, S. Vehicle Characteristic Recognition by Appearance: Computer Vision Methods for Vehicle Make, Color, and License Plate Classification. In Proceedings of the IEEE Pune Section International Conference (PuneCon), Pune, India, 15–17 December 2022; pp. 1–6. [CrossRef]
3. Nazemi, A.; Azimifar, Z.; Shafiee, M.J.; Wong, A. Real-Time Vehicle Make and Model Recognition Using Unsupervised Feature Learning. *IEEE Trans. Intell. Transp. Syst.* **2020**, *21*, 3080–3090. [CrossRef]
4. Hassaballah, M.; Kenk, M.A.; Muhammad, K.; Minaee, S. Vehicle Detection and Tracking in Adverse Weather Using a Deep Learning Framework. *IEEE Trans. Intell. Transp. Syst.* **2021**, *22*, 4230–4242. [CrossRef]
5. Hussain, K.F.; Afifi, M.; Moussa, G. A Comprehensive Study of the Effect of Spatial Resolution and Color of Digital Images on Vehicle Classification. *IEEE Trans. Intell. Transp. Syst.* **2019**, *20*, 1181–1190. [CrossRef]
6. Boukerche, A.; Ma, X. A Novel Smart Lightweight Visual Attention Model for Fine-Grained Vehicle Recognition. *IEEE Trans. Intell. Transp. Syst.* **2022**, *23*, 13846–13862. [CrossRef]
7. Fang, J.; Zhou, Y.; Yu, Y.; Du, S. Fine-Grained Vehicle Model Recognition Using A Coarse-to-Fine Convolutional Neural Network Architecture. *IEEE Trans. Intell. Transp. Syst.* **2017**, *18*, 1782–1792. [CrossRef]
8. Masana, M.; Liu, X.; Twardowski, B.; Menta, M.; Bagdanov, A.D.; van de Weijer, J. Class-Incremental Learning: Survey and Performance Evaluation on Image Classification. *IEEE Trans. Pattern Anal. Mach. Intell.* **2023**, *45*, 5513–5533. [CrossRef]

9. Hafeez, M.A.; Ul-Hasan, A.; Shafait, F. Incremental Learning of Object Detector with Limited Training Data. In Proceedings of the Digital Image Computing: Techniques and Applications (DICTA), Gold Coast, Australia, 29 November–1 December 2021; pp. 1–8. [CrossRef]

10. Zhang, F. Learning Unsupervised Side Information for Zero-Shot Learning. In Proceedings of the International Conference on Signal Processing and Machine Learning (CONF-SPML), Stanford, CA, USA, 14 November 2021; pp. 325–328. [CrossRef]

11. Li, Y.; Kong, D.; Zhang, Y.; Chen, R.; Chen, J. Representation Learning of Remote Sensing Knowledge Graph for Zero-Shot Remote Sensing Image Scene Classification. In Proceedings of the IEEE International Geoscience and Remote Sensing Symposium (IGARSS), Brussels, Belgium, 11–16 July 2021; pp. 1351–1354. [CrossRef]

12. Kezebou, L.; Oludare, V.; Panetta, K.; Agaian, S. Few-Shots Learning for Fine-Grained Vehicle Model Recognition. In Proceedings of the IEEE International Symposium on Technologies for Homeland Security (HST), Boston, MA, USA, 8–9 November 2021; pp. 1–9. [CrossRef]

13. Zhou, F.; Zhang, L.; Wei, W.; Bai, Z.; Zhang, Y. Meta Transfer Learning for Few-Shot Hyperspectral Image Classification. In Proceedings of the IEEE International Geoscience and Remote Sensing Symposium (IGARSS), Brussels, Belgium, 11–16 July 2021; pp. 3681–3684. [CrossRef]

14. Scheirer, W.J.; de Rezende Rocha, A.; Sapkota, A.; Boult, T.E. Toward Open Set Recognition. *IEEE Trans. Pattern Anal. Mach. Intell.* **2013**, *35*, 1757–1772. [CrossRef]

15. Scheirer, W.J.; Jain, L.P.; Boult, T.E. Probability Models for Open Set Recognition. *IEEE Trans. Pattern Anal. Mach. Intell.* **2014**, *36*, 2317–2324. [CrossRef]

16. Rudd, E.M.; Jain, L.P.; Scheirer, W.J.; Boult, T.E. The Extreme Value Machine. *IEEE Trans. Pattern Anal. Mach. Intell.* **2018**, *40*, 762–768. [CrossRef]

17. Ribeiro Mendes Júnior, P.; Boult, T.E.; Wainer, J.; Rocha, A. Open-Set Support Vector Machines. *IEEE Trans. Syst. Man Cybern. Syst.* **2022**, *52*, 3785–3798. [CrossRef]

18. Alfarisy, G.A.F.; Malik, O.A.; Hong, O.W. Quad-Channel Contrastive Prototype Networks for Open-Set Recognition in Domain-Specific Tasks. *IEEE Access* **2023**, *11*, 48578–48592. [CrossRef]

19. Bendale, A.; Boult, T. Towards Open World Recognition. In Proceedings of the IEEE Conference on Computer Vision and Pattern Recognition (CVPR), Boston, MA, USA, 7–12 June 2015; pp. 1893–1902. [CrossRef]

20. Wang, Z.; Salehi, B.; Gritsenko, A.; Chowdhury, K.; Ioannidis, S.; Dy, J. Open-World Class Discovery with Kernel Networks. In Proceedings of the IEEE International Conference on Data Mining (ICDM), Sorrento, Italy, 17–20 November 2020; pp. 631–640. [CrossRef]

21. Han, K.; Rebuffi, S.-A.; Ehrhardt, S.; Vedaldi, A.; Zisserman, A. AutoNovel: Automatically Discovering and Learning Novel Visual Categories. *IEEE Trans. Pattern Anal. Mach. Intell.* **2022**, *44*, 6767–6781. [CrossRef] [PubMed]

22. Han, K.; Vedaldi, A.; Zisserman, A. Learning to Discover Novel Visual Categories via Deep Transfer Clustering. In Proceedings of the IEEE/CVF International Conference on Computer Vision (ICCV), Seoul, Republic of Korea, 27 October–2 November 2019; pp. 8400–8408. [CrossRef]

23. Geng, C.; Chen, S. Collective Decision for Open Set Recognition. *IEEE Trans. Knowl. Data Eng.* **2022**, *34*, 192–204. [CrossRef]

24. Sun, X.; Yang, Z.; Zhang, C.; Ling, K.-V.; Peng, G. Conditional Gaussian Distribution Learning for Open Set Recognition. In Proceedings of the IEEE/CVF Conference on Computer Vision and Pattern Recognition (CVPR), Seattle, WA, USA, 13–19 June 2020; pp. 13477–13486. [CrossRef]

25. Ye, W.; Liu, R.; Li, Y.; Jiao, L. Quantum-inspired evolutionary algorithm for convolutional neural networks architecture search. In Proceedings of the IEEE Congress on Evolutionary Computation (CEC), Glasgow, UK, 19–24 July 2020; pp. 1–8. [CrossRef]

26. Liu, J.; Zhou, S.; Wu, Y.; Chen, K.; Ouyang, W.; Xu, D. Block proposal neural architecture search. *IEEE Trans. Image Process.* **2021**, *30*, 15–25. [CrossRef] [PubMed]

27. Zhou, Y.; Gen, G.G.; Yi, Z. A knee-guided evolutionary algorithm for compressing deep neural networks. *IEEE Trans. Cybern.* **2021**, *51*, 1626–1638. [CrossRef]

28. Operiano, K.R.G.; Iba, H.; Pora, W. Neuroevolution architecture backbone for x−ray object detection. In Proceedings of the IEEE Symposium Series on Computational Intelligence (SSCI), Canberra, Australia, 1–4 December 2020; pp. 2296–2303. [CrossRef]

29. Hassanzadeh, T.; Essam, D.; Sarker, R. 2D to 3D evolutionary deep convolutional neural networks for medical image segmentation. *IEEE Trans. Med. Imaging* **2021**, *40*, 712–721. [CrossRef]

30. Sun, Y.; Xue, B.; Zhang, M.; Yen, G.G. Evolving deep convolutional neural networks for image classification. *IEEE Trans. Evol. Comput.* **2020**, *24*, 394–407. [CrossRef]

31. Zhang, H.; Jin, Y.; Cheng, R.; Hao, K. Efficient evolutionary search of attention convolutional networks via sampled training and node inheritance. *IEEE Trans. Evol. Comput.* **2021**, *25*, 371–385. [CrossRef]

32. Hu, M.; Wang, W.; Liu, L.; Liu, Y. Apenas: An asynchronous parallel evolution based multi-objective neural architecture search. In Proceedings of the IEEE International Conference on Parallel & Distributed Processing with Applications, Big Data & Cloud Computing, Sustainable Computing & Communications, Social Computing & Networking (ISPA/BDCloud/SocialCom/SustainCom), Exeter, UK, 17–19 December 2020; pp. 153–159. [CrossRef]

33. Vargas-Hákim, G.; Mezura-Montes, E.; Acosta-Mesa, H. Hybrid encodings for neuroevolution of convolutional neural networks: A case study. In Proceedings of the Genetic and Evolutionary Computation Conference Companion (GECCO'21), Online, 10–14 July 2021; Association for Computing Machinery: New York, NY, USA, 2021; pp. 1762–1770. [CrossRef]

34. Vargas-Hákim, G.-A.; Mezura-Montes, E.; Acosta-Mesa, H.-G. A Review on Convolutional Neural Network Encodings for Neuroevolution. *IEEE Trans. Evol. Comput.* **2022**, *26*, 12–27. [CrossRef]
35. Khosla, P.; Teterwak, P.; Wang, C.; Sarna, A.; Tian, Y.; Isola, P.; Maschinot, A.; Liu, C.; Krishnan, D. Supervised contrastive learning. *Adv. Neural Inf. Process. Syst.* **2020**, *33*, 18661–18673. [CrossRef]
36. Chen, T.; Kornblith, S.; Norouzi, M.; Hinton, G. A simple framework for contrastive learning of visual representations. In Proceedings of the 37th International Conference on Machine Learning (ICML'20), Online, 13–18 July 2020; pp. 1597–1607.
37. Tafazzoli, F.; Frigui, H.; Nishiyama, K. A Large and Diverse Dataset for Improved Vehicle Make and Model. In Proceedings of the IEEE Conference on Computer Vision and Pattern Recognition Workshops (CVPRW), Honolulu, HI, USA, 21–26 July 2017; pp. 874–881. [CrossRef]
38. Hassan, A.; Ali, M.; Durrani, N.M.; Tahir, M.A. An Empirical Analysis of Deep Learning Architectures for Vehicle Make and Model Recognition. *IEEE Access* **2021**, *9*, 91487–91499. [CrossRef]
39. Kristiani, E.; Yang, C.-T.; Huang, C.-Y. iSEC: An Optimized Deep Learning Model for Image Classification on Edge Computing. *IEEE Access* **2020**, *8*, 27267–27276. [CrossRef]
40. Deb, K.; Pratap, A.; Agarwal, S.; Meyarivan, T. A fast and elitist multiobjective genetic algorithm: NSGA-II. *IEEE Trans. Evol. Comput.* **2002**, *6*, 182–197. [CrossRef]
41. Handl, J.; Knowles, J. An Evolutionary Approach to Multiobjective Clustering. *IEEE Trans. Evol. Comput.* **2007**, *11*, 56–76. [CrossRef]
42. Martínez-Peñaloza, M.; Mezura-Montes, E.; Cruz-Ramírez, N.; Acosta-Mesa, H.; Ríos-Figueroa, H. Improved multi-objective clustering with automatic determination of the number of clusters. *Neural Comput. Appl.* **2017**, *28*, 2255–2275. [CrossRef]
43. Van der Maaten, L.; Hinton, G. Visualizing Data using t-SNE. *J. Mach. Learn. Res.* **2008**, *9*, 2579–2605.

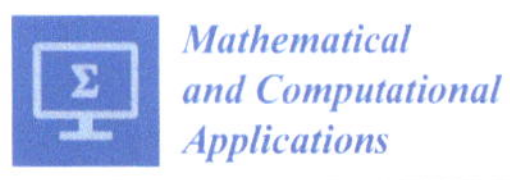
Mathematical and Computational Applications

Article

FE² Computations with Deep Neural Networks: Algorithmic Structure, Data Generation, and Implementation

Hamidreza Eivazi [1,†], Jendrik-Alexander Tröger [2,*,†], Stefan Wittek [1], Stefan Hartmann [2] and Andreas Rausch [1]

[1]   Institute for Software and Systems Engineering, Clausthal University of Technology, 38678 Clausthal-Zellerfeld, Germany; hamidreza.eivazi.kourabbaslou@tu-clausthal.de (H.E.); stefan.wittek@tu-clausthal.de (S.W.); andreas.rausch@tu-clausthal.de (A.R.)

[2]   Institute of Applied Mechanics, Clausthal University of Technology, 38678 Clausthal-Zellerfeld, Germany; stefan.hartmann@tu-clausthal.de

*   Correspondence: jendrik-alexander.troeger@tu-clausthal.de

†   These authors contributed equally to this work.

Abstract: Multiscale FE² computations enable the consideration of the micro-mechanical material structure in macroscopical simulations. However, these computations are very time-consuming because of numerous evaluations of a representative volume element, which represents the microstructure. In contrast, neural networks as machine learning methods are very fast to evaluate once they are trained. Even the DNN-FE² approach is currently a known procedure, where deep neural networks (DNNs) are applied as a surrogate model of the representative volume element. In this contribution, however, a clear description of the algorithmic FE² structure and the particular integration of deep neural networks are explained in detail. This comprises a suitable training strategy, where particular knowledge of the material behavior is considered to reduce the required amount of training data, a study of the amount of training data required for reliable FE² simulations with special focus on the errors compared to conventional FE² simulations, and the implementation aspect to gain considerable speed-up. As it is known, the Sobolev training and automatic differentiation increase data efficiency, prediction accuracy and speed-up in comparison to using two different neural networks for stress and tangent matrix prediction. To gain a significant speed-up of the FE² computations, an efficient implementation of the trained neural network in a finite element code is provided. This is achieved by drawing on state-of-the-art high-performance computing libraries and just-in-time compilation yielding a maximum speed-up of a factor of more than 5000 compared to a reference FE² computation. Moreover, the deep neural network surrogate model is able to overcome load-step size limitations of the RVE computations in step-size controlled computations.

Keywords: multiscale finite element computations; deep neural networks; surrogate modeling; Sobolev training; representative volume element; step-size control

Citation: Eivazi, H.; Tröger, J.-A.; Wittek, S.; Hartmann, S.; Rausch, A. FE² Computations with Deep Neural Networks: Algorithmic Structure, Data Generation, and Implementation. *Math. Comput. Appl.* **2023**, *28*, 91. https://doi.org/10.3390/mca28040091

Academic Editor: Gianluigi Rozza

Received: 17 July 2023
Revised: 4 August 2023
Accepted: 14 August 2023
Published: 16 August 2023

1. Introduction

Nearly all commonly applied engineering materials possess, depending on the detail of investigation, some heterogeneous microstructure, e.g., fiber-reinforced polymers or rolled steel alloys, where the grains can have preferential directions because of the manufacturing process. Since this heterogeneous microstructure can significantly influence the response of these materials to mechanical loading, it is of particular interest to consider the microstructure already in numerical simulations. The development of constitutive models for materials with heterogeneous microstructures is challenging in both aspects, phenomenological constitutive modeling and subsequent experimental calibration. Thus, the so-called FE² method has been developed by [1–6]—to mention only a few—for coupled numerical simulation of structures at macro- and microscale with finite elements. There, in contrast to common finite element computations, a constitutive model is not assigned to an integration point at macroscale. Instead, the stress and consistent tangent quantities are obtained by solving an initial boundary value problem with finite elements on a particular microstructure followed by a numerical homogenization technique. In this context, the microstructure is usually denoted as a representative volume element (RVE). In addition to the aforementioned works, in [7], a comprehensive description of the FE^2-method for the numerical solution of these coupled boundary value problems on different scales is provided. In general, there exist further methods to obtain the response of heterogeneous microstructures, such as Discrete Fourier Transforms or Fast Fourier Transforms; see, for example, [8]. Even the finite cell method is applicable for the homogenization of heterogeneous microstructures; see, for example, [9]. However, in this work, deep neural networks are applied to replace the computationally costly solution of initial boundary value problems at microscale.

Currently, various applications of methods of artificial intelligence exist in the field of solid mechanics. A comprehensive overview of applications in continuum material mechanics is given in [10]. Further reviews are provided in [11,12] for applications in experimental solid mechanics and [13] for material development in additive manufacturing employing machine learning methods. Ref. [14] provides a general introduction to the application of machine learning techniques in material modeling and design of materials. Additionally, in [15], a review and investigation of the ability to apply machine learning in constitutive modeling is provided; however, it is in the context of soils. Most applications of machine learning methods aim to obtain feasible information from huge amounts of data or to increase the speed of particular computations. The source of the data could either be simulations, as in the present work, or directly experimental data, as in the *data-driven mechanics* approach, which was introduced by the aithors of [16], where it is not required to learn the response of constitutive models from simulations.

The application of artificial neural networks for data-based constitutive modeling was originally introduced in [17] and is frequently used in representing the material behavior for finite element simulations since then; see, for example, [18,19]. Recently, different approaches have been published to advance numerical simulations with machine learning methods. An investigation into deep learning surrogate models for accelerating numerical simulations is presented in [20]. Ref. [21] contains a proposal of a combination of physics-based finite element analysis and convolutional neural networks for predicting the mechanical response of materials. In contrast, ref. [22] contains an application of deep learning techniques for extracting rules that are inherent in computational mechanics to propose a new numerical quadrature rule that shows results superior to the well-known Gauss–Legendre quadrature.

Learning material behavior from simulations is generally covered with versatile approaches. In this context, the authors of [23] describe a material modeling framework for hyperelasticity and plasticity, where different architectures of neural networks are employed. Model-free procedures that fit into the data-driven mechanics approach for representing material behavior are described in [24–27], among others. Model-free approaches are suitable, especially for the consideration of elastoplastic material behavior; see [28,29] as well. Artificial neural networks could also be applied for calibrating known constitutive models from experimental data (parameter identification). First attempts are presented in [30,31], whereas in [32], modern deep reinforcement learning techniques are applied for the calibration of history-dependent models. Since the measurement techniques to obtain experimental data have evolved in recent years, modern calibration techniques can consider full-field data, e.g., from digital image correlation; see [33]. Instead of calibrating constitutive models from experimental data with neural networks, where an error is introduced from choosing the constitutive model, the experimental data can be directly employed for discovering the material models from data. This approach is introduced in [34] for hyperelasticity and later on extended to cover elastoplasticity [35] and generalized materials [36]. Similar work with automated discovery of suitable hyperelastic materials is provided in [37], where *constitutive artificial neural networks* are applied, introduced in [38].

Many different machine learning methods are successfully used for multiscale applications in solid mechanics. There, the main objective is to obtain the homogenized response from heterogeneous microstructures. One of the first works in this context is provided in [39], wherein the authors applied neural networks for the homogenization of non-linear elastic materials. Ref. [40] contains a proposal of a data-driven two-scale approach to predict homogenized quantities even for inelastic deformations by drawing on clustering techniques. The ability to replace microscale evaluations with artificial neural networks requires a suitable accuracy of the network after the training process. Regarding this issue, the authors of ref. [41] make use of artificial neural networks as constitutive relation surrogates for nonlinear material behavior. However, based on the evaluation of quality indicators, a reduced-order model can be employed instead of the neural network within an adaptive switching method. The authors of ref. [42] describe the so-called *Structural-Genome-Driven* approach for FE^2 computations of composite structures, wherein even model reduction techniques are applied.

Advanced neural network architectures such as convolutional or recurrent neural networks are regularly applied to predict the homogenized response of microstructures. Here, atomistic data can also be used showing significant acceleration compared to molecular statics [43]. Elastic material behavior is investigated in [44–47]. In ref. [48], significant speed up in homogenization is reached when applying three-dimensional convolutional neural networks in broad ranges of different microstructures, phase fractions, and material parameters. The authors of ref. [49] provide the generalization of data to obtain three-dimensional nonlinear elastic material laws under arbitrary loading conditions. Considering anisotropy, teh authors of ref. [50] predict effective material properties of RVEs with randomly placed and shaped inclusions. The suitability of different machine learning methods for homogenizing polycrystalline materials is studied in [51]. Besides the purely mechanical homogenization approaches, the authors of ref. [52] show that neural networks can be even applied to computational homogenization of electrostatic problems. Moreover, researchers in ref. [53] employ μCT data within a data-driven multiscale framework to study open-cell foam structures.

According to [54], replacing microscale computations in the FE^2 method by surrogate models can be denoted as a *data-driven multiscale finite element method*. The authors of ref. [55] perform multiscale computations with feedforward neural networks and recurrent neural networks for RVEs with inelastic material behavior and further investigate the ability to generalize for unknown loading paths. Researchers in ref. [56] present a hybrid methodology denoted as a model-data-driven one. Therein, the authors apply a combination of conventional constitutive models and a data-driven correction component as a multiscale methodology. Moreover, it is beneficial to incorporate physical knowledge into the development of neural network surrogates. This is achieved, for example, in [57]. The authors propose *thermodynamics-based artificial neural networks* (TANNs) and apply them for multiscale simulations of inelastic lattice structures, while later extending the framework to evolution TANNs [58]. The application of particular physical constraints by using problem-specific invariants as input quantities and the Helmholtz free energy density as output is provided in [59]. The authors provide FE^{ANN} as a data-driven multiscale framework and minimize the number of microscale simulations, which serve as training data, by following an autonomous data mining approach.

Further, probabilistic approaches can be employed while developing the surrogate models; in [60], more accurate results are achieved with Sobolev training [61] compared to regular training for hyperelasticity. In this context, for an extension to multiscale plasticity with geometric machine learning methods, we refer to [62,63]. Elastoplastic solid materials are investigated in [64] using recurrent neural networks and in [65] where the authors employ two separated neural networks for the homogenized stress and tangent information. The authord of ref. [66] apply DeepONet as a surrogate model for the microscale level with two-dimensional elastoplasticity and hyperelasticity. Currently, the authors of ref. [67]

demonstrate the applicability of the encoder/decoder approach for multiscale computations with path-dependent material behavior on the microscale.

Another research track are the so-called *deep material networks* (DMNs), which provide an efficient data-driven multiscale framework to reproduce the homogenized response of RVEs. The introduction of DMNs for two-phase heterogeneous materials is provided in [68] together with an extension to three-dimensional microstructures [69]. The authors of ref. [70] further extend the technique to take into account diverse fiber orientations, applying DMNs for multiscale analysis of composites with full thermo-mechanical coupling [71]. Researchers in ref. [72] employ DMNs with the computation of the tangent operator in a closed form as an output of the network.

The main objective of the present work is to provide a consistent approach for employing deep neural networks (DNN) as surrogate models in step-size controlled multiscale FE^2 computations. As mentioned afore, various publications already deal with embedding artificial neural networks into numerical simulations especially for accelerating computational costly multiscale simulations. A novelty of the present work is that we provide a clear description of the algorithmic structure, which is in general a Multilevel–Newton algorithm (MLNA) that simplifies to a Newton–Raphson algorithm when employing DNN surrogate models. Further, current publications leave out required information; for example, the ways in which the consistent tangent at macroscale integration points is obtained from the microscale information, meaning whether the computations are performed by automatic differentiation, neural network models, or by numerical differentiation. Concerning this objective, we start in Section 2 with an explanation of the underlying equations and the algorithmic structure in FE^2 computations, where we restrict ourselves to small strains and quasi-static problems. Afterwards, two different architectures of neural networks and specific considerations of physical knowledge during the training process are described. Since the amount of training data required to obtain sufficient accuracies in the neural network outputs is of particular interest, this is investigated as well while using regular training and Sobolev training. As another novel contribution, we develop a method for efficiently coupling the different programming codes of the trained neural network and the multiscale finite element code. There, the application of high-performance computing libraries and just-in-time compilation yields significantly higher speed up of the DNN-FE^2 approach in load-step size controlled computations compared to the results presented in the current literature. Furthermore, the DNN surrogate is even able to overcome load-step size limitations that are apparent in FE^2 computations.

The notation in use is defined in the following manner: geometrical vectors are symbolized by $\vec{a}$ and second-order tensors $\mathbf{A}$ by bold-faced Roman letters. In addition, we introduce column vectors and matrices at the global finite element level symbolized by bold-type italic letters A and column vectors and matrices on the local (element) level using bold-type Roman letters $\mathbf{A}$. Further, to distinguish quantities on macroscale and microscale levels, we indicate microscale quantities by $\langle \breve{\cdot} \rangle$. Calligraphic letters $\mathcal{A}$ denote deep neural network surrogate models.

2. Classical FE^2 Computations

In this work, finite elements are employed to perform multiscale computations; see, for example, [4,7]. Hence, only the main equations are recapped, which are necessary to explain the algorithmic structure. In multiscale FE^2 computations, the macro- and microscale levels have to be distinguished regarding the spatial discretization. Here, we restrict ourselves to periodic displacement boundary conditions on the microscale and refer to [4] for other boundary conditions on the microscale. The computation of the system of non-linear equations resulting from the spatial discretization is explained for the Multilevel–Newton algorithm (MLNA), which is here a two-level Newton algorithm. Further, the connection is drawn to embedding deep neural network surrogate models as predictors for homogenized quantities from the microscale in the MLNA.

2.1. Spatial Discretization

In the present work, FE^2 analyses are performed in a quasi-static setting with the restriction to small strains. Thus, no configurations have to be distinguished and we have the symmetric stress tensor $\mathbf{T}(\vec{x}, t)$ and strain tensor $\mathbf{E}(\vec{x}, t) = 1/2\Big(\text{grad}\,\vec{u}(\vec{x}, t) + \text{grad}^T\vec{u}(\vec{x}, t)\Big)$ at positions $\vec{x}$ and time t. $\vec{u}(\vec{x}, t)$ represents the displacement vector. The local balance of linear momentum has to be fulfilled. Here, the weak form is employed, which is also known as the principle of virtual displacements:

$$\pi(t, \mathbf{T}, \delta\vec{u}) := \int_V \delta\mathbf{E}(\vec{x}) \cdot \mathbf{T}(\vec{x}, t)\,\mathrm{d}V - \int_V \delta\vec{u}(\vec{x}) \cdot \rho(\vec{x})\vec{k}\,\mathrm{d}V - \int_A \delta\vec{u}(\vec{x}) \cdot \vec{t}(\vec{x}, t)\,\mathrm{d}A = 0, \quad (1)$$

where $\delta\vec{u}(\vec{x})$ are virtual displacements that are arbitrary but vanish at positions where the displacements $\vec{u}(\vec{x}, t)$ are prescribed. Similarly, $\delta\mathbf{E}(\vec{x}) = 1/2\Big(\text{grad}\,\delta\vec{u}(\vec{x}) + \text{grad}^T\delta\vec{u}(\vec{x})\Big)$ represent virtual strains. Moreover, $\vec{k}$ symbolizes the acceleration of gravity. V and A denote the volume and surface of the material body and $\vec{t}$ are surface tractions. To develop the arising equations of the spatial discretization for three-dimensional continua on both macro- and microscale levels, a consistent matrix notation is followed.

2.1.1. Macroscale

Due to the spatial discretization, volume V and surface A transist into approximations Ω and Γ. Further, $\mathbf{x} \in \Omega$ denote coordinates. Following a Galerkin-based finite element formulation, the ansatz for the displacements and virtual displacements read

$$u^h(\mathbf{x}, t) = \sum_{j=1}^{n_{\text{nodes}}} N_j(\mathbf{x})\mathbf{u}_j(t) = N(\mathbf{x})u(t) + \overline{N}(\mathbf{x})\overline{u}(t) \stackrel{\text{in}\,\textcircled{e}}{=} \mathbf{N}^e(\boldsymbol{\varphi}^e(\mathbf{x}))\mathbf{u}^e(t), \quad (2)$$

$$\delta u^h(\mathbf{x}, t) = \sum_{j=1}^{n_{\text{nodes}}} N_j(\mathbf{x})\delta\mathbf{u}_j = N(\mathbf{x})\delta u \stackrel{\text{in}\,\textcircled{e}}{=} \mathbf{N}^e(\boldsymbol{\varphi}^e(\mathbf{x}))\delta\mathbf{u}^e. \quad (3)$$

$\mathbf{u}_j \in \mathbb{R}^3$ and $\delta\mathbf{u}_j \in \mathbb{R}^3$ denote the macroscopic nodal displacement and virtual nodal displacement vector at node j. The shape functions are denoted by $N_j(\mathbf{x})$, while n_{nodes} corresponds to the number of nodes on a macroscale level. In Equations (2) and (3), it is tacitly assumed that the displacements and virtual displacements are partitioned into unknown and prescribed quantities, i.e., $u \in \mathbb{R}^{n_u}$ are unknown macroscale nodal displacements and $\overline{u} \in \mathbb{R}^{n_p}$ are known (or prescribed) nodal displacements. Analogously, the arbitrary virtual displacements are denoted by $\delta u \in \mathbb{R}^{n_u}$. For the prescribed virtual displacements on the macroscale $\delta\overline{u} = 0$, $\delta\overline{u} \in \mathbb{R}^{n_p}$ holds by definition. Thus, the number of degrees of freedom is $n_a = n_u + n_p$. Further, the transition to a formulation of the displacements on element level in Equations (2) and (3) yields the matrix of shape functions $\mathbf{N}^e \in \mathbb{R}^{3 \times n_u^e}$ within an element, the element nodal displacements $\mathbf{u}^e \in \mathbb{R}^{n_u^e}$, and corresponding virtual element nodal displacements $\delta\mathbf{u}^e \in \mathbb{R}^{n_u^e}$. Here, n_u^e is the number of element nodal degrees of freedom. $\boldsymbol{\xi} = \boldsymbol{\varphi}^e(\mathbf{x}) = \boldsymbol{\chi}^{e-1}(\mathbf{x})$ are the local coordinates in the element domain with the coordinate transformation $\mathbf{x} = \boldsymbol{\chi}^e(\boldsymbol{\xi})$. The assignment between global and element quantities can be formulated as

$$\mathbf{u}^e(t) = \mathbf{Z}^e u(t) + \overline{\mathbf{Z}}^e\,\overline{u}(t), \quad \delta\mathbf{u}^e = \mathbf{Z}^e\delta u. \quad (4)$$

Here, $\mathbf{Z}^e \in \mathbb{R}^{n_u^e \times n_u}$ and $\overline{\mathbf{Z}}^e \in \mathbb{R}^{n_u^e \times n_p}$ are formally introduced incidence matrices (Boolean matrices) which are not programmed but used here for the explanation of the assembling procedure of all element contributions. They assign the global unknown and prescribed nodal displacements to element e. Regarding an explanation of an implementation of these matrices, we refer to [73].

Moreover, the resulting strains and virtual strains read

$$E^h(\mathbf{x},t) = B(\mathbf{x})u(t) + \overline{B}(\mathbf{x})\overline{u}(t) \overset{\text{in}\ \text{\small ⓔ}}{=} \mathbf{B}^e(\varphi^e(\mathbf{x}))\mathbf{u}^e(t) = \mathbf{B}^e(\varphi^e(\mathbf{x}))(Z^e u(t) + \overline{Z}^e \overline{u}(t)), \quad (5)$$

$$\delta E^h(\mathbf{x},t) = B(\mathbf{x})\delta u \overset{\text{in}\ \text{\small ⓔ}}{=} \mathbf{B}^e(\varphi^e(\mathbf{x}))\delta \mathbf{u}^e = \mathbf{B}^e(\varphi^e(\mathbf{x}))Z^e \delta u. \tag{6}$$

Again, a decomposition into known and unknown nodal displacements is employed. Since the strain tensor is symmetric, $\mathbf{E} = \mathbf{E}^T$, the strains can be written in the Voigt notation, i.e., $E^h \in \mathbb{R}^6$ and $\delta E^h \in \mathbb{R}^6$. $B \in \mathbb{R}^{6 \times n_u}$ and $\overline{B} \in \mathbb{R}^{6 \times n_p}$ denote the global strain–displacement matrices on the macroscale level for unknown and prescribed degrees of freedom, respectively, whose mathematical representation is extremely difficult to specify. Thus, the element strain–displacement matrix $\mathbf{B}^e \in \mathbb{R}^{6 \times n_u^e}$ is chosen. Inserting Equations (3) and (6) into the principle of virtual displacements (1) and performing a decomposition of the discretized domain into elements yields the non-linear equations on macroscale

$$g(t, T(t)) := \sum_{e=1}^{n_e} Z^{eT} \left(\sum_{j=1}^{n_G^e} w_j \, \mathbf{B}^{e(j)T} \, \mathbf{T}^{e(j)}(t) \det \mathbf{J}^{e(j)} \right) - \overline{p}(t) = 0, \tag{7}$$

$g \in \mathbb{R}^{n_u}$. n_G^e is the number of integration points in an element on the macroscale. Further, w_j denotes the weighting factors of the spatial integration technique, where here the Gauss-integration is drawn on. Accordingly, ξ_j symbolizes the local coordinates of the integration points. Notation $\langle \cdot \rangle^{e(j)}$ is used to abbreviate quantities of element e at the macroscale integration point j, e.g., $\mathbf{B}^{e(j)} := \mathbf{B}^e(\xi_j)$ for the strain–displacement matrix at integration point ξ_j. Since the coordinates are transformed into a reference domain, $\mathbf{J}^{e(j)} = \partial \chi^e / \partial \xi \big|_{\xi = \xi_j}$ is the Jacobian of the coordinate transformation. Moreover, $\overline{p}(t), \overline{p} \in \mathbb{R}^{n_u}$ represents the equivalent nodal force vector comprising the volume and surface distributed loads:

$$\overline{p}(t) := \int_\Omega N^T(\mathbf{x})\rho(\mathbf{x})\mathbf{k}\,\mathrm{d}\Omega + \int_\Gamma N^T(\mathbf{x})\,\mathbf{t}(\mathbf{x},t)\,\mathrm{d}\Gamma. \tag{8}$$

In Equation (7), $\mathbf{T}^{e(j)} \in \mathbb{R}^6$ are the stresses at a specific integration point j on the macroscale written in the Voigt notation. Usually, these quantities are obtained from the evaluation of particular constitutive models. In contrast, in FE^2 computations, the stresses are computed by a particular homogenization scheme of the microstructure, which is explained in the following section. The macroscopical strains at each integration point read with Equation $(4)_1$

$$\mathbf{E}^{e(j)}(t, u(t)) = \mathbf{B}^{e(j)}\mathbf{u}^e = \mathbf{B}^{e(j)}\left(Z^e u(t) + \overline{Z}^e \,\overline{u}(t)\right). \tag{9}$$

Unfortunately, the principle of virtual displacements does not allow the computation of reaction forces. However, since the macroscale reaction forces are of interest in FE^2 computations, we choose here the Lagrange multiplier method; see [74] and the literature cited therein. Thus, the geometric constraint equation

$$C_c(t, \hat{u}(t)) = \hat{u}(t) - \overline{u}(t) = 0 \tag{10}$$

is introduced, $C_c \in \mathbb{R}^{n_p}$. The prescribed displacements $\overline{u}(t)$ should be identical to the degrees of freedom $\hat{u}(t)$, $\hat{u} \in \mathbb{R}^{n_p}$, representing the degrees of freedom that are initially assumed to be unknown as well. To satisfy constraint Equation (10), the Lagrange multipliers $\lambda \in \mathbb{R}^{n_p}$ are required, which can be interpreted as the negative nodal forces.

The combining of Equations (7) and (10) provides the full system of equations for the discretized weak form of the balance of linear momentum on the macroscale, $g_a \in \mathbb{R}^{n_a}$,

$$g_a(t, \lambda(t), T(t)) := \left\{ \begin{array}{l} \displaystyle\sum_{e=1}^{n_e} Z^{eT} \left(\sum_{j=1}^{n_G^e} w_j \, \mathbf{B}^{e(j)T} \, \mathbf{T}^{e(j)}(t) \det \mathbf{J}^{e(j)} \right) - \overline{p}(t) \\[2em] \displaystyle\sum_{e=1}^{n_e} \overline{Z}^{eT} \left(\sum_{j=1}^{n_G^e} w_j \, \mathbf{B}^{e(j)T} \, \mathbf{T}^{e(j)}(t) \det \mathbf{J}^{e(j)} \right) - \lambda(t) \end{array} \right\} = \mathbf{0}. \qquad (11)$$

Remark 1. *It is worth mentioning that the consideration of reaction forces with the Lagrange multiplier method is performed to obtain a consistent variational formulation. The principle of virtual displacements does not allow the computation of reaction forces since they provide no virtual work (remember that $\delta \overline{\mathbf{u}} = \mathbf{0}$ holds). Thus, another variational principle is required, which is here the Lagrange multiplier method. It is important to state that the Lagrange multipliers do not extend the number of unknowns in the application here, since they can be computed in a post-processing step after the computation of the nodal displacements; see Equation $(11)_2$. Then, the Lagrange multipliers can be interpreted as nodal reaction forces, while considering that, of course, the accuracy of the results depends on the chosen termination criteria for the displacements. As a result, the application of the Lagrange multiplier method bypasses the evaluation of nodal equilibrium to compute the reaction forces at prescribed displacement degrees of freedom. The interested reader is referred to [74] for a detailed description of the method, and further references.*

2.1.2. Microscale

The arising equations from the spatial discretization need to be studied also for the microstructure, which represents the discretized microscale geometry in FE^2 computations and is usually denoted as representative volume element (RVE). In contrast to common finite element simulations, where a constitutive model is evaluated at each integration point, here, the microstructure has to be evaluated.

The discretized weak form of the local equilibrium equation on the microscale can be derived analogously to the macroscale and reads

$$\check{g}_a^{e(j)}(t, u_a(t), \check{u}_a^{e(j)}(t)) = \sum_{\check{e}=1}^{\check{n}_e^{e(j)}} \check{Z}_a^{\check{e}T} \left(\sum_{\check{j}=1}^{\check{n}_G^{\check{e}}} \check{w}_{\check{j}} \, \check{\mathbf{B}}^{\check{e}(\check{j})T} \, \check{\mathbf{T}}^{\check{e}(\check{j})}(t) \det \check{\mathbf{J}}^{\check{e}(\check{j})} \right) = \mathbf{0}. \qquad (12)$$

Some remarks should be made regarding the above equation. First, $\check{u}_a^{e(j)} \in \mathbb{R}^{\check{n}_a^{e(j)}}$ are all displacements in the RVE at integration point j of macroscale element e. $\check{n}_a^{e(j)}$ denotes the number of displacement degrees of freedom on the microscale. It is assumed that all displacements are initially unknown in the RVE. $\check{n}_e^{e(j)}$ defines the number of elements on the microscale, $\check{n}_G^{\check{e}}$ symbolizes the number of microscale integration points per element, and $\check{w}_{\check{j}}$ are the weights of the spatial integration. Matrix $\check{Z}_a^{\check{e}T} \in \mathbb{R}^{\check{n}_u^{\check{e}} \times \check{n}_a^{e(j)}}$ denotes formally the assembling procedure of all element contributions and comprises matrices $\check{Z}^{\check{e}} \in \mathbb{R}^{\check{n}_u^{\check{e}} \times \check{n}_u^{e(j)}}$ and $\overline{Z}^{\check{e}} \in \mathbb{R}^{\check{n}_u^{\check{e}} \times \check{n}_p^{e(j)}}$ for the unknown and prescribed displacement degrees of freedom in the RVE, $\check{n}_u^{e(j)}$ and $\check{n}_p^{e(j)}$, respectively, $\check{Z}_a^{\check{e}T} = [\check{Z}^{\check{e}} \, \overline{Z}^{\check{e}}]$. The strain–displacement matrix on the microscale is defined by $\check{\mathbf{B}}^{\check{e}(\check{j})} \in \mathbb{R}^{6 \times \check{n}_u^{\check{e}}}$. Similarly to the macroscale, $\check{n}_a^{e(j)} = \check{n}_u^{e(j)} + \check{n}_p^{e(j)}$ holds. Moreover, at the microscale level, there are no volume or surface distributed loads, i.e., $\overline{p} = \mathbf{0}$.

In this work, we restrict ourselves to periodic displacement boundary conditions on so-called conform spatial discretizations. We refer to [4] for a detailed description of boundary conditions on the microscale and to [75] regarding periodic displacement boundary conditions on non-conform discretizations. The underlying idea of periodic displacement boundary conditions is that the displacements of nodes, which are positioned on different parts of the surface of the RVE, are coupled. This coupling can be treated as a linear multiple point constraint problem. Thus, we introduce the primary and secondary displacements, $\check{u}_{\mathrm{M}}^{e(j)} \in \mathbb{R}^{\check{n}_{\mathrm{M}}^{e(j)}}$ and $\check{u}_{\mathrm{S}}^{e(j)} \in \mathbb{R}^{\check{n}_{\mathrm{S}}^{e(j)}}$. Here, $\check{n}_{\mathrm{K}}^{e(j)} = \check{n}_{\mathrm{M}}^{e(j)} = \check{n}_{\mathrm{S}}^{e(j)}$ holds for the number of pair-wise coupled displacement degrees of freedom $\check{n}_{\mathrm{K}}^{e(j)}$ and $\check{n}_{\mathrm{P}}^{e(j)} = \check{n}_{\mathrm{S}}^{e(j)}$ for the prescribed degrees of freedom. Since the periodic displacements are applied on the surface of the RVE, the internal (within the volume of the RVE) displacement degrees of freedom are defined as $\check{u}_{\mathrm{V}}^{e(j)} \in \mathbb{R}^{\check{n}_{\mathrm{V}}^{e(j)}}$, where $\check{n}_{\mathrm{u}}^{e(j)} = \check{n}_{\mathrm{V}}^{e(j)} + \check{n}_{\mathrm{M}}^{e(j)}$ holds. Accordingly, the decomposition of the nodal displacement degrees of freedom and the discretized local equilibrium Equation (12) on the microscale

$$\check{u}_{\mathrm{a}}^{e(j)} = \begin{Bmatrix} \check{u}_{\mathrm{V}}^{e(j)} \\ \check{u}_{\mathrm{M}}^{e(j)} \\ \check{u}_{\mathrm{S}}^{e(j)} \end{Bmatrix}, \qquad \check{g}_{\mathrm{a}}^{e(j)}(t, u_{\mathrm{a}}, \check{u}_{\mathrm{a}}^{e(j)}(t)) = \begin{Bmatrix} \check{g}_{\mathrm{V}}^{e(j)}(t, u_{\mathrm{a}}, \check{u}_{\mathrm{V}}^{e(j)}, \check{u}_{\mathrm{M}}^{e(j)}) \\ \check{g}_{\mathrm{M}}^{e(j)}(t, u_{\mathrm{a}}, \check{u}_{\mathrm{V}}^{e(j)}, \check{u}_{\mathrm{M}}^{e(j)}) \\ \check{g}_{\mathrm{S}}^{e(j)}(t, u_{\mathrm{a}}, \check{u}_{\mathrm{V}}^{e(j)}, \check{u}_{\mathrm{M}}^{e(j)}) \end{Bmatrix}, \tag{13}$$

is obtained.

The connection between the macro- and the microscale is achieved with macroscale displacements $u_{\mathrm{a}}(t)$ and microscale displacements $\check{u}_{\mathrm{a}}^{e(j)}(t)$ by specifying constraint

$$\check{C}_{\mathrm{c}}^{e(j)}(\check{u}_{\mathrm{a}}^{e(j)}(t), u_{\mathrm{a}}(t)) = \check{A}_{1}^{e(j)}\check{u}_{\mathrm{a}}^{e(j)}(t) - \check{A}_{2}^{e(j)}\mathbf{E}^{e(j)}(u_{\mathrm{a}}(t)) = \mathbf{0}, \tag{14}$$

with $\check{C}_{\mathrm{c}}^{e(j)} \in \mathbb{R}^{\check{n}_{\mathrm{P}}^{e(j)}}$, $\check{A}_{1}^{e(j)} \in \mathbb{R}^{\check{n}_{\mathrm{K}}^{e(j)} \times \check{n}_{\mathrm{a}}^{e(j)}}$, and $\check{A}_{2}^{e(j)} \in \mathbb{R}^{\check{n}_{\mathrm{K}}^{e(j)} \times 6}$. For the case of periodic displacement boundary conditions, matrices $\check{A}_{1}^{e(j)}$ and $\check{A}_{2}^{e(j)}$ read

$$\check{A}_{1}^{e(j)} = \begin{bmatrix} \underset{(\check{n}_{\mathrm{K}}^{e(j)} \times \check{n}_{\mathrm{V}}^{e(j)})}{\mathbf{0}} & \underset{(\check{n}_{\mathrm{M}}^{e(j)} \times \check{n}_{\mathrm{M}}^{e(j)})}{\check{H}_{\mathrm{M}}^{e(j)}} & \underset{(\check{n}_{\mathrm{S}}^{e(j)} \times \check{n}_{\mathrm{S}}^{e(j)})}{\check{H}_{\mathrm{S}}^{e(j)}} \end{bmatrix} \quad \text{and} \quad \check{A}_{2}^{e(j)} = \check{P}^{e(j)T}, \tag{15}$$

where matrices $\check{H}_{\mathrm{M}}^{e(j)}$ and $\check{H}_{\mathrm{S}}^{e(j)}$ are link-topology matrices that only contain the numbers $0, +1, -1$. $\check{P}^{e(j)} \in \mathbb{R}^{6 \times \check{n}_{\mathrm{K}}^{e(j)}}$ is a matrix that comprises the differences in the corresponding nodal positions. Constraint (14) can be reformulated with $\check{M}^{e(j)} = \check{H}_{\mathrm{S}}^{e(j)-1}\check{H}_{\mathrm{M}}^{e(j)}$, $\check{M}^{e(j)} \in \mathbb{R}^{\check{n}_{\mathrm{S}}^{e(j)} \times \check{n}_{\mathrm{M}}^{e(j)}}$ leading to

$$\check{C}_{\mathrm{c}}^{e(j)}(\check{u}_{\mathrm{a}}^{e(j)}(t), u_{\mathrm{a}}(t)) = \check{M}^{e(j)}\check{u}_{\mathrm{M}}^{e(j)} + \check{u}_{\mathrm{S}}^{e(j)} - \check{H}_{\mathrm{S}}^{e(j)-1}\check{P}^{e(j)T}\mathbf{E}^{e(j)}(t, u(t)). \tag{16}$$

Constraint (16) is, again, enforced with the microscale Lagrange multipliers $\check{\lambda}^{e(j)} \in \mathbb{R}^{\check{n}_{\mathrm{K}}^{e(j)}}$. With decomposition (13), the microscale strain vector $\overset{\times}{\mathbf{E}}^{\check{e}(\check{j})} \in \mathbb{R}^{6}$ of microscale element $\check{e}$ and integration point $\check{j}$ in dependence of the macroscale strains (9) reads

$$\overset{\times}{\mathbf{E}}^{\check{e}(\check{j})}(t, u, \check{u}^{e(j)}) = \check{B}^{\check{e}(\check{j})}\left\{ \check{Z}^{\check{e}}\check{u}^{e(j)} + \check{Z}_{\mathrm{S}}^{\check{e}}\check{H}_{\mathrm{S}}^{e(j)-1}\check{P}^{e(j)T}\mathbf{E}^{e(j)}(t, u(t)) \right\}, \tag{17}$$

where we abbreviate

$$\check{u}^{e(j)} = \begin{Bmatrix} \check{u}_{\mathrm{V}}^{e(j)} \\ \check{u}_{\mathrm{M}}^{e(j)} \end{Bmatrix} \qquad \check{Z}^{\check{e}} = \begin{bmatrix} \check{Z}_{\mathrm{V}}^{\check{e}} & \check{Z}_{\mathrm{M}}^{\check{e}} - \check{Z}_{\mathrm{S}}^{\check{e}}\check{M}^{e(j)} \end{bmatrix}. \tag{18}$$

$\check{u}^{e(j)} \in \mathbb{R}^{\check{n}_{\mathrm{u}}^{e(j)}}$ denotes the assembled displacement vector, while $\check{Z}^{\check{e}} \in \mathbb{R}^{\check{n}_{\mathrm{u}}^{\check{e}} \times \check{n}_{\mathrm{u}}^{e(j)}}$, $\check{Z}_{\mathrm{V}}^{\check{e}} \in \mathbb{R}^{\check{n}_{\mathrm{u}}^{\check{e}} \times \check{n}_{\mathrm{V}}^{e(j)}}$, $\check{Z}_{\mathrm{M}}^{\check{e}} \in \mathbb{R}^{\check{n}_{\mathrm{u}}^{\check{e}} \times \check{n}_{\mathrm{M}}^{e(j)}}$, and $\check{Z}_{\mathrm{S}}^{\check{e}} \in \mathbb{R}^{\check{n}_{\mathrm{u}}^{\check{e}} \times \check{n}_{\mathrm{S}}^{e(j)}}$ represent assignment matrices.

In contrast to the macroscale, on the microscale, constitutive models are applied to describe the mechanical behavior of the materials. Since in this contribution, only elastic materials are studied, the constitutive model has the general form of

$$\check{\mathbf{T}}^{\check{e}(\check{j})} = \check{\mathbf{h}}^{\check{e}(\check{j})}(\check{\mathbf{E}}^{\check{e}(\check{j})}), \tag{19}$$

i.e., the evaluation of algebraic equations yields the stresses $\check{\mathbf{T}}^{\check{e}(\check{j})} \in \mathbb{R}^6$ at the microscale integration points, which are already contained in Equation (12). What remains is the question of how to determine the required macroscale stress $\mathbf{T}^{e(j)}$ from the microscale evaluation. Here, a homogenization procedure is applied that fits into the general form of

$$\mathbf{T}^{e(j)} = \tilde{\mathbf{h}}^{e(j)}(\check{\boldsymbol{\lambda}}^{e(j)}(t)) = \frac{1}{\check{V}^{e(j)}} \check{\mathbf{A}}_2^{e(j)T} \check{\boldsymbol{\lambda}}^{e(j)}(t). \tag{20}$$

In the case of periodic displacement boundary conditions, the secondary displacements $\check{u}_{\mathrm{S}}^{e(j)}$ can be expressed by constraint (16)

$$\check{u}_{\mathrm{S}}^{e(j)} = \check{H}_{\mathrm{S}}^{e(j)-1} \check{P}^{e(j)T} \mathbf{E}^{e(j)}(t, u(t)) - \check{M}^{e(j)} \check{u}_{\mathrm{M}}^{e(j)}. \tag{21}$$

Further, enforcing constraint (16) yields, on the RVE level,

$$\check{g}_{\mathrm{S}}^{e(j)}(t, u_{\mathrm{a}}, \check{u}_{\mathrm{V}}^{e(j)}, \check{u}_{\mathrm{M}}^{e(j)}) - \check{H}_{\mathrm{S}}^{e(j)T} \check{\boldsymbol{\lambda}}^{e(j)} = \mathbf{0}. \tag{22}$$

This allows computation of microscale Lagrange multipliers $\check{\boldsymbol{\lambda}}^{e(j)}$:

$$\check{\boldsymbol{\lambda}}^{e(j)} = \check{H}_{\mathrm{S}}^{e(j)-T} \check{g}_{\mathrm{S}}^{e(j)}(t, u_{\mathrm{a}}, \check{u}_{\mathrm{V}}^{e(j)}, \check{u}_{\mathrm{M}}^{e(j)}). \tag{23}$$

As a result, the homogenized stresses (20) on macroscale element e and integration point j read for periodic displacement boundary conditions:

$$\tilde{\mathbf{h}}^{e(j)}(t, u_{\mathrm{a}}, \check{u}_{\mathrm{V}}^{e(j)}, \check{u}_{\mathrm{M}}^{e(j)}) = \frac{1}{\check{V}^{e(j)}} \check{P}^{e(j)} \check{H}_{\mathrm{S}}^{e(j)-T} \check{g}_{\mathrm{S}}^{e(j)}(t, u_{\mathrm{a}}, \check{u}_{\mathrm{V}}^{e(j)}, \check{u}_{\mathrm{M}}^{e(j)}). \tag{24}$$

2.1.3. General System of Non-Linear Equations

The entire system of equations of FE^2 computations with non-linear elastic material at the microscale level is obtained by formally assembling all independent variables of the RVEs:

$$\check{u}_{\mathrm{a}} = \sum_{e=1}^{n_{\mathrm{e}}} \sum_{j=1}^{n_{\mathrm{G}}^e} Z_{\check{u}_{\mathrm{a}}}^{e(j)T} \check{u}_{\mathrm{a}}^{e(j)}, \qquad \check{\boldsymbol{\lambda}} = \sum_{e=1}^{n_{\mathrm{e}}} \sum_{j=1}^{n_{\mathrm{G}}^e} Z_{\check{\boldsymbol{\lambda}}}^{e(j)T} \check{\boldsymbol{\lambda}}^{e(j)}, \tag{25}$$

$\check{u}_{\mathrm{a}} \in \mathbb{R}^{n_{\mathrm{e}} n_{\mathrm{G}}^e \check{n}_{\mathrm{a}}^{e(j)}}$, $\check{\boldsymbol{\lambda}} \in \mathbb{R}^{n_{\mathrm{e}} n_{\mathrm{G}}^e \check{n}_{\mathrm{K}}^{e(j)}}$, as well as equations

$$\check{g}_{\mathrm{a}}(\check{u}_{\mathrm{a}}, \check{\boldsymbol{\lambda}}) = \sum_{e=1}^{n_{\mathrm{e}}} \sum_{j=1}^{n_{\mathrm{G}}^e} Z_{\check{u}_{\mathrm{a}}}^{e(j)T} \left\{ \check{g}_{\mathrm{a}}^{e(j)}(\check{u}_{\mathrm{a}}^{e(j)}) - \check{A}_1^{e(j)T} \check{\boldsymbol{\lambda}}^{e(j)} \right\}, \tag{26}$$

$$\check{C}_{\mathrm{c}}(u_{\mathrm{a}}, \check{u}_{\mathrm{a}}) = \sum_{e=1}^{n_{\mathrm{e}}} \sum_{j=1}^{n_{\mathrm{G}}^e} Z_{\check{\boldsymbol{\lambda}}}^{e(j)T} \check{C}_{\mathrm{c}}^{e(j)}(u_{\mathrm{a}}, \check{u}_{\mathrm{a}}^{e(j)}). \tag{27}$$

The entire system of non-linear equations is obtained by compiling macroscale Equations (10) and (11) and microscale Equations (26) and (27). The number of equations can be essentially reduced by assuming that constraint (10) is fulfilled after solving the entire system of non-linear equations and by employing Equation (23) on the microscale for periodic displacement boundary conditions. Further, $\check{A}_1^{e(j)}$, as given in Equation $(15)_1$, can be applied. Then, the reduced system of non-linear equations,

$$F(t, y) = \left\{ \begin{array}{c} g(t, u) \\ \check{g}_V(t, u, \check{u}_V, \check{u}_M) \\ \check{g}_M(t, u, \check{u}_V, \check{u}_M) - \check{M}^T \check{g}_S(t, u, \check{u}_V, \check{u}_M) \end{array} \right\} = 0, \tag{28}$$

which is the result of the spatial discretization, has to be solved at each load-step (time-step) with the vector of unknowns

$$y^T = \{u, \check{u}_V, \check{u}_M\}^T. \tag{29}$$

In the equations mentioned above, $\check{u}_V$ and $\check{u}_M$ are the vectors of assembled internal microscale nodal displacements $\check{u}_V^{e(j)}$ and primary nodal displacements $\check{u}_M^{e(j)}$, respectively.

Remark 2. *Further, an important aspect is the time discretization. Since only non-linear elastic material is studied here, Equation (28) represents a purely algebraic system of equations. Nevertheless, time integration methods, such as the Backward–Euler method, can be applied when formally extending Equation (28) with $\dot{t} = 1$ to obtain a system of differential-algebraic equations (DAE), as it is common in finite element computations, where the inelastic material behavior is described by evolution equations for some internal variables [76]. As a result, the application of time integration methods to elastic problems leads to an incremental application of the prescribed loads, which is therefore achieved in the numerical examples of this work. In the case of non-linear elastic material behavior, the load is often applied step-wise, where the previous solution of the nodal displacements is inserted into some Newton-like scheme as starting vector to be close to the solution. Otherwise, problems in the convergence of the iterative scheme are observed. In this sense, the step-wise increase in the load (displacement- or force-controlled) can be interpreted as time integration.*

2.2. Multilevel–Newton Algorithm

What remains is the question of how the system of non-linear Equation (28) is solved in multiscale simulations. Further, to sufficiently embed DNN surrogates, it is important to make clear which parts of the overall computation scheme can be substituted with minimal changes in a finite element program, as it is discussed later on.

There are different approaches to solve the system of non-linear Equation (28). First, the entire system of equations could be solved with the Newton–Raphson method, but this would only be possible for smaller problems due to the extremely large number of equations. An alternative—see [76]—would be to use the Newton–Schur complement, which, on the one hand, requires some intervention in the coding and provision of the derivatives [77]. In traditional approaches, on the other hand, one uses the MLNA considering periodic boundary conditions for the microstructures. Therefore, in order to see what advantage neural networks have here, the Multilevel–Newton algorithm (MLNA) approach is briefly explained.

Since the MLNA, which was originally introduced by [78,79], is frequently applied, we refer especially to [76] regarding the differences between the MLNA and the well-known Newton–Raphson scheme. Thus, only the required equations are recapped here to show the general algorithmic structure of FE^2 computations and the incorporation of DNN surrogate models.

If we interpret the incremental load-control as time integration, the non-linear system (28) has to be evaluated at time t_{n+1}, $t_{n+1} = t_n + \Delta t_n$. Thus, in each time-step, the unknown microscale displacements $\check{u} = \{\check{u}_V, \check{u}_M\}^T$, see Equation $(18)_1$, and the unknown macroscale displacements u are sought. For further treatment of the equations, we also introduce decomposition

$$\check{G} \triangleq \left\{ \begin{array}{c} \check{g}_V \\ \check{g}_M - \check{M}^T \check{g}_S \end{array} \right\}. \tag{30}$$

Thus, the system of non-linear equations

$$G(u, \check{u}) = 0,$$
$$\check{G}(u, \check{u}) = 0 \tag{31}$$

has to be solved.

2.2.1. Multilevel–Newton Algorithm for FE^2 Computations

In the traditional manner, the MLNA is applied to solve Equation (31). The scheme draws on the implicit function theorem, i.e., it is assumed that function $\check{u} = \mathring{u}(u)$ exists. In other words,

$$G(u, \mathring{u}(u)) = 0 \tag{32}$$

has to be solved. The Newton–Raphson method applied to the non-linear system (32) requires in each iteration step the computation of linear system

$$\left[\frac{\partial G}{\partial u} + \frac{\partial G}{\partial \check{u}} \frac{d\mathring{u}}{du} \right] \Delta u = -G(u, \mathring{u}(u)). \tag{33}$$

Here, the iteration index is omitted for brevity. Quantities $d\mathring{u}/du$ and $\check{u} = \mathring{u}(u)$ have to be provided by two additional computational steps, since $\mathring{u}(u)$ is assumed to exist, but its representation is unknown. First,

$$\check{G}(u, \check{u}) = 0 \qquad \leadsto \check{u} \tag{34}$$

is evaluated for a given u, and second, the chain-rule is applied to

$$\check{G}(u, \mathring{u}(u)) = 0 \quad \rightarrow \quad \frac{\partial \check{G}}{\partial u} + \frac{\partial \check{G}}{\partial \check{u}} \frac{d\mathring{u}}{du} = 0 \quad \leadsto \quad \frac{d\mathring{u}}{du}. \tag{35}$$

The entire procedure is shown in Algorithm 1.

In greater detail and with the problem at hand, we proceed as follows. On a microscale, the system of non-linear equations

$$\check{G}(t_{n+1}, u, \check{u}_{\mathrm{V}}, \check{u}_{\mathrm{M}}) = \sum_{e=1}^{n_e} \sum_{j=1}^{n_{\mathrm{G}}^e} Z_{\check{u}}^{e(j)T} \check{G}^{e(j)}(t_{n+1}, u, \check{u}_{\mathrm{V}}^{e(j)}, \check{u}_{\mathrm{M}}^{e(j)}) = 0 \tag{36}$$

has to be be solved for the case of periodic displacement boundary conditions, with

$$\check{G}^{e(j)}(t_{n+1}, u, \check{u}_{\mathrm{V}}^{e(j)}, \check{u}_{\mathrm{M}}^{e(j)}) = \sum_{\check{e}=1}^{\check{n}_e^{e(j)}} \check{Z}^{\check{e}T} \left(\sum_{\check{j}=1}^{\check{n}_{\mathrm{G}}^{\check{e}}} \check{w}_{\check{j}} \, \check{B}^{\check{e}(\check{j})T} \, \check{h}^{\check{e}(\check{j})}(E^{\check{e}(\check{j})}) \det \check{J}^{\check{e}(\check{j})} \right). \tag{37}$$

Here, $\check{Z}^{\check{e}}$ still has representation (18)$_2$, which then leads to the second and third equations of Equation (28). For purely elastic problems, the solution of Equation (36) leads to a linear system on global microscale level within the Newton-iteration step to solve Equation (34),

$$\left[\frac{\partial \check{G}}{\partial \check{u}} \right] \Delta \check{u} = -\check{G}(u, \check{u}), \tag{38}$$

which reads, in detail, as

$$\left[\begin{array}{cc} \dfrac{\partial \check{G}_{\mathrm{V}}}{\partial \check{u}_{\mathrm{V}}} & \dfrac{\partial \check{G}_{\mathrm{V}}}{\partial \check{u}_{\mathrm{M}}} \\ \dfrac{\partial \check{G}_{\mathrm{M}}}{\partial \check{u}_{\mathrm{V}}} - \check{M}^T \left[\dfrac{\partial \check{G}_{\mathrm{S}}}{\partial \check{u}_{\mathrm{V}}} \right] & \dfrac{\partial \check{G}_{\mathrm{M}}}{\partial \check{u}_{\mathrm{M}}} - \check{M}^T \left[\dfrac{\partial \check{G}_{\mathrm{S}}}{\partial \check{u}_{\mathrm{M}}} \right] \end{array} \right] \Bigg|_y \left\{ \begin{array}{c} \Delta \check{u}_{\mathrm{V}} \\ \Delta \check{u}_{\mathrm{M}} \end{array} \right\} = -\left\{ \begin{array}{c} \check{G}_{\mathrm{V}}(y) \\ \check{G}_{\mathrm{M}}(y) - \check{M}^T \check{G}_{\mathrm{S}}(y) \end{array} \right\} \tag{39}$$

with the vector of unknowns $\boldsymbol{y}$ according to Equation (29)$_1$. We apply another relationship between microscale displacements $\breve{\boldsymbol{u}}^{e(j)}$ and the assembled microscale displacements $\breve{\boldsymbol{u}}$, $\breve{\boldsymbol{u}}^{e(j)} = \boldsymbol{Z}_{\breve{u}}^{e(j)}\breve{\boldsymbol{u}}$. The system of linear Equation (38) can be re-written employing the chain rule and applying the decomposition into the macroscale integration point contributions, i.e., contributions of each RVE,

$$\sum_{e=1}^{n_e}\sum_{j=1}^{n_G^e} \boldsymbol{Z}_{\breve{u}}^{e(j)T}\left\{\left[\frac{\partial\breve{\boldsymbol{G}}^{e(j)}}{\partial\breve{\boldsymbol{u}}^{e(j)}}\right]\Delta\breve{\boldsymbol{u}}^{e(j)} + \breve{\boldsymbol{G}}^{e(j)}(\boldsymbol{u},\breve{\boldsymbol{u}}^{e(j)})\right\} = \boldsymbol{0}. \tag{40}$$

Algorithm 1: Multilevel-Newton algorithm for FE^2 computations with periodic displacement boundary conditions on a microscale.

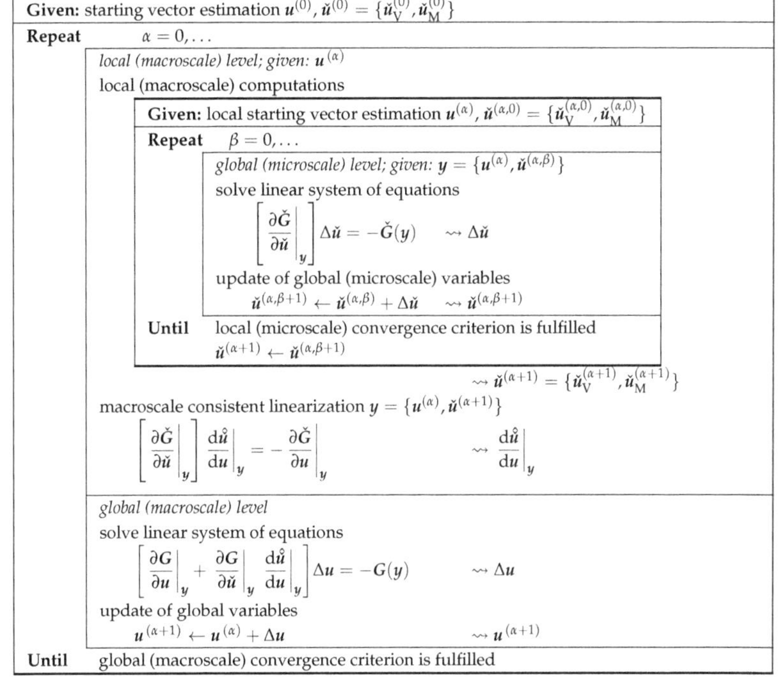

For abbreviation purposes, we introduce the global microscale tangential stiffness matrix

$$\breve{\boldsymbol{K}}^{e(j)} := \frac{\partial\breve{\boldsymbol{G}}^{e(j)}}{\partial\breve{\boldsymbol{u}}^{e(j)}} = \frac{\partial\breve{\boldsymbol{G}}^{e(j)}}{\partial\breve{\boldsymbol{E}}^{\breve{e}(\breve{j})}}\frac{\mathrm{d}\breve{\boldsymbol{E}}^{\breve{e}(\breve{j})}}{\mathrm{d}\breve{\boldsymbol{u}}^{e(j)}}, \tag{41}$$

$\breve{\boldsymbol{K}}^{e(j)} \in \mathbb{R}^{\breve{n}_u^{e(j)}\times\breve{n}_u^{e(j)}}$. Using Equations (17) and (37),

$$\breve{\boldsymbol{K}}^{e(j)} = \sum_{\breve{e}=1}^{\breve{n}_e^{e(j)}} \breve{\boldsymbol{Z}}^{\breve{e}T}\left[\sum_{\breve{j}=1}^{\breve{n}_G^{\breve{e}}} \breve{w}_{\breve{j}}\, \breve{\boldsymbol{B}}^{\breve{e}(\breve{j})T}\underbrace{\left[\frac{\partial\breve{\boldsymbol{h}}^{\breve{e}(\breve{j})}}{\partial\breve{\boldsymbol{E}}^{\breve{e}(\breve{j})}}\right]}_{\breve{\boldsymbol{C}}^{\breve{e}(\breve{j})}}\breve{\boldsymbol{B}}^{\breve{e}(\breve{j})}\,\det\breve{\boldsymbol{J}}^{\breve{e}(\breve{j})}\right]\breve{\boldsymbol{Z}}^{\breve{e}} \tag{42}$$

$$\underbrace{\phantom{\sum_{\breve{j}=1}^{\breve{n}_G^{\breve{e}}} \breve{w}_{\breve{j}}\, \breve{\boldsymbol{B}}^{\breve{e}(\breve{j})T}}}_{\breve{\boldsymbol{k}}^{\breve{e}}}$$

is obtained, where $\check{\mathbf{C}}^{\check{e}(\check{j})} \in \mathbb{R}^{6\times 6}$ denotes the consistent tangent matrix at microscale integration point $\check{j}$ of element $\check{e}$, and $\check{\mathbf{k}}^{\check{e}} \in \mathbb{R}^{\check{n}_{\check{u}}^{\check{e}} \times \check{n}_{\check{u}}^{\check{e}}}$ defines the element stiffness matrix of an element in an RVE. As a result, on a microscale, i.e., each RVE, the system of linear equations

$$\check{K}^{e(j)} \Delta \check{u}^{e(j)} = -\check{G}^{e(j)}(u, \check{u}^{e(j)}) \tag{43}$$

has to be sequentially solved on a global microscale level to reach microscale equilibrium. In other words, the solution of system (43) is repeated until a local convergence criterion is fulfilled. Then, the microscale displacements $\check{u}$ are obtained.

As the next step, the macroscale consistent linearization (35) implies

$$\frac{\partial \check{G}}{\partial \check{u}} \frac{d\mathring{\check{u}}}{du} = -\frac{\partial \check{G}}{\partial u}. \tag{44}$$

These matrices read under consideration of Equations (9), (17) and (42):

$$\frac{\partial \check{G}}{\partial \check{u}} \frac{d\mathring{\check{u}}}{du} = \sum_{e=1}^{n_e} \sum_{j=1}^{n_G^e} Z_{\check{u}}^{e(j)T} \left[\sum_{\check{e}=1}^{\check{n}_e^{e(j)}} \check{Z}^{\check{e}T} \check{k}^{\check{e}} \check{Z}^{\check{e}} \right] \frac{d\check{u}^{e(j)}}{d\mathbf{E}^{e(j)}} \underbrace{\frac{d\hat{\mathbf{E}}^{e(j)}}{du}}_{\mathbf{B}^{e(j)}Z^e}, \tag{45}$$

$$\frac{\partial \check{G}}{\partial u} = \sum_{e=1}^{n_e} \sum_{j=1}^{n_G^e} Z_{\check{u}}^{e(j)T} \left[\sum_{\check{e}=1}^{\check{n}_e^{e(j)}} \check{Z}^{\check{e}T} \check{k}^{\check{e}} \check{Z}_S^{\check{e}} \right] \check{H}_S^{e(j)-1} \check{P}^{e(j)T} \frac{d\hat{\mathbf{E}}^{e(j)}}{du}, \tag{46}$$

where $\partial \check{\mathbf{E}}^{\check{e}(\check{j})}/\partial \mathbf{E}^{e(j)} = \check{\mathbf{B}}^{\check{e}(\check{j})} \check{Z}_S^{\check{e}} \check{H}_S^{e(j)-1} \check{P}^{e(j)T}$ is used in Equation (46). Then, with the two matrices

$$\check{K}^{e(j)} = \sum_{\check{e}=1}^{\check{n}_e^{e(j)}} \check{Z}^{\check{e}T} \check{k}^{\check{e}} \check{Z}^{\check{e}} \quad \text{and} \quad \overline{\check{K}}^{e(j)} = \sum_{\check{e}=1}^{\check{n}_e^{e(j)}} \check{Z}^{\check{e}T} \check{k}^{\check{e}} \check{Z}_S^{\check{e}}, \tag{47}$$

$\overline{\check{K}}^{e(j)} \in \mathbb{R}^{\check{n}_u^{e(j)} \times \check{n}_S^{e(j)}}$, on a global RVE level, the consistent linearization step (44) for each RVE reads

$$\check{K}^{e(j)} \frac{d\check{u}^{e(j)}}{d\mathbf{E}^{e(j)}} = -\overline{\check{K}}^{e(j)} \check{H}_S^{e(j)-1} \check{P}^{e(j)T} \tag{48}$$

in order to compute $d\check{u}^{e(j)}/d\mathbf{E}^{e(j)}$ and, finally, $d\mathring{\check{u}}/du$. Therewith, the local macroscale computations consisting of the two steps, the global microscale level and the macroscale consistent linearization, are finalized.

As a last step in the MLNA, the system of linear equations

$$\left[\frac{\partial G}{\partial u} + \frac{\partial G}{\partial \check{u}} \frac{d\mathring{\check{u}}}{du} \right]\Bigg|_y \Delta u = -G(y), \quad \text{with} \quad y = \left\{ \begin{array}{c} u \\ \check{u} \end{array} \right\}, \tag{49}$$

see Equation (33), has to be solved at a global macroscale level to compute the increment Δu of the macroscale displacements. Here, G is the discretized weak formulation of the equilibrium equation at the macroscale; see Equation (7),

$$G(t_{n+1}, u, \check{u}) = \sum_{e=1}^{n_e} Z^{eT} \left(\sum_{j=1}^{n_G^e} w_j \, \mathbf{B}^{e(j)T} \underbrace{\tilde{h}^{e(j)}(t_{n+1}, u, \check{u}^{e(j)})}_{T^{e(j)}} \det \mathbf{J}^{e(j)} \right) - \bar{p}(t_{n+1}) = \mathbf{0}, \tag{50}$$

in dependence of the homogenized stress state $\mathbf{T}^{e(j)}$ from Equation (20). Analogously to Equation (41), we define the global tangential stiffness matrix

$$K := \frac{\partial G}{\partial u} + \frac{\partial G}{\partial \check{u}} \frac{\mathrm{d}\mathring{u}}{\mathrm{d}u}, \tag{51}$$

$$= \sum_{e=1}^{n_e} \mathbf{Z}^{eT} \left[\sum_{j=1}^{n_G^e} w_j \, \mathbf{B}^{e(j)T} \frac{1}{\check{V}^{e(j)}} \check{P}^{e(j)} \check{H}_{\mathrm{S}}^{e(j)-T} \left[\frac{\partial \check{G}_{\mathrm{S}}^{e(j)}}{\partial u} + \frac{\partial \check{G}_{\mathrm{S}}^{e(j)}}{\partial \check{u}} \frac{\mathrm{d}\mathring{u}}{\mathrm{d}u} \right] \det \mathbf{J}^{e(j)} \right]. \tag{52}$$

$K \in \mathbb{R}^{n_u \times n_u}$, where Equation (20) is already employed in Equation (52). Again, the derivatives can be re-formulated applying the chain rule and the microscale element stiffness matrix $\check{k}^{\check{e}}$ from Equation (42):

$$\frac{\partial \check{G}_{\mathrm{S}}^{e(j)}}{\partial u} = \frac{\partial \check{G}_{\mathrm{S}}^{e(j)}}{\partial \check{\mathbf{E}}^{\check{e}(\check{j})}} \frac{\partial \mathring{\mathbf{E}}^{\check{e}(\check{j})}}{\partial \mathbf{E}^{e(j)}} \frac{\mathrm{d}\mathbf{E}^{e(j)}}{\mathrm{d}u} = \underbrace{\left[\sum_{\check{e}=1}^{\check{n}_e^{e(j)}} \check{\mathbf{Z}}_{\mathrm{S}}^{\check{e}T} \check{k}^{\check{e}} \check{\mathbf{Z}}_{\mathrm{S}}^{\check{e}} \right]}_{\overline{\overline{\check{K}}}^{e(j)}} \check{H}_{\mathrm{S}}^{e(j)-1} \check{P}^{e(j)T} \, \mathbf{B}^{e(j)} \mathbf{Z}^e, \tag{53}$$

$$\frac{\partial \check{G}_{\mathrm{S}}^{e(j)}}{\partial \check{u}} \frac{\mathrm{d}\mathring{u}}{\mathrm{d}u} = \frac{\partial \check{G}_{\mathrm{S}}^{e(j)}}{\partial \check{\mathbf{E}}^{\check{e}(\check{j})}} \frac{\partial \mathring{\mathbf{E}}^{\check{e}(\check{j})}}{\partial \mathring{u}^{e(j)}} \frac{\mathrm{d}\mathring{u}^{e(j)}}{\mathrm{d}\mathbf{E}^{e(j)}} \frac{\mathrm{d}\mathbf{E}^{e(j)}}{\mathrm{d}u} = \left[\sum_{\check{e}=1}^{\check{n}_e^{e(j)}} \check{\mathbf{Z}}_{\mathrm{S}}^{\check{e}T} \check{k}^{\check{e}} \check{\mathbf{Z}}^{\check{e}} \right] \frac{\mathrm{d}\mathring{u}^{e(j)}}{\mathrm{d}\mathbf{E}^{e(j)}} \, \mathbf{B}^{e(j)} \mathbf{Z}^e, \tag{54}$$

with $\overline{\overline{\check{K}}}^{e(j)} \in \mathbb{R}^{\check{n}_{\mathrm{S}}^{e(j)} \times \check{n}_{\mathrm{S}}^{e(j)}}$. Inserting Equations (53) and (54) into Equation (52), the global tangential stiffness matrix reads

$$K = \sum_{e=1}^{n_e} \mathbf{Z}^{eT} \left[\sum_{j=1}^{n_G^e} w_j \, \mathbf{B}^{e(j)T} \mathbf{C}^{e(j)} \mathbf{B}^{e(j)} \det \mathbf{J}^{e(j)} \right] \mathbf{Z}^e. \tag{55}$$

Here, $\mathbf{C}^{e(j)} \in \mathbb{R}^{6 \times 6}$ denotes the consistent tangent matrix at integration point j of the macroscale having the representation

$$\mathbf{C}^{e(j)} = \frac{1}{\check{V}^{e(j)}} \check{P}^{e(j)} \check{H}_{\mathrm{S}}^{e(j)-T} \left[\overline{\overline{\check{K}}}^{e(j)} - \overline{\check{K}}^{e(j)T} \check{K}^{e(j)-1} \overline{\check{K}}^{e(j)} \right] \check{H}_{\mathrm{S}}^{e(j)-1} \check{P}^{e(j)T} \tag{56}$$

with the matrices in Equations (47) and (53) and the application of Equation (48).

2.2.2. Newton Algorithm for FE^2 Computations with DNN Surrogate Models

Obviously, the computations of the non-linear system (34) and the linear system with several right-hand sides (35) within an iterative scheme are very time consuming. Thus, an alternative approach is of particular interest. The basic idea is that the recurrent stress and tangent calculation is learned by a deep neural network and, thus, an efficient evaluation can be achieved. Since we embed DNN surrogate models to accelerate the FE^2 computation, it is important to make clear which quantities are applied as input and as output and what parts of the aforementioned MLNA are replaced by the surrogate.

As mentioned in the introduction, many different architectures of neural networks exist and are regularly applied in the field of computational mechanics. In this work, we draw on common feedforward neural networks to replace the entire local macroscale computations in Algorithm 1. The particular details for the neural networks are given later on; hence, we focus here on the algorithmic structure. During the solution of the boundary value problem, macroscale strains $\mathbf{E}^{e(j)}(t_{n+1}, u(t_{n+1}))$ at integration point j of macroscale element e are provided in dependence of the macroscale displacements u by Equation (9). As explained beforehand, in FE^2 computations, strains $\mathbf{E}^{e(j)}$ serve as an input to compute the displacement boundary conditions (16) under the assumption of periodic displacement degrees of freedom. Thus, strains $\mathbf{E}^{e(j)}$ are input quantities for the DNN surrogate models to predict the homogenized stresses $\mathbf{T}^{e(j)}$ and the consistent tangent matrix $\mathbf{C}^{e(j)}$ with the two surrogate models $\mathcal{T}$ and $\mathcal{C}$,

$$\mathbf{T}^{e(j)} \approx \mathcal{T}(\mathbf{E}^{e(j)}(u); \theta_\mathcal{T}) \quad \text{and} \quad \mathbf{C}^{e(j)} \approx \mathcal{C}(\mathbf{E}^{e(j)}(u); \theta_\mathcal{C}), \tag{57}$$

where $\theta_\mathcal{T}$ and $\theta_\mathcal{C}$ are the parameters concerned of the surrogate models. Here, the notation of the surrogate models $\mathcal{T}$ and $\mathcal{C}$ indicates that the models are evaluated for the strains $\mathbf{E}^{e(j)}$ while parameters $\theta_\mathcal{T}$ and $\theta_\mathcal{C}$ are assumed to be given after sufficient training of the neural network. At first, we introduce two different surrogate models for the stress and consistent tangent prediction. Later on, different realizations of the surrogate models are discussed as well.

The predicted stresses $\mathbf{T}^{e(j)}$ are employed to evaluate the local equilibrium Equation (50), here, of course, without being dependent on $\breve{u}$,

$$G(u) := g(t_{n+1}, u) = \sum_{e=1}^{n_e} \mathbf{Z}^{eT} \left(\sum_{j=1}^{n_G^e} w_j \, \mathbf{B}^{e(j)T} \, \mathcal{T}(\mathbf{E}^{e(j)}(u); \theta_\mathcal{T}) \det \mathbf{J}^{e(j)} \right) - \bar{p}(t_{n+1}) = 0. \tag{58}$$

Again, we omit the iteration indices and the load-step index $n+1$ for brevity. The predicted consistent tangent matrices $\mathbf{C}^{e(j)}$ are assembled into the global stiffness matrix K according to Equation (55),

$$K = \sum_{e=1}^{n_e} \mathbf{Z}^{eT} \left[\sum_{j=1}^{n_G^e} w_j \, \mathbf{B}^{e(j)T} \, \mathcal{C}(\mathbf{E}^{e(j)}(u); \theta_\mathcal{C}) \, \mathbf{B}^{e(j)} \, \det \mathbf{J}^{e(j)} \right] \mathbf{Z}^e. \tag{59}$$

As a result, when following the DNN-FE^2 approach for multiscale FE^2 computations, only the solution of the linear system of equations

$$K \, \Delta u = -G(u) \tag{60}$$

is necessary on a global macroscale level in each iteration. The entire Newton algorithm for FE^2 computations with DNN surrogate models and non-linear elastic material on a microscale is provided in Algorithm 2.

Algorithm 2: Newton algorithm for FE^2 computations following the DNN-FE^2 approach.

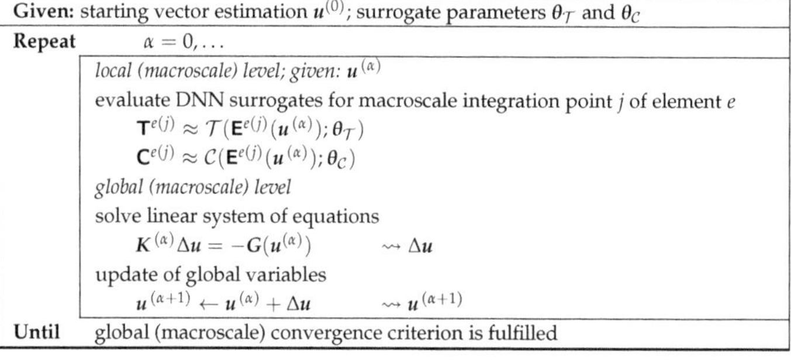

Given:	starting vector estimation $u^{(0)}$; surrogate parameters $\theta_\mathcal{T}$ and $\theta_\mathcal{C}$	
Repeat	$\alpha = 0, \ldots$	
	local (macroscale) level; given: $u^{(\alpha)}$	
	evaluate DNN surrogates for macroscale integration point j of element e	
	$\quad \mathbf{T}^{e(j)} \approx \mathcal{T}(\mathbf{E}^{e(j)}(u^{(\alpha)}); \theta_\mathcal{T})$	
	$\quad \mathbf{C}^{e(j)} \approx \mathcal{C}(\mathbf{E}^{e(j)}(u^{(\alpha)}); \theta_\mathcal{C})$	
	global (macroscale) level	
	solve linear system of equations	
	$\quad K^{(\alpha)} \Delta u = -G(u^{(\alpha)})$	$\leadsto \Delta u$
	update of global variables	
	$\quad u^{(\alpha+1)} \leftarrow u^{(\alpha)} + \Delta u$	$\leadsto u^{(\alpha+1)}$
Until	global (macroscale) convergence criterion is fulfilled	

It should be emphasized that the explained algorithm for DNN-FE^2 simulations in Algorithm 2 only holds for elastic problems. The mapping between macroscale strains and homogenized stress and consistent tangent changes essentially when applying viscous or path-dependent materials, such as plasticity or viscoplasticity, which is not discussed here.

3. Deep Neural Networks

In this section, a brief introduction is provided to deep neural networks and state-of-the-art frameworks for the implementation of these learning methodologies. First, a fully connected deep neural network—also known as a multilayer perceptron (MLP)—is considered. An MLP consists of a consecutive repetition of so-called layers. Each layer contains a set of nodes, so-called neurons, which are densely connected to the nodes of the preceding and succeeding layers. A deep neural network (DNN) is a neural network with multiple layers between the input and output layers which are the so-called hidden layers. Data sample $\mathbf{x}$ in space $\chi \subset \mathbb{R}^n$ and the corresponding target output $\mathbf{y}$ in space $\psi \subset \mathbb{R}^m$ are considered. Then, the objective of a deep neural network is to learn the mapping, $\mathcal{F} : \chi \rightarrow \psi$, from the data by minimizing a scalar-valued loss function $L(\mathcal{F}(\mathbf{x}; \boldsymbol{\theta}), \mathbf{y})$ for all the samples in the training data set, where $\boldsymbol{\theta} \in \mathbb{R}^{n_\theta}$ represents the trainable parameters of the network. To this end, the data are processed through each layer i as

$$\boldsymbol{\zeta}^{(i)} = \boldsymbol{\varphi}^{(i)}(\mathbf{W}^{(i)}\boldsymbol{\zeta}^{(i-1)} + \mathbf{b}^{(i)}), \quad i = 1, \ldots, n_{\text{layer}}, \tag{61}$$

where $\boldsymbol{\zeta}^{(i-1)} \in \mathbb{R}^{p^{(i)}}$ and $\boldsymbol{\zeta}^{(i)} \in \mathbb{R}^{q^{(i)}}$ are the input and output of the ith layer with the number of neurons $p^{(i)}$ in the previous layer and $q^{(i)}$ neurons in the current layer. Further, $\boldsymbol{\zeta}^{(0)} = \mathbf{x}$ holds. $\mathbf{W}^{(i)} \in \mathbb{R}^{q^{(i)} \times p^{(i)}}$ represents a weighting matrix and $\mathbf{b}^{(i)} \in \mathbb{R}^{q^{(i)}}$ is the bias vector. $\boldsymbol{\varphi}^{(i)} : \mathbb{R}^{q^{(i)}} \rightarrow \mathbb{R}^{q^{(i)}}$ symbolizes the element-wise applied activation function in layer i. Parameters $\boldsymbol{\theta}$ of the network are determined by applying a gradient-descent optimization technique for minimizing the loss function on the training data set. The updates of the parameters are obtained as $\Delta\boldsymbol{\theta} = \eta\, \partial\mathcal{L}/\partial\boldsymbol{\theta}$ where η denotes the learning rate. The gradient of the loss function with respect to the trainable parameters can be obtained using automatic differentiation (AD) [80]. All the neural networks discussed in this study were developed applying machine learning software frameworks developed by Google Research called TensorFlow [81] and JAX [82].

Automatic differentiation (AD), also known as *algorithmic differentiation* or "auto-diff" (automatic differentiation), is a family of methods for evaluating the derivatives of numeric functions expressed as computer programs efficiently and accurately through the accumulation of values during code execution. AD has an extensive application in machine learning and also well-established use cases in computational fluid dynamics [83], atmospheric sciences [84], and engineering design optimization [85]. In the field of computational solid mechanics, see [86] and the literature cited therein. The idea behind AD is to break down the function into its elementary operations and compute the derivative of each operation using symbolic rules of differentiation. This means that instead of relying on numerical approximations or finite differences to compute the derivative, AD can provide exact derivatives with machine precision. To do this, AD keeps track of the derivative values at each stage of the computation applying a technique called forward or reverse mode differentiation. This allows AD computing the derivative of the overall composition of the function by combining the derivatives of the constituent operations through the chain rule. The benefit of AD is that it can be applied to a wide range of computer programs, allowing for the efficient and accurate computation of derivatives. This makes it a powerful tool for scientific computing, optimization, and machine learning, where derivatives are needed for tasks such as gradient descent, optimization, and training of neural networks. AD techniques include forward and reverse accumulation modes. Forward-mode AD is efficient for functions $f : \mathbb{R} \rightarrow \mathbb{R}^m$, while for cases $f : \mathbb{R}^n \rightarrow \mathbb{R}^m$ where $n \gg m$, AD in its reverse accumulation mode is preferred [80]. For state-of-the-art deep learning models, n can be as large as millions or billions. In this research work, we utilized reverse-mode AD for the training of the neural networks and also for obtaining the Jacobian of the outputs with respect to the inputs. This should be demonstrated for both applied frameworks, TensorFlow and JAX. If one considers a batch of input vectors $\mathbf{x}$ and the corresponding outputs $\mathbf{y}$, then the Jacobian matrix $\mathbf{J}$ can be easily computed in batch mode using AD via TensorFlow and JAX according to Algorithm 3.

Algorithm 3: Computing the Jacobian matrix **J** of function f via reverse mode AD in TensorFlow and JAX frameworks for a batch of samples **x**.

TensorFlow:

```
def Jacobian(f, x):
    with tf.GradientTape() as tape:
        tape.watch(x)
        y = f(x)
    return tape.batch_jacobian(y, x)
J = Jacobian(f, x)
```

JAX:

```
Jacobian = jax.vmap(jax.jacrev(f))
J = Jacobian(x)
```

3.1. Deep Neural Networks as Surrogate Models for Local RVE Computations

In the MLNA described in Section 2.2.1, the computations on local macroscale level are very expensive to perform. Thus, the objective is to develop a data-driven surrogate model for substituting the local macroscale computations with deep neural networks. To this end, the macroscale strains $\mathbf{E}^{e(j)}(t_{n+1})$ at each integration point ξ_j and time (load-step) t_{n+1} are taken as the input and macroscale stresses $\mathbf{T}^{e(j)}(t_{n+1})$ and the consistent tangent matrix $\mathbf{C}^{e(j)}(t_{n+1})$ are provided as the output of the surrogate model. In the following, the FE^2 analysis is performed in a quasi-static setting with the restriction to small strains. For the sake of simplicity of the notation and for a two-dimensional set-up, we refer to the input of the surrogate model as $\overline{\mathbf{E}} = \{E_{11}, E_{22}, E_{12}\}^T$, $\overline{\mathbf{E}} \in \mathbb{R}^3$, and to the outputs as $\overline{\mathbf{T}} = \{T_{11}, T_{22}, T_{12}\}^T$, $\overline{\mathbf{T}} \in \mathbb{R}^3$, and $\overline{\mathbf{C}} = \{C_{11}, C_{12}, C_{13}, C_{21}, C_{22}, C_{23}, C_{31}, C_{32}, C_{33}\}^T$, $\overline{\mathbf{C}} \in \mathbb{R}^9$. Here, it should be mentioned that we do not employ the symmetry of the consistent tangent matrix due to the application of AD, where we compute the Jacobian matrix of the neural network containing the partial derivatives of each element of $\overline{\mathbf{T}}$ with respect to each element of the input $\overline{\mathbf{E}}$. Thus, we apply a soft symmetry constraint to the Jacobian matrix of the neural network by the data.

The inputs of the surrogate model are computed using an MPI (message passing interface) parallelized FORTRAN code. We employ FORPy [87], a library for FORTRAN-Python interoperability, to perform the data communications between FORTRAN and Python codes in an efficient and parallel manner. In particular, we load the required Python libraries and the DNN models only once and conduct the RVE computations in parallel, which leads to a considerable speed-up. The obtained outputs from the RVE surrogate model are passed to the FORTRAN code for further computations.

3.2. Training and Validation Datasets

Since the FE^2 framework in Section 2 is derived for the case of small strains, we consider a domain of application for our surrogate model with the upper and lower bounds of $E_{i,\min} = -0.04$ and $E_{i,\max} = 0.04$, respectively, where E_i represents the ith component of the strain input $\overline{\mathbf{E}}$. A dataset is generated by imposing different strain inputs to an RVE and computing the corresponding stress components and consistent tangent matrices. We utilized Latin hypercube sampling (LHS) [88] to efficiently sample from the input space. To generate the data, we consider two global symmetries in the input space:

$$T_{12}(E_{11}, E_{22}, -E_{12}) = -T_{12}(E_{11}, E_{22}, E_{12}), \tag{62}$$

$$T_{11}(-E_{11}, -E_{22}, E_{12}) = -T_{11}(E_{11}, E_{22}, E_{12}),$$
$$T_{22}(-E_{11}, -E_{22}, E_{12}) = -T_{22}(E_{11}, E_{22}, E_{12}). \tag{63}$$

It is important to mention that the assumed symmetries can be employed as long as the materials in the RVE show no tension–compression asymmetry, anisotropy, or rate- or path-dependent behavior, i.e., the reduction in the input space is not applicable for more complex behavior such as plasticity or anisotropic behavior. Thus, the data are generated using the numerical solver for one quarter of the input space according to the region marked by blue in Figure 1.

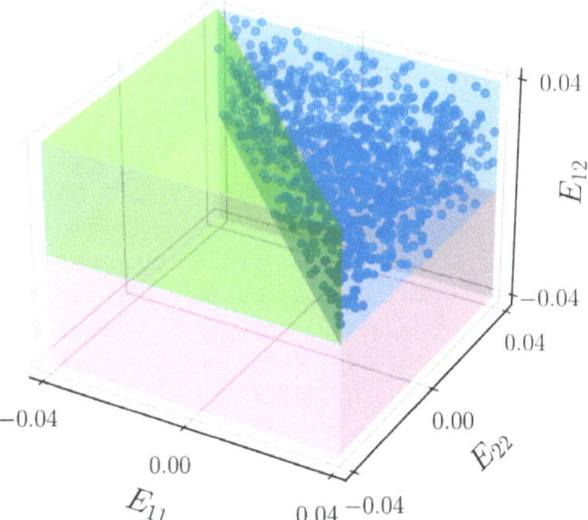

Figure 1. Domain of application for the surrogate model. The two global symmetries in the input space, Equations (62) and (63), are marked using pink and green colors, respectively. Blue dots show a subset of the sampled data points using LHS.

The dataset is augmented by transforming the generated data through the aforementioned global symmetries (62) and (63). This leads to a reduction in the computation time required for the preparation of the training data. After generating the dataset, it is decomposed into 80% for training and 20% for validation. Later on, in Section 4.2, the effect of the size of the training dataset on the accuracy of the final solution obtained from the DNN-based FE^2 simulation is investigated. It should be noted that the DNN models are tested by conducting DNN-FE^2 simulations and comparing the obtained solutions with those of the reference FE^2 simulations.

3.3. Architecture and Training Process

As it was mentioned in Sections 2.2 and 3.1, the surrogate model for the local RVE takes $\overline{\mathbf{E}} \in \mathbb{R}^3$ as the input and predicts $\overline{\mathbf{T}} \in \mathbb{R}^3$ and $\overline{\mathbf{C}} \in \mathbb{R}^9$ as the outputs. The obtained stress components and the consistent tangent matrix are assembled at the global finite element level for computation of the next global iteration in the Newton–Raphson method. In this work, two model architectures for our DNN-based surrogate models are considered, both are developed based on MLPs. In the first architecture, two separate neural networks $\mathcal{T}$ and $\mathcal{C}$ are implemented that map $\overline{\mathbf{E}}$ to $\overline{\mathbf{T}}$ and $\overline{\mathbf{C}}$,

$$\overline{\mathbf{T}} = \mathcal{T}(\overline{\mathbf{E}}; \boldsymbol{\theta}_\mathcal{T}), \quad \overline{\mathbf{C}} = \mathcal{C}(\overline{\mathbf{E}}; \boldsymbol{\theta}_\mathcal{C}), \tag{64}$$

where $\boldsymbol{\theta}_\mathcal{T}$ and $\boldsymbol{\theta}_\mathcal{C}$ represent the trainable parameters of deep neural network, $\mathcal{T}$ and $\mathcal{C}$, respectively. The notation in use indicates that the deep neural networks are evaluated for strain inputs $\overline{\mathbf{E}}$ with the given parameters after training the neural network. Throughout the article, this architecture is denoted as NN–2. However, this architecture does not consider that the consistent tangent matrix is the functional matrix of the stress components with respect to the strains,

$$\overline{\mathbf{C}} = \begin{bmatrix} C_{ij} \end{bmatrix} \quad \text{with} \quad C_{ij} = \frac{\partial T_i}{\partial E_j}, \quad i, j = 1, 2, 3, \tag{65}$$

where T_i and E_j are the corresponding entries in $\overline{\mathbf{T}}$ and $\overline{\mathbf{E}}$, respectively. Thus, this is taken into account in the second architecture by computing $\overline{\mathbf{C}}$ as the output of the Jacobian function $\mathcal{T}'$ as

$$\overline{\mathbf{T}} = \mathcal{T}(\overline{\mathbf{E}}; \theta_\mathcal{T}), \quad \overline{\mathbf{C}} = \frac{\partial \mathcal{T}}{\partial \overline{\mathbf{E}}} := \mathcal{T}'(\overline{\mathbf{E}}; \theta_\mathcal{T}), \tag{66}$$

where $\mathcal{T}'$ is obtained by applying reverse mode AD on the deep neural network surrogate $\mathcal{T}$, which is parameterized with trainable parameters $\theta_\mathcal{T}$. This architecture is denoted as NN–AD. Moreover, this approach is known as the so-called Sobolev training [61] in which both the target and its derivative with respect to the input are considered for supervised learning. Particular explanations regarding the application of Sobolev training in multiscale simulations are provided in [63], while the method is also employed in [60]. By optimizing the parameters of neural networks to approximate not only the function's outputs but also the function's derivatives, the model can encode additional information about the target function within its parameters. Therefore, the quality of the predictions, the data efficiency, and generalization capabilities of the learned neural network can be improved.

In the following, we provide a detailed discussion on the data pre-processing, model training, model selection, and hyperparameter tuning.

3.3.1. Data Pre-Processing

We perform a standardization step on both input and outputs of the model to obtain efficient training of the networks using the statistics of the training dataset. A training dataset $\mathbb{D}_{\text{train}} = \left[\overline{\mathbf{E}}, \overline{\mathbf{T}}, \overline{\mathbf{C}}\right]_{\text{train}}$ is considered with its mean and standard deviation over the samples as vectors μ and σ, respectively. For training of the NN–2 model, the input and the outputs are standardized independently with their means and standard deviations,

$$\tilde{V}_i = \frac{V_i - \mu_i}{\sigma_i}, \tag{67}$$

where V_i represents the ith component of $\overline{\mathbf{E}}$ or $\overline{\mathbf{T}}$ with the mean and standard deviation μ_i and σ_i, respectively. In contrast, for the NN–AD model, the consistent tangent matrix $\overline{\mathbf{C}}$ should be scaled consistently with the scaling of $\overline{\mathbf{E}}$ and $\overline{\mathbf{T}}$ so that their relationship is preserved. Therefore, scaling (67) is performed for the NN–AD model for the strains and stresses, while the components of $\overline{\mathbf{C}}$ are scaled as

$$\tilde{C}_{ij} = \frac{\partial \tilde{T}_i}{\partial \tilde{E}_j} = \frac{\partial \tilde{T}_i}{\partial T_i} \frac{\partial T_i}{\partial E_j} \frac{\partial E_j}{\partial \tilde{E}_j} = \frac{\sigma_j}{\sigma_i} C_{ij}, \quad i, j = 1, 2, 3. \tag{68}$$

3.3.2. Training

In the following, a detailed discussion of the training process of all DNNs implemented in this research work is provided. We utilize an extended version of the stochastic gradient descent algorithm, known as *Adam* [89], for optimizing the parameters of the network during the training process. The weights and biases of the DNN are initialized using the Glorot uniform algorithm [90] and zero initialization, respectively. All the neural networks are trained for 4000 epochs with an exponential decay of learning rate of

$$\eta = \eta_{\text{initial}}\, \gamma^{(\text{current step}/\text{decay step})}, \tag{69}$$

where η represents the learning rate. $\eta_{\text{initial}} = 10^{-3}$ is the initial learning rate and $\gamma = 0.1$ is the decay rate. Here, a decay step of 1000 is employed. The decay of the learning rate is applied according to Equation (69) every 1000 epochs to obtain a staircase behavior. For different sizes of the training dataset, the batch size is set such that 100 batches are obtained in order to have the same number of training updates for different sizes of the training dataset. The mean squared error (MSE) is utilized as the loss function. For the NN–2 model, the loss for a sample can be obtained as

$$L_{\mathcal{T}}(\mathcal{T}(\overline{\mathbf{E}};\boldsymbol{\theta}_{\mathcal{T}}),\overline{\mathbf{T}}^{\text{ref}}) = \frac{1}{3}\sum_{i=1}^{3}(\tilde{T}_i^{\text{ref}} - \tilde{T}_i^{\text{pred}}(\overline{\mathbf{E}};\boldsymbol{\theta}_{\mathcal{T}}))^2, \tag{70}$$

$$L_{\mathcal{C}}(\mathcal{C}(\overline{\mathbf{E}};\boldsymbol{\theta}_{\mathcal{C}}),\overline{\mathbf{C}}^{\text{ref}}) = \frac{1}{9}\sum_{i=1}^{9}(\tilde{C}_{ij}^{\text{ref}} - \tilde{C}_{ij}^{\text{pred}}(\overline{\mathbf{E}};\boldsymbol{\theta}_{\mathcal{C}}))^2, \tag{71}$$

where $L_{\mathcal{T}}$ and $L_{\mathcal{C}}$ indicate the loss for the mapping $\mathcal{T}$ and $\mathcal{C}$, respectively. Here, $\tilde{T}_i^{\text{ref}}$ denotes the reference stress value and $\tilde{T}_i^{\text{pred}}$ is the prediction of the neural network. Accordingly, $\tilde{C}_i^{\text{ref}}$ is the reference value in the consistent tangent and $\tilde{C}_i^{\text{pred}}$ is the corresponding prediction.

The loss for a data sample for the NN–AD architecture is computed as

$$L(\mathcal{T}(\overline{\mathbf{E}};\boldsymbol{\theta}_{\mathcal{T}}),\overline{\mathbf{T}}^{\text{ref}},\overline{\mathbf{C}}^{\text{ref}}) = \alpha\, L_{\mathcal{T}}(\mathcal{T}(\overline{\mathbf{E}};\boldsymbol{\theta}_{\mathcal{T}}),\overline{\mathbf{T}}^{\text{ref}}) + \beta\, L_{\mathcal{T}'}(\mathcal{T}'(\overline{\mathbf{E}};\boldsymbol{\theta}_{\mathcal{T}}),\overline{\mathbf{C}}^{\text{ref}}) \tag{72}$$

$$= \alpha\,\frac{1}{3}\sum_{i=1}^{3}(\tilde{T}_i^{\text{ref}} - \tilde{T}_i^{\text{pred}}(\overline{\mathbf{E}};\boldsymbol{\theta}_{\mathcal{T}}))^2 + \beta\,\frac{1}{9}\sum_{i=1}^{9}(\tilde{C}_i^{\text{ref}} - \tilde{C}_i^{\text{pred}}(\overline{\mathbf{E}};\boldsymbol{\theta}_{\mathcal{T}}))^2, \tag{73}$$

where α and β are the weighting coefficients for the two components of the loss. In all the cases, the loss for a batch of data is calculated by taking the average of the per-sample losses in the batch.

3.3.3. Model Selection

During the training process of each model, we track the validation loss and save the parameters of the model which lead to the lowest validation loss as the best model parameters. This helps to avoid overfitting of our deep neural networks. As mentioned earlier, the data are decomposed randomly into 80% for training and 20% for validation.

3.3.4. Hyperparameter Tuning

Hyperparameters in machine learning are the parameters that are defined by the user. Their values are set before starting the learning process of the model, such as number of neurons and hidden layers. The values of the hyperparameters remain unchanged during the training process and the following prediction. In machine learning applications, it is important to set the hyperparameters of a model such that the best performance is obtained regarding both prediction and generalization. Here, we perform hyperparameter tuning using a simple grid search algorithm to optimize the model performance on the validation dataset. For this experiment, a dataset with the size of $N_{\mathbb{D}} = 10^5$ is selected. We investigate three hyperparameters, i.e., the number of hidden layers N_h, the number of neurons per each hidden layer N_n, and the activation function φ, and carry out the hyperparameter tuning for the NN–2 architecture. Further, the same hyperparameters are employed for the NN–AD architecture for the sake of comparability. Moreover, for the NN–AD model, the weighting coefficients of the two components of the loss, α and β, are studied as well.

Results of the hyperparameter tuning for the number of hidden layers N_h, the number of neurons per each hidden layer N_n, and the activation function φ are reported in Appendix A. We observe that a model with eight hidden layers, 128 neurons per each hidden layer, and a *swish* activation function leads to $L_{\mathcal{T}}^{\text{val}} = 3.52 \times 10^{-8}$ and $L_{\mathcal{C}}^{\text{val}} = 2.84 \times 10^{-7}$. Moreover, our results show that increasing the model complexity to more than the aforementioned values would not lead to a significant gain in the model accuracy. Thus, we select these model hyperparameters for further analysis.

Other hyperparameters for the NN–AD model are the weighting coefficients α and β of the components of the loss, i.e., $L_{\mathcal{T}}$ and $L_{\mathcal{T}'}$. Here, the effect of the weighting on the obtained validation losses is investigated. The results are reported in Table 1.

Table 1. Effect of the weighting coefficients α and β for the components of the loss (72) on the performance of the NN–AD model.

(α, β)	$L_{\mathcal{T}}^{\text{val}}$	$L_{\mathcal{T}'}^{\text{val}}$
$(1, 0.01)$	4.84×10^{-8}	1.35×10^{-6}
$(1, 1)$	2.20×10^{-8}	8.85×10^{-8}
$(1, 100)$	3.55×10^{-8}	2.97×10^{-8}

We choose $\alpha = 1$ for all the cases and change β from 0.01 to 100. It can be observed that having a small β may lead to an imbalanced training where a difference of almost two orders of magnitude between validation losses $\mathcal{L}_{\mathcal{T}}$ and $\mathcal{L}_{\mathcal{T}'}$ exists. However, a β of one or larger results in a more balanced training leading to validation losses with nearly the same scale. According to the results of this study, we select a model with $\alpha = 1$ and $\beta = 100$ for further analysis.

4. Numerical Experiments

In this section, the DNN-FE^2 approach for the simulation of two canonical test cases in computational solid mechanics is investigated, i.e., an L-profile and Cook's membrane, and compare the results with those of FE^2 reference simulation. The numerical experiments are performed using both architectures, NN-2 as well as NN-AD, to provide a detailed discussion regarding the accuracy and efficiency of the simulations. The results are reported for the accuracy of the simulations, required time for model development (e.g., computational time needed for training), time of numerical simulation, number of load steps, and the total number of global iterations required for reaching the convergence of the FE^2 simulation. To this end, the absolute percentage error

$$\epsilon = \frac{|V^{\text{ref}} - V^{\text{pred}}|}{\langle |V^{\text{ref}}| \rangle} \times 100 \tag{74}$$

is utilized, where V indicates any component of stress or strain tensors $\overline{\mathbf{T}}$ and $\overline{\mathbf{E}}$, respectively. To avoid the division by zero, the absolute mean of the reference solution on the global grid is employed as the denominator, where $\langle \cdot \rangle$ shows the ensemble average.

All DNN models are trained on an NVIDIA RTX A2000 Laptop GPU with CUDA 12.1. The DNN-FE^2 simulations are performed on an 11th Gen Intel(R) Core(TM) i7-11850H @ 2.50GHz CPU with 16 threads. In contrast, the FE^2 reference simulations are conducted on a second-Gen Intel(R) Xeon(R) Silver 4216 @ 2.10GHz CPU with 16 processes and one thread per process. The speed-up gain is calculated by dividing the total time of computation required by the FE^2 reference simulation by that of the DNN-FE^2 simulation.

4.1. Problem Setup

For the numerical experiments, we restrict ourselves to two-dimensional test cases, where a plane strain case is always assumed. The representative volume element (RVE) under consideration is chosen as a commonly applied geometry in the mechanical analysis of composite materials; see Figure 2.

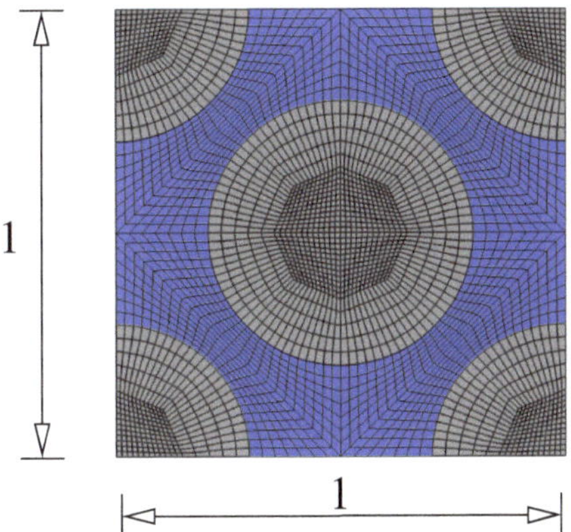

Figure 2. Geometry of the RVE (dimensions in mm) used as microstructure in the numerical experiments with fibers (grey) and matrix material (blue).

Here, the fiber volume fraction of the applied RVE is approximately 55%. The fibers are assumed to behave linearly, in an isotropic elastic manner with bulk modulus K_{f} and shear modulus G_{f}; see Table 2.

Table 2. Material parameters for elastic fiber and non-linear elastic matrix material in the RVE.

K_{f} N mm^{-2}	G_{f} N mm^{-2}	K_{m} N mm^{-2}	α_1 N mm^{-2}	α_2 -
4.35×10^4	2.99×10^4	4.78×10^3	5.0×10^1	6.0×10^{-2}

In contrast, the matrix material is modeled with a non-linear elastic material behavior, which is extracted from an originally viscoplastic constitutive model where the shear modulus is deformation-depending; see [91]. The particular stress–strain relation reads

$$\mathbf{T} = K_{\mathrm{m}}(\mathrm{tr}\,\mathbf{E})\mathbf{I} + G_{\mathrm{m}}(\mathbf{E}^{\mathrm{D}})\mathbf{E}^{\mathrm{D}} \qquad \text{with} \qquad G_{\mathrm{m}}(\mathbf{E}^{\mathrm{D}}) = \frac{\alpha_1}{\alpha_2 + ||\mathbf{E}^{\mathrm{D}}||_2}. \tag{75}$$

The material parameters for the non-linear elastic material are the bulk modulus K_{m} and the parameters α_1 and α_2; see Table 2. The spatial discretization of the RVE is achieved with $\check{n}_{\mathrm{e}}^{e(j)} = 3456$ eight-noded quadrilateral elements and $\check{n}_{\mathrm{nodes}}^{e(j)} = 10,561$ nodes.

The application of DNN surrogate models in multiscale simulations is studied for two macroscale test cases—L-profile and Cook's membrane; see Figure 3.

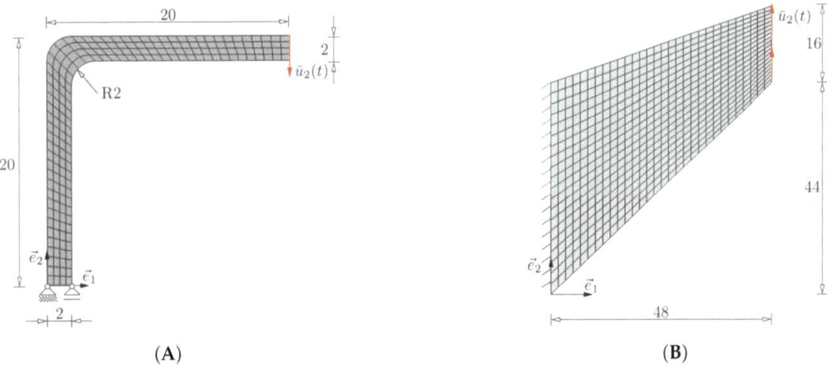

Figure 3. Spatial discretization and boundary conditions for macroscale test cases. (**A**) L-profile; (**B**) Cook's membrane.

The spatial discretization is achieved with eight-noded quadrilateral elements. As a result, $n_G^e = 9$ integration points are present in each macroscale element, which means that $n_e n_G^e$ calls of the RVE in Figure 2 are necessary in each global Newton iteration of a FE^2 computation. The L-profile is spatially discretized with $n_e = 200$ elements and $n_{\text{nodes}} = 709$ nodes. For the Cook's membrane, $n_e = 600$ elements and $n_{\text{nodes}} = 1901$ nodes are used. The L-profile has a prescribed displacement boundary condition on the top right edge with $\bar{u}_2(t) = -3\,\text{mm s}^{-1}\,t$. In contrast, the Cook's membrane is fixed on the left edge and has an applied displacement boundary condition $\bar{u}_2(t) = 2\,\text{mm s}^{-1}\,t$ on the right edge.

The initial time-step size of both numerical examples is $\Delta t_0 = 1 \times 10^{-3}\,\text{s}$, whereas the simulation is performed for $t \subset [0,1]$. The time discretization is achieved with the Backward–Euler method; see also Remark 1 regarding the time discretization for purely elastic problems. Here, the time-step size Δt is not fixed but chosen based on the number of Newton iterations N_{iter} and the time-step size of the current step Δt_n,

$$\Delta t_{\text{new}} = \Delta t_n \times \begin{cases} f_{\max} & \text{if } N_{\text{iter}} \leq 5, \\ f_{\min} & \text{if } N_{\text{iter}} > 15, \\ 1 & \text{if } N_{\text{iter}} > 5 \text{ and } N_{\text{iter}} \leq 15. \end{cases} \tag{76}$$

In this work, the quantities $f_{\max} = 1.2$ and $f_{\min} = 0.3$ are chosen. The termination criteria of the global Newton iteration are applied as

$$||\Delta u|| \leq \text{tol}_u \quad \text{and} \quad ||G(u)|| \leq \text{tol}_G. \tag{77}$$

It should be mentioned that the applied tolerance values are rarely reported in current literature to DNN-FE coupling in multiscale applications, which makes it difficult to draw comparisons. In this work, tolerances $\text{tol}_u = 1 \times 10^{-6}$ and $\text{tol}_G = 1 \times 10^{-3}$ are chosen.

4.2. Investigation on the Size of Dataset

Next, the effect of increasing the size of the dataset on the performance of the DNN models as well as the efficiency and accuracy of the DNN-FE^2 simulations are assessed. Different sizes of datasets, i.e., $N_{\mathbb{D}} \in \{10^3, 10^4, 10^5, 10^6, 4 \times 10^6\}$, are considered, while each dataset is generated according to the explanations in Section 3.2. Note that in all the cases, 80% of the samples in the dataset are used for training and 20% for validation. Results are reported for both NN–2 and NN–AD architectures for having a comprehensive comparison. It should also be noted that the efficient size of the dataset depends on the complexity of the model and the mapping that must be learned. Here, results are reported for NN architectures containing eight hidden layers with 128 neurons per each hidden layer and *swish* as the activation function, which are selected based on the results of hyperparameter tuning; see Section 3.3.4. Figure 4 illustrates the lowest $L_{\mathcal{T}}^{\text{val}}$ obtained during the training of the DNNs using different sizes of the dataset. It can be seen that increasing the size of the dataset from 10^3 to 10^5 leads to a significant reduction in $L_{\mathcal{T}}^{\text{val}}$. However, improvements in the performance of the model are not significant when further increasing $N_{\mathbb{D}}$. We also report the required time of training in Figure 4. The training time t_{train} is normalized by the computational time $t_{\text{comp,Cook}}^{\text{FE}^2}$ needed for FE^2 simulation of the Cook's membrane,

$$t_{\text{rel,train}} = \frac{t_{\text{train}}}{t_{\text{comp,Cook}}^{\text{FE}^2}}, \tag{78}$$

to offer an insight into the cost of developing an NN-based surrogate model for the RVE in comparison with the FE^2 reference simulation. As it is expected, increasing $N_{\mathbb{D}}$ results in an increase in the required training time. It can be observed that even for the largest dataset ($N_{\mathbb{D}} = 4 \times 10^6$), the training time is much shorter than the computational time of the FE^2 simulation. For the dataset with $N_{\mathbb{D}}$ of 10^5, only 1.39% of the computational time of the FE^2 simulation is needed for the training of the NN–AD model.

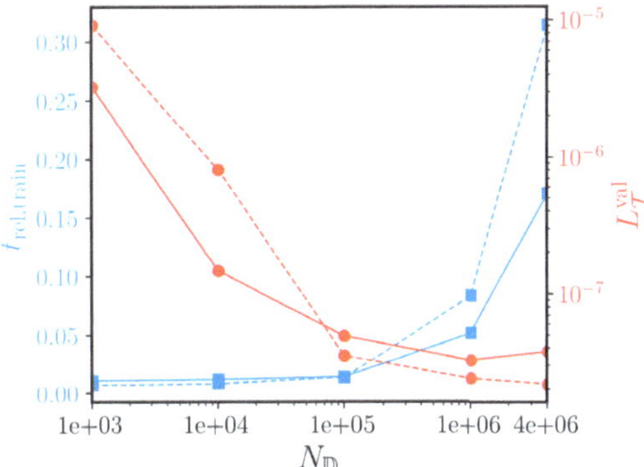

Figure 4. Influence of the size of the dataset $N_{\mathbb{D}}$ on training time $t_{\text{rel,train}}$ (78) and validation loss $L_{\mathcal{T}}^{\text{val}}$ (70) (dashed lines corresponds to NN–AD architecture and solid lines to NN–2 architecture).

4.3. Numerical Simulations

The numerical results obtained from DNN-FE^2 simulations are reported in this section for the L-profile and Cook's membrane test cases and compared with the FE^2 reference simulations regarding accuracy and efficiency. The accuracy of the simulations is estimated by computing the mean and standard deviation of the absolute percentage error, defined in Equation (74), ϵ_{mean} and ϵ_{std}, respectively, over all the components of $\overline{\mathbf{E}}$ and $\overline{\mathbf{T}}$ and over all the global grid points. The speed-up gain is computed by dividing the total time of computation for the FE^2 reference simulation by that of the DNN-FE^2 simulation. Results are reported for different sizes of the dataset $N_{\mathbb{D}}$ and for both architectures NN–2 as well as NN–AD architectures. We refer to the models trained on datasets with different sizes as model–∗ where ∗ × 10^3 shows the size of the dataset.

4.3.1. L-Profile

Results for the simulation of the L-profile test case are summarized in Table 3.

Table 3. Results of the DNN-FE^2 simulation of the L-profile for different sizes of the training/validation dataset.

Model	$N_{\mathbb{D}}$	ϵ_{mean} (%)	ϵ_{std} (%)	Speed-Up	N_{iter}	N_t
NN–2–1	1×10^3	8.79	10.8	232×	104	32
NN–AD–1	1×10^3	4.68	8.52	443×	79	30
NN–2–10	1×10^4	3.20	4.87	246×	101	32
NN–AD–10	1×10^4	0.42	0.69	400×	86	30
NN–2–100	1×10^5	2.48	3.68	254×	97	32
NN–AD–100	1×10^5	0.21	0.40	462×	73	30
NN–2–1000	1×10^6	1.59	2.62	251×	97	31
NN–AD–1000	1×10^6	0.20	0.37	452×	76	30
NN–2–4000	4×10^6	1.87	2.89	236×	103	32
NN–AD–4000	4×10^6	0.15	0.30	462×	73	30

It is evident that both DNN surrogate models, NN–2–1 and NN–AD–1, which were trained on a dataset with only 1000 samples, are accurate enough for achieving the convergence of the FE^2 simulation. This is also the case for the NN–2–10 model leading to ϵ_{mean} and ϵ_{std} of 3.20% and 4.87%, respectively. Employing the NN–AD–10 model as the DNN surrogate model leads to the convergence of the simulation and provides very accurate results with ϵ_{mean} and ϵ_{std} of 0.42% and 0.69%, respectively. All the DNN-based models trained on larger datasets lead to the convergence of the simulation. Our results show the superior performance of the NN–AD models in comparison with the NN–2 models in all the numerical experiments. For instance, the NN–2–100 model results in ϵ_{mean} and ϵ_{std} of 2.48% and 3.68%, respectively, while the NN–AD–100 model provides more accurate results with ϵ_{mean} and ϵ_{std} of 0.21% and 0.40%, respectively. Moreover, we observe that the NN–AD architecture is more efficient regarding the required size of the dataset where the NN–AD–10 model performs better than the NN–2–4000 model.

Apart from accuracy aspects, the speed-up gain obtained from the DNN-FE^2 simulation is of particular interest. In Table 3, it can be observed that a speed-up of 400$\times$ can be obtained from the NN–AD–10 model. This huge speed-up gain shows the excellent potential of the DNN-FE^2 approach for fast and accurate multiscale simulations of solid materials. Our results show that the NN–AD–4000 model leads to the best performance regarding the accuracy, speed-up, and the required number of iterations. In general, we can observe that the NN–AD architecture is more efficient than the NN–2 architecture regarding the speed-up gain where, for instance, the NN–AD–100 model obtains a speed-up of 462$\times$ against 254$\times$ of the NN–2–100 model. The models of both architectures require a quite similar number of time-steps N_t, which are here load-steps. The lesser number of load-steps for the NN-AD architecture results from the fewer number of Newton iterations, which leads to slightly higher load-step sizes according to Equation (76). Moreover, it is evident that the speed-up of the NN-AD architecture is higher than for the NN-2 architecture. This is caused, on the one hand, by the lesser number of global Newton iterations N_{iter} because of the higher prediction accuracy of the consistent tangent matrix. On the other hand, in our implementation, the NN-AD model consisting of one feedforward neural network and the backpropagation step for AD is faster to evaluate than the NN-2 model, which comprises two different feedforward neural networks.

Figures 5 and 6 depict the results obtained from the DNN-FE^2 simulation using the NN–AD–100 model as the DNN surrogate model in comparison with that of the FE^2 reference simulation for all the components of $\overline{\mathbf{E}}$ and $\overline{\mathbf{T}}$, respectively. The reference solution is illustrated on the left panel, the DNN-FE^2 solution is in the middle, and the absolute percentage error ϵ is on the right. It can be observed in Figure 5 that for the normal components of the strain tensor, E_{11} and E_{22}, a maximum absolute percentage error of 1.03% is achieved which shows excellent performance of our DNN-FE^2 approach. For the shear strain E_{12}, the error is slightly higher where a maximum absolute percentage error of 4.38% is obtained. The same conclusion can be drawn from Figure 6, where for the normal components of the stress tensor T_{11} and T_{22}, the maximum percentage errors are 2.12% and 1.02%, respectively, while for the shear stress, T_{12}, the maximum percentage error is slightly higher and is equal to 4.26%.

Moreover, Figure 7 shows the distribution of the absolute percentage error in the solution for the components of $\overline{\mathbf{T}}$ (up) and $\overline{\mathbf{E}}$ (bottom) obtained from the DNN-FE^2 simulation using the NN–AD–100 model on all the integration points of the L-profile test case. The green area and the marked percentage indicate the samples with less than 1% of error and their population proportion. We observe that excellent simulation results are obtained where an absolute percentage error of less than 1% is acquired for most of the samples. For instance, it can be observed that 94.06% of the samples have an error of less than 1% in the solution for the shear stress T_{12}.

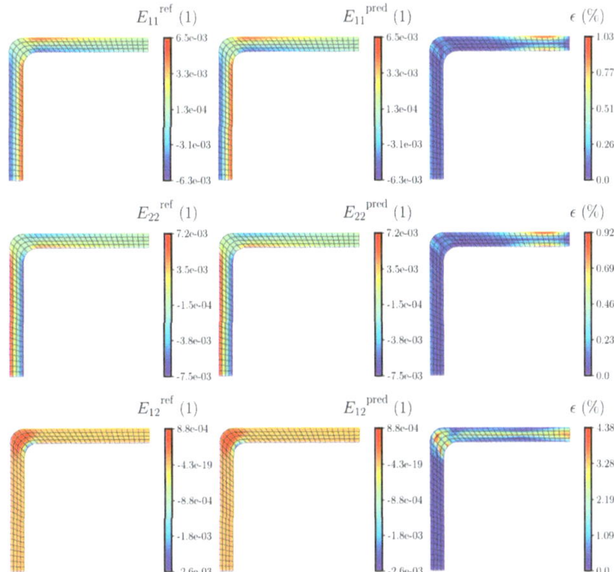

Figure 5. Reference data (**left**) and results obtained from DNN-FE^2 simulation (**middle**) with NN–AD–100 model as well as error measure (74) (**right**) for the components of strain tensor $\overline{\mathbf{E}}$.

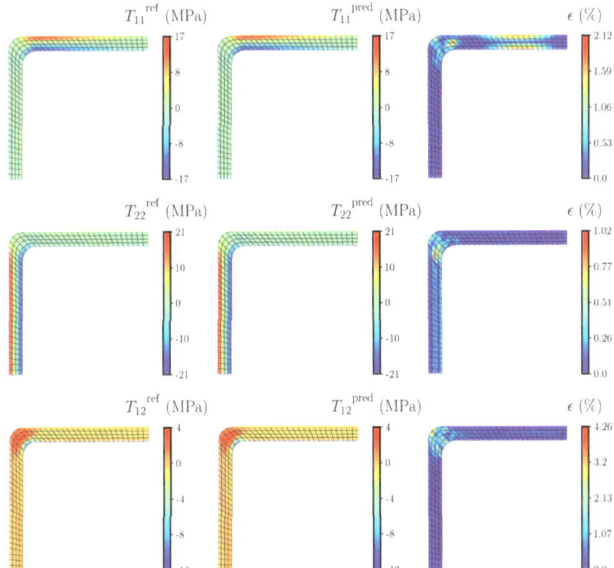

Figure 6. Reference data (**left**) and results obtained from DNN-FE^2 simulation (**middle**) with NN–AD–100 model as well as error measure (74) (**right**) for the components of stress tensor $\overline{\mathbf{T}}$.

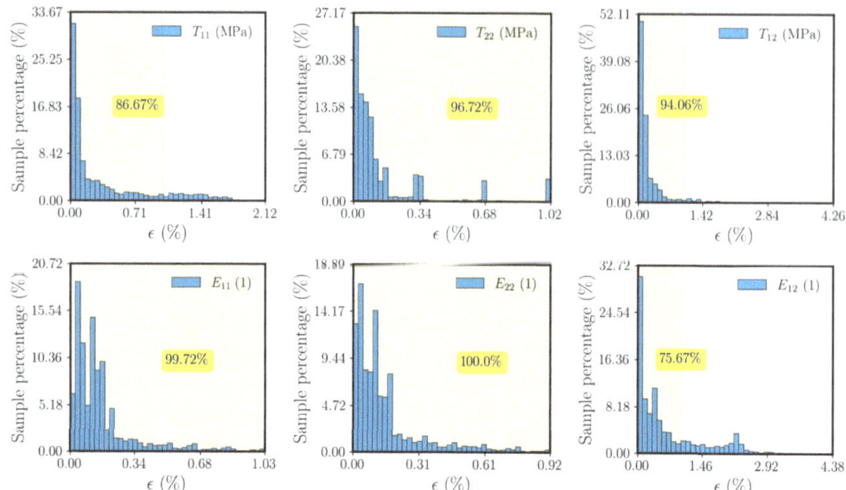

Figure 7. Histograms of the error (74) for the L-profile when applying the NN–AD-100 model. The top and bottom panels illustrate the error for the components of the stress and strain tensors, $\overline{\mathbf{T}}$ and $\overline{\mathbf{E}}$, respectively.

4.3.2. Cook's Membrane

Furthermore, we apply the DNN-FE^2 approach for the simulation of Cook's membrane test case and compare the obtained solution with that of the reference FE^2. Table 4 summarizes the results for models based on the NN–2 and NN–AD architectures trained on datasets with different sizes.

Table 4. Results of the DNN-FE^2 simulation of the Cook's membrane different for sizes of the training/validation dataset.

Model	$N_{\mathbb{D}}$	$\epsilon_{\mathbf{mean}}$ (%)	$\epsilon_{\mathbf{std}}$ (%)	Speed-Up	N_{iter}	N_t
NN–2–1	1×10^3	0.68	0.89	242×	123	32
NN–AD–1	1×10^3	0.60	0.71	527×	85	30
NN–2–10	1×10^4	0.31	0.44	292×	104	32
NN–AD–10	1×10^4	0.09	0.13	542×	84	30
NN–2–100	1×10^5	0.19	0.26	287×	104	32
NN–AD–100	1×10^5	0.02	0.02	554×	82	30
NN–2–1000	1×10^6	0.13	0.16	286×	105	32
NN–AD–1000	1×10^6	0.03	0.06	575×	79	30
NN–2–4000	4×10^6	0.12	0.17	291×	103	32
NN–AD–4000	4×10^6	0.01	0.01	611×	73	30

The NN–2–1 and NN–AD–1 models provide a converged solution with less than 1% of error. This is also the case for the NN–2–10 and NN–2–100. Utilizing NN–AD–10 and NN–AD–100 models leads to the convergence of the simulation with excellent accuracy. We obtain ϵ_{mean} and ϵ_{std} of 0.02% using the NN–AD–100 model, which shows the excellent capability of the NN–AD architecture for surrogate modeling of the local macroscale computations. The results show that the NN–AD architecture outperforms the NN–2 architecture in all the tests where, similar to the results obtained for the L-profile, the errors ϵ_{mean} and ϵ_{std} are almost an order of magnitude lower.

The results are also reported regarding the computational efficiency of the proposed framework in Table 4 for the second numerical experiment of Cook's membrane; see Figure 3B. It can be observed that the NN–AD–10 model obtains a speed-up gain of 542× for this example. The results show that increasing the size of the dataset leads to a more efficient simulation with a lesser number of iterations and lower computational time. The NN–AD–4000 model provides a speed-up of 611× and leads to the convergence of the simulation in 30 time-steps. Compared to the L-profile test case, higher speed-up gain is achieved for the Cook's membrane. This is due to the fact that the number of elements, which require microscale computations, is three times higher than that of the L-profile. This suggests that utilizing the DNN-FE^2 approach for more expensive computations, e.g., three-dimensional problems, could even lead to a higher speed-up gain.

Figures 8 and 9 show the results obtained from the DNN-FE^2 simulation employing the NN–AD–100 model as the DNN surrogate model in comparison with that of the FE^2 reference simulation for all the components of $\overline{\mathbf{E}}$ and $\overline{\mathbf{T}}$, respectively. It can be observed that for all the components of strain and stress tensors, very accurate results can be obtained. For normal strains E_{11} and E_{22} and the shear strain E_{12}, the maximum absolute percentage errors are 0.14%, 0.16%, and 0.18%, respectively. Moreover, for the stress components T_{11}, T_{22}, and T_{12} the maximum errors are equal to 0.23%, 0.38%, and 0.10%.

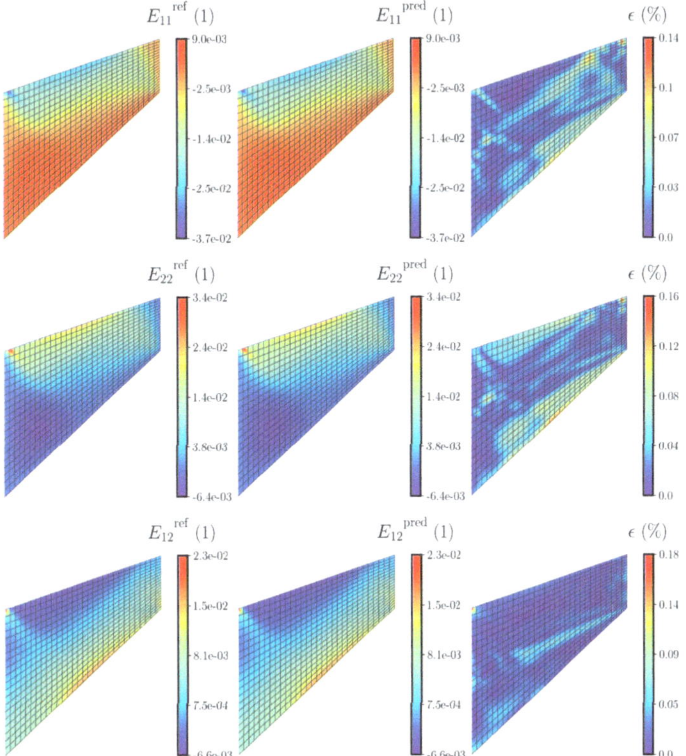

Figure 8. Reference data (**left**) and results obtained from DNN-FE^2 simulation (**middle**) with NN–AD–100 model as well as error measure (74) (**right**) for the components of strain tensor $\overline{\mathbf{E}}$.

We also report the distribution of the absolute percentage errors over all the integration points for Cook's membrane test case in Figure 10.

For this case, the green area and the marked percentage indicate the samples with less than 0.1% of error and their population proportion. It can be seen that for most of the sample points, an error of less than 0.1% has been obtained. Our results show the excellent capability of the NN–AD models for very accurate and efficient FE^2 simulations.

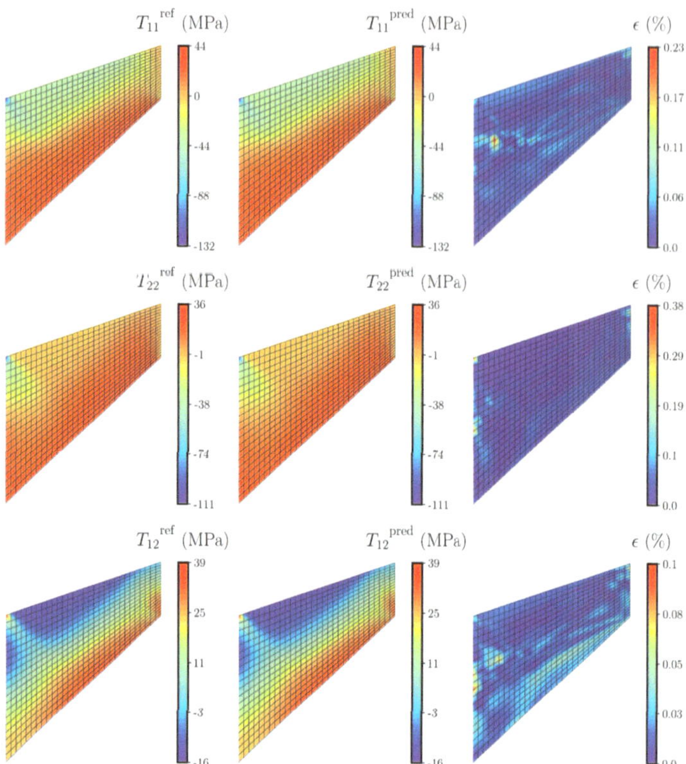

Figure 9. Reference data (**left**) and results obtained from DNN-FE^2 simulation (**middle**) with NN–AD–100 model as well as error measure (74) (**right**) for the components of stress tensor $\overline{\mathbf{T}}$.

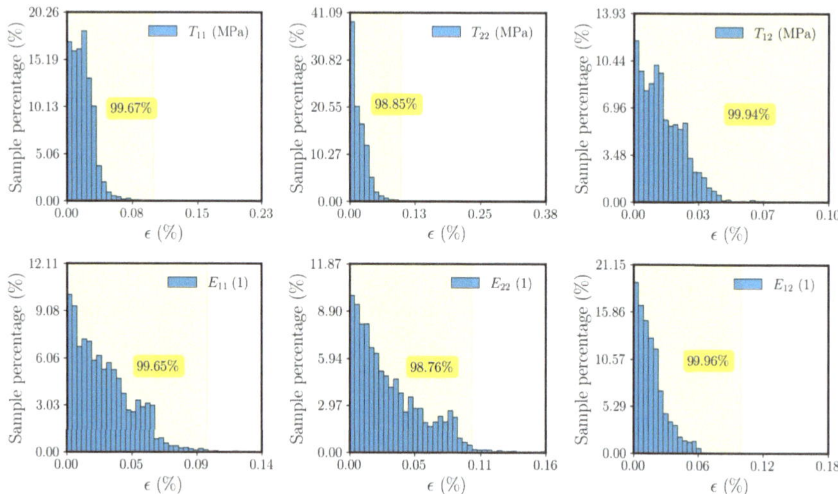

Figure 10. Histograms of the error (74) for the Cook's membrane when applying the NN–AD-100 model. The top and bottom panels illustrate the error for the components of the stress and strain tensors, $\overline{\mathbf{T}}$ and $\overline{\mathbf{E}}$, respectively.

4.4. Load-Step Size Behavior

Since FE^2 computations usually require small initial time-step sizes Δt_0, the application of certain time-step control schemes is reasonable. In general, the determination of the time-step size Δt_{new} for the following time-step depending on the current time-step size Δt_n can be achieved via different methods. In this contribution, we consider the number of (global) Newton iterations N_{iter} for the load step-size control; see Equation (76).

The step-size behavior, when using a step-size control based on global Newton iterations, is shown for our two numerical experiments in Figure 11.

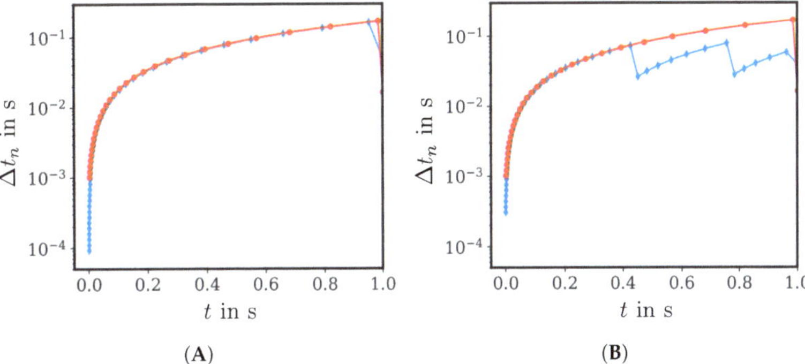

Figure 11. Step-size behavior of DNN-FE^2 simulations (red) with NN–AD–100 model and FE^2 reference simulation (blue), only accepted time-step sizes are shown. (**A**) L-profile; (**B**) Cook's membrane.

For the L-profile, the overall step-size behavior is quite similar for both the FE^2 reference simulation and with embedding the DNN surrogate model NN–AD–100; see Figure 11A. However, it is evident that the initial time-step size, which is chosen as $\Delta t_0 = 10^{-3}$ s, is suitable for the surrogate model, but not for the FE^2 reference simulation as the time-step size is rejected multiple times and convergence is initially reached at $\Delta t = 9 \times 10^{-5}$ s. The rejections of the initial time-step size, which represents here the initially applied load-step, result from divergence at the global macroscale level.

A similar behavior is observed for the Cook's membrane and shown in Figure 11B. The first accepted time-step size is $\Delta t = 3 \times 10^{-4}$ s, whereas the DNN surrogate already shows convergence at $\Delta t_0 = 10^{-3}$ s. In contrast to the L-profile, the step-size behavior shows significant differences. For the reference FE^2 simulation of the Cook's membrane, a certain limit exists, where the step size decreases because of failures in the local-level computations of the MLNA (RVE computations), i.e., the applied load leads to certain limitations in the step-size behavior of the RVEs. However, the DNN-FE^2 computation with the embedded DNN surrogate model is successfully converging even for higher step sizes. The different step-size behaviors for L-profile and Cook's membrane are obtained due to different magnitudes of the strains at each macroscale integration point that result from the loading conditions in Figure 3; see the results in Figures 5 and 8 as well.

As a result, the application of DNN surrogate models is not only possible for load-step size controlled FE^2 computations, but it also leads to certain advantages. On the one hand, higher initial load-step sizes are possible and, on the other hand, certain limitations in the load-step size can be overcome and thus larger step sizes can be applied compared to classical FE^2 computations.

5. Speed-Up with JAX and Just-in-Time Compilation

JAX [82] is a Python library developed by Google Research for high-performance numerical computing. It utilizes an updated version of *Autograd* [92] for automatic differentiation of native Python and NumPy functions. JAX supports reverse-mode differentiation as well as forward-mode differentiation, and the two can be composed arbitrarily to any order. Moreover, JAX uses XLA (accelerated linear algebra) [93] to compile and run NumPy programs on GPUs and TPUs (tensor processing units), which is performed by just-in-time (JIT) compilation and execution of the calls. JAX also allows just-in-time compilation of user-defined Python functions into XLA-optimized kernels using a one-function application programming interface, *jit*. Compilation and automatic differentiation can be composed arbitrarily, so one can express sophisticated algorithms and obtain maximal performance without leaving Python. These properties allow the implementation of our NN–AD architecture for RVE surrogate modeling efficiently using JAX. In the following, we discuss just-in-time compilation and its application in our DNN-FE^2 simulation framework in combination with FORPy [87].

5.1. Just-in-Time Compilation

Just-in-time (JIT) compilation is a technique used in modern programming languages to improve the performance of code execution at runtime. With JIT compilation, the code is compiled from a high-level language into machine code at the moment it is needed, rather than ahead of time. This allows a more efficient use of resources and can lead to significant performance improvements, especially for applications that require repeated execution of the same code. JIT compilers work by analyzing the code being executed and dynamically generating optimized machine code that is tailored to the specific execution context. In particular, when a program is executed, the JIT compiler analyzes the code being executed and identifies hot spots or sections of code that are frequently executed. These sections of code are then compiled into machine code and stored in memory for future use. The next time the same section of code is executed, the JIT compiler can use the pre-compiled machine code instead of interpreting the code again. This leads to significant performance improvements, as the program spends less time interpreting code and more time executing the machine code.

The JIT compiler in JAX is based on XLA, a domain-specific compiler that optimizes numerical computations for modern hardware architectures. With JAX, users can write a Python code that looks like a NumPy code but runs much faster on specialized hardware. This makes JAX an ideal library for scientific computing, machine learning, and other high-performance computing tasks. In addition to JIT compilation, JAX also provides tools for distributed computing and parallelization, making it a versatile library for a wide range of applications.

5.2. Speed-Up with JAX and JIT

JIT compilation is of interest in the DNN-FE^2 approach since a repeated execution of the surrogate model in every iteration and for every integration point occurs. Thus, the JIT compilation of the prediction function, which is called from the FORTRAN finite element code through FORPy, allows for more efficient use of resources and can lead to significant performance improvements. To this end, we developed the NN–AD–100 model using JAX and a neural network library and ecosystem for JAX called Flax [94]. Further, the prediction function is compiled using `jax.jit` transformation.

The results are reported in Table 5, where we compare the computational efficiency of our TensorFlow and JAX implementations. It should be noted that we utilize the same set of hyperparameters and similar training processes for both implementations. It can be seen that the required time of training is shorter for the JAX implementation, and it is equal to 9.49×10^{-3} of the computational time required for FE^2 simulation of the Cook's membrane. Moreover, we gain a significant speed-up from the JAX implementation in comparison with the TensorFlow implementation. The speed-up gain for the L-profile and Cook's membrane test cases are, respectively, equal to $4629\times$ and $5853\times$ for JAX and $462\times$ and $554\times$ for TensorFlow. To the best of the authors' knowledge, our JAX implementation provides the highest speed-up in the context of DNN-FE^2 simulations for non-linear elastic material behavior in the literature.

Table 5. Comparison of JAX and TensorFlow implementations of the surrogate model NN–AD–100 regarding the computational efficiency.

Framework	$t_{\text{rel,train}}$	Speed-Up for L-Profile	Speed-Up for Cook's Membrane
TensorFlow	1.39×10^{-2}	$462\times$	$554\times$
JAX	9.49×10^{-3}	$4629\times$	$5853\times$

6. Conclusions

In the present work, a DNN-FE^2 approach is explained in detail to significantly accelerate multiscale FE^2 simulations. In general, the algorithmic structure of FE^2 computations is a Multilevel-Newton algorithm, even for the case of purely elastic material behavior without internal variables. The main source of computational costs are the local macroscale computations, which include the numerous computations of representative volume elements. Thus, in the DNN-FE^2 approach, we replace the local macroscale computations by drawing on a deep neural network surrogate model, which is very fast to evaluate after sufficient training. Here, it turns out that using automatic differentiation and Sobolev training to obtain the consistent tangent information is superior to an approach with two deep neural networks for the prediction of stresses and consistent tangent regarding data efficiency and prediction accuracy. Moreover, in step-size-controlled computations, the deep neural network surrogates are able to overcome certain step-size limitations of the FE^2 reference computations. For the Cook's membrane as a particular example in this contribution, we achieve a speed-up factor of more than 5000 compared to a FE^2 reference simulation when using just-in-time compilation techniques together with an efficient coupling between different programming codes using the FORPy library. The main advantage of the explained DNN-FE^2 approach is that it can be easily implemented to the existing finite element codes since just the evaluation of a surrogate model for each macroscale integration point has to be considered.

Author Contributions: Conceptualization, investigation and methodology, H.E., J.-A.T. and S.W.; writing—original draft preparation, H.E. and J.-A.T.; writing—review and editing, S.W. and S.H.; supervision, S.H. and A.R. All authors have read and agreed to the published version of the manuscript.

Funding: HE's research was conducted within the Research Training Group CircularLIB, supported by the Ministry of Science and Culture of Lower Saxony with funds from the program zukunft.niedersachsen of the Volkswagen Foundation. We further acknowledge support by Open Access Publishing Fund of Clausthal University of Technology.

Institutional Review Board Statement: Not applicable.

Data Availability Statement: All the codes employed for developing the DNN-based surrogate models are released as open-source in the GitHub-repository https://github.com/HamidrezaEiv/FE2-Computations-With-Deep-Neural-Networks (accessed on 16 July 2023).

Conflicts of Interest: The authors declare no conflict of interest.

Abbreviations

The following abbreviations are used in this manuscript:

DNN	Deep neural network
FE	Finite element
RVE	Representative volume element
TANN	Thermodynamics-based artificial neural network
DMN	Deep material network
NN	Neural network
MLNA	Multilevel Newton algorithm
DAE	Differential-algebraic equations
AD	Automatic differentiation

MPI   Message passing interface
LHS   Latin hypercube sampling
XLA   Accelerated linear algebra
JIT   Just-in-time

Appendix A. Hyperparameter Tuning

We conduct hyperparameter tuning using a grid search algorithm to optimize the performance of the model on the validation dataset. For this purpose, we employ a dataset with a size of $N_{\mathbb{D}} = 10^5$. Our attention is directed towards three specific hyperparameters: the number of hidden layers N_h, the number of neurons per hidden layer N_n, and the choice of the activation function φ. We apply hyperparameter tuning to the NN–2 architecture. Results obtained from different sizes of the neural network (number of hidden layers $N_h \times$ number of neurons per hidden layer N_n) with *swish* activation function are reported in Table A1.

Table A1. Summary of the results obtained for training and validation losses and the required time of training for different sizes of the NN–2–100 model. Results are reported for models with *swish* activation function.

$N_h \times N_n$	$L_{\mathcal{T}}^{\text{train}}$	$L_{\mathcal{T}}^{\text{val}}$	$L_{\mathcal{C}}^{\text{train}}$	$L_{\mathcal{C}}^{\text{val}}$	$t_{\text{rel,train}}$
64×2	1.06×10^{-6}	1.06×10^{-6}	1.35×10^{-4}	1.33×10^{-4}	6.97×10^{-3}
64×4	1.49×10^{-7}	1.51×10^{-7}	3.58×10^{-6}	3.68×10^{-6}	8.36×10^{-3}
64×8	1.02×10^{-7}	1.04×10^{-7}	9.27×10^{-7}	9.35×10^{-7}	1.17×10^{-2}
64×16	1.66×10^{-7}	1.70×10^{-7}	3.93×10^{-7}	4.29×10^{-7}	1.81×10^{-2}
128×2	1.13×10^{-6}	1.11×10^{-6}	1.71×10^{-4}	1.67×10^{-4}	7.80×10^{-3}
128×4	1.14×10^{-7}	1.15×10^{-7}	9.47×10^{-7}	1.03×10^{-6}	9.19×10^{-3}
128×8	5.05×10^{-8}	4.96×10^{-8}	2.58×10^{-7}	3.10×10^{-7}	1.45×10^{-2}
128×16	5.22×10^{-8}	5.30×10^{-8}	2.19×10^{-7}	3.35×10^{-7}	2.08×10^{-2}
256×2	1.55×10^{-6}	1.54×10^{-6}	1.60×10^{-4}	1.58×10^{-4}	7.90×10^{-3}
256×4	7.95×10^{-8}	8.02×10^{-8}	5.08×10^{-7}	5.85×10^{-7}	1.18×10^{-2}
256×8	7.10×10^{-8}	7.00×10^{-8}	1.01×10^{-7}	1.67×10^{-7}	1.75×10^{-2}
256×16	5.19×10^{-8}	5.18×10^{-8}	3.35×10^{-7}	1.76×10^{-7}	3.37×10^{-2}

The training and the validation losses are reported for both mappings, $\mathcal{T}$ and $\mathcal{C}$, which are selected during the training process as the best models based on the lowest validation loss. Also, the relative time of training $t_{\text{rel,train}}$ is outlined. It is evident that the lowest validation loss $L_{\mathcal{T}}^{\text{val}} = 4.96 \times 10^{-8}$ is obtained from a deep neural network with eight hidden layers and 128 neurons per hidden layer, where the corresponding validation loss $L_{\mathcal{C}}^{\text{val}} = 3.10 \times 10^{-7}$ is obtained. The results show that networks with a larger number of parameters lead to losses with the same order of magnitude and do not show a considerable improvement while requiring more time for training. Therefore, we select $N_h = 8$ and $N_n = 128$ for our analysis, see Table A1.

Moreover, Figure A1 illustrates the influence of the choice of activation function on the learning curves for both mappings, $\mathcal{T}$ and $\mathcal{C}$.

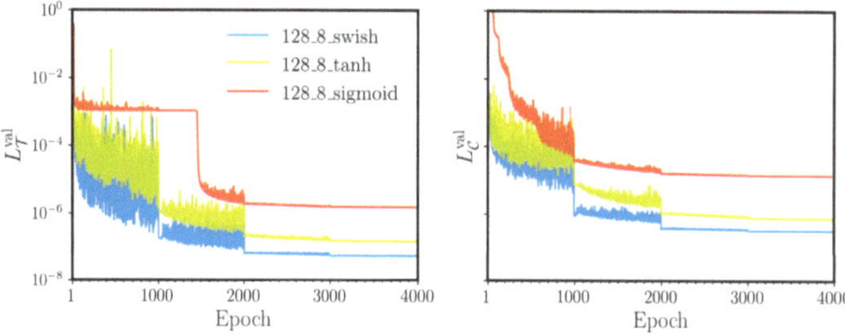

Figure A1. Influence of the choice of activation function on the learning process; $L_{\mathcal{T}}^{\text{val}}$ (**left**) and $L_{\mathcal{C}}^{\text{val}}$ (**right**). Results are reported for models containing 8 hidden layers and 128 neurons per each hidden layer.

Results are reported for deep neural networks with $N_h = 8$ and $N_n = 128$. Similar influence in all the other cases with different sizes of the neural network can be observed. Further, it can be seen that the *swish* activation function performs better than *sigmoid* and *tanh*. It should be noted that rectified linear unit (ReLU) activation function is not used since the mapping requires to be continuously differentiable, especially for the NN–AD architecture.

References

1. Smit, R.J.; Brekelmans, W.M.; Meijer, H.E. Prediction of the mechanical behavior of nonlinear heterogeneous systems by multi-level finite element modeling. *Comput. Methods Appl. Mech. Eng.* **1998**, *155*, 181–192. [CrossRef]
2. Feyel, F. Multiscale FE^2 elastoviscoplastic analysis of composite structures. *Comput. Mater. Sci.* **1999**, *16*, 344–354. [CrossRef]
3. Kouznetsova, V.; Brekelmans, W.A.M.; Baaijens, F.P.T. An approach to micro-macro modeling of heterogeneous materials. *Comput. Mech.* **2001**, *27*, 37–48. [CrossRef]
4. Miehe, C.; Koch, A. Computational micro-to-macro transitions of discretized microstructures undergoing small strains. *Arch. Appl. Mech.* **2002**, *72*, 300–317. [CrossRef]
5. Miehe, C. Computational micro-to-macro transitions for discretized micro-structures of heterogeneous materials at finite strains based on the minimization of averaged incremental energy. *Comput. Methods Appl. Mech. Eng.* **2003**, *192*, 559–591. [CrossRef]
6. Kouznetsova, V.; Geers, M.G.D.; Brekelmans, W.A.M. Multi-scale second-order computational homogenization of multi-phase materials: A nested finite element solution strategy. *Comput. Methods Appl. Mech. Eng.* **2004**, *193*, 5525–5550. [CrossRef]
7. Schröder, J. A numerical two-scale homogenization scheme: The FE^2-method. In *Plasticity and Beyond: Microstructures, Crystal-Plasticity and Phase Transitions*; Schröder, J., Hackl, K., Eds.; Springer: Vienna, Austria, 2014; pp. 1–64.
8. Kochmann, J.; Wulfinghoff, S.; Reese, S.; Mianroodi, J.R.; Svendsen, B. Two-scale FE–FFT-and phase-field-based computational modeling of bulk microstructural evolution and macroscopic material behavior. *Comput. Methods Appl. Mech. Eng.* **2016**, *305*, 89–110. [CrossRef]
9. Düster, A.; Sehlhorst, H.G.; Rank, E. Numerical homogenization of heterogeneous and cellular materials utilizing the finite cell method. *Comput. Mech.* **2012**, *50*, 413–431. [CrossRef]
10. Bock, F.E.; Aydin, R.C.; Cyron, C.J.; Huber, N.; Kalidindi, S.R.; Klusemann, B. A Review of the Application of Machine Learning and Data Mining Approaches in Continuum Materials Mechanics. *Front. Mater.* **2019**, *6*, 110. [CrossRef]
11. Brodnik, N.; Muir, C.; Tulshibagwale, N.; Rossin, J.; Echlin, M.; Hamel, C.; Kramer, S.; Pollock, T.; Kiser, J.; Smith, C.; et al. Perspective: Machine learning in experimental solid mechanics. *J. Mech. Phys. Solids* **2023**, *173*, 105231. [CrossRef]
12. Jin, H.; Zhang, E.; Espinosa, H.D. Recent Advances and Applications of Machine Learning in Experimental Solid Mechanics: A Review. *arXiv* **2023**, arXiv:2303.07647.
13. Johnson, N.; Vulimiri, P.; To, A.; Zhang, X.; Brice, C.; Kappes, B.; Stebner, A. Invited review: Machine learning for materials developments in metals additive manufacturing. *Addit. Manuf.* **2020**, *36*, 101641. [CrossRef]
14. Kumar, S.; Kochmann, D.M. What Machine Learning Can Do for Computational Solid Mechanics. In *Current Trends and Open Problems in Computational Mechanics*; Aldakheel, F., Hudobivnik, B., Soleimani, M., Wessels, H., Weißenfels, C., Marino, M., Eds.; Springer International Publishing: Cham, Switzerland, 2022; pp. 275–285.
15. Zhang, P.; Yin, Z.Y.; Jin, Y.F. State-of-the-Art Review of Machine Learning Applications in Constitutive Modeling of Soils. *Arch. Comput. Methods Eng.* **2021**, *28*, 3661–3686. [CrossRef]
16. Kirchdoerfer, T.; Ortiz, M. Data-driven computational mechanics. *Comput. Methods Appl. Mech. Eng.* **2016**, *304*, 81–101. [CrossRef]
17. Ghaboussi, J.; Garrett, J.H.; Wu, X. Knowledge-Based Modeling of Material Behavior with Neural Networks. *J. Eng. Mech.* **1991**, *117*, 132–153. [CrossRef]
18. Lefik, M.; Schrefler, B. Artificial neural network as an incremental non-linear constitutive model for a finite element code. *Comput. Methods Appl. Mech. Eng.* **2003**, *192*, 3265–3283. [CrossRef]
19. Hashash, Y.M.A.; Jung, S.; Ghaboussi, J. Numerical implementation of a neural network based material model in finite element analysis. *Int. J. Numer. Methods Eng.* **2004**, *59*, 989–1005. [CrossRef]
20. Deshpande, S.; Sosa, R.I.; Bordas, S.P.A.; Lengiewicz, J. Convolution, aggregation and attention based deep neural networks for accelerating simulations in mechanics. *Front. Mater.* **2023**, *10*, 1128954. [CrossRef]
21. Yao, H.; Gao, Y.; Liu, Y. FEA-Net: A physics-guided data-driven model for efficient mechanical response prediction. *Comput. Methods Appl. Mech. Eng.* **2020**, *363*, 112892. [CrossRef]
22. Oishi, A.; Yagawa, G. Computational mechanics enhanced by deep learning. *Comput. Methods Appl. Mech. Eng.* **2017**, *327*, 327–351. [CrossRef]
23. Huang, D.; Fuhg, J.N.; Weißenfels, C.; Wriggers, P. A machine learning based plasticity model using proper orthogonal decomposition. *Comput. Methods Appl. Mech. Eng.* **2020**, *365*, 113008. [CrossRef]
24. Nguyen, L.T.K.; Keip, M.A. A data-driven approach to nonlinear elasticity. *Comput. Struct.* **2018**, *194*, 97–115. [CrossRef]
25. Stainier, L.; Leygue, A.; Ortiz, M. Model-free data-driven methods in mechanics: Material data identification and solvers. *Comput. Mech.* **2019**, *64*, 381–393. [CrossRef]
26. Eggersmann, R.; Kirchdoerfer, T.; Reese, S.; Stainier, L.; Ortiz, M. Model-Free Data-Driven Inelasticity. *Comput. Methods Appl. Mech. Eng.* **2019**, *350*, 81–99. [CrossRef]

27. González, D.; Chinesta, F.; Cueto, E. Thermodynamically consistent data-driven computational mechanics. *Contin. Mech. Thermodyn.* **2019**, *31*, 239–253. [CrossRef]
28. Ciftci, K.; Hackl, K. Model-free data-driven simulation of inelastic materials using structured data sets, tangent space information and transition rules. *Comput. Mech.* **2022**, *70*, 425–435. [CrossRef]
29. Eghbalian, M.; Pouragha, M.; Wan, R. A physics-informed deep neural network for surrogate modeling in classical elasto-plasticity. *Comput. Geotech.* **2023**, *159*, 105472. [CrossRef]
30. Huber, N.; Tsakmakis, C. Determination of constitutive properties from sperical indentation data using neural networks, Part I: The case of pure kinematic hardening in plasticity laws. *J. Mech. Phys. Solids* **1999**, *47*, 1569–1588. [CrossRef]
31. Huber, N.; Tsakmakis, C. Determination of constitutive properties from sperical indentation data using neural networks, Part II: Plasticity with nonlinear and kinematic hardening. *J. Mech. Phys. Solids* **1999**, *47*, 1589–1607. [CrossRef]
32. Villarreal, R.; Vlassis, N.; Phan, N.; Catanach, T., Jones, R.; Trask, N.; Kramer, S.; Sun, W. Design of experiments for the calibration of history-dependent models via deep reinforcement learning and an enhanced Kalman filter. *Comput. Mech.* **2023**, *72*, 95–124. [CrossRef]
33. Hamel, C.M.; Long, K.N.; Kramer, S.L.B. Calibrating constitutive models with full-field data via physics informed neural networks. *Strain* **2022**, *59*, e12431. [CrossRef]
34. Flaschel, M.; Kumar, S.; De Lorenzis, L. Unsupervised discovery of interpretable hyperelastic constitutive laws. *Comput. Methods Appl. Mech. Eng.* **2021**, *381*, 113852. [CrossRef]
35. Flaschel, M.; Kumar, S.; De Lorenzis, L. Discovering plasticity models without stress data. *NPJ Comput. Mater.* **2022**, *8*, 91. [CrossRef]
36. Flaschel, M.; Kumar, S.; De Lorenzis, L. Automated discovery of generalized standard material models with EUCLID. *Comput. Methods Appl. Mech. Eng.* **2023**, *405*, 115867. [CrossRef]
37. Linka, K.; Kuhl, E. A new family of Constitutive Artificial Neural Networks towards automated model discovery. *Comput. Methods Appl. Mech. Eng.* **2023**, *403*, 115731. [CrossRef]
38. Linka, K.; Hillgärtner, M.; Abdolazizi, K.P.; Aydin, R.C.; Itskov, M.; Cyron, C.J. Constitutive artificial neural networks: A fast and general approach to predictive data-driven constitutive modeling by deep learning. *J. Comput. Phys.* **2021**, *429*, 110010. [CrossRef]
39. Le, B.A.; Yvonnet, J.; He, Q.C. Computational homogenization of nonlinear elastic materials using neural networks. *Int. J. Numer. Methods Eng.* **2015**, *104*, 1061–1084. [CrossRef]
40. Liu, Z.; Bessa, M.; Liu, W.K. Self-consistent clustering analysis: An efficient multi-scale scheme for inelastic heterogeneous materials. *Comput. Methods Appl. Mech. Eng.* **2016**, *306*, 319–341. [CrossRef]
41. Fritzen, F.; Fernández, M.; Larsson, F. On-the-Fly Adaptivity for Nonlinear Twoscale Simulations Using Artificial Neural Networks and Reduced Order Modeling. *Front. Mater.* **2019**, *6*, 75. [CrossRef]
42. Yang, J.; Xu, R.; Hu, H.; Huang, Q.; Huang, W. Structural-Genome-Driven computing for thin composite structures. *Compos. Struct.* **2019**, *215*, 446–453. [CrossRef]
43. Mianroodi, J.; Rezaei, S.; Siboni, N.; Xu, B.X.; Raabe, D. Lossless multi-scale constitutive elastic relations with artificial intelligence. *NPJ Comput. Mater.* **2022**, *8*, 67. [CrossRef]
44. Gupta, A.; Bhaduri, A.; Graham-Brady, L. Accelerated multiscale mechanics modeling in a deep learning framework. *Mech. Mater.* **2023**, *184*, 104709. [CrossRef]
45. Nguyen-Thanh, V.M.; Trong Khiem Nguyen, L.; Rabczuk, T.; Zhuang, X. A surrogate model for computational homogenization of elastostatics at finite strain using high-dimensional model representation-based neural network. *Int. J. Numer. Methods Eng.* **2020**, *121*, 4811–4842. [CrossRef]
46. Aldakheel, F.; Elsayed, E.S.; Zohdi, T.I.; Wriggers, P. Efficient multiscale modeling of heterogeneous materials using deep neural networks. *Comput. Mech.* **2023**, *72*, 155–171. [CrossRef]
47. Kim, S.; Shin, H. Data-driven multiscale finite-element method using deep neural network combined with proper orthogonal decomposition. *Eng. Comput.* **2023**. [CrossRef]
48. Eidel, B. Deep CNNs as universal predictors of elasticity tensors in homogenization. *Comput. Methods Appl. Mech. Eng.* **2023**, *403*, 115741. [CrossRef]
49. Yang, H.; Guo, X.; Tang, S.; Liu, W.K. Derivation of heterogeneous material laws via data-driven principal component expansions. *Comput. Mech.* **2019**, *64*, 365–379. [CrossRef]
50. Rao, C.; Liu, Y. Three-dimensional convolutional neural network (3D-CNN) for heterogeneous material homogenization. *Comput. Mater. Sci.* **2020**, *184*, 109850. [CrossRef]
51. Reimann, D.; Nidadavolu, K.; ul Hassan, H.; Vajragupta, N.; Glasmachers, T.; Junker, P.; Hartmaier, A. Modeling Macroscopic Material Behavior With Machine Learning Algorithms Trained by Micromechanical Simulations. *Front. Mater.* **2019**, *6*, 181. [CrossRef]
52. Göküzüm, F.S.; Nguyen, L.T.K.; Keip, M.A. An Artificial Neural Network Based Solution Scheme for Periodic Computational Homogenization of Electrostatic Problems. *Math. Comput. Appl.* **2019**, *24*, 40. [CrossRef]
53. Korzeniowski, T.F.; Weinberg, K. Data-driven finite element computation of open-cell foam structures. *Comput. Methods Appl. Mech. Eng.* **2022**, *400*, 115487. [CrossRef]
54. Xu, R.; Yang, J.; Yan, W.; Huang, Q.; Giunta, G.; Belouettar, S.; Zahrouni, H.; Zineb, T.B.; Hu, H. Data-driven multiscale finite element method: From concurrence to separation. *Comput. Methods Appl. Mech. Eng.* **2020**, *363*, 112893. [CrossRef]

55. Li, B.; Zhuang, X. Multiscale computation on feedforward neural network and recurrent neural network. *Front. Struct. Civ. Eng.* **2020**, *14*, 1285–1298. [CrossRef]
56. Fuhg, J.N.; Böhm, C.; Bouklas, N.; Fau, A.; Wriggers, P.; Marino, M. Model-data-driven constitutive responses: Application to a multiscale computational framework. *Int. J. Eng. Sci.* **2021**, *167*, 103522. [CrossRef]
57. Masi, F.; Stefanou, I. Multiscale modeling of inelastic materials with Thermodynamics-based Artificial Neural Networks (TANN). *Comput. Methods Appl. Mech. Eng.* **2022**, *398*, 115190. [CrossRef]
58. Masi, F.; Stefanou, I. Evolution TANN and the identification of internal variables and evolution equations in solid mechanics. *J. Mech. Phys. Solids* **2023**, *174*, 105245. [CrossRef]
59. Kalina, K.A.; Linden, L.; Brummund, J.; Kästner, M. FE^{ANN}: An efficient data-driven multiscale approach based on physics-constrained neural networks and automated data mining. *Comput. Mech.* **2023**, *71*, 827–851. [CrossRef]
60. Feng, N.; Zhang, G.; Khandelwal, K. Finite strain FE^2 analysis with data-driven homogenization using deep neural networks. *Comput. Struct.* **2022**, *263*, 106742. [CrossRef]
61. Czarnecki, W.M.; Osindero, S.; Jaderberg, M.; Swirszcz, G.; Pascanu, R. Sobolev Training for Neural Networks. In *Advances in Neural Information Processing Systems*; Guyon, I., Luxburg, U.V., Bengio, S., Wallach, H., Fergus, R., Vishwanathan, S., Garnett, R., Eds.; Curran Associates, Inc.: Red Hook, NY, USA, 2017; Volume 30.
62. Vlassis, N.N.; Sun, W. Sobolev training of thermodynamic-informed neural networks for interpretable elasto-plasticity models with level set hardening. *Comput. Methods Appl. Mech. Eng.* **2021**, *377*, 113695. [CrossRef]
63. Vlassis, N.N.; Sun, W. Geometric learning for computational mechanics Part II: Graph embedding for interpretable multiscale plasticity. *Comput. Methods Appl. Mech. Eng.* **2023**, *404*, 115768. [CrossRef]
64. Ghavamian, F.; Simone, A. Accelerating multiscale finite element simulations of history-dependent materials using a recurrent neural network. *Comput. Methods Appl. Mech. Eng.* **2019**, *357*, 112594. [CrossRef]
65. Drosopoulos, G.A.; Stavroulakis, G.E. Data-Driven Computational Homogenization Using Neural Networks: FE2-NN Application on Damaged Masonry. *J. Comput. Cult. Herit.* **2021**, *14*, 1–19. [CrossRef]
66. Yin, M.; Zhang, E.; Yu, Y.; Karniadakis, G.E. Interfacing finite elements with deep neural operators for fast multiscale modeling of mechanics problems. *Comput. Methods Appl. Mech. Eng.* **2022**, *402*, 115027. [CrossRef] [PubMed]
67. Rocha, I.; Kerfriden, P.; van der Meer, F. Machine learning of evolving physics-based material models for multiscale solid mechanics. *Mech. Mater.* **2023**, *184*, 104707. [CrossRef]
68. Liu, Z.; Wu, C.; Koishi, M. A deep material network for multiscale topology learning and accelerated nonlinear modeling of heterogeneous materials. *Comput. Methods Appl. Mech. Eng.* **2019**, *345*, 1138–1168. [CrossRef]
69. Liu, Z.; Wu, C. Exploring the 3D architectures of deep material network in data-driven multiscale mechanics. *J. Mech. Phys. Solids* **2019**, *127*, 20–46. [CrossRef]
70. Gajek, S.; Schneider, M.; Böhlke, T. An FE–DMN method for the multiscale analysis of short fiber reinforced plastic components. *Comput. Methods Appl. Mech. Eng.* **2021**, *384*, 113952. [CrossRef]
71. Gajek, S.; Schneider, M.; Böhlke, T. An FE-DMN method for the multiscale analysis of thermomechanical composites. *Comput. Mech.* **2022**, *69*, 1–27. [CrossRef]
72. Nguyen, V.D.; Noels, L. Micromechanics-based material networks revisited from the interaction viewpoint; robust and efficient implementation for multi-phase composites. *Eur. J. Mech. A Solids* **2022**, *91*, 104384. [CrossRef]
73. Hughes, T.J.R. *The Finite Element Method*; Prentice-Hall: Englewood Cliffs, NJ, USA, 1987.
74. Hartmann, S.; Quint, K.J.; Hamkar, A.W. Displacement control in time-adaptive non-linear finite-element analysis. *ZAMM J. Appl. Math. Mech.* **2008**, *88*, 342–364. [CrossRef]
75. Nguyen, V.D.; Béchet, E.; Geuzaine, C.; Noels, L. Imposing periodic boundary condition on arbitrary meshes by polynomial interpolation. *Comput. Mater. Sci.* **2012**, *55*, 390–406. [CrossRef]
76. Hartmann, S. A remark on the application of the Newton-Raphson method in non-linear finite element analysis. *Comput. Mech.* **2005**, *36*, 100–116. [CrossRef]
77. Lange, N.; Hütter, G.; Kiefer, B. An efficient monolithic solution scheme for FE^2 problems. *Comput. Methods Appl. Mech. Eng.* **2021**, *382*, 113886. [CrossRef]
78. Rabbat, N.B.G.; Sangiovanni-Vincentelli, A.L.; Hsieh, H.Y. A Multilevel Newton Algorithm with Macromodeling and Latency for the Analysis of Large-Scale Nonlinear Circuits in the Time Domain. *IEEE Trans. Circuits Syst.* **1979**, *26*, 733–740. [CrossRef]
79. Hoyer, W.; Schmidt, J.W. Newton-Type Decomposition Methods for Equations Arising in Network Analysis. *ZAMM Z. Angew. Math. Und Mech.* **1984**, *64*, 397–405. [CrossRef]
80. Baydin, A.G.; Pearlmutter, B.A.; Radul, A.A.; Siskind, J.M. Automatic Differentiation in Machine Learning: A Survey. *J. Mach. Learn. Res.* **2017**, *18*, 5595–5637.
81. Abadi, M.; Barham, P.; Chen, J.; Chen, Z.; Davis, A.; Dean, J.; Devin, M.; Ghemawat, S.; Irving, G.; Isard, M.; et al. Tensorflow: A system for large-scale machine learning. In Proceedings of the 12th USENIX Symposium on Operating Systems Design and Implementation, Savannah, GA, USA, 2–4 November 2016; Volume 16, pp. 265–283.
82. Bradbury, J.; Frostig, R.; Hawkins, P.; Johnson, M.J.; Leary, C.; Maclaurin, D.; Necula, G.; Paszke, A.; VanderPlas, J.; Wanderman-Milne, S.; et al. JAX: Composable Transformations of Python+NumPy Programs. 2018. Available online: https://news.ycombinator.com/item?id=22812312 (accessed on 16 July 2023).

83. Müller, J.D.; Cusdin, P. On the performance of discrete adjoint CFD codes using automatic differentiation. *Int. J. Numer. Methods Fluids* **2005**, *47*, 939–945. [CrossRef]
84. Charpentier, I.; Ghemires, M. Efficient adjoint derivatives: Application to the meteorological model meso-nh. *Optim. Methods Softw.* **2000**, *13*, 35–63. [CrossRef]
85. Chandrasekhar, A.; Sridhara, S.; Suresh, K. AuTO: A framework for Automatic differentiation in Topology Optimization. *Struct. Multidiscip. Optim.* **2021**, *64*, 4355–4365. [CrossRef]
86. Rothe, S.; Hartmann, S. Automatic Differentiation for stress and consistent tangent computation. *Arch. Appl. Mech.* **2015**, *85*, 1103–1125. [CrossRef]
87. Rabel, E.; Rüger, R.; Govoni, M.; Ehlert, S. Forpy: A library for Fortran-Python interoperability. Available online: https://github.com/ylikx/forpy (accessed on 16 July 2023).
88. McKay, M.D.; Beckman, R.J.; Conover, W.J. A Comparison of Three Methods for Selecting Values of Input Variables in the Analysis of Output from a Computer Code. *Technometrics* **1979**, *21*, 239–245.
89. Kingma, D.P.; Ba, J. Adam: A Method for Stochastic Optimization. *arXiv* **2017**, arXiv:1412.6980.
90. Glorot, X.; Bengio, Y. Understanding the difficulty of training deep feedforward neural networks. In Proceedings of the Thirteenth International Conference on Artificial Intelligence and Statistics, Chia Laguna Resort, Sardinia, Italy, 13–15 May 2010; Volume 9, pp. 249–256.
91. Hartmann, S. A thermomechanically consistent constitutive model for polyoxymethylene: Experiments, material modeling and computation. *Arch. Appl. Mech.* **2006**, *76*, 349–366. [CrossRef]
92. Maclaurin, D.; Duvenaud, D.; Adams, R.P. Autograd: Effortless gradients in numpy. In Proceedings of the ICML 2015 AutoML Workshop, Paris, France, 11 July 2015; Volume 238.
93. Sabne, A. XLA: Compiling Machine Learning for Peak Performance. 2020. Available online: https://research.google/pubs/pub50530/ (accessed on 16 July 2023).
94. Heek, J.; Levskaya, A.; Oliver, A.; Ritter, M.; Rondepierre, B.; Steiner, A.; van Zee, M. Flax: A Neural Network Library and Ecosystem for JAX. 2023. Available online: https://github.com/google/flax (accessed on 16 July 2023).

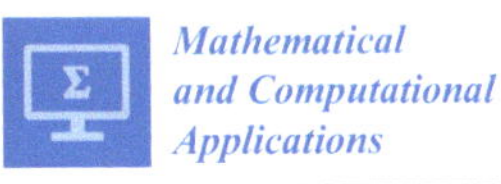

Mathematical and Computational Applications

Article

Conservation Laws and Symmetry Reductions of the Hunter–Saxton Equation via the Double Reduction Method

Molahlehi Charles Kakuli [1,2,*], **Winter Sinkala** [1] **and Phetogo Masemola** [2]

[1] Department of Mathematical Sciences and Computing, Faculty of Natural Sciences, Walter Sisulu University, Private Bag X1, Mthatha 5117, South Africa; wsinkala@wsu.ac.za

[2] School of Mathematics, University of the Witwatersrand, Johannesburg 2000, South Africa; phetogo.masemola@wits.ac.za

* Correspondence: ckakuli@wsu.ac.za; Tel.: +27-047-502-2295

Abstract: This study investigates via Lie symmetry analysis the Hunter–Saxton equation, an equation relevant to the theoretical analysis of nematic liquid crystals. We employ the multiplier method to obtain conservation laws of the equation that arise from first-order multipliers. Conservation laws of the equation, combined with the admitted Lie point symmetries, enable us to perform symmetry reductions by employing the double reduction method. The method exploits the relationship between symmetries and conservation laws to reduce both the number of variables and the order of the equation. Five nontrivial conservation laws of the Hunter–Saxton equation are derived, four of which are found to have associated Lie point symmetries. Applying the double reduction method to the equation results in a set of first-order ordinary differential equations, the solutions of which represent invariant solutions for the equation. While the double reduction method may be more complex to implement than the classical method, since it involves finding Lie point symmetries and deriving conservation laws, it has some advantages over the classical method of reducing PDEs. Firstly, it is more efficient in that it can reduce the number of variables and order of the equation in a single step. Secondly, by incorporating conservation laws, physically meaningful solutions that satisfy important physical constraints can be obtained.

Keywords: double reduction; Hunter–Saxton equation; lie symmetry analysis; conservation law; invariant solution

Citation: Kakuli, M.C.; Sinkala, W.; Masemola, P. Conservation Laws and Symmetry Reductions of the Hunter–Saxton Equation via the Double Reduction Method. *Math. Comput. Appl.* **2023**, *28*, 92. https://doi.org/10.3390/mca28050092

Academic Editor: Tasawar Hayat

Received: 10 July 2023
Revised: 11 August 2023
Accepted: 21 August 2023
Published: 22 August 2023

1. Introduction

In this research article, we focus on the Hunter–Saxton equation, a mathematical model described by the partial differential equation (PDE),

$$(u_t + uu_x)_x = \tfrac{1}{2}u_x^2, \tag{1}$$

which arises as an Euler–Lagrange equation of a variational principle in the study of a nonlinear wave equation for the director field of a nematic liquid crystal [1]. Equation (1) has attracted significant attention from researchers, prompting numerous studies on it and its derivatives. These investigations have often employed Lie symmetry analysis to explore various properties of the equations and, in certain instances, to uncover solutions.

Nadjafikhah and Ahangari [2] determined the Lie point symmetries of the equation and used the symmetries to find conservation laws and conduct symmetry reductions of the equation. An optimal system of one-dimensional subalgebras of the symmetry algebra of the Hunter–Saxton equation was also constructed. San et al. [3] investigated a modified version of the Hunter–Saxton equation, a third-order nonlinear PDE. Their work featured the utilization of Ibragimov's nonlocal conservation method to derive conservation laws for the equation. Liu and Zhao [4] undertook the study of a generalized two-component Hunter–Saxton system of equations. They determined similarity variables and executed symmetry reductions for this new generalized system, leading to the discovery of some exact solutions of the system. Yao et al. [5] tackled the periodic Hunter–Saxton equation, introducing a variable coefficient into the generalized equation. They succeeded in finding exact solutions for specific selections of the variable coefficient by employing the classical approach to finding invariant solutions. Johnpillai and Khalique [6] also used Lie symmetry analysis to find exact solutions for yet another generalized version of the Hunter–Saxton equation.

In line with the research outlined above, our study is dedicated to examining the symmetry reductions of the Hunter–Saxton equation, utilizing the double reduction method. Our objectives encompass the identification of Lie point symmetries, the determination of conservation laws through the multiplier method, and the application of the double reduction method to achieve symmetry reductions. This research serves as a valuable addition to the existing body of work on the Hunter–Saxton equation, while also contributing insights into the double reduction method in the search for solutions of PDEs. It must be noted that the double reduction routine we adopt in this article is based on the generalized approach proposed by Bokhari et al. [7], which can be used to study PDEs such as those studied in [8–10], of dimension higher than $1 + 1$.

The double reduction method, introduced by Sjöberg [11,12], is a technique for solving PDEs based on the use of Lie symmetries and conservation laws. For a $(1 + 1)$ PDE of order q, the double reduction theory allows for the reduction in the PDE to an ODE of order $q - 1$, provided that the PDE possesses a conservation law and an associated symmetry. Generalizations of the double reduction method have been proposed to handle higher-dimensional PDEs and systems of PDEs [7,13,14]. Anco and Gandarias [15] have introduced a further generalization of the double reduction method to handle partial differential equations (PDEs) with $n \geq 2$ independent variables and a symmetry algebra of dimension at least $n - 1$. In their work [15], they present an algorithm for identifying all symmetry-invariant conservation laws that reduce to first integrals for the corresponding ordinary differential equation (ODE) governing symmetry-invariant solutions of the PDE.

Moreover, Anco and Gandarias [15] propose an improved formulation for assessing the symmetry invariance of conservation laws by utilizing multipliers. This refined formulation enables the direct derivation of symmetry-invariant conservation laws, eliminating the need to first obtain conservation laws and subsequently verify their invariance.

The subsequent sections of this paper are structured as follows: Section 2 provides an overview of the necessary preliminaries and outlines the fundamental principles of the double reduction theorem. In Section 3, we calculate the Lie point symmetries and conservation laws for the Hunter–Saxton equation, determining which conservation laws are associated with symmetries. Section 4 focuses on executing symmetry reductions for the Hunter–Saxton equation. Finally, in Section 5, we present our concluding remarks.

2. Fundamentals of the Double Reduction Theorem

In this section, we present the double reduction routine for a qth-order $(q \geq 1)$ partial differential equation with n independent variables $x = (x^1, x^2, \ldots, x^n)$ and one dependent variable $u = u(x)$, namely

$$F(x, u, u_{(1)}, u_{(2)}, \ldots, u_{(q)}) = 0, \tag{2}$$

where $u_{(q)}$ denotes the collection $\{u_q\}$ of qth-order partial derivatives. In this connection, we first present the following well-known definitions and results (see, e.g., [7,16–19]).

1. The total derivative operator with respect to x^i is

$$D_i = \frac{\partial}{\partial x^i} + u_i \frac{\partial}{\partial u} + u_{ij} \frac{\partial}{\partial u_j} + \cdots, \quad i = 1, 2, \ldots, n, \tag{3}$$

where u_i denotes the derivative of u with respect to x^i. Similarly, u_{ij} denotes the derivative of u with respect to x^i and x^j.

2. An n-tuple $T = (T^1, T^2, \ldots, T^n)$, $\quad i = 1, 2, \ldots, n$, such that

$$D_i T^i = 0 \tag{4}$$

holds for all solutions of (2) is known as a conservation law of (2).

3. Multiplier Λ for Equation (2) is a non-singular function on the solution space of (2) with the property

$$D_i T^i = \Lambda E \tag{5}$$

for arbitrary function $u(x^1, x^2, \ldots, x^n)$.

4. The determining equations for multipliers are obtained by taking the variational derivative

$$\frac{\delta}{\delta u}(\Lambda E) = 0, \tag{6}$$

where the Euler operator $\delta/\delta u$ is defined by

$$\frac{\delta}{\delta u} = \frac{\partial}{\partial u} - D_i \frac{\partial}{\partial u_i} + D_{ij} \frac{\partial}{\partial u_{ij}} - D_{ijk} \frac{\partial}{\partial u_{ijk}} + \cdots . \tag{7}$$

5. A Lie symmetry of (2) with infinitesimal generator $X = \xi_i \partial x_i + \eta \partial u$ is said to be associated with a conserved law (4) if the symmetry and the conservation law satisfy the relations [16]

$$\left[T^i, X \right] = X\left(T^i \right) + T^i D_j \xi^j - T^j D_j \xi^i, \quad i = 1, \ldots, n. \tag{8}$$

Suppose that the PDE (2) admits a Lie point symmetry with infinitesimal generator $X = \xi_i \partial x_i + \eta \partial u$ that is associated with a conservation law $D_i T^i = 0$. The following steps constitute the routine of the double reduction method:

I. Find similarity variables $\tilde{x}_i, i = 1, 2, \ldots, n$ and w,

$$
\begin{aligned}
\tilde{x}_i &= \tilde{x}_i\left(x^1, x^2, \ldots, x^n \right), \quad i = 1, 2, \ldots, n \\
w(\tilde{x}_1, \ldots, \tilde{x}_{n-1}) &= \omega\left(x^1, x^2, \ldots, x^n \right) u,
\end{aligned}
$$

such that in these variables $X = \dfrac{\partial}{\partial \tilde{x}_n}$.

II. Find inverse canonical coordinates

$$
\begin{aligned}
x^i &= x^i(\tilde{x}_1, \tilde{x}_2, \ldots, \tilde{x}_n), \quad i = 1, 2, \ldots, n \\
u\left(x^1, x^2, \ldots, x^n \right) &= \psi(\tilde{x}_1, \tilde{x}_2, \ldots, \tilde{x}_n) w.
\end{aligned}
$$

III. Write partial derivatives of u in terms of the similarity variables.

IV. Construct matrices A and A^{-1} as follows:

$$
A = \begin{pmatrix}
\tilde{D}_1 x_1 & \tilde{D}_1 x_2 & \cdots & \tilde{D}_1 x_n \\
\tilde{D}_2 x_1 & \tilde{D}_2 x_2 & \cdots & \tilde{D}_2 x_n \\
\vdots & \vdots & \vdots & \vdots \\
\tilde{D}_n x_1 & \tilde{D}_n x_2 & \cdots & \tilde{D}_n x_n
\end{pmatrix}, \quad
A^{-1} = \begin{pmatrix}
D_1 \tilde{x}_1 & D_1 \tilde{x}_2 & \cdots & D_1 \tilde{x}_n \\
D_2 \tilde{x}_1 & D_2 \tilde{x}_2 & \cdots & D_2 \tilde{x}_n \\
\vdots & \vdots & \vdots & \vdots \\
D_n \tilde{x}_1 & D_n \tilde{x}_2 & \cdots & D_n \tilde{x}_n
\end{pmatrix}.
$$

V. Write components T^i of the conserved vector in terms of the similarity variables as follows:

$$\begin{pmatrix} \widetilde{T}^1 \\ \widetilde{T}^2 \\ \vdots \\ \widetilde{T}^n \end{pmatrix} = J\left(A^{-1}\right)^T \begin{pmatrix} T^1 \\ T^2 \\ \vdots \\ T^n \end{pmatrix}, \tag{9}$$

where $J = \det(A)$. Note that $T^1, \ldots, T^n$ in (9) are easily expressed in terms of the similarity variables in light of II and III.

VI. The reduced conservation law becomes

$$D_1\widetilde{T}^1 + D_2\widetilde{T}^2 + \cdots + D_{n-1}\widetilde{T}^{n-1} = 0. \tag{10}$$

3. Symmetries and Conservation Laws of the Hunter–Saxton Equation

The Hunter–Saxton Equation (1) is a $(1+1)$ PDE with two independent variables $x = (x^1, x^2) = (t, x)$ and one dependent variable $u = u(t, x)$. It admits the following four symmetries:

$$\begin{aligned} X_1 &= x\frac{\partial}{\partial x} + u\frac{\partial}{\partial u} & X_2 &= \frac{\partial}{\partial t} \\ X_3 &= t\frac{\partial}{\partial t} + x\frac{\partial}{\partial x} & X_4 &= t^2\frac{\partial}{\partial t} + 2tx\frac{\partial}{\partial x} + 2x\frac{\partial}{\partial u}. \end{aligned} \tag{11}$$

The symmetries are easily computed using MathLie, the symmetry-finding package for Mathematica [20] developed by G. Baumann [21]. We use the multiplier approach to derive conservation laws for the Hunter–Saxton Equation (1). We seek first-order multipliers

$$\Lambda = \Lambda(x, t, u, u_x, u_t) \tag{12}$$

of (1), for which the determining equation according to (6) is

$$\frac{\delta}{\delta u}\left[\Lambda\left((u_t + uu_x)_x - \frac{1}{2}ux^2\right)\right] = 0, \tag{13}$$

where the standard Euler operator $\delta/\delta u$, as defined in (7), is

$$\frac{\delta}{\delta u} = \frac{\partial}{\partial u} - D_t\frac{\partial}{\partial u_t} - D_x\frac{\partial}{\partial u_x} + D_t^2\frac{\partial}{\partial u_{tt}} + D_x^2\frac{\partial}{\partial u_{xx}} + D_xD_t\frac{\partial}{\partial u_{tx}} - \cdots, \tag{14}$$

and total derivative operators D_t and D_x using (3) are

$$D_t = \frac{\partial}{\partial t} + u_t\frac{\partial}{\partial u} + u_{tt}\frac{\partial}{\partial u_t} + u_{tx}\frac{\partial}{\partial u_x} + \cdots,$$

$$D_x = \frac{\partial}{\partial x} + u_x\frac{\partial}{\partial u} + u_{xx}\frac{\partial}{\partial u_x} + u_{tx}\frac{\partial}{\partial u_t} + \cdots.$$

The determining equation for the multiplier Λ after expansion takes the following form:

$$\Omega_0 + u_{tt}\Omega_1 + u_{tx}\Omega_2 + (u_{tx})^2\Omega_3 + u_{xx}\Omega_4 + u_{xx}u_{tt}\Omega_5 = 0, \tag{15}$$

where

$$\begin{aligned}
\Omega_0 &= u_x \Lambda_{tu} - \frac{1}{2}u_x^2 \Lambda_{tu_t} + \Lambda_{tx} - \frac{1}{2}u_x^3 \Lambda_{uu_x} - \frac{1}{2}u_x^2 u_t \Lambda_{uu_t} + uu_x^2 \Lambda_{uu} + u_x u_t \Lambda_{uu} \\
&\quad + 2uu_x \Lambda_{xu} + u_t \Lambda_{xu} + u\Lambda_{xx} - \frac{1}{2}u_x^2 \Lambda_{xu_x} + \frac{3u_x^2 \Lambda_u}{2} + u_x \Lambda_x, \\
\Omega_1 &= u_x \Lambda_{uu_t} - \frac{1}{2}u_x^2 \Lambda_{u_t u_t} + \Lambda_{xu_t}, \\
\Omega_2 &= 2uu_x \Lambda_{uu_t} + 2u\Lambda_{xu_t} - u_x^2 \Lambda_{u_t u_x} + 2\Lambda_u, \\
\Omega_3 &= u\Lambda_{u_t u_t} - \Lambda_{u_t u_x}, \\
\Omega_4 &= \Lambda_{tu_x} - u\Lambda_{tu_t} + u_t \Lambda_{uu_x} + u\Lambda_{xu_x} + uu_x \Lambda_{uu_x} - uu_t \Lambda_{uu_t} - \frac{1}{2}u_x^2 \Lambda_{u_x u_x} \\
&\quad + 2u\Lambda_u - u_x \Lambda_{u_x} - u_t \Lambda_{u_t} + \Lambda, \\
\Omega_5 &= \Lambda_{u_t u_x} - u\Lambda_{u_t u_t}.
\end{aligned}$$

The multiplier determining Equation (15) splits with respect to different combinations of the derivatives u_{xx}, u_{tx} and u_{tt} yielding an overdetermined linear system of equations for the multiplier. The system of equations was solved using Mathematica [20] to obtain

$$\Lambda = u_t \left(\delta_2 + \delta_3 t - \frac{\delta_1 t^2}{2} \right) + u_x x(\delta_3 - \delta_1 t) + \delta_1 x + \delta_4 u_x + \frac{\delta_5}{u_x^2}, \tag{16}$$

where δ_i, $i = 1, 2, \ldots, 5$, are arbitrary constants. From (5) and (16), we obtain

$$\left[(u_t + uu_x)_x - \frac{1}{2}ux^2 \right] \left[u_t \left(\delta_2 + \delta_3 t - \frac{\delta_1 t^2}{2} \right) + u_x x(\delta_3 - \delta_1 t) \right.$$
$$\left. + \delta_1 x + \delta_4 u_x + \frac{\delta_5}{u_x^2} \right] = D_t T^t + D_x T^x, \tag{17}$$

where

$$\begin{aligned}
T^t &= u_x^2 \left(u\left(\frac{\delta_1 t^2}{4} - \frac{\delta_2}{2} - \frac{\delta_3 t}{2} \right) + x\left(\frac{\delta_3}{2} - \frac{\delta_1 t}{2} \right) + \frac{\delta_4}{2} \right) - \frac{\delta_5}{u_x} + \phi_2(x) \\
&\quad + u_x(\delta_1 x - \delta_1 tu + \phi_1(u)), \\
T^x &= u_t^2 \left(\frac{\delta_2}{2} + \frac{\delta_3 t}{2} - \frac{\delta_1 t^2}{4} \right) + uu_x^2 \left(x\left(\frac{\delta_3}{2} - \frac{\delta_1 t}{2} \right) + \frac{\delta_4}{2} \right) - \frac{\delta_5 u}{u_x} + \frac{3\delta_5 x}{2} \\
&\quad + u_x \left(uu_t \left(\delta_2 + \delta_3 t - \frac{\delta_1 t^2}{2} \right) + \delta_1 ux \right) + u_t(\delta_1 tu - \phi_1(u)) + \phi_3(t)
\end{aligned}$$

for arbitrary functions $u(t, x)$. When $u(t, x)$ is a solution of Equation (1), the left hand side of (17) vanishes and we obtain conservation laws of the Hunter–Saxton Equation (1) for which the conserved vectors (T_i^1, T_i^2), $i = 1, 2, \ldots, 5$, are given by

$$\begin{aligned}
T_1^1 &= u_x \left(ux - \frac{1}{2}t^2 uu_t \right) - \frac{t^2 u_t^2}{4} - \frac{1}{2}tuu_x^2 x + u_t(tu - \phi_1(u)) + \phi_3(t), \\
T_1^2 &= u_x^2 \left(\frac{t^2 u}{4} - \frac{tx}{2} \right) + u_x(x - tu + \phi_1(u)) + \phi_2(x),
\end{aligned}$$

$$\begin{aligned}
T_2^1 &= \phi_3(t) + uu_x u_t - u_t \phi_1(u) + \frac{u_t^2}{2}, \\
T_2^2 &= u_x \phi_1(u) + \phi_2(x) - \frac{uu_x^2}{2},
\end{aligned}$$

$$T_3^1 = tuu_xu_t + \frac{tu_t^2}{2} + \phi_3(t) + \frac{1}{2}uu_x^2x - u_t\phi_1(u),$$

$$T_3^2 = u_x^2\left(\frac{x}{2} - \frac{tu}{2}\right) + u_x\phi_1(u) + \phi_2(x),$$

$$T_4^1 = \phi_3(t) - u_t\phi_1(u) + \frac{uu_x^2}{2},$$

$$T_4^2 = \phi_2(x) + u_x\phi_1(u) + \frac{u_x^2}{2},$$

$$T_5^1 = \phi_3(t) - u_t\phi_1(u) - \frac{u}{u_x} + \frac{3x}{2},$$

$$T_5^2 = \phi_2(x) + u_x\phi_1(u) - \frac{1}{u_x}.$$

According to (8), symmetry X is associated with conservation law $D_tT^t + D_xT^x = 0$ if the following formula is satisfied:

$$X\begin{pmatrix} T^t \\ T^x \end{pmatrix} - \begin{pmatrix} D_t\xi^t & D_x\xi^t \\ D_t\xi^x & D_x\xi^x \end{pmatrix}\begin{pmatrix} T^t \\ T^x \end{pmatrix} + (D_t\xi^t + D_x\xi^x)\begin{pmatrix} T^t \\ T^x \end{pmatrix} = 0. \tag{18}$$

It turns out that the association of symmetries and conservation laws of (1) is obtained in the following cases:

$$\kappa_1(X_1 + 2X_3) + \kappa_2X_2 \rightarrow \begin{cases} T_2^1 = \frac{u_t^2}{2} - \frac{\delta_1 u_t}{u} + uu_xu_t \\ T_2^2 = \frac{\delta_1 u_x}{u} + \frac{\delta_3}{x} - \frac{uu_x^2}{2} \end{cases},$$

$$\kappa_1(X_1 + X_3) + \kappa_2X_2 \rightarrow \begin{cases} T_4^1 = \frac{uu_x^2}{2} - \frac{\delta_1 u_t}{u} \\ T_4^2 = \frac{\delta_1 u_x}{u} + \frac{\delta_3}{x} + \frac{u_x^2}{2} \end{cases},$$

$$\kappa_1\left(X_1 - \frac{X_3}{2}\right) + \kappa_2X_2 \rightarrow \begin{cases} T_5^1 = \frac{\delta_2}{2\kappa_2 - \kappa_1 t} - \frac{\delta_1 u_t}{u} - \frac{u}{u_x} + \frac{3x}{2} \\ T_5^2 = \frac{\delta_1 u_x}{u} + \frac{\delta_3}{x} - \frac{1}{u_x} \end{cases},$$

$$X_3 \rightarrow \begin{cases} T_3^1 = \frac{\delta_1}{t} + tuu_xu_t + \frac{tu_t^2}{2} + \frac{1}{2}uu_x^2x - u_t\phi_1(u) \\ T_3^2 = \frac{\delta_2}{x} + u_x^2\left(\frac{x}{2} - \frac{tu}{2}\right) + u_x\phi_1(u) \end{cases}.$$

It is important to observe that among the five computed conservation laws, we identified associated Lie point symmetries for only four. Notably, the conservation law T_1 lacks any associated Lie point symmetry of the Hunter–Saxton equation.

4. Double Reduction of the Hunter–Saxton Equation

4.1. Double Reduction of (1) by $\langle \kappa_1(X_1 + 2X_3) + \kappa_2X_2 \rangle$

We transform the generator $Z = \kappa_1(X_1 + 2X_3) + \kappa_2X_2$ to its canonical form $Y = 0\frac{\partial}{\partial r} + \frac{\partial}{\partial s} + 0\frac{\partial}{\partial w}$. Therefore, canonical coordinates $r = r(t, x)$, $s = s(t, x)$ and $w = w(t, x, u)$ must be found such that $Z(r) = 0$, $Z(s) = 1$ and $Z(w) = 0$. While the coordinates r and w are obtained from invariants of Z, the coordinate s may be determined by inspection. More systematically, it can be obtained from an invariant $J = v - s(x, y)$ of the extended operator $Z + \partial_v$, where v is an auxiliary variable [19]. We obtain

$$r = \frac{x}{(2\kappa_1 t + \kappa_2)^{3/2}}, \quad s = \frac{\ln x}{3\kappa_1}, \quad w = \frac{u}{\sqrt{2\kappa_1 t + \kappa_2}}, \quad \kappa_1 \neq 0, \tag{19}$$

where $w = w(r)$. Inverse canonical coordinates follow from (19) and are given by

$$t = \frac{e^{2\kappa_1 s} - \kappa_2 r^{2/3}}{2\kappa_1 r^{2/3}}, \quad x = e^{3\kappa_1 s}, \quad u = \frac{we^{\kappa_1 s}}{r^{1/3}}. \tag{20}$$

Computing A and $\left(A^{-1}\right)^T$, we obtain

$$A = \begin{pmatrix} D_r t & D_r x \\ D_s t & D_s x \end{pmatrix} = \begin{pmatrix} -\dfrac{e^{2\kappa_1 s}}{3\kappa_1 r^{5/3}} & 0 \\ \dfrac{e^{2\kappa_1 s}}{r^{2/3}} & 3e^{3\kappa_1 s}\kappa_1 \end{pmatrix}$$

and

$$\left(A^{-1}\right)^T = \begin{pmatrix} D_t r & D_x r \\ D_t s & D_x s \end{pmatrix} = \begin{pmatrix} -3e^{-2\kappa_1 s}\kappa_1 r^{5/3} & e^{-3\kappa_1 s}r \\ 0 & \dfrac{e^{-3\kappa_1 s}}{3\kappa_1} \end{pmatrix}.$$

The partial derivatives of u from (20) are given by

$$\begin{aligned}
u_t &= \kappa_1 \sqrt[3]{r}\, e^{-\kappa_1 s}(w - 3rw_r), \quad u_x = r^{2/3}w_r e^{-2\kappa_1 s}, \\
u_{tx} &= -\kappa_1 r^{4/3}e^{-4\kappa_1 s}(3rw_{rr} + 2w_r), \\
u_{xx} &= r^{5/3}w_{rr}e^{-5\kappa_1 s}.
\end{aligned} \tag{21}$$

The reduced conserved form is given by

$$\begin{pmatrix} T_2^r \\ T_2^s \end{pmatrix} = J\left(A^{-1}\right)^T \begin{pmatrix} T_2^t \\ T_2^x \end{pmatrix}, \tag{22}$$

where $J = \det(A) = -\dfrac{e^{5\kappa_1 s}}{r^{5/3}}$. By substituting (20) and (21) into (22), we obtain

$$\begin{aligned}
T_2^r &= \delta_1 \kappa_1 + 3\delta_3 \kappa_1 + 3\kappa_1^2 rww_r - \frac{9}{2}\kappa_1^2 r^2 w_r^2 - \frac{\kappa_1^2 w^2}{2} + \frac{3}{2}\kappa_1 rww_r^2 - \kappa_1 w^2 w_r, \\
T_2^s &= w_r\left(\kappa_1 w - \frac{\delta_1}{w} - \frac{w^2}{3r}\right) + \frac{\delta_1}{3r} - \frac{\kappa_1 w^2}{6r} + w_r^2\left(w - \frac{3\kappa_1 r}{2}\right),
\end{aligned} \tag{23}$$

where the reduced conserved form satisfies

$$D_r T_2^r = 0. \tag{24}$$

From (23) and (24), we have

$$3\kappa_1^2 rww_r - \frac{9}{2}\kappa_1^2 r^2 w_r^2 - \frac{\kappa_1^2 w^2}{2} + \frac{3}{2}\kappa_1 rww_r^2 - \kappa_1 w^2 w_r = k,$$

where k is an arbitrary constant.

4.2. Double Reduction of (1) by $\langle \kappa_1(X_1 + X_3) + \kappa_2 X_2 \rangle$

Canonical coordinates determined from $\langle \kappa_1(X_1 + X_3) + \kappa_2 X_2 \rangle$ are

$$r = \frac{x}{(\kappa_1 t + \kappa_2)^2}, \quad s = \frac{\ln x}{2\kappa_1}, \quad w = \frac{u}{\sqrt{x}}, \quad \kappa_1 \neq 0, \tag{25}$$

where $w = w(r)$, and the inverse canonical coordinates are given by

$$t = -\frac{\kappa_2 \sqrt{r} - e^{\kappa_1 s}}{\kappa_1 \sqrt{r}}, \quad x = e^{2\kappa_1 s} \quad u = we^{\kappa_1 s}. \tag{26}$$

Therefore, the partial derivatives of u from (26) are given by

$$\begin{aligned}
u_t &= -2\kappa_1 r^{3/2}w_r, \quad u_x = \frac{1}{2}e^{-\kappa_1 s}(2rw_r + w), \\
u_{tx} &= -e^{-2\kappa_1 s}\kappa_1 r^{3/2}(2rw_{rr} + 3w_r, \\
u_{xx} &= -\frac{1}{4}e^{-3\kappa_1 s}(w - 4r(rw_{rr} + w_r)).
\end{aligned} \tag{27}$$

As for A and $\left(A^{-1}\right)^T$, we obtain

$$A = \begin{pmatrix} D_r t & D_r x \\ D_s t & D_s x \end{pmatrix} = \begin{pmatrix} -\dfrac{e^{\kappa_1 s}}{2\kappa_1 r^{3/2}} & 0 \\ \dfrac{e^{\kappa_1 s}}{\sqrt{r}} & 2e^{2\kappa_1 s}\kappa_1 \end{pmatrix},$$

and

$$\left(A^{-1}\right)^T = \begin{pmatrix} D_t r & D_x r \\ D_t s & D_x s \end{pmatrix} = \begin{pmatrix} -2e^{-\kappa_1 s}\kappa_1 r^{3/2} & e^{-2\kappa_1 s} r \\ 0 & \dfrac{e^{-2\kappa_1 s}}{2\kappa_1} \end{pmatrix}.$$

Therefore, from

$$\begin{pmatrix} T_4^r \\ T_4^s \end{pmatrix} = J\left(A^{-1}\right)^T \begin{pmatrix} T_4^t \\ T_4^x \end{pmatrix}, \tag{28}$$

where $J = \det(A) = -\dfrac{e^{3\kappa_1 s}}{r^{3/2}}$, we obtain

$$T_4^r = \delta_1 \kappa_1 + 2\delta_3 \kappa_1 + \kappa_1 r^2 w_r^2 + \kappa_1 r w w_r + \frac{\kappa_1 w^2}{4} - \frac{1}{2} r^{3/2} w w_r^2 - \frac{w^3}{8\sqrt{r}} - \frac{1}{2}\sqrt{r} w^2 w_r,$$

$$T_4^s = -\frac{\delta_1 w_r}{w} - \frac{w^3}{16\kappa_1 r^{3/2}} - \frac{w^2 w_r}{4\kappa_1 \sqrt{r}} - \frac{\sqrt{r} w w_r^2}{4\kappa_1}. \tag{29}$$

From the reduced conservation law $D_r T_4^r = 0$, we obtain

$$\kappa_1 r^2 w_r^2 + \kappa_1 r w w_r + \frac{\kappa_1 w^2}{4} - \frac{1}{2} r^{3/2} w w_r^2 - \frac{w^3}{8\sqrt{r}} - \frac{1}{2}\sqrt{r} w^2 w_r = k,$$

where k is an arbitrary constant.

4.3. Double Reduction of (1) by $\left\langle \kappa_1 \left(X_1 - \frac{X_3}{2} \right) + \kappa_2 X_2 \right\rangle$

Canonical coordinates determined from $\left\langle \kappa_1 \left(X_1 - \frac{X_3}{2} \right) + \kappa_2 X_2 \right\rangle$ are

$$r = x(2\kappa_2 - \kappa_1 t), \quad s = \frac{2\ln x}{\kappa_1}, \quad w = \frac{u}{x^2}, \quad \kappa_1 \neq 0, \tag{30}$$

where $w = w(r)$, and the inverse canonical coordinates are given by

$$t = \frac{2\kappa_2 - r e^{-\frac{1}{2}\kappa_1 s}}{\kappa_1}, \quad x = e^{\frac{\kappa_1 s}{2}}, \quad u = w e^{\kappa_1 s} \tag{31}$$

Therefore, the partial derivatives of u from (31) are given by

$$u_t = -\kappa_1 w_r e^{\frac{3\kappa_1 s}{2}}, \quad u_x = e^{\frac{\kappa_1 s}{2}}(r w_r + 2w),$$
$$u_{tx} = -\kappa_1 e^{\kappa_1 s}(r w_{rr} + 3w_r), \tag{32}$$
$$u_{xx} = r(r w_{rr} + 4w_r) + 2w.$$

Therefore,

$$A = \begin{pmatrix} D_r t & D_r x \\ D_s t & D_s x \end{pmatrix} = \begin{pmatrix} -\dfrac{e^{-\frac{1}{2}\kappa_1 s}}{\kappa_1} & 0 \\ \dfrac{1}{2}e^{-\frac{1}{2}\kappa_1 s} r & \dfrac{1}{2}e^{\frac{\kappa_1 s}{2}}\kappa_1 \end{pmatrix}$$

and

$$\left(A^{-1}\right)^T = \begin{pmatrix} D_t r & D_x r \\ D_t s & D_x s \end{pmatrix} = \begin{pmatrix} -e^{\frac{\kappa_1 s}{2}}\kappa_1 & e^{-\frac{1}{2}\kappa_1 s} r \\ 0 & \dfrac{2e^{-\frac{1}{2}\kappa_1 s}}{\kappa_1} \end{pmatrix}.$$

Therefore, from

$$\begin{pmatrix} T_5^r \\ T_5^s \end{pmatrix} = J\left(A^{-1}\right)^T \begin{pmatrix} T_5^t \\ T_5^x \end{pmatrix}, \tag{33}$$

where $J = \det(A) = -\frac{1}{2}$, we obtain

$$T_5^r = \frac{2\kappa_1(2\delta_1 r w_r + 4\delta_1 w + \delta_3 r w_r + 2\delta_3 w - 1) - 2\delta_2(r w_r + 2w) - r(3r w_r + 4w)}{4r w_r + 8w},$$

$$T_5^s = -\frac{2\delta_1\kappa_1 r^2 w_r^2 + 4\delta_1\kappa_1 r w w_r + 2\delta_2 r w w_r + 4\delta_2 w^2 + 3r^2 w w_r + 4r w^2}{2\kappa_1 r^2 w w_r + 4\kappa_1 r w^2}. \tag{34}$$

From the reduced conservation law $D_r T_5^r = 0$, we obtain

$$\frac{2\kappa_1(2\delta_1 r w_r + 4\delta_1 w + \delta_3 r w_r + 2\delta_3 w - 1) - 2\delta_2(r w_r + 2w) - r(3r w_r + 4w)}{4r w_r + 8w} = k,$$

where k is an arbitrary constant.

4.4. Double Reduction of (1) by $\langle X_3 \rangle$

Canonical coordinates determined from X_3 are

$$r = \frac{x}{t}, \quad s = \ln x \quad w = u, \tag{35}$$

where $w = w(r)$, and the inverse canonical coordinates are given by

$$t = \frac{e^s}{r}, \quad x = e^s \quad u = w. \tag{36}$$

Therefore, the partial derivatives of u from (36) are given by

$$u_t = -r^2 e^{-s} w_r, \quad u_x = r e^{-s} w_r,$$
$$u_{tx} = -r^2 e^{-2s}(r w_{rr} + w_r), \tag{37}$$
$$u_{xx} = r^2 e^{-2s} w_{rr}.$$

As for A and $\left(A^{-1}\right)^T$, we obtain

$$A = \begin{pmatrix} D_r t & D_r x \\ D_s t & D_s x \end{pmatrix} = \begin{pmatrix} -\frac{e^s}{r^2} & 0 \\ \frac{e^s}{r} & e^s \end{pmatrix},$$

and

$$\left(A^{-1}\right)^T = \begin{pmatrix} D_t r & D_x r \\ D_t s & D_x s \end{pmatrix} = \begin{pmatrix} -e^{-s} r^2 & e^{-s} r \\ 0 & e^{-s} \end{pmatrix}.$$

Therefore, from

$$\begin{pmatrix} T_3^r \\ T_3^s \end{pmatrix} = J\left(A^{-1}\right)^T \begin{pmatrix} T_3^t \\ T_3^x \end{pmatrix}, \tag{38}$$

where $J = \det(A) = -\frac{e^{2s}}{r^2}$, we obtain

$$T_3^r = \delta_2 - \delta_1,$$
$$T_3^s = \frac{1}{2} w_r^2 (w - r) - \frac{\delta_1}{r} - w_r \phi_1(w). \tag{39}$$

It is remarkable that in this case, because T_3^r in (39) is simply a constant, the reduced conservation law $D_r T_3^r = 0$ does not result in an ODE that can be solved for w. Therefore, no invariant solution arises via the double reduction method from the association of X_3 and the conservation law T_3.

5. Concluding Remarks

In this paper, a study of the Hunter–Saxton equation using Lie symmetry analysis was presented. Symmetry reductions of the equation were carried out by employing the generalized approach to double reduction theory proposed by Bokhari et al. [7]. By utilizing the multiplier method, nontrivial conservation laws for the Hunter–Saxton equation were derived. These conservation laws, along with the Lie point symmetries of the equation, were employed to perform symmetry reductions via the double reduction method.

Through the analysis, a set of first-order ODEs was obtained, whose solutions represent invariant solutions for the Hunter–Saxton equation. Out of the five nontrivial conservation laws constructed, it was observed that only four had associated Lie point symmetries according to the definition provided by Kara and Mahomed [16]. The conservation law T_1 did not have any linear combination of symmetries associated with it. Additionally, it is noteworthy that despite the conservation law T_3 having an associated Lie point symmetry, X_3, the application of the double reduction method in this case did not yield a symmetry reduction of the Hunter–Saxton equation. This outcome could be attributed to the "collapse" of the first integral, which was expected to represent a reduced ODE for the PDE but instead resulted in a constant value.

Author Contributions: Conceptualization, M.C.K. and W.S.; methodology, M.C.K., W.S. and P.M.; software, M.C.K. and W.S.; validation, W.S. and P.M.; formal analysis, M.C.K., W.S. and P.M.; writing—original draft preparation, M.C.K. and W.S.; writing—review and editing, M.C.K., W.S. and P.M. All authors have read and agreed to the published version of the manuscript.

Funding: This research received no external funding.

Acknowledgments: The authors would like to thank the Directorate of Research Development and Innovation of Walter Sisulu University for continued support.

Conflicts of Interest: The authors declare no conflict of interest.

References

1. Hunter, J.K.; Saxton, R. Dynamics of director fields. *SIAM J. Appl. Math.* **1991**, *51*, 1498–1521. [CrossRef]
2. Nadjafikhah, M.; Ahangari, F. Symmetry Analysis and Conservation Laws for the Hunter–Saxton Equation. *Commun. Theor. Phys.* **2013**, *59*, 335–348. [CrossRef]
3. San, S.; Yaşar, E. On the Conservation Laws and Exact Solutions of a Modified Hunter–Saxton Equation. *Adv. Math. Phys.* **2014**, *2014*, 349059. [CrossRef]
4. Liu, W.; Zhao, Y. Lie symmetry reductions and exact solutions to a generalized two-component Hunter–Saxton system. *AIMS Math.* **2021**, *6*, 1087–1100.
5. Yao, S.W.; Gulsen, S.; Hashemi, M.S.; Inc, M.; Bicer, H. Periodic Hunter–Saxton equation parametrized by the speed of the Galilean frame: Its new solutions, Nucci's reduction, first integrals and Lie symmetry reduction. *Results Phys.* **2023**, *47*, 106370. [CrossRef]
6. Johnpillai, A.G.; Khalique, C.M. Symmetry Reductions, Exact Solutions, and Conservation Laws of a Modified Hunter–Saxton Equation. *Abstr. Appl. Anal.* **2013**, *2013*, 204746. [CrossRef]
7. Bokhari, A.H.; Al-Dweik, A.Y.; Zaman, F.D.; Kara, A.H.; Mahomed, F.M. Generalization of the Double Reduction Theory. *Nonlinear Anal. Real World Appl.* **2010**, *11*, 3763–3769. [CrossRef]
8. Zhao, Z.; He, L. Lie symmetry, nonlocal symmetry analysis, and interaction of solutions of a $(2+1)$-dimensional KdV–mKdV equation. *Theor. Math. Phys.* **2021**, *206*, 142–162. [CrossRef]
9. Zhao, Z.; Yue, J.; He, L. New type of multiple lump and rogue wave solutions of the $(2+1)$-dimensional Bogoyavlenskii–Kadomtsev–Petviashvili equation. *Appl. Math. Lett.* **2022**, *133*, 108294. [CrossRef]
10. Zhao, Z.; He, L.; Wazwaz, A.M. Dynamics of lump chains for the BKP equation describing propagation of nonlinear waves. *Chin. Phys. B* **2023**, *32*, 040501. [CrossRef]
11. Sjöberg, A. Double Reduction of PDEs from the Association of Symmetries with Conservation Laws with Applications. *Appl. Math. Comput.* **2007**, *184*, 608–616. [CrossRef]
12. Sjöberg, A. On Double Reductions from Symmetries and Conservation Laws. *Nonlinear Anal. Real World Appl.* **2009**, *10*, 3472–3477. [CrossRef]
13. Özkan, Y.S. Double reduction of second order Benjamin–Ono equation via conservation laws and the exact solutions. *Balıkesir Üniversitesi Fen Bilimleri Enstitüsü Dergisi* **2021**, *23*, 210–223.
14. Morris, R.; Kara, A.H. Double reductions/analysis of the Drinfeld–Sokolov–Wilson equation. *Appl. Math. Comput.* **2013**, *219*, 6473–6483. [CrossRef]

15. Anco, S.C.; Gandarias, M.L. Symmetry Multi-reduction Method for Partial Differential Equations with Conservation Laws. *Commun. Nonlinear Sci. Numer. Simul.* **2020**, *91*, 105349. [CrossRef]
16. Kara, A.; Mahomed, F. Relationship between symmetries and conservation laws. *Int. J. Theor. Phys.* **2000**, *39*, 23–40. [CrossRef]
17. Steeb, W.H.; Strampp, W. Diffusion equations and Lie and Lie–Bäcklund transformation groups. *Phys. A Stat. Mech. Appl.* **1982**, *114A*, 95–99. [CrossRef]
18. Kara, A.; Mahomed, F. Action of Lie–Bäcklund Symmetries on Conservation Laws. In Proceedings of the International Conference on Modern Group Analysis VII, Nordfjordeid, Norway, 30 June–5 July 1997.
19. Olver, P.J. *Applications of Lie Groups to Differential Equations*, 2nd ed.; Springer: New York, NY, USA, 1993.
20. Wolfram Research, Inc. *Mathematica, Version 9.0, Wolfram Research*; Wolfram Research, Inc.: Champaign, IL, USA, 2012.
21. Baumann, G. *Symmetry Analysis of Differential Equations with Mathematica*; Telos/Springer: New York, NY, USA, 2000.

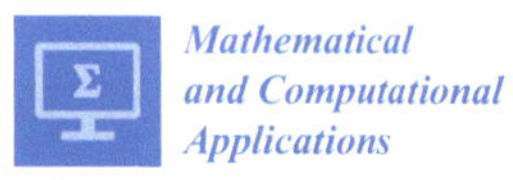
Mathematical and Computational Applications

Article

Exploring the Potential of Mixed Fourier Series in Signal Processing Applications using One-Dimensional Smooth Closed-Form Functions with Compact Support: A Comprehensive Tutorial

Carlos-Iván Páez-Rueda *, Arturo Fajardo, Manuel Pérez, German Yamhure and Gabriel Perilla

Department of Electronic Engineering, Pontificia Universidad Javeriana, Carrera 7 #40-62, Bogota 110311, Colombia; fajardoa@javeriana.edu.co (A.F.); manuel.perez@javeriana.edu.co (M.P.); gyamhure@javeriana.edu.co (G.Y.); gabriel.perilla@javeriana.edu.co (G.P.)
* Correspondence: paez.carlos@javeriana.edu.co

Abstract: This paper studies and analyzes the approximation of one-dimensional smooth closed-form functions with compact support using a mixed Fourier series (i.e., a combination of partial Fourier series and other forms of partial series). To explore the potential of this approach, we discuss and revise its application in signal processing, especially because it allows us to control the decreasing rate of Fourier coefficients and avoids the Gibbs phenomenon. Therefore, this method improves the signal processing performance in a wide range of scenarios, such as function approximation, interpolation, increased convergence with quasi-spectral accuracy using the time domain or the frequency domain, numerical integration, and solutions of inverse problems such as ordinary differential equations. Moreover, the paper provides comprehensive examples of one-dimensional problems to showcase the advantages of this approach.

Keywords: function reconstruction; Fourier series; Gibbs phenomenon; convergence acceleration; exponential accuracy

MSC: 42A16; 42A20; 41A10

Citation: Páez-Rueda, C.-I.; Fajardo, A.; Pérez, M.; Yamhure, G.; Perilla, G. Exploring the Potential of Mixed Fourier Series in Signal Processing Applications using One-Dimensional Smooth Closed-Form Functions with Compact Support: A Comprehensive Tutorial. *Math. Comput. Appl.* **2023**, *28*, 93. https://doi.org/10.3390/mca28050093

Academic Editor: Gianluigi Rozza

Received: 26 July 2023
Revised: 11 August 2023
Accepted: 14 August 2023
Published: 1 September 2023

1. Introduction

The Fourier series of a function with compact support, denoted by $g:[0,T] \to \mathbb{R}$, has been the cornerstone of several modern applications through harmonic analysis and Fourier synthesis. On the one hand, harmonic analysis allows the study of the original function or phenomenon through the superposition of simpler trigonometric functions. This analysis represents a wide branch of study in mathematics [1–3] and is used in many applications in physics [4–6], engineering [7–9], medicine [10–12], and music [13]. Fourier synthesis, on the other hand, uses a linear combination of basis functions to approximate the original function [14], which has many applications in boundary value problems [15–17], data interpolation [18–21], and compression [22].

The Fourier series has several advantages for representing a function with compact support (i.e., non-periodic function). First, it has been extensively studied, and many well-known analytical and numerical properties can be applied [1,2]. Second, unlike other series based on local information, such as the Taylor series or the Spline series, the Fourier series does not require the use of high-order derivatives, and their coefficients are calculated by well-conditioned algorithms. Finally, unlike other series based on non-trigonometric basis functions using an inner product, such as orthogonal polynomials, the Fourier series offers the advantage of reducing the computational cost of obtaining the coefficients by employing the FFT (fast Fourier transform). However, despite those advantages, it is well-known that its performance degrades when the equivalent periodic function (denoted by

$\bar{g}{:}\mathbb{R} \to \mathbb{R}$) loses its smooth property. The first shortcoming is the presence of unacceptable oscillations (i.e., ringing artifacts) in the approximation, which are generally known as the Gibbs (or Gibbs–Wilbraham) phenomenon [23]. The second shortcoming is slow convergence because the magnitude of Fourier coefficients is $O(|k|^{-1})$ [24] or $O(|k|^{-2})$ [25], which makes it difficult to obtain a suitable representation using few coefficients for many purposes, such as data compression or fast solvers of inverse problems. To address these drawbacks, several techniques have been proposed over the last several years. They can be classified into averaging and filtering techniques [26–56], polynomial techniques [57–69], and discontinuity subtraction techniques [70–86].

The Windowing technique [26] is possibly the most important filtering technique for harmonic analysis, in which the function is artificially smoothed with a relevant distortion cost for small bandwidths. Averaging and filtering techniques are more diverse for synthesis applications. For instance, the Fejér's arithmetic mean method removes the Gibbs oscillations [27–29]. Similarly, the Lanczos sigma approximation [30–32], Mollifiers [33–38], and other averaging methods [39,40] and filters [41–44] can reduce ringing artifacts, too. Moreover, they can be combined with special wavelets [45–50] with the same purpose. Furthermore, these techniques can be merged with Fourier extension methods, in which artificial and convenient information in an extended interval $t \in [-a, T + a]$ allows for a reduction in undesirable phenomena in $t \in [0, T]$ [51–56]. Despite their success, averaging and filtering techniques have several drawbacks due to the artificial modification and slow convergence of the Fourier coefficients.

The main polynomial methodology for synthesis application may be the spectral reprojection approach [57], in which Fourier coefficients are reprojected onto other basis functions conformed by polynomials. For instance, the Gegenbauer polynomials [58–60] and general polynomials using inverse methods [61,62] are successful in removing the Gibbs phenomenon. In the same direction, Chebyshev polynomials produce a strongly nonuniform distribution of points with good performance for interpolation [63,64]. However, the solution loses simplification by using non-equidistant data. Similarly, other related techniques can address these concerns, such as Padé approximations [65–67], convergence acceleration, and inverse methods [68,69]. Although polynomial techniques reduce or remove ringing artifacts from arbitrary Fourier coefficients, they have some drawbacks due to their complexities or ill-conditioned solutions.

Discontinuity subtraction techniques are employed to separate the discontinuities in the original function, yielding a more convenient Fourier representation. To the best of our knowledge, the concept of removing discontinuities using polynomials was first introduced by Russian works in the 1900s. For instance, A. N. Krylov proposed the method of Acad using a piecewise linear polynomial [70], ([71], p. 79) and A. S. Maliev proposed a strengthened convergence method using high-order piecewise polynomials using a Fourier extension method ([71], p. 86). Those approaches were generalized by C. Lanczos in ([72], p. 98), using quasi-Bernoulli polynomials, denoted by $B_m(t)$, in the same domain of the function. The Maliev–Lanczos approach has enormous potential because it avoids the Gibbs phenomenon and allows for generating Fourier coefficients with convergence $O(|k|^{-M})$. Despite the fact that these works demonstrated that the Fourier series can achieve accelerated convergence for smooth functions, there was little scientific discussion for almost three decades about these methods [77,78]. The Lanczos approach emerged again in the 1990s in the works of K. S. Eckhoff [79–81], who proposed the reconstruction of piecewise smooth functions with M jumps by solving a linear system of M equations using quasi-Bernoulli polynomials and the spectral domain. Over the last two decades, several Armenian researchers have made significant contributions to the Lanczos approach. For instance, the works of A. Nersessian and A. Poghosyan addressed the main issue of some alternatives to the quasi-Bernoulli series in [67,87–92], such as the quasi-polynomial series, the Fourier–Pade series, the trigonometric interpolations series, and the quasi-polynomial Pade series. Similarly, A. Nersessian studied a framework based on a biorthogonal system and adaptive algorithms with a strong potential for accelerating the convergence of Fourier series due to an over-convergence phenomenon [93–96]. Furthermore, some simplifications and applications of Eckhoff algorithm have been studied by A. Poghosyan et al. in [83–86,97], such as its application to two-dimensional functions, the simplification of the minimization problem, and the study of trigonometric interpolations series. Finally, several researchers around the world, who are not fully discussed in this introduction due to space limitations, have also contributed to this technique [82,98–101]. For example, B. Adcock provided a comprehensive discussion and evaluation of several related techniques in [99], and D. Batenkov proposed a novel decimated Eckhoff's algorithm in [101].

On the other hand, Fourier series have been widely used to solve differential problems, such as ODEs, PDEs, and eigenvalues [14,102,103]. In particular, we found that, unrelated to the previous state of the art, P. Roache proposed the method of "reduction to periodicity" in [104] for solving differential equations in fluid dynamics [105,106]. That method applies the discontinuity subtraction technique using simple polynomials in a normalized domain (i.e., $\phi(t) = \sum_{k=0}^{M} a_k \cdot t^k, \forall t \in [0,1]$), where the coefficients are chosen to produce a smooth periodic residual error, and therefore, the solution increases the convergence and removes ringing artifacts using the Fourier approach.

The Maliev–Lanczos approach to approximating closed-form smooth functions has four disadvantages in applied problems. First, the method requires explicit knowledge of the function's derivative at the edges of the interval. However, for continuous-time applications where the closed-form function is known, this requirement does not cause setbacks because derivative operators may be easily computed using the chain rule. These continuous-time applications include the solutions to direct and inverse problems using linear operators (i.e., $\mathscr{L}\{\cdot\} : g \to h$ such that $\mathscr{L}\{\alpha \cdot g_1(t) + \beta \cdot g_2(t)\} = \alpha \cdot \mathscr{L}\{g_1(t)\} + \beta \cdot \mathscr{L}\{g_2(t)\}$). Second, Fourier coefficients of the original function must be determined because the method is supported by the Fourier framework. For common functions, these coefficients are not always known in closed form. Third, the evaluation of quasi-Bernoulli polynomials requires an iterative algorithm that increases addition and product operations and, therefore, increases computation time and sensitivity to rounding errors in the operations $\mathscr{L}\{B_m(t)\}$. Finally, it would seem that the approach using quasi-Bernoulli polynomials is the best framework for accelerated convergence because they are directly found from integration by parts using the integral definition of Fourier coefficients.

Despite the advances of the last decades, the Maliev–Lanczos approach is not widely used or recognized as one would expect in continuous-time problems involving closed-form smooth functions in applied mathematics, physics, and engineering. For example, this technique receives minimal attention as an alternative to the Taylor series for non-analytic smooth functions on a closed interval. Unfortunately, the Maliev–Lanczos approach is rarely mentioned in engineering textbooks, despite the fact that the Fourier series and Taylor series are fundamental tools for a wide range of problems and applications (e.g., Riemann integration, integral equations, and boundary or initial value problems). Perhaps this is due to the aforementioned shortcomings, as well as a lack of useful information and discussions, canonical examples, or practical applications.

The contributions of this publication are categorized as follows. The first category involves making scientific contributions to the Maliev–Lanczos approach. This includes advancements, improvements, and novel insights related to the approach, which are listed below:

1. We prove that the Roache approach (i.e., using simple polynomials derived from the residual error framework) is spectrally equivalent to the Lanczos approach (i.e., using quasi-Bernoulli polynomials derived from an integration-by-parts framework) for any harmonic different from zero (i.e., $\forall k \in \mathbb{Z} - \{0\}$), where simple polynomial coefficients are determined by a low-cost backward algorithm. Although both approaches have the same complexity when using linear operators based on integrals or derivatives, simple polynomials are easier to manipulate using these operators, and they reduce rounding errors by lowering addition and multiplication operations.

2. We propose a reprojection method that allows the transformation from Fourier series coefficients to mixed Fourier series coefficients. We introduce the term "mixed Fourier series" to designate the summation of series derived from a smooth periodic residual error, wherein one of the series is the Fourier series. This method allows for recovering convergence $O(|k|^{-M})$ using standard Fourier coefficients obtained from any smooth function. The proposed method has the advantage of avoiding the temporal information of $g{:}[0, T] \to \mathbb{R}$. Therefore, it has the potential to be particularly useful for native spectral applications (e.g., solving differential equations with spectral methods).

3. By employing the Maliev–Lanczos approach and leveraging the residual error framework, we introduce and evaluate a novel sub-harmonic mixed Fourier series. This new series demonstrates enhanced performance and versatility in approximating wide-band or pass-band functions compared to the quasi-Bernoulli series. It is worth noting that the Maliev–Lanczos approach presents a set of continuity-based constraints that can be applied to any series complementing the Fourier series. Moreover, the conditions for achieving accelerated convergence can be readily obtained using the residual error framework.

The second category focuses on utilizing case studies as canonical examples to inspire and encourage non-specialists to apply the Maliev–Lanczos approach to real-life problems. The contributions pertaining to this category are listed below:

1. We discuss several examples of common smooth functions whose approximations using polynomials and trigonometric series exhibit several well-known adverse phenomena, such as the Gibbs phenomenon, Runge's phenomenon, spectral leakage, and non-convergence by a non-analytic point or a limited region of convergence (using the Taylor series), which are successfully represented by the mixed Fourier series. The results demonstrate the potential of the Maliev–Lanczos approach in the approximation of the usual smooth functions in applied problems, even outperforming, in several scenarios, the Taylor series, orthogonal polynomials, and Chebyshev polynomials using nonuniform sampling.

2. We illustrate the application of the mixed Fourier series with linear operators. In particular, we solve a common direct problem in applied mathematics (Numerical Riemann Integration) and a common inverse problem in fluid dynamics (Poisson's equation). In both examples, we show the benefit of employing simple polynomials, and we illustrate fast convergence without the Gibbs phenomenon.

3. We evaluate the use of a mixed evaluation (i.e., a combination of closed-form derivatives and the DFT approach) to find the mixed Fourier series of functions without closed-form Fourier coefficients. In that case, we show that the DFT reflects the property $O(|k|^{-M})$ for smooth functions, which allows accelerated discrete Fourier processing. Therefore, this approach has a huge potential for a wide range of practical situations.

4. Finally, we show in detail the methodology used to define a new mixed Fourier series using the residual error framework. Additionally, the versatility of this new series is demonstrated through several examples.

The rest of this article is divided as follows. Section 2 develops the continuous-time theory, and Section 3 discusses several continuous-time examples and applications. Finally, the last sections present open challenges and future work (Section 4), followed by conclusions (Section 5).

2. Continuous-Time Theory

2.1. Fourier Series Fundamentals

Let $g:[0, T] \to \mathbb{R}$ be a function with compact support. Traditionally, the partial Fourier series of $g(t)$ is given by [3], p. 62

$$S_N\{g; t\} := \sum_{k=-N}^{N} G_k \cdot e^{2\pi i \cdot t \cdot k f_0}, \tag{1}$$

where $f_0 = \frac{1}{T}$, and the Fourier coefficients are

$$G_k := \frac{1}{T} \cdot \int_0^T g(t) \cdot e^{-2\pi i \cdot t \cdot k f_0} dt, \forall k \in \mathbb{Z}. \tag{2}$$

This paper makes the assumption that $g(t)$ is a well-behaved (i.e., not pathological [107]) function such that $\lim_{N \to \infty} S_N\{g; t\} \to g(t), \forall t \in [0, T]$ using a well-defined concept of convergence, such as point-wise, uniform, or based on the 2-norm (i.e., $\|\cdot\|_2$) [1,2]. Moreover, the periodic equivalent function (denoted by $\bar{g} : \mathbb{R} \to \mathbb{R}$) is defined from the partial Fourier series by

$$\bar{g}(t) := \lim_{N \to \infty} S_N\{g; t\}, \forall t \in \mathbb{R}. \tag{3}$$

Definition 1. *Let $C^0[0, T]$ be the set of continuous functions on $[0, T]$. Let $C^K[0, T]$ be the set of continuous functions with K-times continuously differential properties on $[0, T]$, where we use only the right-hand derivative definition at $t = 0$ and the left-hand derivative definition at $t = T$.*

Even though $g(t)$ has a smooth property, defined by $g \in C^K[0, T]$, the Fourier series could be inefficient for analysis or synthesis applications because the periodic equivalent function could have discontinuities caused by the edges of the interval (e.g., $g(0^+) \neq g(T^-) \Rightarrow \bar{g}(mT^+) \neq \bar{g}(mT^-), \forall m \in \mathbb{Z}$). As a result, the periodic equivalent function is usually inconvenient for a Fourier representation because it lacks a smooth property (i.e., $\bar{g} \notin C^K(\mathbb{R})$).

2.2. Mixed Fourier Series

The disadvantages of the Fourier series representation could be solved using the linear combination given by

$$g(t) := P_M\{g; t\} + r(t), \forall t \in [0, T], \tag{4}$$

where the function $r(t)$ means the residual error between $g(t)$ and an arbitrary partial series $P_M\{g; t\}$. If a convenient form of $P_M\{g; t\}$ is chosen, then an equivalent periodic residual error (i.e., $\bar{r} : \mathbb{R} \to \mathbb{R}$) with suitable properties for a Fourier series representation can be obtained. Therefore, as part of the method, we design $P_M\{g; t\}$ such that $\bar{r} \in C^M(\mathbb{R})$ for some $0 \leq M \leq K$. As a result, the partial mixed Fourier series, defined by

$$g_{N,M}(t) := P_M\{g; t\} + R_{N,M}\{r; t\}, \forall t \in [0, T] \tag{5}$$

$$= P_M\{g; t\} + \sum_{k=-N}^{N} R_{k,M}^g \cdot e^{2\pi i \cdot t \cdot k f_0}, \forall t \in [0, T], \tag{6}$$

has greater potential for processing applications because the partial Fourier series $R_{N,M}\{r; t\}$ avoids the Gibbs phenomenon with a better decreasing rate of their Fourier coefficients, where

$$R_{k,M}^g = \frac{1}{T} \cdot \int\limits_0^T \{g(t) - P_M\{g;t\}\} \cdot e^{-2\pi i \cdot t \cdot k f_0} dt. \tag{7}$$

The new Fourier coefficient (i.e., $R_{k,M}^g$) is the kth harmonic of the residual error formed between the original function $g(t)$ and the hypothesis $P_M\{g;t\}$.

Lastly, in this paper, in order to simplify the results, we study a simple polynomial series,

$$P_M\{g;t\} := \sum_{m=1}^{M+1} P_{m,M}^g \cdot \left(\frac{t}{T}\right)^m, \forall t \in [0,T] \tag{8}$$

with coefficients $P_{m,M}^g$, $M \in \mathbb{N} \cup \{0\}$ and $0 \leq M \leq K$. This polynomial series is mostly equivalent to the one proposed by Roache in [104], where we utilize an arbitrary compact interval $[0,T]$ and we avoid the use of the coefficient $P_{0,M}^g$.

The mixed Fourier series defined in (5) is a combination of functions without a Fourier series representation (i.e., polynomial series are not mandatory in $P_M\{g;t\}$) and standard trigonometric functions (i.e., using kf_0-harmonics in $R_{N,M}\{r;t\}$) with constants $P_{m,M}^g$ and $R_{k,M}^g$, respectively. As we prove with the theory, and we show with several study cases, the mixed Fourier series representation can substantially enhance the processing of $g(t)$ with low-cost of implementation and storage.

2.3. Polynomial Coefficients in Closed Form

To find the general coefficients of $P_M\{g;t\}$, we first study the methodology for the cases $M \in \{0,1,2\}$.

2.3.1. Case $M = 0$

Using (4) and (8) with $M = 0$, we obtain the edges

$$g(0) = r(0), \tag{9}$$
$$g(T) = P_{1,0}^g + r(T). \tag{10}$$

If $\tilde{r} \in C^0(\mathbb{R})$, then $r(T) = r(0)$. As a result, solving (9) and (10), we obtain

$$P_{1,0}^g = g(T) - g(0). \tag{11}$$

In particular, the trivial case $g(T) = g(0)$ has the trivial representation $P_{1,0}^g = 0$. Finally, we approximate $g(t)$ by means of $g_{N,0}(t)$ using (5)–(8).

2.3.2. Case $M = 1$

Using (4) and (8) with $M = 1$, we obtain the edges

$$g(0) = r(0), \tag{12}$$
$$g(T) = P_{1,1}^g + P_{2,1}^g + r(T). \tag{13}$$

Let $g^{(m)}(t)$ be the mth derivative of $g(t)$, or $\frac{d^m}{dt^m} g(t)$, where $g^{(0)}(t) := g(t)$. If $g, r \in C^1[0,T]$, then we obtain

$$g^{(1)}(t) = \frac{1}{T} P_{1,1}^g + \frac{2}{T^2} P_{2,1}^g \cdot t + r^{(1)}(t), \forall t \in [0,T] \tag{14}$$

with the edges

$$g^{(1)}(0) = \frac{1}{T}P^g_{1,1} + r^{(1)}(0), \tag{15}$$

$$g^{(1)}(T) = \frac{1}{T}P^g_{1,1} + \frac{2}{T}P^g_{2,1} + r^{(1)}(T). \tag{16}$$

If $\bar{r} \in \mathcal{C}^1(\mathbb{R})$, then $r^{(1)}(T) = r^{(1)}(0)$ and $r(T) = r(0)$. Solving (15)–(16) and (12)–(13), we obtain

$$P^g_{2,1} = \frac{T}{2}\{g^{(1)}(T) - g^{(1)}(0)\}, \tag{17}$$

$$P^g_{1,1} = \{g(T) - g(0)\} - P^g_{2,1}. \tag{18}$$

We want to note that the coefficients of $M = 0$ are the same coefficients of $M = 1$ with $P^g_{2,1} = 0$. Finally, we approximate $g(t)$ through $g_{N,1}(t)$ using (5)–(8).

2.3.3. Case $M = 2$

Using (4) and (8) with $M = 2$, and $g, r \in \mathcal{C}^2[0, T]$, we obtain the edges

$$g(0) = r(0), \tag{19}$$

$$g(T) = P^g_{1,2} + P^g_{2,2} + P^g_{3,2} + r(T), \tag{20}$$

$$g^{(1)}(0) = \frac{1}{T}P^g_{1,2} + r^{(1)}(0), \tag{21}$$

$$g^{(1)}(T) = \frac{1}{T}P^g_{1,2} + 2\frac{1}{T}P^g_{2,2} + 3\frac{1}{T}P^g_{3,2} + r^{(1)}(T), \tag{22}$$

$$g^{(2)}(0) = 2\frac{1}{T^2}P^g_{2,2} + r^{(2)}(0), \tag{23}$$

$$g^{(2)}(T) = 2\frac{1}{T^2}P^g_{2,2} + 6\frac{1}{T^2}P^g_{3,2} + r^{(2)}(T). \tag{24}$$

Solving (23)–(24), (21)–(22), and (19)–(20) with $\bar{r} \in \mathcal{C}^2(\mathbb{R})$, we obtain

$$P^g_{3,2} = \frac{T^2}{6}\{g^{(2)}(T) - g^{(2)}(0)\}, \tag{25}$$

$$P^g_{2,2} = \frac{T}{2}\{g^{(1)}(T) - g^{(1)}(0)\} - \frac{3}{2}P_{3,2}, \tag{26}$$

$$P^g_{1,2} = \{g(T) - g(0)\} - P_{2,2} - P_{3,2}. \tag{27}$$

Again, we want to note that the coefficients of $M = 1$ are the same coefficients of $M = 2$ with $P^g_{3,2} = 0$. Finally, we approximate $g(t)$ through $g_{N,2}(t)$ using (5)–(8).

2.3.4. General Case: Arbitrary $M \in \mathbb{N} + \{0\}$ Such That $M \le K$

If $g \in C^K[0, T]$, then the polynomial coefficients $P^g_{m,M}$ in closed form can be determined from (8) by the property

$$g^{(k)}(t) = r^{(k)}(t) + \frac{1}{T^k} \cdot \sum_{m=k}^{M+1} \alpha_{k+1,m} \cdot P^g_{m,M} \cdot \left(\frac{t}{T}\right)^{m-k}, \forall k \in \{M, M-1, \cdots, 1\}, \forall t \in [0, T], \tag{28}$$

where

$$\alpha_{k,m} = \frac{m!}{(m - k + 1)!}, \forall 1 \le k \le m. \tag{29}$$

If we design $P_M\{g; t\}$ such that $\bar{r} \in C^M(\mathbb{R})$, then the boundaries

$$r^{(k)}(T) = r^{(k)}(0) \tag{30}$$

are mandatory for any $0 \leq k \leq M \leq K$. From those boundaries and (28), the unknown constants can be easily obtained using the backward algorithm derived from

$$P^g_{M+1,M} = \frac{1}{(M+1)!} F^g_{M'}$$
(31)

$$P^g_{k,M} = \frac{1}{k!} F^g_{k-1} - \frac{1}{k!} \cdot \sum_{m=k+1}^{M+1} \alpha_{k,m} \cdot P^g_{m,M'} \; \forall k \in \{M, M-1, \cdots, 1\},$$
(32)

where

$$F^g_k := T^k \cdot \{g^{(k)}(T) - g^{(k)}(0)\}.$$
(33)

Corollary 1. *If* $g^{(M)}(T) = g^{(M)}(0)$, *then* $P^g_{M+1,M} = 0$ *and* $P^g_{k,M} = P^g_{k,M-1}, \forall k = \{M, \cdots, 1\}$.

Proof. Trivial from (31)–(33). □

2.4. Fourier Coefficients in Closed Form

We have two ways to determine $R^g_{k,M}$ in closed form. With that objective, we first present the following lemmas.

Lemma 1. *If* $h_0(t) := 1$ *and* $h_m(t) := \frac{1}{T^m} t^m, \forall m \in \mathbb{N}$, *then*

$$H_{k,m} := \frac{1}{T} \int_0^T h_m(t) \cdot e^{-2\pi i \cdot t \cdot k f_0} dt = -\frac{1}{2\pi i \cdot k} + \frac{m \cdot H_{k,m-1}}{2\pi i \cdot k}$$
(34)

for $\forall m \in \mathbb{N}$ *and* $\forall k \in \mathbb{Z} - \{0\}$, *where* $H_{k,0} = 0, \forall k \in \mathbb{Z} - \{0\}$.

Proof. First, we note that

$$H_{k,0} = \frac{1}{T} \int_0^T h_0(t) \cdot e^{-2\pi i \cdot t \cdot k f_0} dt = 0, \forall k \in \mathbb{Z} - \{0\}$$

and $\frac{d}{dt} h_m(t) = \frac{m}{T^m} t^{m-1} = \frac{m}{T} \frac{1}{T^{m-1}} t^{m-1} = \frac{m}{T} \cdot h_{m-1}(t), \forall m \in \mathbb{N}$. If we use integration by parts in $h_m(t)$, then

$$\begin{aligned}
H_{k,m} &= \frac{1}{T} \int_0^T h_m(t) \cdot e^{-2\pi i \cdot t \cdot k f_0} dt \\
&= \frac{1}{T} \frac{e^{-2\pi i \cdot t \cdot k f_0}}{(-2\pi i \cdot k f_0)} h_m(t) \Big|_{t=0}^{t=T} - \frac{1}{T} \int_0^T \frac{d}{dt} h_m(t) \frac{e^{-2\pi i \cdot t \cdot k f_0}}{-2\pi i \cdot k f_0} dt \\
&= -\frac{1}{2\pi i \cdot k} + \frac{m}{2\pi i \cdot k f_0} \cdot \frac{1}{T^2} \cdot \int_0^T h_{m-1}(t) e^{-2\pi i \cdot t \cdot k f_0} dt \\
&= -\frac{1}{2\pi i \cdot k} + \frac{m \cdot H_{k,m-1}}{2\pi i \cdot k}, \forall k \in \mathbb{Z} - \{0\}.
\end{aligned}$$

□

Lemma 2. *If* $h_m(t) = \frac{1}{T^m} t^m, \forall m \in \mathbb{N}$, *then*

$$H_{k,m} = -\sum_{n=1}^m \frac{m!}{(m-n+1)!} \frac{1}{(2\pi i \cdot k)^n}, \forall k \in \mathbb{Z} - \{0\}.$$
(35)

Proof. If we use (34) with $m = 1$, then

$$H_{k,1} = -\frac{1}{2\pi i \cdot k} + \frac{1}{2\pi i \cdot k} H_{k,0} = -\frac{1}{2\pi i \cdot k}, \forall k \in \mathbb{Z} - \{0\}.$$

For the general case $m \geq 2$, we have

$$
\begin{aligned}
H_{k,m} &= -\frac{1}{2\pi i \cdot k} + \frac{m}{2\pi i \cdot k} H_{k,m-1} \\
&= -\frac{1}{2\pi i \cdot k} + \frac{m}{2\pi i \cdot k}\left\{-\frac{1}{2\pi i \cdot k} + \frac{m-1}{2\pi i \cdot k} H_{k,m-2}\right\} \\
&\;\;\vdots \\
&= -\frac{1}{2\pi i \cdot k} - \frac{m}{(2\pi i \cdot k)^2} - \frac{m(m-1)}{(2\pi i \cdot k)^3} - \cdots - \frac{m!}{(2\pi i \cdot k)^m}
\end{aligned}
$$

for $\forall k \in \mathbb{Z} - \{0\}$. $\quad\square$

Therefore, using (7) and Lemma 2, we obtain $R_{k,M}^g$ by means of

$$
R_{k,M}^g = G_k - \sum_{m=1}^{M+1} P_{m,M}^g \cdot H_{k,m}, \forall k \in \mathbb{Z} - \{0\}. \tag{36}
$$

This equation has the simplification of Corollary 2, which is very useful for low-order values of M.

Corollary 2 (First Method). *If we use (4) and (8) such that $\bar{r} \in C^M(\mathbb{R})$ for $M \in \mathbb{N} \cup \{0\}$, then their Fourier coefficients are*

$$
R_{k,M}^g = G_k + \sum_{m=1}^{M+1} \frac{F_{m-1}^g}{(2\pi i \cdot k)^m}, \forall k \in \mathbb{Z} - \{0\}, M \geq 0 \tag{37}
$$

$$
= R_{k,M-1}^g + f_0^{-M} \cdot \frac{\{g^{(M)}(T) - g^{M)}(0)\}}{(2\pi i \cdot k)^{M+1}}, \forall k \in \mathbb{Z} - \{0\}, M \geq 1. \tag{38}
$$

Proof. We obtain (37) by replacing (31)–(32) in (36). The simplification is straightforward from its definition. $\quad\square$

For instance, for $M \in \{0, 1, 2\}$, we obtain the following Fourier coefficients:

$$
R_{k,0}^g = G_k + \frac{\{g(T) - g(0)\}}{2\pi i \cdot k}, \forall k \in \mathbb{Z} - \{0\}, \tag{39}
$$

$$
R_{k,1}^g = R_{k,0}^g + f_0^{-1} \cdot \frac{\{g^{(1)}(T) - g^{(1)}(0)\}}{(2\pi i \cdot k)^2}, \forall k \in \mathbb{Z} - \{0\}, \tag{40}
$$

$$
R_{k,2}^g = R_{k,1}^g + f_0^{-2} \cdot \frac{\{g^{(2)}(T) - g^{(2)}(0)\}}{(2\pi i \cdot k)^3}, \forall k \in \mathbb{Z} - \{0\}. \tag{41}
$$

On the other hand, if $g^{(M)}(t)$ and their Fourier coefficients are easy to calculate, then we can use the simplification of Lemma 3 and Corollary 3.

Lemma 3. *If $\bar{r} \in C^M(\mathbb{R})$ for $M \in \mathbb{N} \cup \{0\}$, then their Fourier coefficients for $\forall k \in \mathbb{Z} - \{0\}$ are*

$$
R_{k,M}^g = \frac{1}{(2\pi i \cdot k f_0)^M} \cdot \frac{1}{T} \int_0^T r^{(M)}(t) \cdot e^{-2\pi i \cdot t \cdot k f_0} dt. \tag{42}
$$

Proof. For more detail, see [108–111]. $\quad\square$

Corollary 3 (Second Method). *If we use (4) and (8) such that $\bar{r} \in \mathcal{C}^M(\mathbb{R})$ for $M \in \mathbb{N} \cup \{0\}$, then their Fourier coefficients are*

$$R_{k,M}^g = \frac{f_0^{-M}}{(2\pi i \cdot k)^M}\left(D_{k,M}^g + \frac{\{g^{(M)}(T) - g^{(M)}(0)\}}{2\pi i \cdot k}\right) \forall k \in \mathbb{Z} - \{0\}, \tag{43}$$

where

$$D_{k,M}^g = \frac{1}{T} \cdot \int_0^T g^{(M)}(t) \cdot e^{-2\pi i \cdot t \cdot k f_0} dt, \forall k \in \mathbb{Z} - \{0\}. \tag{44}$$

Proof. For $\bar{r} \in \mathcal{C}^0(\mathbb{R})$, we obtain the same result as in (39) by

$$R_{k,0}^g = D_{k,0}^g + \frac{\{g(T) - g(0)\}}{2\pi i \cdot k}, \forall k \in \mathbb{Z} - \{0\}$$

$$= G_k + \frac{\{g(T) - g(0)\}}{2\pi i \cdot k}, \forall k \in \mathbb{Z} - \{0\}.$$

For $\bar{r} \in \mathcal{C}^M(\mathbb{R})$ such that $M \in \mathbb{N}$, we obtain the expression using (4), (42), and (44) in

$$R_{k,M}^g = \frac{1}{(2\pi i \cdot k f_0)^M}\left(D_{k,M}^g - \frac{1}{T}\int_0^T \frac{d^M}{dt^M}P_M\{g,t\} \cdot e^{-2\pi i \cdot t \cdot k f_0} dt\right)$$

$$= \frac{f_0^{-M}}{(2\pi i \cdot k)^M}\left(D_{k,M}^g - \frac{1}{T^M}F_M^g \cdot H_{k,1}\right), \forall k \in \mathbb{Z} - \{0\}$$

$$= \frac{f_0^{-M}}{(2\pi i \cdot k)^M}\left(D_{k,M}^g + \frac{\{g^{(M)}(T) - g^{(M)}(0)\}}{2\pi i \cdot k}\right), \forall k \in \mathbb{Z} - \{0\}.$$

$\square$

Finally, we obtain the case $k = 0$ by definition:

$$R_{0,M}^g := \frac{1}{T} \cdot \int_0^T r(t)dt = \frac{1}{T} \cdot \int_0^T g(t)dt - \frac{1}{T} \cdot \int_0^T P_M\{g;t\}dt$$

$$= G_0 - \sum_{m=1}^{M+1} \frac{1}{m+1} \cdot P_{m,M}^g. \tag{45}$$

2.5. Enhanced Continuous-Time Processing

The potential of the mixed Fourier series approach is supported by the following well-known theorem [108–111].

Theorem 1. *If $\bar{r} \in \mathcal{C}^M(\mathbb{R})$ for $M \in \mathbb{N} \cup \{0\}$, then their Fourier coefficients are bounded for $\forall k \in \mathbb{Z} - \{0\}$ by*

$$\left|R_{k,M}^g\right| \leq D_{max} \cdot |k|^{-M}, \tag{46}$$

where

$$D_{max} = \frac{1}{(2\pi f_0)^M} \cdot \sup_{\forall t \in [0,T]}\left|r^{(M)}(t)\right|. \tag{47}$$

Proof. If $\bar{r} \in \mathcal{C}^M(\mathbb{R})$ for $M \in \mathbb{N} \cup \{0\}$, then we obtain the inequality using (42) for $\forall k \in \mathbb{Z} - \{0\}$ by

$$\left| R_{k,M}^g \right| = \frac{1}{|k|^M} \cdot \frac{1}{(2\pi f_0)^M} \cdot \frac{1}{T} \left| \int_0^T r^{(M)}(t) \cdot e^{-2\pi i \cdot t \cdot k f_0} dt \right|$$

$$\leq \frac{1}{|k|^M} \cdot \frac{1}{(2\pi f_0)^M} \cdot \frac{1}{T} \int_0^T \left| r^{(M)}(t) \right| dt$$

$$\leq \frac{1}{|k|^M} \cdot \frac{1}{(2\pi f_0)^M} \cdot \sup_{\forall t \in [0,T]} \left| r^{(M)}(t) \right| = D_{max} \cdot |k|^{-M}.$$

$\square$

Therefore, if we design (4) such that $\bar{r} \in C^M(\mathbb{R})$, then the two major drawbacks of the partial Fourier series of $g \in C^K[0,T]$ are solved. First, $R_{N,M}\{r;t\}$ does not have the Gibbs phenomenon because $\bar{r} \in C^0(\mathbb{R})$. Second, the decreasing rates of their coefficients are controlled toward $O(|k|^{-M})$ because D_{max} is bounded by the Boundedness Theorem.

Moreover, the mixed Fourier series allows the use of the linear property because both summations are linear, as summarized in the following corollary.

Corollary 4 (Superposition Property). *Let $v \in C^K[0,T]$ be a function with mixed Fourier series given by $P_{k,M}^v$ and $R_{k,M}^v$. Let $w \in C^K[0,T]$ be a function with mixed Fourier series given by $P_{k,M}^w$ and $R_{k,M}^w$. Consequently, $g \in C^K[0,T]$ obtained by $g(t) = \alpha \cdot v(t) + \beta \cdot w(t)$ has mixed Fourier series given by $P_{k,M}^g = \alpha \cdot P_{k,M}^v + \beta \cdot P_{k,M}^w$ and $R_{k,M}^g = \alpha \cdot R_{k,M}^v + \beta \cdot R_{k,M}^w$.*

2.6. Relation with the Maliev–Lanczos Approach

In [71], p. 86, A.S. Maliev proposed using a Fourier extension method through $g_e : [-\pi, \pi] \to \mathbb{R}$ to enhance the processing of $g : [0, \pi] \to \mathbb{R}$, where $g_e(t), \forall t \in [-\pi, 0]$ is represented by polynomials based on its continuity properties utilizing the edge information $g^{(m)}(0)$ and $g^{(m)}(\pi)$. Although we acknowledge this fundamental concept, we do not delve into that idea in this paper because it involves artificially increasing the domain (i.e., it can be a major issue for some applications), and it implies increasing the complexity of Fourier estimation (i.e., the Fourier approach goes from fundamental period T to fundamental period $2T$, which implies an increase in the frequency resolution from f_0 to $\frac{1}{2}f_0$). However, it should be noted that Maliev's approach can improve the performance of (5) at the expense of doubling the number of unknown variables (or doubling the number of samples) for a fixed bandwidth.

On the other hand, in [72], p. 98, C. Lanczos simplified Maliev's works by using quasi-Bernoulli polynomials without changing the domain. This approach for a partial series, using the Lanczos nomenclature, is defined by $g, h_p : [-1, 1] \to \mathbb{R}$ such that

$$h_p(t) \approx g(t) - \frac{1}{2} \sum_{m=0}^{p} \{ g^{(m)}(1) - g^{(m)}(-1) \} \cdot B_{m+1}(t) - \frac{1}{2} \int_{-1}^{1} g(t) dt, \tag{48}$$

where

$$B_{m+1}(t) = \frac{t^m}{m!} - b_2 \frac{t^{m-2}}{(m-2)!} + b_4 \frac{t^{m-4}}{(m-4)!} - \cdots \tag{49}$$

and

$$\frac{2t}{e^t - e^{-t}} = 1 - b_2 t^2 + b_4 t^4 - b_6 t^6 + \cdots. \tag{50}$$

The function $B_{m+1}(t)$ is closely related to Bernoulli polynomials ([72], p. 106 and p. 109), and it takes, for example, the following values: $B_1(t) = t$, $B_2(t) = \frac{1}{2}t^2 - \frac{1}{6}$,

$B_3(t) = \frac{1}{6}t^3 - \frac{1}{6}t$, $B_4(t) = \frac{1}{24}t^4 - \frac{1}{12}t^2 + \frac{7}{360}$. From Lanczos's works, it is easy to deduce that the Fourier coefficients of (48) are

$$H_{k,p} = \begin{cases} 0 & ,k=0 \\ G_k + \frac{1}{2} \cdot \sum_{m=0}^{p} \{g^{(m)}(1) - g^{(m)}(-1)\} \cdot \frac{(-1)^k}{(i\pi k)^{m+1}} & ,\forall k \in \mathbb{Z} - \{0\} \end{cases}. \tag{51}$$

Conclusively, we find that $R_{k,M}^{g}$ has the same Fourier coefficients $H_{k,p}$ for $\forall k \in \mathbb{Z} - \{0\}$ when we replace the function $B_{m+1}(t)$ in (48) by $\frac{T^m}{2^m} \cdot B_{m+1}(\frac{2}{T} \cdot t - 1)$, defined in a new domain $t \in [0, T]$, with new boundaries $g^{(m)}(T)$, $g^{(m)}(0)$, and $p = M$. Therefore, the simple polynomial series simplifies (48) using a different perspective based on the residual error framework. Because consecutive derivatives or integrals are easy to evaluate using simple polynomials, our result facilitates the application of $g_{N,M}(t)$ through many linear operators. Because we intend to use $M \ll N$, the numerical values of $P_{m,M}^{g}$ calculated by the backward algorithm derived from (31)–(33) have a low computational cost.

2.7. A Simple Reprojection Method: Using Standard Closed-Form Fourier Coefficients to Define a Mixed Fourier Series

Because the Fourier series is widely used in signal theory, determining a mixed Fourier series (i.e., $P_{m,M}^{g}$ and $R_{k,M}^{g}$) from standard closed-form Fourier coefficients (i.e., G_k) may be necessary in some cases, either to avoid the Gibbs phenomenon or to improve convergence for a fixed number of harmonics. In this subsection, we briefly discuss that methodology for $M = 0$.

If $g \in C^0[0, T]$, then

$$g_{N,0}(t) = F_0^g \cdot \left(\frac{t}{T}\right) + \sum_{k=-N, k\neq 0}^{N} \{G_k + \frac{F_0^g}{2\pi i \cdot k}\} \cdot e^{2\pi i \cdot t \cdot k f_0} + R_{0,0}^g \tag{52}$$

may exist for $\forall t \in [0, T]$. As a result of the continuity property, the approximation $g_{N,0}(t)$ must have the property $g_{N,0}(\Delta) \approx g_{N,0}(0)$ for a small and convenient value Δ. Therefore, we find

$$F_0^g \approx \frac{\sum_{k=-N, k\neq 0}^{N} G_k \cdot \{1 - e^{2\pi i \cdot \Delta \cdot k f_0}\}}{\left(\frac{\Delta}{T}\right) - \sum_{k=-N, k\neq 0}^{N} \frac{1}{2\pi i \cdot k} \{1 - e^{2\pi i \cdot \Delta \cdot k f_0}\}} \tag{53}$$

from (52). Because the approximation $g_{N,0}(t)$ has the bandwidth Nf_0, we can select any $0 < \Delta \leq \frac{1}{4} \frac{1}{Nf_0}$ in order to model the discontinuity without ringing artifacts.

Example 1. *Let $g:[0,2] \to \mathbb{R}$ be a test function with closed-form Fourier coefficients given by*

$$G_k = \frac{-16\,i\pi k^3 - 16k^2 + 36\,i\pi k - 36}{\pi^2(16k^4 - 72k^2 + 81)}, \forall k \in \mathbb{Z}. \tag{54}$$

Figure 1 shows $g_{N,0}(t)$ using (53) with $\Delta = \frac{1}{4}\frac{1}{Nf_0}$ and $\Delta = \frac{1}{16}\frac{1}{Nf_0}$, where, by reference, the test function is $g(t) = t \cdot \cos(\frac{3}{2} \cdot \pi \cdot t), \forall t \in [0, 2]$. As the figure makes clear, the approximation of F_0^g allows the recovery of convergence $O(|k|^{-2})$ for $R_{k,0}^g$ from G_k without ambiguities. As a result, the Gibbs phenomenon is removed.

In the other cases (i.e., $M \geq 1$), we can repeat a similar procedure using the partial Maclaurin Series of $g_{N,M}^{(m)}(t) \perp m \in \{0, \cdots, M-1\}$ at $t = \Delta$, and $g_{N,M}^{(M)}(\Delta) \approx g_{N,M}^{(M)}(0)$. For instance, if $M = 1$, then we obtain $g_{N,1}(\Delta) \approx g_{N,1}(0) + \frac{1}{1!}g_{N,1}^{(1)}(0) \cdot \Delta$ and $g_{N,1}^{(1)}(\Delta) \approx g_{N,1}^{(1)}(0)$, which results in a system of two equations that can be solved using standard matrix techniques. In conclusion, this method is a low-complexity alternative to spectral reprojection methods [58–60] for removing the Gibbs phenomenon of $g \in C^K[0, T]$ because the mixed Fourier series has the potential to improve convergence (i.e., not just remove the Gibbs phenomenon) from the original Fourier coefficients using a straightforward and simple procedure.

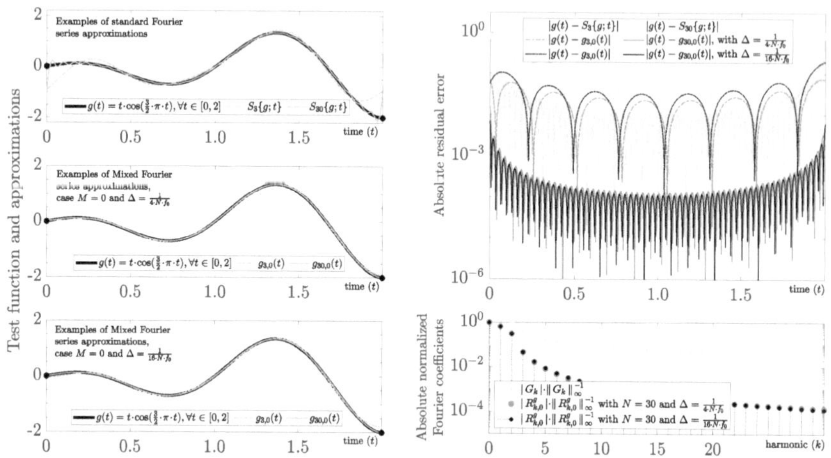

Figure 1. Example of removing Gibbs phenomenon from closed-form Fourier coefficients (G_k) with a bandwidth Nf_0.

3. Continuous-Time Examples and Applications

3.1. A Different Perspective for Convergent Series of Functions

By Weierstrass's approximation theorem ([112], §14.08), a function $h \in \mathcal{C}^0[0, T]$ can be uniformly approximated by polynomials as closely as desired. Because polynomial series are important in signal theory, we link our approach with that perspective as follows.

Definition 2. *If* $g, r, h_m : [0, T] \to \mathbb{R}$, $P_M\{g; t\} := \sum_{m=1}^{M+1} P_{m,M}^g \cdot h_m(t)$ *and* $g(t) = P_M\{g; t\} + r(t)$, *then a predisposed series of* $g \in \mathcal{C}^0[0, T]$ *is* $P_M\{g; t\}$ *such that* $\lim_{M\to\infty} P_M\{g\} \to g$.

Consequently, a predisposed series is a linear combination of functions (e.g., polynomials) designed to have direct convergence toward $g(t)$ (i.e., a series predisposed to converge directly). A complementary interpretation of this definition is obtained by analyzing the periodic functions using harmonic analysis. If $\bar{r}(t)$ is the equivalent periodic residual error resulting from the periodic extension of $g(t) - P_M\{g; t\}$, then a predisposed series is obtained when $P_M\{g; t\}$ allows that $\lim_{M\to\infty} \bar{r}(t) \to 0$ using point-wise or uniform convergence. Naturally, the concept of convergence can be generalized in a weak sense (i.e., using the "almost anywhere" concept) by other norms, such as $\|\cdot\|_2$. Predisposed series, such as sequences of polynomials based on orthogonalization or useful solutions to ordinary differential equations, are difficult to find or build because $\lim_{M\to\infty} \bar{r}(t) \to 0$ is a hard constraint (i.e., difficult to achieve with a potentially slow convergence rate), with relevant challenges in updating their coefficients at a low computational cost.

The Maliev–Lanczos approach implies a different class of convergence. We design $P_M\{g; t\}$, with a reasonably low order, for some particular application and function $g : [0, T] \to \mathbb{R}$, such that $\bar{r} \in \mathcal{C}^M(\mathbb{R})$. Afterward, we find a harmonic approximation for a "nonzero residual error" through $R_{k,M}^g$. Therefore, the convergence proposal is indirect in the sense that $\bar{r}(t) \neq 0$, with several advantages. First, ringing artifacts are removed because we can control the type of convergence; for example:

Corollary 5 (Convergence Everywhere). *If* $g \in \mathcal{C}^1[0, T]$ *and* $P_1\{g; t\}$ *are given by (8) with coefficients (31)–(33), then the Fourier series of* $r(t) = g(t) - P_1\{g; t\}, \forall t \in [0, T]$ *has uniform convergence.*

Proof. In the methodology proposed, we design $P_1\{g;t\}$ such that $\bar{r} \in C^1(\mathbb{R})$. Therefore, $\bar{r}(t)$ has Fourier series with uniform convergence [109,110]. Consequently, $r(t)$ has uniform convergence, too. $\square$

Second, the bandwidth of $\bar{r}(t)$ becomes more compacted for base-band functions because $|R^g_{k,M}|$ is $O(|k|^{-M})$. Conclusively, low-order harmonics of $R_{N,M}\{r;t\}$ will provide good approximations, where its discrete-time signal will have less aliasing. Furthermore, we could use (4) through linear operators without ambiguities because $P_M\{g;t\}$ is well-defined for many linear operators, and $\bar{r} \in C^M(\mathbb{R})$ does not have the Gibbs phenomenon, with a small enough bandwidth for many practical applications.

3.2. Canonical Examples of Approximation Using Closed-Form Smooth Functions

This subsection discusses typical and well-known closed-form smooth functions approximated by $g_{N,M}(t)$ using a reasonably small value of M to clearly explain the methodology and encourage the use of the mixed Fourier series in more complex problems.

3.2.1. Generic Sawtooth Function

We define this function by

$$g(t) = \alpha + \beta \cdot t, \forall t \in [0,T],$$

where $\alpha, \beta \in \mathbb{R}$. The coefficients of $S_N\{g;t\}$ are

$$G_k = \begin{cases} \alpha + \frac{1}{2} \cdot T \cdot \beta & ,k = 0 \\ -\frac{T \cdot \beta}{2\pi i \cdot k} & ,\forall k \in \mathbb{Z} - \{0\} \end{cases}. \tag{55}$$

Using $M = 0$, we obtain $g(0) = \alpha$, $g(T) = \alpha + \beta \cdot T$. Consequently, $P^g_{1,0} = \beta \cdot T$ and

$$P_0\{g;t\} = \beta \cdot t, \forall t \in [0,T], \tag{56}$$

$$R^g_{k,0} = \alpha \cdot \mathrm{sinc}(k) = \begin{cases} \alpha & ,k = 0 \\ 0 & ,\text{elsewhere} \end{cases}. \tag{57}$$

As expected, we obtain the best possible scenario because only two coefficients (i.e., $P^g_{1,0} = \beta \cdot T$ and $R^g_{0,0} = \alpha$) are necessary to model this function without errors. From Corollary 1, we obtain $P_{M \geq 0}\{g;t\} = \beta \cdot t, \forall t \in [0,T]$.

3.2.2. Power Function

We define this function by

$$g(t) = t^m, \forall t \in [0,T],$$

where $m \in \mathbb{N} - \{1\}$. The coefficients of $S_N\{g;t\}$ are

$$G_k = \begin{cases} \frac{1}{m+1} \cdot T^m & ,k = 0 \\ T^m \cdot H_{k,m} & ,\text{elsewhere} \end{cases}. \tag{58}$$

We are interested in this case because many functions can have partial Taylor approximations. As a result, the ability to approximate polynomials with the mixed series can arise as a relevant question. This function produces ringing artifacts in $S_N\{g;t\}$ for higher values of $m \cdot T^{m-1}$ caused by the change in amplitude and slope at edges. In Figure 2, we show the Fourier series for $m = 5$ and $T = 2$ using $N = 3$ and $N = 30$.

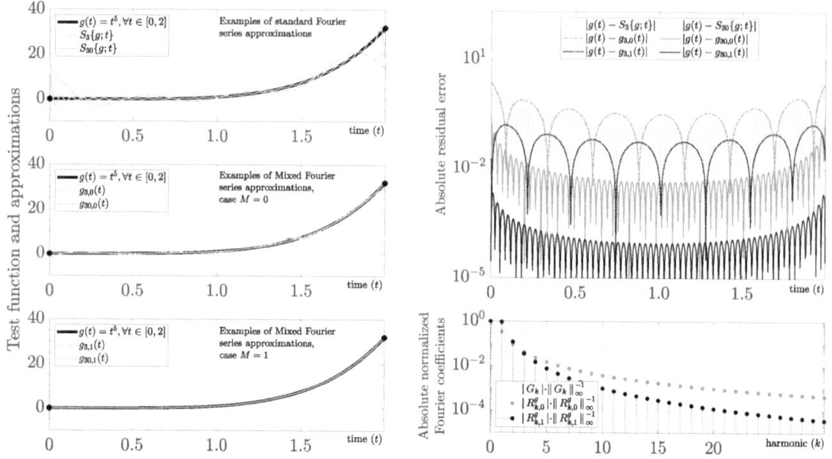

Figure 2. Evaluation of $g(t) = t^5, \forall t \in [0,2]$.

In this case, the polynomial constants (31)–(33) use

$$g^{(k)}(t) = \begin{cases} \frac{m!}{(m-k)!} \cdot t^{m-k} & ,k \le m, \forall t \in [0,T] \\ 0 & ,k > m, \forall t \in [0,T] \end{cases}.$$ (59)

For instance, for $m = 5$ and $T = 2$, we obtain the following polynomials:

$$P_0\{g;t\} = 16t, \forall t \in [0,2],$$ (60)

$$P_1\{g;t\} = 20t^2 - 24t, \forall t \in [0,2],$$ (61)

and using (37) and (45), we obtain the Fourier coefficients

$$R_{k,0} = \begin{cases} -\frac{32}{3} & ,k = 0 \\ 2^5 \cdot H_{k,5} + 2^5 \frac{1}{2\pi i k} & ,\text{elsewhere} \end{cases},$$ (62)

$$R_{k,1} = \begin{cases} \frac{8}{3} & ,k = 0 \\ R_{k,0} - 2^5 \frac{5}{(2\pi k)^2} & ,\text{elsewhere} \end{cases}.$$ (63)

As we show in Figure 2, the approximations $g_{N,0}(t)$ and $g_{N,1}(t)$ do not have the Gibbs phenomenon, and we obtain control of the decreasing rate of the Fourier coefficients. For instance, we find that the 30th harmonic has $|G_{30}| \cdot \|G_k\|_\infty^{-1} \sim 3.1 \times 10^{-2}$ and $|R^g_{30,M}| \cdot \|R^g_{k,M}\|_\infty^{-1} \sim 3.5 \times 10^{-5}$ for $M = 1$ in this example. As a result, the increase in resolution using the mixed Fourier series is nearly cubic (29.47 dB) for that harmonic. If $M \ge m - 1$, then we obtain again the best possible scenario because $P_{M \ge m-1}\{g;t\} = t^m, \forall t \in [0,T]$. Applying superposition, it follows that arbitrary polynomials with degree D have an exact representation when $M \ge D - 1$. We emphasize, however, that by using a low-order value of M in the mixed Fourier series, we can avoid using high-order derivatives.

3.2.3. Exponential Function

We define this function by

$$g(t) = e^{\alpha \cdot t}, \forall t \in [0,T],$$ (64)

where $\alpha \in \mathbb{R} - \{0\}$. The coefficients of $S_N\{g;t\}$ are

$$G_k = \frac{e^{T\alpha} - 1}{T\alpha - 2i\pi k}, \forall k \in \mathbb{Z}. \tag{65}$$

This function is of our interest because it could increase or decrease its values very fast. In Figure 3, we show the Fourier series for $\alpha = -4$ and $T = 2$ using $N = 3$ and $N = 30$.

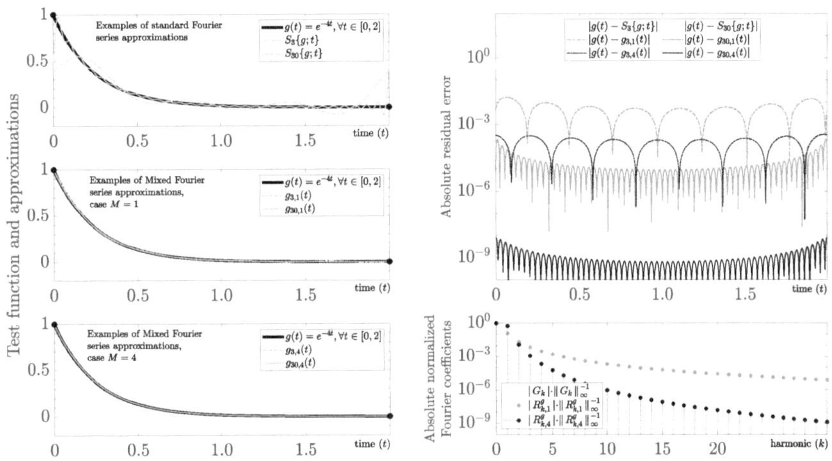

Figure 3. Evaluation of $g(t) = e^{-4t}, \forall t \in [0,2]$.

In this case, the polynomial constants (31)–(33) use

$$g^{(k)}(t) = \alpha^k \cdot e^{\alpha \cdot t}, \forall t \in [0, T]. \tag{66}$$

For instance, for $\alpha = -4$ and $T = 2$, we obtain the following polynomials:

$$P_1\{g;t\} = -\left(e^{-8} - 1\right)\left(t^2 - \frac{5}{2}t\right), \forall t \in [0, 2], \tag{67}$$

$$P_4\{g;t\} = \left(e^{-8} - 1\right)\sum_{m=1}^{5} \phi_m \cdot t^m, \forall t \in [0, 2], \tag{68}$$

where $\phi_1 = \frac{209}{90}$, $\phi_2 = -\frac{31}{3}$, $\phi_3 = \frac{124}{9}$, $\phi_4 = -\frac{20}{3}$, and $\phi_5 = \frac{16}{15}$. On the other hand, the Fourier coefficients $R_{k,M}$ can be calculated efficiently using Corollary 3. In particular, for this example, we find

$$R_{k,1} = \begin{cases} -\frac{31}{24}\left(e^{-8} - 1\right) & ,k = 0 \\ \frac{T\alpha}{(2\pi i \cdot k)}\left(\frac{e^{T\alpha} - 1}{T\alpha - 2i\pi k} + \frac{e^{T\alpha} - 1}{2\pi i \cdot k}\right) & ,\text{elsewhere} \end{cases} \tag{69}$$

$$R_{k,4} = \begin{cases} -\frac{209}{360}\left(e^{-8} - 1\right) & ,k = 0 \\ \frac{T^4\alpha^4}{(2\pi i \cdot k)^4}\left(\frac{e^{T\alpha} - 1}{T\alpha - 2i\pi k} + \frac{e^{T\alpha} - 1}{2\pi i \cdot k}\right) & ,\text{elsewhere} \end{cases} \tag{70}$$

As Figure 3 makes clear, we are able to recover an approximation of this function with $\|g(t) - g_{3,4}(t)\|_\infty \sim 10^{-8}$ using 30 harmonics (i.e., $N = 30$). From a practical point of view, for $M = 4$, we obtain a worst absolute error of $\sim 10^{-3}$ and $\sim 10^{-6}$ using only 3 and 10 harmonics, respectively. For instance, we find that the 30th harmonic has $|G_{30}| \cdot \|G_k\|_\infty^{-1} \sim 4.2 \times 10^{-2}$ and $|R_{30,M}^g| \cdot \|R_{k,M}^g\|_\infty^{-1} \sim 1.2 \times 10^{-9}$ for $M = 4$ in this example. As a result, the increase in resolution using the mixed Fourier series is almost 75.44 dB for that harmonic.

3.2.4. Base-Band Cosine Function

We define this function by

$$g(t) = \cos(2\pi \frac{\beta}{T} \cdot t), \forall t \in [0, T],$$

where $0 < \beta < 1$. The coefficients of $S_N\{g; t\}$ are

$$G_k = \frac{1}{2}e^{-i\pi(k-\beta)}\text{sinc}(k - \beta) + \frac{1}{2}e^{-i\pi(k+\beta)}\text{sinc}(k + \beta). \tag{71}$$

This function is of our interest because it allows us to study the spectrum leakage in trigonometric base-band functions. As Figure 4 makes clear, even though the Fourier series obtains a small Gibbs phenomenon for $\beta = 0.9$ and $T = 2$, the spectrum leakage could be relevant for many practical applications.

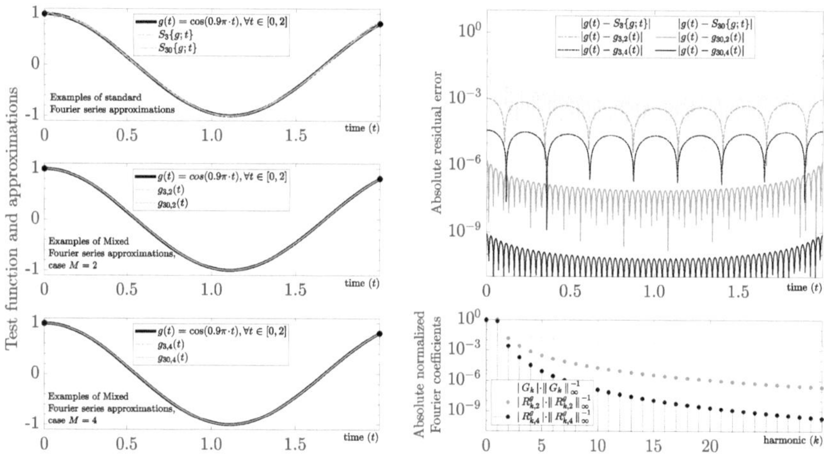

Figure 4. Evaluation of $g(t) = cos(0.9\pi \cdot t), \forall t \in [0, 2]$.

In this case, the polynomial constants (31)–(33) use

$$g^{(k)}(t) = (2\pi\frac{\beta}{T})^k \begin{cases} (-1)^{\frac{k}{2}+\frac{1}{2}} \cdot \sin(2\pi\frac{\beta}{T} \cdot t), \forall k \in \{1, 3, 5, \cdots\}, \forall t \in [0, T] \\ (-1)^{\frac{k}{2}} \cdot \cos(2\pi\frac{\beta}{T} \cdot t) \quad, \forall k \in \{0, 2, 4, \cdots\}, \forall t \in [0, T] \end{cases}' \tag{72}$$

where the Fourier coefficients $R^g_{k,M}$ can be calculated efficiently using Corollary 3. As an example, for $\beta = 0.9$ and $T = 2$, we find the following Fourier coefficients:

$$R^g_{k,2} = \begin{cases} 0.268535523631802 & , k = 0 \\ -\left(\frac{0.9}{ik}\right)^2\left(G_k + \frac{\{\cos(1.8\pi)-1\}}{2\pi ik}\right) & , \text{elsewhere} \end{cases}' \tag{73}$$

$$R^g_{k,4} = \begin{cases} 0.416158284137357 & , k = 0 \\ \left(\frac{0.9}{ik}\right)^4\left(G_k + \frac{\{\cos(1.8\pi)-1\}}{2\pi ik}\right) & , \text{elsewhere} \end{cases}. \tag{74}$$

As expected, Figure 4 shows a reduction in spectral leakage by increasing M. In contrast to the spectral distortion caused by improving amplitude-based frequency discrimination using the Windowing technique, the mixed Fourier series improves amplitude-based frequency discrimination only by increasing M (i.e., without adding artificial distortion). For instance, using a criteria of 10^{-2} in amplitude-based frequency discrimination, $|G_k| \cdot \|G_k\|_\infty^{-1} \geq 10^{-2}$ requires eight harmonics in Figure 4. In contrast, $|R^g_{k,M}| \cdot \|R^g_{k,M}\|_\infty^{-1} \geq 10^{-2}$ requires only three and two harmonics using $M = 2$ and $M = 4$, respectively. Using more selective criteria of 10^{-3} in this example, the standard Fourier approach requires 65 harmonics, and the mixed Fourier series requires 4 and 3 harmonics using $M = 2$ and $M = 4$, respectively. Nonetheless, a redefinition of $P_M\{g;t\}$ is required to apply this technique directly to carrier detection (i.e., $\beta \gg 1$) because $P_M\{g;t\}$ composed only of polynomials is a base-band function.

3.3. Comparison with Selected State-of-the-Art Techniques

In this subsection, we compare the performance of the mixed Fourier series with other types of series in a variety of scenarios using convenient test functions. Although the term "convenient function" is debatable, we define it as a function that is demanding enough for trigonometric and polynomial basis functions on $[0, T]$ and, at the same time, has a simple mathematical structure that allows us to avoid debating its influence on the numerical implementation. For this reason, we start our comparison with the exponential function previously studied in Section 3.2.3.

Figure 5 shows the performance of the most common averaging and filtering techniques for $N = 10$ [27,32], where $F_N\{g;t\}$ is the partial Fejér's series (i.e., Fejér's arithmetic mean method), given by

$$F_N\{g;t\} := \frac{1}{N+1} \sum_{m=0}^{N} S_m\{g;t\}, \tag{75}$$

and $\sigma_{N,M}\{g;t\}$ is a partial Fourier series using a particular σ_l-filter with the Mth order, given by

$$\sigma_{N,M}\{g;t\} := \sum_{k=-N}^{N} G_k \cdot e^{2\pi i \cdot t \cdot k f_0} \cdot \sigma_l^{M+1}\left(k \cdot N^{-1}\right). \tag{76}$$

For example, the standard Lanczos filter (also known as σ-approximation) is given by $\sigma_1(x) = \mathrm{sinc}(x) := \frac{\sin(\pi x)}{\pi x}$, the Raised cosine filter is given by $\sigma_2(x) := \frac{1}{2}\{1 + \cos(\pi x)\}$, and the Sharpened Raised cosine filter is given by $\sigma_3(x) := \sigma_2^4(x) \cdot \{35 - 84 \cdot \sigma_2(x) + 70 \cdot \sigma_2^2(x) - 20 \cdot \sigma_2^3(x)\}$. As can be concluded from a cursory examination of Figure 5, we obtain better performance using $g_{N,0}(t)$. Because the average and convolution operators are smooth operators, methods based on them converge more slowly than $g_{N,M\geq 1}(t)$.

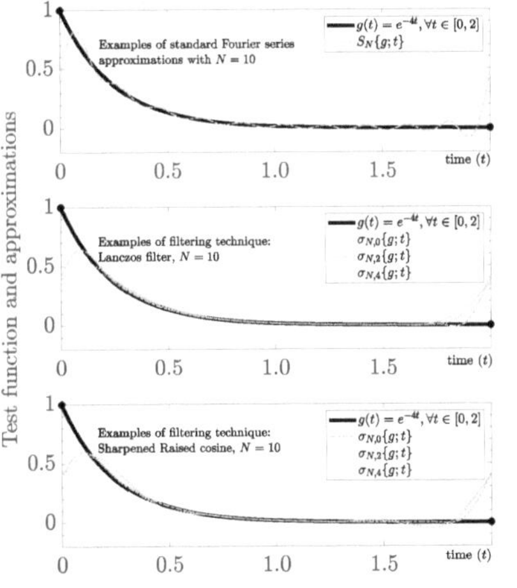
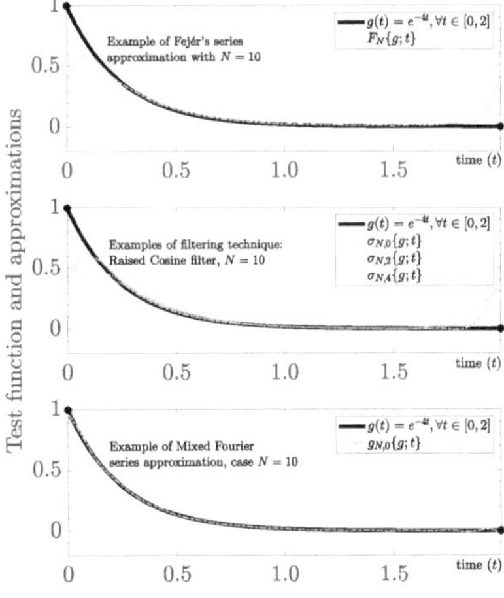

Figure 5. Example of removing Gibbs phenomenon using averaging and filtering techniques and mixed Fourier series for $g(t) = e^{-4t}, \forall t \in [0,2]$.

On the other hand, a comparison between mixed Fourier series and orthogonal polynomials is also pertinent. Figure 6 shows the absolute residual error using the Legendre orthogonal polynomials, denoted by $W_M^{Leg}\{g;t\}$, and the Chebyshev orthogonal polynomials, denoted by $W_M^{Che}\{g;t\}$, both defined on $[0,T]$. As the figure makes clear, both approximations have good performance for $M = 10$, where $\|g(t) - W_M^{Leg}\{g;t\}\|_\infty \simeq 9.2 \times 10^{-6}$ and $\|g(t) - W_M^{Che}\{g;t\}\|_\infty \simeq 3.1 \times 10^{-6}$. We are interested in determining some fair comparatives because the mixed Fourier series has two degrees of freedom (i.e., N and M). For instance, we obtain $\|g(t) - g_{N,10}(t)\|_\infty \leq 2.7 \times 10^{-7}$ for $N \geq 3$, which implies that we can improve both orthogonal polynomials by combining simple polynomials with the same degree and a few harmonics from the residual. As another example, we find $\|g(t) - g_{10,M}(t)\|_\infty \leq 1.6 \times 10^{-6}$ for $M \geq 4$, which implies that the same number of unknown harmonics can also improve both orthogonal polynomials using low-order derivatives from the edges. Finally, we obtain $\|g(t) - g_{6,5}(t)\|_\infty \simeq 2.7 \times 10^{-6}$ such that $\min\{M + N\}$, which implies that the mixed Fourier series has a better performance using at least 19 unknown variables (i.e., $M + 2N + 2$ unknown variables) versus the 11 unknown variables (i.e., $M + 1$ unknown variables) from the orthogonal polynomials. In summary, the mixed Fourier series outperforms orthogonal polynomials in several ways. First, we have an additional degree of freedom that has a significant impact on convergence. Second, we have less computational complexity because the interior product is more simple and computationally efficient using the Fourier approach (i.e., G_k can be defined in terms of an inner product using the same framework of orthogonal polynomials). Third, our approach implies uniform sampling, which simplifies the numerical implementation using the DFT$\{\cdot\}$. Finally, we also find a quasi-spectral accuracy because $|R_{N,M}^g| \leq D_{max} \cdot |N|^{-M}$, where $M \leq K$ and D_{max} is (47).

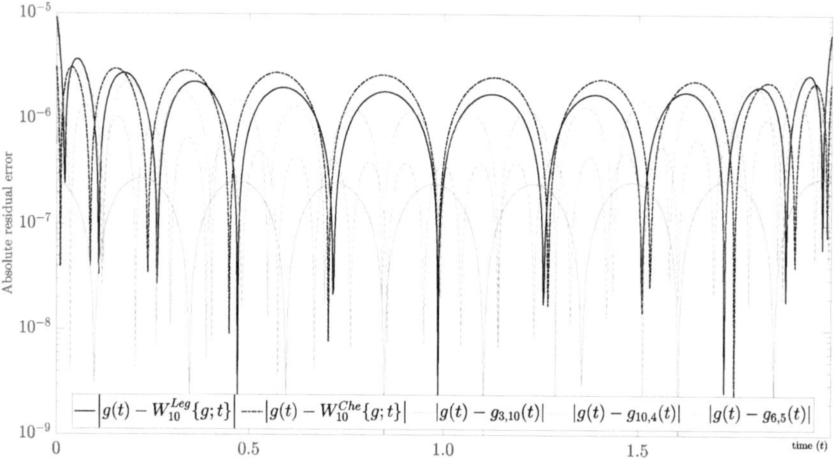

Figure 6. Absolute residual error between $g(t) = e^{-4t}, \forall t \in [0,2]$, and the approximations using orthogonal polynomials and mixed Fourier series.

Another critical situation to discuss is when the Taylor series of $g(t)$ at t_0, defined by $T_M\{g;t;t_0\} := \sum_{k=0}^{M} \frac{1}{k!} g^{(k)}(t_0)(t - t_0)^k$, cannot converge in the whole domain or at some point of the domain (e.g., $g(t)$ is a smooth function but a non-analytic function at some $t_0 \in [0,T]$). For instance, the case $g(t) = \ln(t + 1), \forall t \in [0,2]$, allows us to discuss a typical example where the Taylor series cannot converge in the whole domain using $t_0 = 0$ because its residual has a region of convergence $|t| < 1$, as we show in Figure 7. As happens in this example, the mixed Fourier series can be found by using a mixed evaluation (i.e., a combination of closed-form and numerical evaluations). On the one hand, we can usually find $P^g_{m,M}$ from closed-form derivatives, as shown in this example by

$$g^{(k)}(t) = \begin{cases} \ln(t+1) & k = 0, \forall t \in [0,T] \\ -\dfrac{(-1)^k \cdot (k-1)!}{(t+1)^k} & k \in \mathbb{N}, \forall t \in [0,T] \end{cases} \tag{77}$$

On the other hand, because closed-form Fourier coefficients are relatively uncommon for many well-known functions, we can approximate $G_k \approx \mathcal{T}_N\{\hat{G}_k\}$ and $R^g_{k,M} \approx \mathcal{T}_N\{\hat{R}^g_{k,M}\}$ using the Discrete Fourier Transform (DFT) of $2N+1$ points through

$$\mathcal{T}_N\{\hat{G}_k\} := \frac{1}{2N+1} \cdot \begin{cases} \hat{G}_0 & , k = 0 \\ \hat{G}_k & , \forall k \in [1,N] \\ \hat{G}_{2N+k+1}, & \forall k \in [-N,-1] \end{cases}, \tag{78}$$

where

$$\hat{G}_k = \mathrm{DFT}\{g(t_n)\}^{2N}_{n=0} := \sum_{n=0}^{2N} g(t_n) \cdot e^{-\frac{2\pi}{2N+1} i \cdot kn}, \forall k \in [0,2N] \tag{79}$$

and

$$\hat{R}^g_{k,M} := \mathrm{DFT}\{g(t_n) - P_M\{g; t_n\}\}^{2N}_{n=0} \tag{80}$$

using the uniform samples $t_n = h \cdot n, \forall n \in \{0,1,\cdots,2N\}$, and $h = \frac{T}{2N+1}$. According to the Sampling Theory [113,114], this approach converges by increasing N because the aliasing from the discrete-time model is removed when $N \to \infty$ for $g \in \mathcal{C}^K[0,T]$. In particular, because $\ln(\cdot)$ does not have closed-form Fourier coefficients, we can use (80) to obtain $g_{N,M}(t)$. As shown in Figure 7, the mixed evaluation allows us to obtain a convergent approximation for increasing values of M and N. Similarly, another relevant case study is given by $g(t) = e^{-1/t^2}, \forall t \in (0,2]$ and $g(0) = 0$ because it is a typical smooth function with non-analytic behavior at $t_0 = 0$ (i.e., caused by $g^{(k)}(0) = 0, \forall k \in \mathbb{N} \cup \{0\}$). As we show in Figure 7, we also obtain a convergent mixed Fourier series for increasing values of M and N.

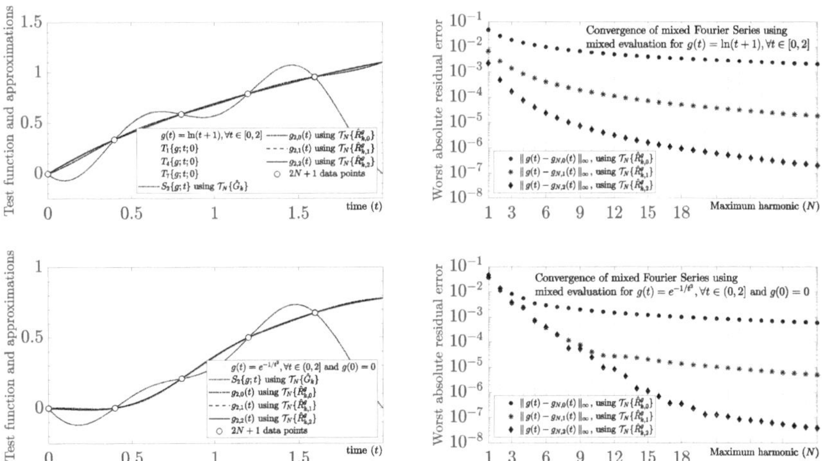

Figure 7. Smooth functions with anomalous Taylor series behavior, and mixed Fourier series using mixed evaluation (i.e., derivatives calculated using the closed form, and Fourier coefficients approximated by the DFT).

Lastly, it is pertinent to evaluate this mixed evaluation with special cases, for example, when the test function exhibits both Gibbs and Runge's phenomena. In particular, we propose the study case $g(t) = \tanh(\alpha \cdot t - \frac{1}{2}\alpha \cdot T), \forall t \in [0, T]$, using $\alpha = 4$ and $T = 2$. As shown in Figure 8, we obtain Runge's phenomenon by using a partial interpolating polynomial series $L_{2N+1}\{g; t\} = \sum_{m=0}^{2N+1} \alpha_m \cdot (\frac{t}{T})^m$ and uniform matching nodes $t_n = h \cdot n, \forall n \in \{0, 1, \cdots, 2N+1\}$. On the other hand, we obtain the Gibbs phenomenon by using a partial Fourier interpolating series $S_N\{g; t\} = \sum_{k=-N}^{N} G_k \cdot e^{2\pi i \cdot t \cdot k f_0}$, where G_k is approximated by $\mathcal{T}_N\{\hat{G}_k\}$ with uniform matching nodes $t_n = h \cdot n, \forall n \in \{0, 1, \cdots, 2N\}$. One well-known solution for this situation is obtained by using nonuniform sampling, for example, with the Chebyshev interpolating function $T_{2N+1}\{g; t\} = \sum_{m=0}^{2N+1} \beta_m \cdot T_m(\frac{2}{T}t - 1)$ and the Chebyshev–Gauss–Lobatto (CGL) matching nodes given by $t_l := \frac{T}{2}\{1 + \cos(\frac{l \cdot \pi}{2N+1})\}$, $\forall l \in \{0, \cdots, 2N+1\}$ [115]. In this paper, we propose a different solution obtained by $g_{N,1}(t)$ using uniform matching nodes $t_n = h \cdot n, \forall n \in \{0, 1, \cdots, 2N\}$, $\mathcal{T}_N\{\hat{R}_{k,1}^g\}$, and

$$P_1\{g; t\} = 2 \cdot \tanh(\frac{1}{2}\alpha \cdot T) \cdot \left(\frac{t}{T}\right), \forall t \in [0, T]. \tag{81}$$

Despite the reduced performance due to the mixed evaluation, Figure 8 shows that $g_{N,1}(t)$ has the best performance without Gibbs and Runge's phenomena, outperforming the Chebyshev interpolating function. For instance, we obtain an absolute residual error around $\|g(t) - g_{7,1}(t)\|_\infty \sim 1.5 \times 10^{-4}$ using seven harmonics (i.e., $N = 7$). If we increase α, then more harmonics (and samples) will be necessary for a good approximation because the test function increases its bandwidth.

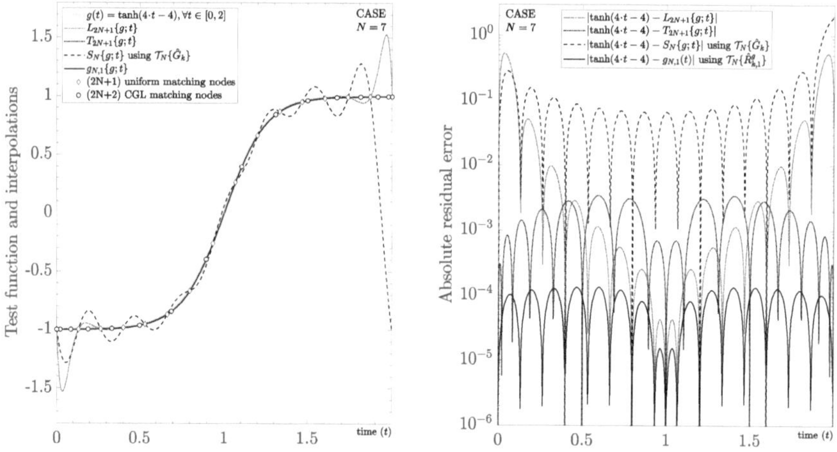

Figure 8. Interpolation of a closed-form function affected by Gibbs phenomenon and Runge's phenomenon simultaneously.

Although the Legendre, Chebyshev, or other modern interpolating series may perform better for other smooth functions, we have shown in several situations that the mixed Fourier series allows us to interpolate a closed-form function without anomalous phenomena caused by convergence. Unlike interpolating methods such as Legendre and Chebyshev, the method for finding $g_{N,M}(t)$ is well conditioned because the DFT is well conditioned and the polynomial constants are found using a low-cost backward algorithm. In summary, we can derive a mixed Fourier series from a closed-form evaluation of $g(t)$ or from a mixed evaluation of $g(t)$, where the new series may outperform common signal presentations with low complexity and well-conditioned methodologies. The Maliev–Lanczos approach has two degrees of freedom, which allows quasi-spectral accuracy. Moreover, it has a simple method based on uniform sampling with convergence everywhere for $M \geq 1$, which allows us to avoid Gibbs and Runge phenomena for $g \in C^{K \geq 1}[0, T]$.

3.4. A Canonical Direct Problem: Numerical Riemann Integration of Closed-Form Smooth Functions

Numerical integration using uniform samples has several advantages because of its computational simplicity. Because the magnitude of Fourier coefficients of $\tilde{r} \in C^M(\mathbb{R})$ has the property $O(|k|^{-M})$, it is a reasonable hypothesis that $r(t)$ allows better numerical integration using the Newton–Cotes quadrature rules [116]. Therefore, we propose the numerical integration of $g(t)$ using $r(t)$ by means of

$$I = \int_0^T g(t)dt = \int_0^T (P_M\{g;t\} + r(t))dt$$
$$\approx T \cdot \sum_{m=1}^{M+1} \frac{1}{m+1} \cdot P_{m,M}^g + \sum_{n=0}^{2N+1} w_n \cdot r(t_n), \tag{82}$$

where $h = T/(2N+1)$, $t_n = n \cdot h$, $r(t_n) = g(t_n) - P_M\{g;t_n\}$, and w_n denotes the weights for a particular quadrature rule [116].

As a case of study, we evaluate $I = \int_0^3 e^{-t^2} dt$ in Figure 9 by means of the absolute relative error of the integral defined by $\eta := \left|1 - \frac{I_{approx}}{I_{exact}}\right|$. We compare (82) with

$$\int_0^T g(t)dt \approx \sum_{n=0}^{2N+1} w_n \cdot g(t_n) \tag{83}$$

using the left rectangular rule (i.e., left Riemann sums), the trapezoidal rule, and Simpson's rule. As we show in this example, the simple left rectangular rule (i.e., $w_n = h, \forall n \in [0, 2N]$ and $w_{2N+1} = 0$) obtains a higher performance. For instance, Figure 9 shows that using that simplest integration scheme, the evaluation of (82) only requires $N > 10$ with $M = 3$ for a typical relative integration error of 1×10^{-10}. This result makes sense using the Fourier framework because $r(t_n)$ has less aliasing than its counterpart $g(t_n)$.

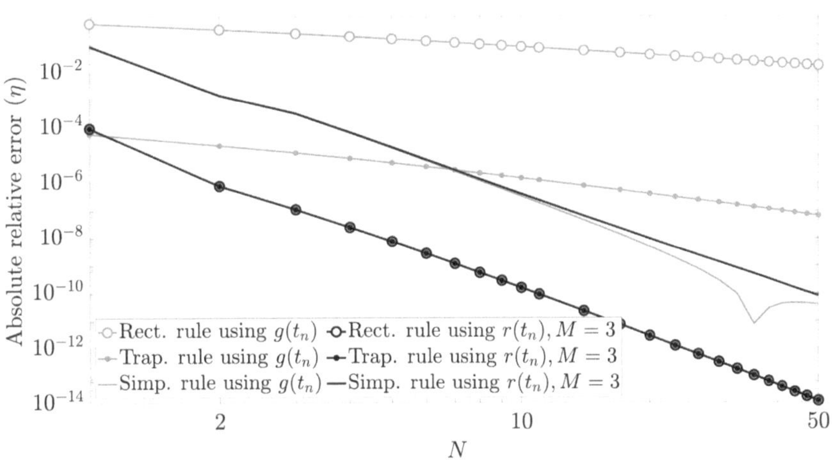

Figure 9. Numerical evaluation of $\int_0^3 e^{-t^2} dt$ using Newton–Cotes quadrature rules.

3.5. A Canonical Inverse Problem: Solution of a Boundary Value Problem (BVP) Using Standard Closed-Form Fourier Coefficients

The mixed Fourier series emerged when we were analyzing the solution of Poisson's equation in one dimension with a Dirichlet boundary condition using the Fourier series [117]. Our approach to solving that problem is as follows.

Let $x \in C^2[0, T]$ be an unknown function such that

$$\frac{d^2}{dt^2} x(t) = \lambda \cdot y(t), \forall t \in [0, T],$$ (84)

$$\text{s.t. } x(0), x(T) \in \mathbb{R}$$ (85)

where we use the right-hand derivative definition at $t = 0$, the left-hand derivative definition at $t = T$, and $\lambda \in \mathbb{R}$.

If we assume that $y \in C^0[0, T]$ has a partial Fourier series given by

$$S_N\{y; t\} = \sum_{k=-N}^{N} Y_k \cdot e^{2\pi i \cdot t \cdot k f_0}, \forall t \in [0, T],$$ (86)

such that $|Y_k|$ is $O(|k|^{-1})$, then the partial series solution of this problem is

$$x_{N,1}(t) = \sum_{k=-N}^{N} R_{k,1}^x \cdot e^{2\pi i \cdot t \cdot k f_0} + \sum_{m=1}^{2} P_{m,1}^x \cdot \left(\frac{t}{T}\right)^m, \forall t \in [0, T].$$ (87)

Replacing $x_{N,1}(t)$ and $S_N\{y; t\}$ in (84), and using the boundaries and the orthogonality of the harmonics, we obtain

$$R_{k,1}^x = \frac{\lambda}{(2\pi i \cdot k f_0)^2} \cdot Y_k, \forall k \in \{\pm 1, \cdots, \pm N\},$$ (88)

$$R_{0,1}^x = x(0) - 2 \cdot \sum_{k=1}^{N} \text{Re}\{R_{k,1}^x\},$$ (89)

$$P_{2,1}^x = \frac{\lambda}{2} \cdot T^2 \cdot Y_0,$$ (90)

$$P_{1,1}^x = \{x(T) - x(0)\} - P_{2,1}^x.$$ (91)

This result is always convergent because $y(t)$ is bounded by the Boundedness Theorem, $|Y_k|$ has a decreasing rate $O(|k|^{-1})$ [24], and $|R_{k,1}^x|$ has a decreasing rate $O(|k|^{-3})$. Therefore, $\sum_{k=1}^{\infty} \text{Re}\{R_{k,1}^x\}$ is bounded, too.

In particular, if $x(t) = \sin(t), \forall t \in [0, \frac{\pi}{2}]$, and $\lambda = -1$, then $y(t) = \sin(t), \forall t \in [0, \frac{\pi}{2}]$, and

$$Y_k = \frac{2}{\pi} \cdot \frac{4ik - 1}{16k^2 - 1}.$$ (92)

In this case, we can see that there are two types of series for the same function $x(t) = y(t) = \sin(t), \forall t \in [0, \frac{\pi}{2}]$. The first is the standard partial Fourier series using the coefficients (92), which have a decreasing rate $O(|k|^{-1})$, and they produce ringing artifacts, as shown in Figure 10. The second is a mixed Fourier series using particular coefficients with $x(0) = 0$ and $x(T) = 1$ obtained from the BVP by (88)–(91). The mixed series includes Fourier coefficients with a decreasing rate $O(|k|^{-3})$, and they do not produce ringing artifacts, as shown in Figure 10. These characteristics motivated us to develop an in-depth analysis of this series and its applications in the framework of signal processing, which is described in this paper.

On the other hand, the solution using the mixed series has better accuracy compared with the standard Finite Difference Method (FDM) [118] given by

$$\begin{bmatrix} -2 & 1 & & & \\ 1 & -2 & 1 & & \\ & \ddots & \ddots & \ddots & \\ & & 1 & -2 & 1 \\ & & & 1 & -2 \end{bmatrix} \begin{bmatrix} x_{2N}(t_1) \\ x_{2N}(t_2) \\ \vdots \\ x_{2N}(t_{2N-1}) \\ x_{2N}(t_{2N}) \end{bmatrix} = \lambda \cdot h^2 \begin{bmatrix} y(t_1) \\ y(t_2) \\ \vdots \\ y(t_{2N-1}) \\ y(t_{2N}) \end{bmatrix} - \begin{bmatrix} x(0) \\ 0 \\ \vdots \\ 0 \\ x(T) \end{bmatrix},$$ (93)

where $h = T/(2N+1)$ and $t_n = n \cdot h, \forall n \in [1, 2N]$. Furthermore, the complexity to find the unknown constants in both cases is $O(N)$ because the solution of (93) using the tridiagonal matrix algorithm is $O(N)$, the evaluation of (88) or (89) is $O(N)$, and the evaluation of (90) or (91) is $O(1)$. Figure 10 shows the FDM solution using linear interpolation, too.

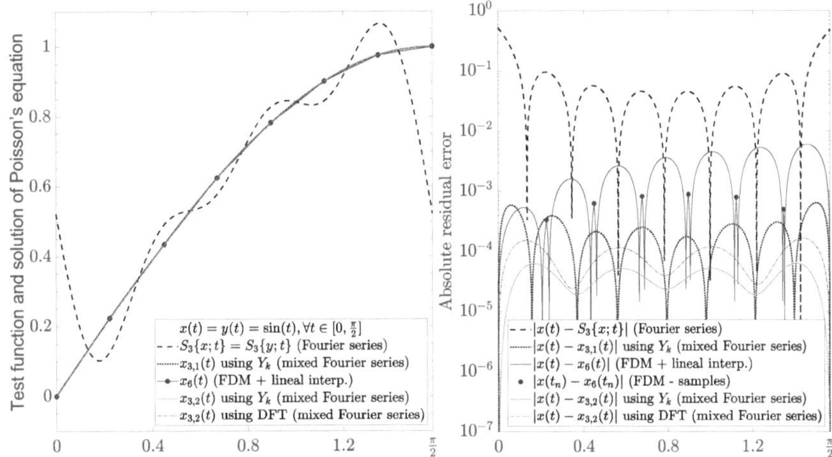

Figure 10. Evaluation of Poisson's equation with Dirichlet boundary condition using mixed Fourier series.

Finally, we can still improve the accuracy of Poisson's Equation solution for any $y \in C^0[0, T]$ using $y_{N,0}(t)$ and $x_{N,2}(t)$, where

$$R_{k,0}^y = \begin{cases} Y_0 - \frac{1}{2}\{y(T) - y(0)\} & k = 0 \\ Y_k + \frac{\{y(T) - y(0)\}}{2\pi i \cdot k} & \text{elsewhere} \end{cases}, \tag{94}$$

$$R_{k,2}^x = \frac{\lambda}{(2\pi i \cdot k f_0)^2} \cdot R_{k,0}^y, \forall k \in \{\pm 1, \cdots, \pm N\}, \tag{95}$$

$$R_{0,2}^x = x(0) - 2 \cdot \sum_{k=1}^{N} \text{Re}\{R_{k,2}^x\}, \tag{96}$$

$$P_{3,2}^x = \frac{\lambda}{6} \cdot T^2 \cdot (y(T) - y(0)), \tag{97}$$

$$P_{2,2}^x = \frac{\lambda}{2} \cdot T^2 \cdot R_{0,0}^y, \tag{98}$$

$$P_{1,2}^x = x(T) - x(0) - P_{2,2}^x - P_{3,2}^x. \tag{99}$$

Figure 10 also shows the solution using $x(T) = y(T) = 1$ and $x(0) = y(0) = 0$, where $\|x(t) - x_{3,2}(t)\|_\infty \sim 6 \times 10^{-5}$ is obtained using only three harmonics (i.e., $N = 3$).

3.6. A Canonical Inverse Problem: Solution of a Boundary Value Problem (BVP) Using the DFT

In the absence of direct knowledge of the Fourier coefficients of $y(t)$, we can use the approximation based on the DFT, given by $\mathcal{T}_N\{\hat{R}_{k,M}^y\}$, without ambiguities at the edges of the interval because we always obtain $\bar{r}_y \in C^M(\mathbb{R})$ using the Maliev–Lanczos approach. Furthermore, $P_{m,M}^y$ is obtained from $y(t)$, and the coefficients $P_{m,M+L}^x$ and $R_{k,M+L}^x$ are obtained from $P_{m,M}^y$ and $R_{k,M}^y$ using the boundaries.

In particular, the case $M = 0$ is always relevant because we simplify the formulation without derivatives. For example, in Figure 10, we compare Poisson's solution using $\mathcal{T}_N\{\hat{R}_{k,M}^y\}$ with (95)–(99). As we show, even though the solution loses accuracy in comparison to the theoretical value $R_{k,0}^y$, the residual error is still acceptable in comparison to the other solutions.

3.7. Toward an Ideal Sampling Theorem for Truncated Continuous-Time Functions

Let $g_T : \mathbb{R} \to \mathbb{R}$ be a truncated function defined by

$$g_T(t) = \begin{cases} g(t) & , \forall t \in [0, T] \\ 0 & , \text{elsewhere} \end{cases}. \tag{100}$$

From the Fourier analysis, it is well known that truncated functions are not band-limited. As a result, sampling that function may result in relevant aliasing when the sampling frequency is reasonably close to twice the usual bandwidth (BW) definitions, such as half power bandwidth or first null bandwidth. Using the ideal sampling theorem [113,114], the number of instantaneous samples required to rebuild $g_T : \mathbb{R} \to \mathbb{R}$ using a Fourier approach is asymptotic, and it is given by $\frac{T}{T_s} = \frac{f_s}{f_0} \gg \frac{2BW}{f_0}$. This result implies many samples to rebuild the truncated function for high-resolution applications. The mixed Fourier series provides a method for quantifying the finite number of instantaneous samples required to rebuild $g_T(t)$ through $g(t)$ such that $g \in \mathcal{C}^M[0, T], \forall M \in \mathbb{N}$. The procedure can be argued as follows:

1. If $\lim_{M \to \infty} \left| P_{M+1,M}^g \right| \to 0$, then $\exists M_0 \geq 0$ such that $\lim_{M \to \infty} P_M\{g; t\} = P_{M_0}\{g; t\} + \epsilon_1(t)$, where $\sup_{\forall t \in [0,T]} |\epsilon_1(t)|$ can be as small as desired.

2. If $\lim_{k, M \to \infty} \left| R_{k,M}^g \right| \to 0$, then $\exists N_0 \geq 0$ such that $\lim_{N, M \to \infty} R_{NM}\{r; t\} = R_{N_0}\{r; t\} + \epsilon_2(t)$, where $\sup_{\forall t \in [0,T]} |\epsilon_2(t)|$ can be as small as desired. The bandwidth of $\bar{r}(t)$ with this approach is $BW = N_0 f_0$.

3. Conclusively, if both previous limits converge to zero, then $\lim_{N, M \to \infty} g_{NM}(t) = P_{M_0}\{g; t\} + R_{N_0}\{r; t\} + \epsilon(t)$, where $\sup_{\forall t \in [0,T]} |\epsilon(t)|$ can be as small as desired.

Therefore, $2M_0 + 2$ instantaneous samples from the edges (i.e., $g(0), g(T), \cdots, g_0^{(M_0)}(0), g^{(M_0)}(T))$ are required to obtain $P_{M_0}\{g, t\}$, and $2N_0 + 1$ instantaneous samples related to the periodic residual error (i.e., $r(t_i) = g(t_i) - P_{M_0}\{g; t_i\}$, where $t_0 = 0, \cdots, t_{2N_0} = T$) are required to obtain $R_{N_0}\{r; t\}$ by means of DFT. Conclusively, we require at least $2M_0 + 2N_0 + 1$ different instantaneous samples from $g(t)$ and its derivatives to rebuild its form in a finite interval $[0, T]$ with an error as small as desired.

Example 2 (Numerical case). *We studied the exponential function with $\alpha = -4$ and $T = 2$ in Section 3.2.3 using the Fourier series and the mixed Fourier series with $M = 1$ and $M = 4$. First, the Fourier series in this example does not converge to zero using $\sup_{\forall t \in [0,T]} |g(t) - S_N\{g; t\}|$ because it has the Gibbs phenomenon. In contrast, the mixed Fourier series converges with $\sup_{\forall t \in [0,T]} |g(t) - g_{N,4}\{g; t\}| \leq 10^{-11}$ using $N \geq 74$. As a result, our approach for $M = 4$ requires at least 149 samples (i.e, $2N + 1$) to estimate the Fourier coefficients numerically using the DFT and 10 samples (i.e., $2M + 2$) of the kth derivatives at the edges to determine the constants $P_{1,4}^g, \cdots, P_{5,4}^g$. If we make the same calculation using the Taylor series, then this example requires $M \geq 38$ for $t_0 = 0$, $M \geq 23$ for $t_0 = \frac{T}{2} = 1$, and $M \geq 34$ for $t_0 = T = 2$. As a result, in the best of those cases, the Taylor series with the same error requires 24 samples (i.e., $M + 1$) of the kth derivatives at $t_0 = \frac{T}{2} = 1$. Nevertheless, the cases $M = 1$ and $M = 4$ using the mixed Fourier series were only included in Section 3.2.3 to compare low-order convergences. If $M := [M_m] = [1 \ \ 2 \ \ 3 \ \ 4 \ \ 5 \ \ 6 \ \ 7 \ \ 8 \ \ 9 \ \ 10 \ \ 11 \ \ 12 \ ]$, then the mixed Fourier series converges with the same criterion for $N := [N_m] = [26{,}000 \ \ 2800 \ \ 380 \ \ 74 \ \ 54 \ \ 32 \ \ 20 \ \ 16 \ \ 11 \ \ 9 \ \ 8 \ \ 7]$. As can be seen, the number of harmonics does not change significantly when $M_m > 8$. In fact, we want to emphasize the limiting factor $N_m + M_m \approx 19$ for $M_m > 8$. The major advantage of this approach is given by avoiding the information of higher-order derivatives at one point in the exchange of instantaneous samples in the whole domain. The absolute residual errors for some cases are shown in Figure 11.*

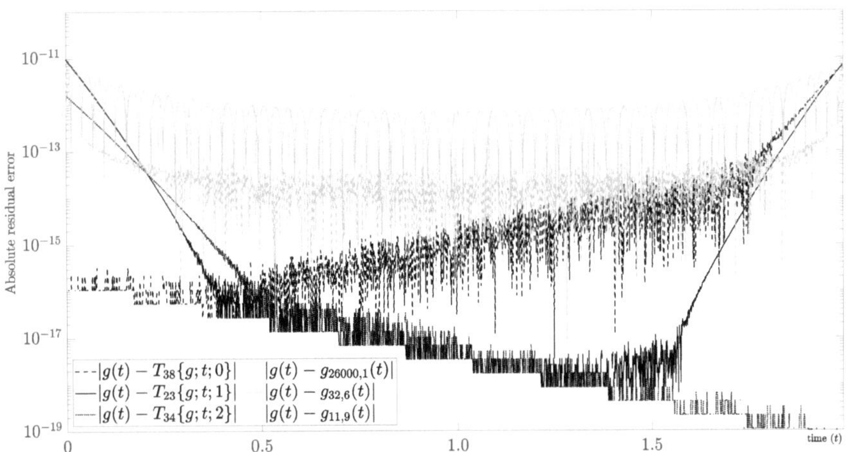

Figure 11. Convergence example for $g(t) = e^{-4t}, \forall t \in [0,2]$, using Taylor and mixed Fourier series.

Example 3 (Theoretical case). *The exponential function (64) has the following properties:* $g^{(M)}(t) = \alpha^M g(t)$, $g \in C^M[0,T]$ *for* $\forall M \in \mathbb{N} \cup \{0\}$, *and* $G_k = \frac{e^{T\alpha}-1}{T\alpha - 2i\pi k}$, $\forall k \in \mathbb{Z}$. *This function has* $P_{M+1,M}^g = \frac{1}{(M+1)!}(T\alpha)^M\{e^{T\alpha} - 1\}$ *using (31). Therefore, we can choose a finite* $M_0 > T|\alpha|$ *such that* P_{M_0+1,M_0}^g *is as small as desired because the factorial grows faster than polynomials and exponentials. On the other hand, we find that* $R_{k,M}^g = \left(\frac{T\alpha}{2\pi i \cdot k}\right)^M \left(\frac{\{e^{T\alpha}-1\}}{T\alpha - 2i\pi k} + \frac{\{e^{T\alpha}-1\}}{2\pi i \cdot k}\right)$ *for* $\forall k \neq 0$ *using (43). Therefore, if* $N_0 > T \cdot |\alpha|$, *then* $\lim_{M\to\infty}\left|R_{N_0,M}^g\right| \to 0$. *Conclusively, using the mixed Fourier series, the number of different instantaneous samples required to rebuild* $g(t) = e^{\alpha t}, \forall t \in [0,T]$, *with an error as small as desired is* $2M_0 + 2N_0 + 1$, *where* M_0, N_0 *are convenient and bounded constants such that* $M_0, N_0 > T|\alpha|$.

3.8. Canonical Example of a Non-Polynomial Mixed Fourier Series: The Sub-Harmonic Case

The methodology utilized in Section 2.2 to obtain the polynomial mixed Fourier series based on the smooth periodic residual error is a framework for defining any mixed Fourier series. In this subsection, we illustrate this methodology to find a novel mixed Fourier series with a non-polynomial form.

Let

$$P_M\{g;t\} = \sum_{m=1}^{\frac{M+1}{2}} A_{m,M}^g \cdot \cos(2\pi \hat{f}_m t) + \sum_{m=1}^{\frac{M+1}{2}} B_{m,M}^g \cdot \sin(2\pi \hat{f}_m t), \forall t \in [0,T] \qquad (101)$$

be a sub-harmonic partial series, where $M \in \mathbb{N}$ is an odd number, $\hat{f}_m \in \mathbb{R} - \{k \cdot f_0\}, \forall k \in \mathbb{Z}$, such that $0 < \hat{f}_1 < \cdots < \hat{f}_{(M+1)/2}$, and $A_{m,M}^g, B_{m,M}^g \in \mathbb{R}$. Let

$$H_{N,M}\{g;t\} := P_M\{g;t\} + \sum_{k=-N}^{N} R_{k,M}^g \cdot e^{2\pi i \cdot t \cdot k f_0}, \forall t \in [0,T] \qquad (102)$$

be the sub-harmonic mixed Fourier series, where

$$R_{k,M}^g = G_k + \sum_{m=1}^{\frac{M+1}{2}} A_{m,M}^g \cdot \frac{ik \cdot C_m - \rho_m \cdot S_m}{2\pi(\rho_m^2 - k^2)} + \sum_{m=1}^{\frac{M+1}{2}} B_{m,M}^g \cdot \frac{ik \cdot S_m + \rho_m \cdot C_m}{2\pi(\rho_m^2 - k^2)}, \forall k \in \mathbb{Z}, \qquad (103)$$

and $\rho_m := T \cdot \hat{f}_m$, $C_m := \cos(2\pi\rho_m) - 1$, and $S_m := \sin(2\pi\rho_m)$.

Because we assume the property $g, r \in C^M[0, T]$, the derivative

$$g^{(k)}(t) = r^{(k)}(t) + \sum_{m=1}^{\frac{M+1}{2}} A_{m,M}^g \cdot \frac{d^k}{dt^k} \cos(2\pi \hat{f}_m t) + \sum_{m=1}^{\frac{M+1}{2}} B_{m,M}^g \cdot \frac{d^k}{dt^k} \sin(2\pi \hat{f}_m t) \tag{104}$$

exists for $\forall t \in [0, T]$ and $\forall k \in \{0, 1, \cdots, M\}$. Because we design $P_M\{g; t\}$ such that the equivalent periodic residual error has a smooth property derived from $r^{(k)}(0) = r^{(k)}(T)$, the unknown sub-harmonic coefficients (i.e., $A_{m,M}^g$ and $B_{m,M}^g$) can be obtained by

$$g^{(k)}(T) - g^{(k)}(0) = \sum_{m=1}^{\frac{M+1}{2}} A_{m,M}^g \cdot \left(\frac{d^k}{dt^k} \cos(2\pi \hat{f}_m t)|_{t=T} - \frac{d^k}{dt^k} \cos(2\pi \hat{f}_m t)|_{t=0} \right)$$
$$+ \sum_{m=1}^{\frac{M+1}{2}} B_{m,M}^g \cdot \left(\frac{d^k}{dt^k} \sin(2\pi \hat{f}_m t)|_{t=T} - \frac{d^k}{dt^k} \sin(2\pi \hat{f}_m t)|_{t=0} \right) \tag{105}$$

for $\forall k \in \{0, 1, \cdots, M\}$. For example, if $M = 1$, then we obtain

$$\begin{bmatrix} C_1 & S_1 \\ -\rho_1 \cdot S_1 & \rho_1 \cdot C_1 \end{bmatrix} \begin{bmatrix} A_{1,1}^g \\ B_{1,1}^g \end{bmatrix} = \begin{bmatrix} F_0^g \\ (2\pi)^{-1} \cdot F_1^g \end{bmatrix}, \tag{106}$$

and if $M = 3$, then we obtain

$$\begin{bmatrix} C_1 & C_2 & S_1 & S_2 \\ -\rho_1 \cdot S_1 & -\rho_2 \cdot S_2 & \rho_1 \cdot C_1 & \rho_2 \cdot C_2 \\ \rho_1^2 \cdot C_1 & \rho_2^2 \cdot C_2 & \rho_1^2 \cdot S_1 & \rho_2^2 \cdot S_2 \\ -\rho_1^3 \cdot S_1 & -\rho_2^3 \cdot S_2 & \rho_1^3 \cdot C_1 & \rho_2^3 \cdot C_2 \end{bmatrix} \begin{bmatrix} A_{1,3}^g \\ A_{2,3}^g \\ B_{1,3}^g \\ B_{2,3}^g \end{bmatrix} = \begin{bmatrix} F_0^g \\ (2\pi)^{-1} \cdot F_1^g \\ -(2\pi)^{-2} \cdot F_2^g \\ -(2\pi)^{-3} \cdot F_3^g \end{bmatrix}, \tag{107}$$

where F_k^g is (33). The matrix formulation for an arbitrary odd case is easily generalized from (105)–(107).

The sub-harmonic mixed Fourier series could have a better performance and versatility than the polynomial mixed Fourier series in several scenarios because it is phenomenologically related to the Fourier basis functions (i.e., it is interpreted literally as a better spectral resolution for some harmonics). Additionally, we can select the sub-harmonics following some special profile for any band-base or pass-band functions. For instance, we can use a uniform distribution (e.g., sub-frequencies $\hat{f}_m \in \{\frac{1}{3}f_0, \frac{2}{3}f_0\}$ or sub-harmonics $\rho_m \in \{\frac{1}{3}, \frac{2}{3}\}$), a non-self-interfering but equally spaced distribution (e.g., sub-frequencies $\hat{f}_m \in \{\frac{1}{6}f_0, \frac{1}{3}f_0\}$ or sub-harmonics $\rho_m \in \{\frac{1}{6}, \frac{1}{3}\}$), or a particular logarithm distribution (e.g., sub-frequencies $\hat{f}_m \in \{\frac{1}{9}f_0, \frac{1}{3}f_0\}$ or sub-harmonics $\rho_m \in \{\frac{1}{9}, \frac{1}{3}\}$) for any band-base function using $M = 3$. Many others are possible depending on the characteristic of $g(t)$ or the conditioning of the matrix resulting from (105).

Although a comprehensive examination of all the characteristics and applications of this new mixed Fourier series is beyond the scope of this paper, we will cover some of them briefly below. First, we obtain the same performance as the polynomial mixed Fourier series when $T \cdot \hat{f}_{(M+1)/2} \ll 1$ because

$$\cos(2\pi \hat{f}_m t) \approx \sum_{k=0}^{(M+1)/2} \frac{(-1)^k}{(2k)!} (2\pi \hat{f}_m t)^{2k}, \forall t \in [0, T], \tag{108}$$

$$\sin(2\pi \hat{f}_m t) \approx \sum_{k=0}^{(M-1)/2} \frac{(-1)^k}{(2k+1)!} (2\pi \hat{f}_m t)^{2k+1}, \forall t \in [0, T], \tag{109}$$

form a non-normalized polynomial mixed Fourier series for $\forall m \in \{1, \cdots, (M+1)/2\}$. The absolute residual errors for $M = 1$ and $M = 3$ using several sub-harmonic profiles, the polynomial mixed Fourier series, and the Fourier series for the exponential test function are compared in Figure 12.

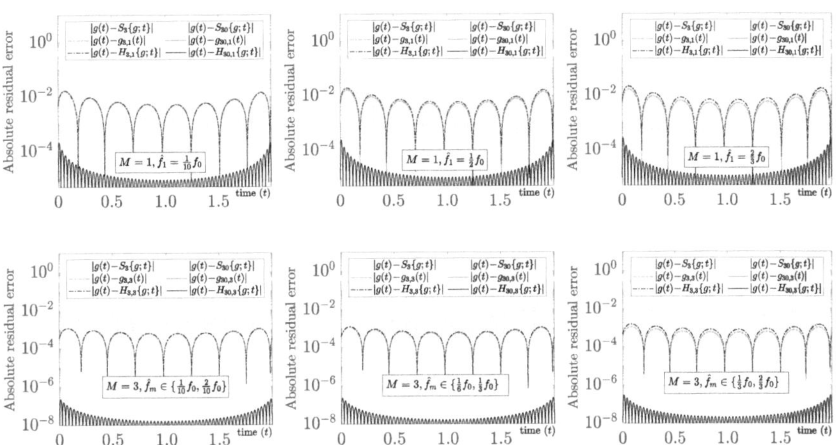

Figure 12. Sub-harmonic mixed Fourier series evaluation using the test function $g(t) = e^{-4t}$, $\forall t \in [0, 2]$.

This new series can be used with base-band functions with a wide-band characteristic, where the main information is influenced by many different harmonics. For example, the test function formed by a base-band frequency sweep given by

$$g(t) = \sin(2\pi \cdot e^{K_f f_0 \cdot t} - 2\pi \cdot K_f f_0 t), \forall t \in [0, T] \tag{110}$$

has considerable spectral information in the instantaneous frequencies

$$f_{ins}(t) := \frac{1}{2\pi} \frac{d}{dt}(2\pi \cdot e^{K_f f_0 \cdot t} - 2\pi \cdot K_f \cdot f_0 t) = K_f f_0 \cdot (e^{K_f f_0 \cdot t} - 1), \forall t \in [0, T]. \tag{111}$$

For narrow-band applications (e.g., $0 < K_f \leq 1$), it is well known that the Chebyshev interpolating function using nonuniform samples achieves greater accuracy for these kind of functions. However, the ill conditioning of that solution for wide-band applications (e.g., $K_f > 1$) produces relevant errors for many applications (e.g., inverse problems). For instance, Figure 13 shows the absolute relative error using the Chebyshev interpolating function with CGL matching nodes for $N \in \{7, 10, 13\}$ and $K_f = 1.6$. The same figure shows that the polynomial mixed Fourier series, where $g_{N \geq 140,1}(t)$, $g_{N \geq 28,3}(t)$, and $g_{N \geq 17,5}(t)$, improve the Chebyshev results using uniform samples and a mixed evaluation (i.e., with $\mathcal{T}_N\{\hat{R}^g_{k,M}\}$). Because the instantaneous frequencies are in the interval $[0\,\text{Hz}, 3.16\,\text{Hz}]$ and the spectral resolution using the Fourier series is $f_0 = \frac{1}{2}\,\text{Hz}$, sub-frequencies $\hat{f}_m \in \{0.75\,\text{Hz}, 1.25\,\text{Hz}, 1.75\,\text{Hz}, 2.25\,\text{Hz}, 2.75\,\text{Hz}, 3.25\,\text{Hz}\}$ would contribute relevant information to reduce the bandwidth of the test function. As Figure 13 makes clear, the sub-harmonic approach using a mixed evaluation improves polynomial approaches for $H_{N \geq 23,1}\{g; t\}$ using $\hat{f}_1 = 3.25\,\text{Hz}$ and for $H_{N \geq 17,3}\{g; t\}$ using $\hat{f}_m \in \{2.75\,\text{Hz}, 3.25\,\text{Hz}\}$. This finding, however, can be improved by performing a local search for the best sub-harmonics in this specific study situation.

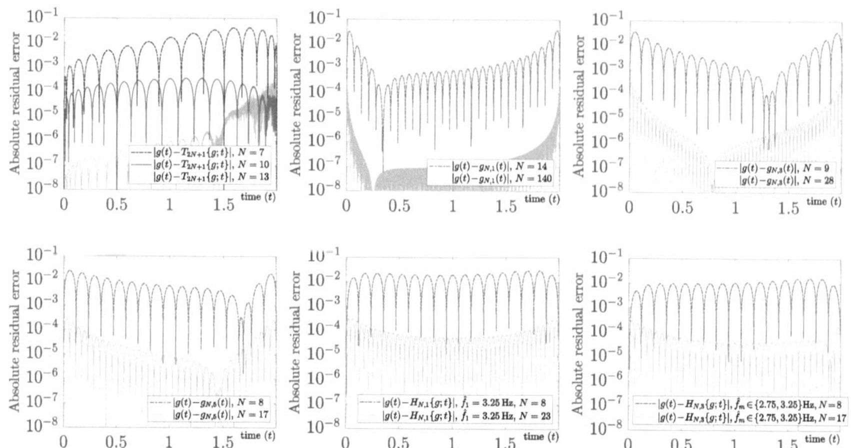

Figure 13. Sub-harmonic mixed Fourier series evaluation using the test function $g(t) = \sin(2\pi \cdot e^{K_f \cdot \frac{t}{2}} - \pi K_f \cdot t), \forall t \in [0,2]$ and $K_f = 1.6$.

Finally, the approximation increases its spectral discrimination when $\hat{f}_m \to L \cdot f_0$, where $L \in \mathbb{Z}$. As a result, pass-band functions can obtain a better spectral discrimination around their main frequencies. For instance, if we assume the test function

$$g(t) = \cos(2\pi \cdot 8.1t + 2.2) + 30\cos(2\pi \cdot 10.3t + 3.7) + \cos(2\pi \cdot 13.2t), \forall t \in [0,2], \quad (112)$$

then the spectral leakage does not allow the Fourier series (or the DFT) to obtain a good discrimination of their estimated carriers using the fundamental frequency $f_0 = \frac{1}{2}$ (i.e., $8\,\text{Hz}, 10.5\,\text{Hz}$, and $13\,\text{Hz}$, which are harmonics $16, 21$, and 26). As illustrated in Figure 14, harmonic 16 is not detectable from the magnitude spectrum using G_k, and it is quite difficult to recognize the three fundamental carries. On the other hand, the polynomial mixed Fourier series improves its accuracy, and it eliminates the Gibbs phenomenon. However, it has low performance, and the magnitude spectrum using $R_{k,M}^g$ also has poor spectral discrimination for pass-band functions because the convergence $O(|k|^{-M})$ lowers high-frequency information, as seen in Figure 14. In contrast, if we use $M = 1$ with $\hat{f}_1 = \frac{(20+21)}{2} f_0 = 20.5 \times f_0 = 10.25\,\text{Hz}$, where harmonics 20 and 21 were obtained from the two major and adjunct harmonics of $|G_k|$, then $|R_{k,1}^g|$ has a better discrimination of carriers using the sub-harmonic mixed Fourier series and increases its approximation accuracy for $N > 45$, at the same time. Although $P_M\{g;t\}$ and $S_N\{r;t\}$ are not orthogonal, the sub-harmonic coefficients can be shown simultaneously in the magnitude spectrum for $\hat{f}_1 > f_0$ through $|C_{m,M}^g| := \sqrt{(A_{m,M}^g)^2 + (B_{m,M}^g)^2}$ because they can be interpreted in the same way as a standard Fourier coefficients, which are the peak amplitudes of the trigonometric basis function at the sub-frequencies $\hat{f}_m$ (or sub-harmonic ρ_m). Continuing this process repeatedly with $M = 2$, we propose $\hat{f}_2 = \frac{(26+27)}{2} \cdot f_0 = 26.5 \times f_0 = 13.25\,\text{Hz}$, where the harmonics were chosen using $|R_{k,1}^g|$ based on its highest next relevance and lowest spectral selectivity. Conclusively, the magnitude spectrum based on $R_{k,3}^g$ allows for a good discrimination of the three carriers with a sub-harmonic resolution and a significant improvement in accuracy for $N > 35$, at the same time. Better results can be obtained by using other strategies, such as the least-square-error approach or a low-cost local search for $\hat{f}_m$. This mixed Fourier series removes the distortion (or loss of spectral information) caused by other approaches, such as the Windowing technique, while simultaneously combining all information into a single spectral diagram.

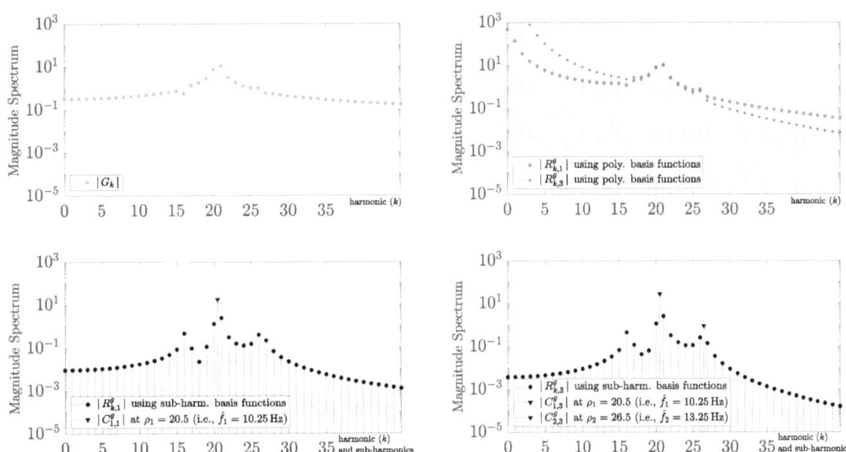

Figure 14. Pass-band application using sub-harmonic mixed Fourier series.

4. Open Challenges and Future Work

Although the mixed Fourier series was discussed and analyzed with several study cases, it is necessary to explore its application to other signal processing problems. In fact, if we replace the original function by its residual error (i.e., $r = g - P_M\{g\}$) in any processing technique, then the technique will process a function with a more compacted spectrum. Therefore, it is reasonable to assume a priori that linear processing techniques will perform better for a fixed bandwidth.

The discrete-time case of this framework requires much more discussion because we found several ways to find the unknown constants. In addition, it is necessary to study the fundamental ambiguity caused by the loss of the sample $g(T)$ using the usual Digital Signal Processing framework (i.e., by taking only N samples, denoted by $t_0, \cdots, t_{N-1}$). For the same reason, more analysis is needed to efficiently integrate the FFT$\{\cdot\}$ into this approach. Furthermore, it is necessary to evaluate and modify our results for noisy discrete-time signals because the constants $P^g_{m,M}$ are derivative-dependent, and thus, the performance may be sensitive to their discrete estimations. However, this technique has a promising future in that circumstance due to the use of the least-square-method or modern noise-robust differentiators.

From Section 3.7, it is seems that a wide class of smoothness functions with compact support $g : [0, T] \to \mathbb{R}$ can be approximated everywhere by a polynomial function with a finite degree (i.e., with a finite value of M) plus a periodic band-limited function (i.e., with a finite value of N) as closely as desired. Nevertheless, more discussion and research on that or related topics are required because it has several consequences for sampling limits for continuous and piecewise functions. For instance, following the Fourier approach, the number of samples to rebuild a pulse function with a duration $0 < T_1 < T$ is an asymptotically large number. Using a piecewise mixed Fourier series, it requires only four samples to rebuild that function with $M = 0$.

In future work, we will research the methods and applications of mixed Fourier series for piecewise continuous functions, and we will apply the Ideal Sampling Theorem to extend our findings to the discrete-time case. In addition, a comprehensive comparison will be made with other modern methods, such as the spectral reprojection method. Although our work is currently limited to one-dimensional problems, we aspire to encourage the exploration of this approach in high-dimensional scenarios.

5. Conclusions

This paper discusses and extends the Maliev–Lanczos approach for processing continuous-time functions with compact support. In contrast to the Taylor series or the Fourier series, the mixed Fourier series uses local and global information. A convenient partial series contains local information about the derivatives at the edges of the interval, whereas the Fourier series contains global information about the remainders throughout the whole domain. The mixed Fourier series avoids the Gibbs phenomenon, and it allows uniform convergence for functions with a bounded continuous first derivative in a closed interval. With the inclusion of $M + 1$ real constants related to simple polynomials computed by a backward algorithm, a major improvement in the error of the approximation is found using N harmonics because the magnitudes of new Fourier coefficients have convergence $O(|k|^{-M})$. In fact, the results evidence that the improvement is better than $O(|k|^{-2-M})$ using common smoothness functions. Similarly, its application in numerical integration shows high performance (e.g., absolute relative error better than 10^{-10}) with a low number of samples using the simple left rectangular rule. On the other hand, in the case of interpolation, we found that the hyperbolic tangent test function (which exhibits Runge's and Gibbs phenomena) can be well represented with $M = 1$, outperforming the Chebyshev interpolation technique using nonuniform sampling. Furthermore, we found that by solving $M + 1$ linear equations, the Fourier series of smooth functions may be easily reprojected to the polynomial mixed Fourier series without using time-domain information (i.e., without derivatives). Several additional canonical examples, applications, and discussions were presented throughout the paper, demonstrating a relevant improvement in the processing of smooth functions.

Author Contributions: Conceptualization, C.-I.P.-R.; methodology, C.-I.P.-R., A.F., M.P., G.Y. and G.P.; software, C.-I.P.-R.; validation, C.-I.P.-R., A.F., M.P., G.Y. and G.P.; formal analysis, C.-I.P.-R., A.F., M.P., G.Y. and G.P.; investigation, C.-I.P.-R.; resources, C.-I.P.-R., A.F., M.P., G.Y. and G.P.; data curation, C.-I.P.-R., A.F., M.P., G.Y. and G.P.; writing—original draft preparation, C.-I.P.-R. and A.F.; writing—review and editing, M.P., G.Y. and G.P.; visualization, C.-I.P.-R.; supervision, C.-I.P.-R.; project administration, C.-I.P.-R. and A.F.; funding acquisition, C.-I.P.-R., A.F., M.P., G.Y. and G.P. All authors have read and agreed to the published version of the manuscript.

Funding: The APC was funded by the Pontificia Universidad Javeriana.

Institutional Review Board Statement: Not applicable.

Informed Consent Statement: Not applicable.

Data Availability Statement: The corresponding author can provide access to the source code and data used in this study upon request.

Acknowledgments: The authors would like to thank the Electronics Department and Electronics Laboratory of the Pontificia Universidad Javeriana for providing the required resources to conduct this study.

Conflicts of Interest: The authors declare no conflict of interest.

References

1.	Zygmund, A. *Trigonometric Series*, 3rd ed.; Cambridge University Press: Cambridge, UK, 2003. [CrossRef]
2.	Allen, R.; Mills, D. *Signal Analysis: Time, Frequency, Scale, and Structure*; Wiley-IEEE Press: Hoboken, NJ, USA, 2004. [CrossRef]
3.	Knapp, A.W. *Basic Real Analysis*; Birkhäuser: Basel, Switzerland, 2005. [CrossRef]
4.	Cho, C.H.; Chen, C.Y.; Chen, K.C.; Huang, T.W.; Hsu, M.C.; Cao, N.P.; Zeng, B.; Tan, S.G.; Chang, C.R. Quantum Computation: Algorithms and Applications. *Chin. J. Phys.* **2021**, *72*, 248–269. [CrossRef]
5.	Bao, S.; Cao, J.; Wang, S. Vibration Analysis of Nanorods by the Rayleigh-Ritz Method and Truncated Fourier Series. *Results Phys.* **2019**, *12*, 327–334. [CrossRef]
6.	Paez-Rueda, C.; Bustamante-Miller, R. Novel Computational Approach to Solve Convolutional Integral Equations: Method of Sampling for One Dimension. *Ing. Univ.* **2019**, *23*, 1–32. [CrossRef]
7.	Sokhal, S.; Ram Verma, S. A Fourier Wavelet Series Solution of Partial Differential Equation Through the Separation of Variables Method. *Appl. Math. Comput.* **2021**, *388*, 125480. [CrossRef]

8. Gurpinar, E.; Sahu, R.; Ozpineci, B. Heat Sink Design for WBG Power Modules Based on Fourier Series and Evolutionary Multi-Objective Multi-Physics Optimization. *IEEE Open J. Power Electron.* **2021**, *2*, 559–569. [CrossRef]

9. Acero, J.; Lope, I.; Carretero, C.; Burdío, J.M. Analysis and Modeling of the Forces Exerted on the Cookware in Induction Heating Applications. *IEEE Access* **2020**, *8*, 131178–131187. [CrossRef]

10. Momose, A. X-ray Phase Imaging Reaching Clinical Uses. *Phys. Med.* **2020**, *79*, 93–102. [CrossRef]

11. Katiyar, R.; Gupta, V.; Pachori, R.B. FBSE-EWT-Based Approach for the Determination of Respiratory Rate from PPG Signals. *IEEE Sens. Lett.* **2019**, *3*, 7001604. [CrossRef]

12. Tripathy, R.K.; Bhattacharyya, A.; Pachori, R.B. A Novel Approach for Detection of Myocardial Infarction from ECG Signals of Multiple Electrodes. *IEEE Sens. J.* **2019**, *19*, 4509–4517. [CrossRef]

13. Lostanlen, V.; Andén, J.; Lagrange, M. Fourier at the Heart of Computer Music: From Harmonic Sounds to Texture. *Comptes Rendus Phys.* **2019**, *20*, 461–473. [CrossRef]

14. Canuto, C.G.; Hussaini, M.Y.; Quarteroni, A.; Zang, T.A. *Spectral Methods: Fundamentals in Single Domains*; Scientific Computation; Springer: Berlin/Heidelberg, Germany, 2010. [CrossRef]

15. Chawde, D.P.; Bhandakkar, T.K. Mixed Boundary Value Problems in Power-law Functionally Graded Circular Annulus. *Int. J. Press. Vessel. Pip.* **2021**, *192*, 104402. [CrossRef]

16. Nie, G.; Hu, H.; Zhong, Z.; Chen, X. A Complex Fourier Series Solution for Free Vibration of Arbitrary Straight-sided Quadrilateral Laminates with Variable Angle Tows. *Mech. Adv. Mater. Struct.* **2022**, *29*, 1081–1096. [CrossRef]

17. Chen, Q.; Du, J. A Fourier Series solution for the Transverse Vibration of Rotating Beams with Elastic Boundary Supports. *Appl. Acoust.* **2019**, *155*, 1–15. [CrossRef]

18. Zhang, M.Y.; Hu, D.Y.; Yang, C.; Shi, W.; Liao, A.H. An Improvement of the Generalized Discrete Fourier Series Based Patch Near-field Acoustical Holography. *Appl. Acoust.* **2021**, *173*, 107711. [CrossRef]

19. Cheng, D.; Kou, K.I. Multichannel Interpolation of Nonuniform Samples with Application to Image Recovery. *J. Comput. Appl. Math.* **2020**, *367*, 112502. [CrossRef]

20. Cheng, D.; Kou, K.I. FFT Multichannel Interpolation and Application to Image Super-resolution. *Signal Process.* **2019**, *162*, 21–34. [CrossRef]

21. Brooks, E.B.; Thomas, V.A.; Wynne, R.H.; Coulston, J.W. Fitting the Multitemporal Curve: A Fourier Series Approach to the Missing Data Problem in Remote Sensing Analysis. *IEEE Trans. Geosci. Remote Sens.* **2012**, *50*, 3340–3353. [CrossRef]

22. Jayasankar, U.; Thirumal, V.; Ponnurangam, D. A Survey on Data Compression Techniques: From the Perspective of Data Quality, Coding Schemes, Data Type and Applications. *J. King Saud Univ. Comput. Inf. Sci.* **2021**, *33*, 119–140. [CrossRef]

23. Hewitt, E.; Hewitt, R.E. The Gibbs-Wilbraham Phenomenon: An Episode in Fourier Analysis. *Arch. Hist. Exact Sci.* **1979**, *21*, 129–160. [CrossRef]

24. Reade, J.B. On the Order of Magnitude of Fourier Coefficients. *SIAM J. Math. Anal.* **1986**, *17*, 469–476. [CrossRef]

25. Jackson, D. The Convergence of Fourier Series. *Am. Math. Mon.* **1934**, *41*, 67–84. [CrossRef]

26. Harris, F. On the Use of Windows for Harmonic Analysis with the Discrete Fourier Transform. *Proc. IEEE* **1978**, *66*, 51–83. [CrossRef]

27. Jerri, A.J. *The Gibbs Phenomenon in Fourier Analysis, Splines and Wavelet Approximations*; Mathematics and Its Applications; Springer: New York, NY, USA, 1998. [CrossRef]

28. Lanczos, C. *Applied Analysis*; Dover Publications: Mineola, NY, USA, 2013. [CrossRef]

29. Torcal-Milla, F.J. A Simple Approach to the Suppression of the Gibbs Phenomenon in Diffractive Numerical Calculations. *Optik* **2021**, *247*, 167921. [CrossRef]

30. Hamming, R. *Numerical Methods for Scientists and Engineers*, 2nd ed.; Dover: Mineola, NY, USA, 1987.

31. Jerri, A.J. Lanczos-Like σ-Factors for Reducing the Gibbs Phenomenon in General Orthogonal Expansions and Other Representations. *J. Comput. Anal. Appl.* **2000**, *2*, 111–127. [CrossRef]

32. Yun, B.I.; Rim, K.S. Construction of Lanczos Type Filters for the Fourier Series Approximation. *Appl. Numer. Math.* **2009**, *59*, 280–300. [CrossRef]

33. Murio, D.A. *The Mollification Method and the Numerical Solution of Ill-Posed Problems*; Wiley-Interscience: Hoboken, NJ, USA, 1993. [CrossRef]

34. Tadmor, E.; Tanner, J. Adaptive Mollifiers for High Resolution Recovery of Piecewise Smooth Data from its Spectral Information. *Found. Comput. Math.* **2002**, *2*, 155–189. [CrossRef]

35. Tadmor, E.; Tanner, J. Adaptive Filters for Piecewise Smooth Spectral Data. *IMA J. Numer. Anal.* **2005**, *25*, 635–647. [CrossRef]

36. Tanner, J. Optimal Filter and Mollifier for Piecewise Smooth Spectral Data. *Math. Comput.* **2006**, *75*, 767–790. [CrossRef]

37. Tadmor, E. Filters, Mollifiers and the Computation of the Gibbs Phenomenon. *Acta Numer.* **2007**, *16*, 305–379. [CrossRef]

38. Piotrowska, J.; Miller, J.M.; Schnetter, E. Spectral Methods in the Presence of Discontinuities. *J. Comput. Phys.* **2019**, *390*, 527–547. [CrossRef]

39. Yun, B.I.; Kim, H.C.; Rim, K.S. An Averaging Method for the Fourier Approximation to Discontinuous functions. *Appl. Math. Comput.* **2006**, *183*, 272–284. [CrossRef]

40. Duman, O. Generalized Cesàro Summability of Fourier Series and its Applications. *Constr. Math. Anal.* **2021**, *4*, 135–144. [CrossRef]

41. Arrowood, J.; Smith, M. Gibbs Phenomenon Suppression Using Fir Time-Varying Filter Banks. In Proceedings of the Digital Signal Processing Workshop, Utica, IL, USA, 13–16 September 1992; pp. 2.1.1–2.1.2. [CrossRef]
42. Gelb, A.; Gottlieb, S. The Resolution of the Gibbs Phenomenon for Fourier Spectral Methods. In *Advances in the Gibbs Phenomenon*; Sampling Publishing: Potsdam, NY, USA, 2007.
43. Yun, B.I. A Weighted Averaging Method for Treating Discontinuous Spectral Data. *Appl. Math. Lett.* **2012**, *25*, 1234–1239. [CrossRef]
44. Ruijter, M.; Versteegh, M.; Oosterlee, C.W. On the Application of Spectral Filters in a Fourier Option Pricing Technique. *J. Comput. Financ.* **2015**, *19*, 75–106. [CrossRef]
45. Walter, G.G.; Shim, H.T. Gibbs' Phenomenon for Sampling Series and What to do About it. *J. Fourier Anal. Appl.* **1988**, *4*, 357–375. [CrossRef]
46. Song, R.; Liang, Y.; Wang, X.; Qi, D. Elimination of Gibbs Phenomenon in Computational Information based on the V-system. In Proceedings of the 2007 2nd International Conference on Pervasive Computing and Applications, Birmingham, UK, 26–27 July 2007; pp. 337–341. [CrossRef]
47. Greene, N. Inverse Wavelet Reconstruction for Resolving the Gibbs Phenomenon. *Int. J. Circuits Syst. Signal Process.* **2008**, *2*, 73–77.
48. Morita, T.; Sato, K.i. Mollification of the Gibbs Phenomena Using Orthogonal Wavelets. In Proceedings of the 2011 International Conference on Multimedia Technology, Hangzhou, China, 26–28 July 2011; pp. 6441–6444. . [CrossRef]
49. Ding, Y.; Selesnick, I.W. Artifact-Free Wavelet Denoising: Non-convex Sparse Regularization, Convex Optimization. *IEEE Signal Process. Lett.* **2015**, *22*, 1364–1368. [CrossRef]
50. Lombardini, R.; Acevedo, R.; Kuczala, A.; Keys, K.P.; Goodrich, C.P.; Johnson, B.R. Higher-Order Wavelet Reconstruction/Differentiation Filters and Gibbs Phenomena. *J. Comput. Phys.* **2016**, *305*, 244–262. [CrossRef]
51. Pan, C. Gibbs Phenomenon Removal and Digital Filtering Directly Through the Fast Fourier Transform. *IEEE Trans. Signal Process.* **2001**, *49*, 444–448. [CrossRef]
52. Boyd, J.P. A Comparison of Numerical Algorithms for Fourier Extension of the First, Second, and Third Kinds. *J. Comput. Phys.* **2002**, *178*, 118–160. [CrossRef]
53. De Ridder, F.; Pintelon, R.; Schoukens, J.; Verheyden, A. Reduction of the Gibbs Phenomenon Applied on Nonharmonic Time Base Distortions. *IEEE Trans. Instrum. Meas.* **2005**, *54*, 1118–1125. [CrossRef]
54. Huybrechs, D. On the Fourier Extension of Nonperiodic Functions. *SIAM J. Numer. Anal.* **2010**, *47*, 4326–4355. [CrossRef]
55. Adcock, B.; Huybrechs, D. On the Resolution Power of Fourier Extensions for Oscillatory Functions. *J. Comput. Appl. Math.* **2014**, *260*, 312–336. [CrossRef]
56. Geronimo, J.; Liechty, K. The Fourier Extension Method and Discrete Orthogonal Polynomials on an Arc of the Circle. *Adv. Math.* **2020**, *365*, 107064. [CrossRef]
57. Gelb, A.; Tanner, J. Robust Reprojection Methods for the Resolution of the Gibbs phenomenon. *Appl. Comput. Harmon. Anal.* **2006**, *20*, 3–25. [CrossRef]
58. Gottlieb, D.; Shu, C.W.; Solomonoff, A.; Vandeven, H. On the Gibbs Phenomenon I: Recovering Exponential Accuracy from the Fourier Partial Sum of a Nonperiodic Analytic Function. *J. Comput. Appl. Math.* **1992**, *43*, 81–98. [CrossRef]
59. Gelb, A. A Hybrid Approach to Spectral Reconstruction of Piecewise Smooth Functions. *J. Sci. Comput.* **2000**, *15*, 293–322. [CrossRef]
60. Shizgal, B.D.; Jung, J.H. Towards the Resolution of the Gibbs Phenomena. *J. Comput. Appl. Math.* **2003**, *161*, 41–65. [CrossRef]
61. Jung, J.H.; Shizgal, B.D. Generalization of the Inverse Polynomial Reconstruction Method in the Resolution of the Gibbs Phenomenon. *J. Comput. Appl. Math.* **2004**, *172*, 131–151. [CrossRef]
62. Chen, X.; Jung, J.H.; Gelb, A. Finite Fourier Frame Approximation Using the Inverse Polynomial Reconstruction Method. *J. Sci. Comput.* **2018**, *76*, 1127–1147. [CrossRef]
63. Boyd, J.P. *Chebyshev and Fourier Spectral Methods*, 2nd revised ed.; Dover Publications: Mineola, NY, USA, 2001.
64. Pan, J.; Li, H. A New Collocation Method using Near-minimal Chebyshev Quadrature Nodes on a Square. *Appl. Numer. Math.* **2020**, *154*, 104–128. [CrossRef]
65. Driscoll, T.; Fornberg, B. A Padé-based Algorithm for Overcoming the Gibbs Phenomenon. *Numer. Algorithms* **2001**, *26*, 77–92. [CrossRef]
66. Beckermann, B.; Matos, A.C.; Wielonsky, F. Reduction of the Gibbs Phenomenon for Smooth Functions with Jumps by the ε-algorithm. *J. Comput. Appl. Math.* **2008**, *219*, 329–349. [CrossRef]
67. Nersessian, A.; Poghosyan, A.; Barkhudaryan, R. Convergence Acceleration for Fourier Series. *J. Contemp. Math. Anal.* **2006**, *41*, 39–51.
68. Brezinski, C. Extrapolation Algorithms for Filtering Series of Functions, and Treating the Gibbs Phenomenon. *Numer. Algorithms* **2004**, *36*, 309–329. [CrossRef]
69. Pasquetti, R. On Inverse Methods for the Resolution of the Gibbs Phenomenon. *J. Comput. Appl. Math.* **2004**, *170*, 303–315. [CrossRef]
70. Krylov, A.N. *On Approximate Calculations, Lectures Delivered in 1906*; Tipolitography of Birkenfeld: St. Petersburg, Russia, 1907. (In Russian)

71. Kantorovich, L.V.; Krylov, V. *Approximate Methods of Higher Analysis*, 3rd ed.; Interscience Publishers Inc.: New York, NY, USA, 1964. [CrossRef]
72. Lanczos, C. *Discourse on Fourier Series*; Hafner: New York, NY, USA, 1966. [CrossRef]
73. Banerjee, N.S.; Geer, J.F. *Exponential Approximations Using Fourier Series Partial Sums*; Technical Report; ICASE, NASA Langley Research Center: Hampton, VA, USA, 1997.
74. Rim, K.S.; Yun, B.I. Gibbs Phenomenon Removal by Adding Heaviside Functions. *Adv. Comput. Math.* **2013**, *38*, 683–699. [CrossRef]
75. Yun, B.I. Improving Fourier Partial Sum Approximation for Discontinuous Functions Using a Weight Function. *Abstr. Appl. Anal.* **2017**, *2017*, 1364914. [CrossRef]
76. Wangüemert-Pérez, J.G.; Godoy-Rubio, R.; Ortega-Moñux, A.; Molina-Fernández, I. Removal of the Gibbs Phenomenon and its Application to Fast-Fourier-Transform-based mode Solvers. *J. Opt. Soc. Am. A* **2007**, *24*, 3772–3780. [CrossRef]
77. Jones, W.B.; Hardy, G. Accelerating Convergence of Trigonometric Approximations. *Math. Comput.* **1970**, *24*, 547–560. [CrossRef]
78. Lyness, J.N. Computational Techniques Based on the Lanczos Representation. *Math. Comput.* **1974**, *28*, 81–123. [CrossRef]
79. Eckhoff, K.S. Accurate and Efficient Reconstruction of Discontinuous Functions from Truncated Series Expanstions. *Math. Comput.* **1993**, *61*, 745–763. [CrossRef]
80. Eckhoff, K.S. Accurate Reconstructions of Functions of Finite Regularity from Truncated Fourier Series Expansions. *Math. Comput.* **1995**, *64*, 671–690. [CrossRef]
81. Eckhoff, K.S. On a High Order Numerical Method for Functions with Singularities. *Math. Comput.* **1998**, *67*, 1063–1088. [CrossRef]
82. Li, W. Alternative Fourier Series Expansions with Accelerated Convergence. *Appl. Math.* **2016**, *7*, 1824–1845. [CrossRef]
83. Barkhudaryan, A.; Barkhudaryan, R.; Poghosyan, A. Asymptotic Behavior of Eckhoff's Method for Fourier Series Convergence Acceleration. *Anal. Theory Appl.* **2007**, *23*, 228–242. [CrossRef]
84. Poghosyan, A. Asymptotic Behavior of the Krylov-lanczos Interpolation. *Anal. Appl.* **2009**, *7*, 199–211. [CrossRef]
85. Poghosyan, A. Asymptotic Behavior of the Eckhoff Approximation in Bivariate Case. *Anal. Theory Appl.* **2012**, *28*, 329–362. [CrossRef]
86. Poghosyan, A. On an Autocorrection Phenomenon of the Eckhoff Interpolation. *Aust. J. Math. Anal. Appl.* **2012**, *9*, 1–31. [CrossRef]
87. Nersessian, A.; Poghosyan, A. Accelerating the Convergence of Trigonometric Series. *Cent. Eur. J. Math.* **2006**, *4*, 435–448. [CrossRef]
88. Poghosyan, A.V.; Poghosyan, L. On a Pointwise Convergence of Quasi-Periodic-Rational Trigonometric Interpolation. *Int. J. Anal.* **2014**, *2014*, 249513. [CrossRef]
89. Poghosyan, A.; Bakaryan, T. Optimal Rational Approximations by the Modified Fourier Basis. *Abstr. Appl. Anal.* **2018**, *2018*, 1705409. [CrossRef]
90. Poghosyan, A.; Poghosyan, L.; Barkhudaryan, R. On some quasi-periodic approximations. *Armen. J. Math.* **2020**, *12*, 1–27. [CrossRef]
91. Poghosyan, A.; Poghosyan, L.; Barkhudaryan, R. On the Convergence of the Quasi-periodic Approximations on a Finite Interval. *Armen. J. Math.* **2021**, *13*, 1–44. [CrossRef]
92. Nersessian, A.; Poghosyan, A. On a Rational Linear Approximation of Fourier Series for Smooth Functions. *J. Sci. Comput.* **2006**, *26*, 111–125. [CrossRef]
93. Nersessian, A. On an Over-Convergence Phenomenon for Fourier series. *Armen. J. Math.* **2018**, *10*, 1–21; Correction in *Armen. J. Math.* **2019**, *11*, 1–2. [CrossRef]
94. Nersessian, A. Fourier Tools are Much More Powerful than Commonly Thought. *Lobachevskii J. Math.* **2019**, *40*, 1122–1131. [CrossRef]
95. Nersessian, A. *Operator Theory and Harmonic Analysis*; Springer Proceedings in Mathematics and Statistics; Chapter On Some Fast Implementations of Fourier Interpolation; Springer: Berlin/Heidelberg, Germany, 2021; pp. 463–477. [CrossRef]
96. Nersessian, A. Acceleration of Convergence of Fourier Series Using the Phenomenon of Over-Convergence. *Armen. J. Math.* **2022**, *14*, 1–31. [CrossRef]
97. Nersessian, A.; Poghosyan, A. The convergence acceleration of two-dimensional Fourier interpolation. *Armen. J. Math.* **2008**, *1*, 50–63.
98. Baszenski, G.; Delvos, F.; Tasche, M. A United Approach to Accelerating Trigonometric Expansions. *Comput. Math. Appl.* **1995**, *30*, 33–49. [CrossRef]
99. Adcock, B. Modified Fourier Expansions: Theory, Construction and Applications. Ph.D. Thesis, Trinity Hall, University of Cambridge, Cambridge, UK, 2010. [CrossRef]
100. Batenkov, D.; Yomdin, Y. Algebraic Fourier Reconstruction of Piecewise Smooth Functions. *Math. Comput.* **2012**, *81*, 277–318. [CrossRef]
101. Batenkov, D. Complete Algebraic Reconstruction of Piecewise-smooth Functions from Fourier Data. *Math. Comput.* **2015**, *84*, 2329–2350. [CrossRef]
102. Trefethen, L.N. *Spectral Methods in MATLAB*; Society for Industrial and Applied Mathematics: Philadelphia, PA, USA, 2000.
103. Shen, J.; Tang, T.; Wang, L.L. *Spectral Methods: Algorithms, Analysis and Applications*; Springer Series in Computational Mathematics 41; Springer Publishing Company Incorporated: New York, NY, USA, 2011.

104. Roache, P.J. A Pseudo-spectral FFT Technique for Non-periodic Problems. *J. Comput. Phys.* **1978**, *27*, 204–220. [CrossRef]
105. Lee, H.N. An Alternate Pseudospectral Model for Pollutant Transport, Diffusion and Deposition in the atmosphere. *Atmos. Environ.* **1981**, *15*, 1017–1024. [CrossRef]
106. Biringen, S.; Kao, K.H. On the Application of Pseudospectral FFT Techniques to Non-periodic Problems. *Int. J. Numer. Methods Fluids* **1989**, *9*, 1235–1267. [CrossRef]
107. Kleiner, I. Evolution of the Function Concept: A Brief Survey. *Coll. Math. J.* **1989**, *20*, 282–300. [CrossRef]
108. Katznelson, Y. *An Introduction to Harmonic Analysis*; Cambridge University Press: Cambridge, UK, 2004. [CrossRef]
109. Grafakos, L. *Classical Fourier Analysis*, 3rd ed.; Graduate Texts in Mathematics 249; Springer: New York, NY, USA, 2014. [CrossRef]
110. Tveito, A.; Winther, R. *Introduction to Partial Differential Equations: A Computational Approach*; Springer: Berlin/Heidelberg, Germany, 2005; Volume 29. [CrossRef]
111. Friesecke, G. *Lectures on Fourier Analysis*; University of Warwick: Coventry, UK, 2007.
112. Jeffreys, H.; Jeffreys, B. *Methods of Mathematical Physics*, 3rd ed.; Cambridge Mathematical Library, Cambridge University Press: Cambridge, UK, 2000. [CrossRef]
113. Unser, M. Sampling-50 Years After Shannon. *Proc. IEEE* **2000**, *88*, 569–587. [CrossRef]
114. Vaidyanathan, P. Generalizations of the Sampling Theorem: Seven Decades After Nyquist. *IEEE Trans. Circuits Syst. Fundam. Theory Appl.* **2001**, *48*, 1094–1109. [CrossRef]
115. Xu, K. The Chebyshev Points of the First Kind. *Appl. Numer. Math.* **2016**, *102*, 17–30. [CrossRef]
116. Press, W.H.; Teukolsky, S.A.; Vetterling, W.T.; Flannery, B.P. *Numerical Recipes: The Art of Scientific Computing*, 3rd ed.; Cambridge University Press: Cambridge, UK, 2007. [CrossRef]
117. Sköllermo, G. A Fourier Method for the Numerical Solution of Poisson's Equation. *Math. Comput.* **1975**, *29*, 697. [CrossRef]
118. Leveque, R. *Finite Difference Methods for Ordinary and Partial Differential Equations: Steady-State and Time-Dependent Problems*; Classics in Applied Mathematics; SIAM, Society for Industrial and Applied Mathematics: Philadelphia, PA, USA, 2007. [CrossRef]

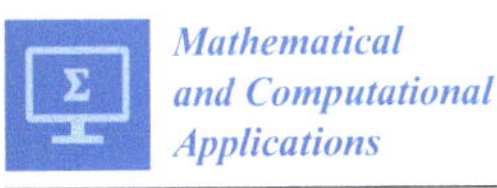
Mathematical and Computational Applications

Article

New Quality Measures for Quadrilaterals and New Discrete Functionals for Grid Generation

Guilmer Ferdinand González Flores * and Pablo Barrera Sánchez

Department of Mathematics, Faculty of Sciences, National Autonomous University of Mexico, Mexico City 04510, Mexico
* Correspondence: guilmerg@ciencias.unam.mx; Tel.: +52-55-5622-4928

Abstract: In this paper, we review some grid quality metrics and define some new quality measures for quadrilateral elements. The curved elements are not discussed. Usually, the maximum value of a quality measure corresponds to the minimum value of the energy density over the grid. We also define new discrete functionals, which are implemented as objective functions in an optimization-based method for quadrilateral grid generation and improvement. These functionals are linearly combined with a discrete functional whose domain has an infinite barrier at the boundary of the set of unfolded grids to preserve convex grid cells in each step of the optimization process.

Keywords: mesh generation; quality measure; aspect ratio; quality improvement

Citation: González Flores, G.F.; Barrera Sánchez, P. New Quality Measures for Quadrilaterals and New Discrete Functionals for Grid Generation. *Math. Comput. Appl.* **2023**, *28*, 95. https://doi.org/10.3390/mca28050095

Academic Editor: Oliver Schütze

Received: 1 June 2023
Revised: 12 August 2023
Accepted: 23 August 2023
Published: 9 September 2023

1. Introduction

Research on mesh generation in Computer Graphics, Scientific Visualization and Computational Field Simulations has led to a substantial number of methods within the last six decades. An exhaustive description of this field is beyond the scope of this paper; one can refer to the many surveys available, see, e.g., Thompson et al. [1] and Lo [2]. However, that mesh generation in regions in 2D and 3D is a central task, used in numerical methods for the solution of partial differential equations, using finite difference, finite element and finite volume methods.

There is also a special interest in studying meshes formed by triangular elements. Our interest here is to generate structured meshes with quadrilateral elements; however, all of the discussion can be applied to unstructured meshes. The simplest way to generate a structured mesh is via interpolation of the boundaries, but it is difficult to ensure that the mesh thus obtained is a convex one. In 2010, Barrera et al. [3] provided a review of some functionals and conditions that guarantee the existence of optimal meshes that are convex over irregular planar regions.

Our interest now is to improve the mesh quality via controlling the shape of the elements. The improvement of mesh quality can be carried out in two ways:

Clean-up. This consists of the elimination, insertion and reconnection of nodes to eliminate the worst elements. Some authors call that this procedure topological optimization in the sense that the connectivity of the nodes is removed to obtain an optimal configuration.

Smoothing. This consists of node repositioning without changing the connectivity of the elements.

In both cases, the goal is to obtain a quality mesh with a low number of distorted elements. To achieve this goal for a quadrilateral, it is neccesary to define an ad hoc quality measure.

Definition 1 (González [4])**.** *We say that a real-valued function $\mu(Q)$ over a quadrilateral Q is a quality measure in the sense of Field-Oddy if it*

(1) *Has the ability to detect degenerated elements;*
(2) *Is bounded and continuous;*
(3) *Is independent of scale;*
(4) *Is normalized;*
(5) *Is invariant under rigid transformations.*

For practical purposes, it is convenient to define an acceptability interval $[\mu_0, 1]$ for the quality measure, i.e., when a quadrilateral has a suitable shape, outside of this interval, we say that the quadrilateral does not have the desired shape. The acceptability interval are defined empirically for each quality measure.

In this paper, we are interested in identifying the shape of the cells and in quantifying the distortion of a quadrilateral when it is not a square or a rectangle. In the remainder of this paper, we will discuss the most used quality measures for rectangles and then propose new quality measures.

The remainder of this paper is structured as follows. The following, background section presents the most used quality measures for rectangles based on angles. Section 3 then presents some new quality measures based on geometric properties. Section 4 then presents some classical global quality metrics and proposes a statistical analysis of all elements of the mesh. Section 5 then presents some concepts to grid quality improvement using quality measures. Finally in the Section 6 then presents some new quality discrete functionals for improvement of the mesh.

2. Background

Following the ideas behind the quality measures for triangles, it is straightforward to define some figures that measure the shape of quadrilaterals. One of these is the aspect ratio, which is defined by comparing to the ideal case when the quadrilateral is a rectangle; it represents the ratio of the largest to the smallest sides. An estimator for this ratio was discussed in 1987 by Robinson [5]. The idea is to associate a rectangle with the convex quadrilateral: a rectangle passing through the midpoints of the sides of the quadrilateral, see Figure 1.

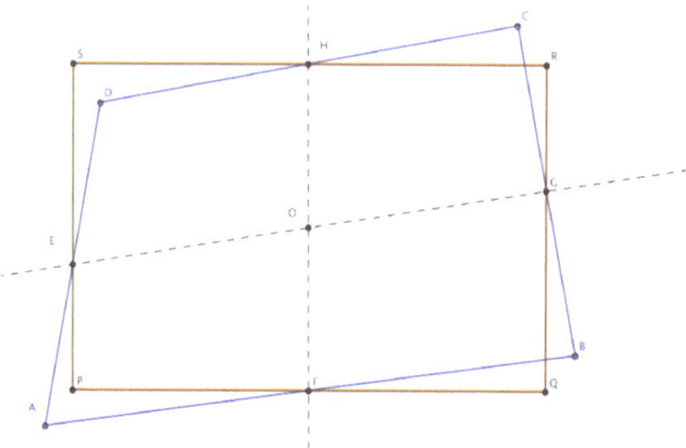

Figure 1. *PQRS* is the rectangle associated by Robinson to quadrilateral *ABCD*.

This idea is usual in continuum mechanics. Robinson proposed a practical method of calculating it by means of bilinear mapping between the unit square and the quadrilateral

$$
x \;=\; e_1 + e_2\xi + e_3\eta + e_4\xi\eta, \tag{1}
$$
$$
y \;=\; f_1 + f_2\xi + f_3\eta + f_4\xi\eta, \tag{2}
$$

where the e and f coefficients are realated to the nodal coordinates x_1 to x_4 and y_1 to y_4 by

$$
\begin{aligned}
e_1 &= \tfrac{1}{4}(x_1 + x_2 + x_3 + x_4) & e_2 &= \tfrac{1}{4}(-x_1 + x_2 + x_3 - x_4) \\
e_3 &= \tfrac{1}{4}(-x_1 - x_2 + x_3 + x_4) & e_4 &= \tfrac{1}{4}(x_1 - x_2 + x_3 - x_4) \\
f_1 &= \tfrac{1}{4}(y_1 + y_2 + y_3 + y_4) & f_2 &= \tfrac{1}{4}(-y_1 + y_2 + y_3 - y_4) \\
f_3 &= \tfrac{1}{4}(-y_1 - y_2 + y_3 + y_4) & f_4 &= \tfrac{1}{4}(y_1 - y_2 + y_3 - y_4)
\end{aligned}
\tag{3}
$$

The meaning of the coefficients in Equation (3) is now shown in [5]. Which yields

$$
\text{aspect ratio} = \max\left\{\frac{e_2}{f_3}, \frac{f_3}{e_2}\right\}.
$$

The associated rectangle has sides, which are parallel to the coordinate axes and pass through the midpoints of the sides of the quadrilateral. In spite of its simplicity, this analytic representation is not satisfactory since it depends on an orthogonal coordinate system. In 2000, Field [6] reviewed this definition and suggested calculating the aspect ratio of Robinson and orthogonalizing the main axes, and proposed a quality measure to detect squares.

In 1989, Lo [7] reviewed the classical quality measure for triangles $T(a, b, c)$ with side lengths l_1, l_2 and l_3,

$$
g_i = 4\sqrt{3}\,\frac{\text{area}(T_i)}{l_1^2 + l_2^2 + l_3^2}
\tag{4}
$$

which attains its optimum value in equilateral triangles, and again proposed calculating each one of those values over the four T_i triangles, which are defined by the sides and diagonals of a quadrilateral, but reordering these quantities in such a way that

$$
g_1 \leq g_2 \leq g_3 \leq g_4,
\tag{5}
$$

and using

$$
\mu 1(Q) = \frac{g_1 g_2}{g_3 g_4},
\tag{6}
$$

as a quality measure. The maximum value is 1 and it is obtained for rectangles. This is a quality measure because it is continuous, bounded and identifies degenerate and even non-convex quadrilaterals. The measure that Lo uses for triangles T is the reciprocal of the number of conditions of a linear mapping $\mu(T) = 1/\kappa_2(T)$, which Knupp [8] used in 2001 to measure the distortion of the elements. Locally, Lo's measure may have more critical points, which can be far from representing a rectangle.

Another measure of quality for quadrilaterals is described by van Rens et al. [9] as follows: compute the inner angles θ_k and define

$$
\mu_2(Q) = \prod_{k=1}^{4}\left(1 - \left|\frac{\frac{\pi}{2} - \theta_k}{\frac{\pi}{2}}\right|\right).
\tag{7}
$$

This function is continuous, dimensionless and $0 \leq \mu_2(Q) \leq 1$. One can see that $\mu_2(Q) = 0$ if Q is a triangle and $\mu_2(Q) = 1$ only if Q is a rectangle.

In 2012, Remacle et al. [10] described the Blossom-Quad algorithm to construct a non-structured mesh with quadrilateral elements obtained from a previous triangulation and used a cost function to produce a quality mesh. They used

$$
\mu_3(Q) = \max\left\{1 - \frac{2}{\pi}\max_k\left\{\left|\frac{\pi}{2} - \theta_k\right|\right\}, 0\right\},
\tag{8}
$$

and observed that if the value of this function is 1, Q is a perfect quadrilateral, and it is 0 if any of the angles are greater than or equal to π, i.e., when the quadrilateral degenerates into a triangle or is nonconvex. This function is also a quality measure.

As noted, unlike other measures for rectangles we have discussed up to this point, the last two ones do not depend either on the shape of the quadrilaterals or the aspect ratio or proportion of their sides; they only measure how near or far away a quadrilateral is from being a rectangle using only the internal angles.

Another function based on the inner angles was proposed by Wu [11]. This author used the same idea as Lo: to order the inner angles θ_i so that $\theta_1 \leq \theta_2 \leq \theta_3 \leq \theta_4$ and define

$$\mu_4(Q) = \frac{\theta_1 \theta_2}{\theta_3 \theta_4}. \tag{9}$$

This function reaches its maximum value of 1 on rectangles. However, this is not a good measure in the sense of Field-Oddy, since it is not capable of detecting degenerate quadrilaterals.

3. New Quality Measures

In the previous section, we had reviewed some measures that characterize rectangles and also pointed out some intervals of acceptability to decide if a quadrilateral is close to the desired shape. However, which rectangle is it close to?

3.1. Quality Measure of Rectangles

A very interesting problem in computational geometry is the following: given a cloud of points, calculate the rectangle of the minimum area that contains them. It is known that this problem can be raised directly on the convex hull of the cloud of points, and therefore the problem can be regarded as calculating a rectangle of minimum area that contains a convex polygon.

We propose the use of the rectangle of minimum area to define a distortion measure of the quadrilateral in the sense that it measures how close or far a quadrilateral Q is from being a rectangle, Figure 2.

Example 1. *For the quadrilateral $A(-6.84, 7.5)$, $B(-10, -4)$, $C(11.81, -1.38)$ and $D(9.27, 11.94)$, the rectangle of minimum area is $A'BCD'$ with an aspect ratio of 1.62. See Figure 2.*

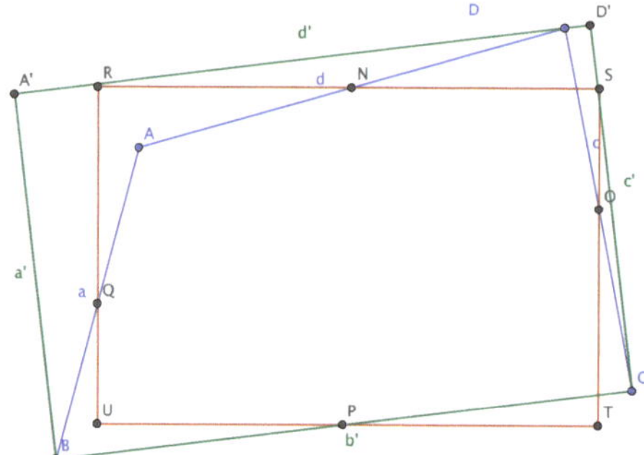

Figure 2. $A'BCD'$ is the rectangle of minimum area for quadrilateral $ABCD$ and $RUTS$ is the rectangle associated by Robinson for Example 1.

On one hand, the cell area a_c, is less than the rectangle area; that is, $a_c \leq a_R$, and it is easy to see that

$$\frac{2a_c - a_R}{a_R} \leq 1;$$

see Lassak [12] for the proof. The quotient thus defined reaches its maximum value of 1 on rectangles. Therefore, we propose the value

$$\mu_{r1}(Q) = \frac{2a_c - a_R}{a_R} \tag{10}$$

as a new quality measure to characterize rectangles.

One must note that $\mu_{r1}(Q)$ is a good quality measure according to Field-Oddy, and $\mu_{r1}(Q) = 0$ if Q is a triangle.

Another good quality measure in this sense is

$$\mu_{r2}(Q) = \frac{2a_-}{a_R}, \tag{11}$$

where

$$a_- = \min\{a_1, a_2, a_3, a_4\} \tag{12}$$

and a_i are the area of the four triangles defined by taking the four vertices of a quadrilateral into groups of three.

3.2. New Aspect Ratio

Using the rectangle of minimum area for Q, we propose the use of the ratio of the largest to the smallest side as the aspect ratio. This measure is invariant under rigid and scaling transformations. This measure is better than Robinson's aspect ratio. It is easy to construct an example for which the Robinson *aspect ratio* is 1 but the quadrilateral is distorted following the next example.

Example 2. *For the quadrilateral A(3.53, 10.21), B(−10, −4), C(11.81, −1.38) and D(9.27, 11.94), Robinson's* aspect ratio *is 1.00 but using the rectangle of minimum area $NBCD$, the* aspect ratio *is 1.62. See Figure 3.*

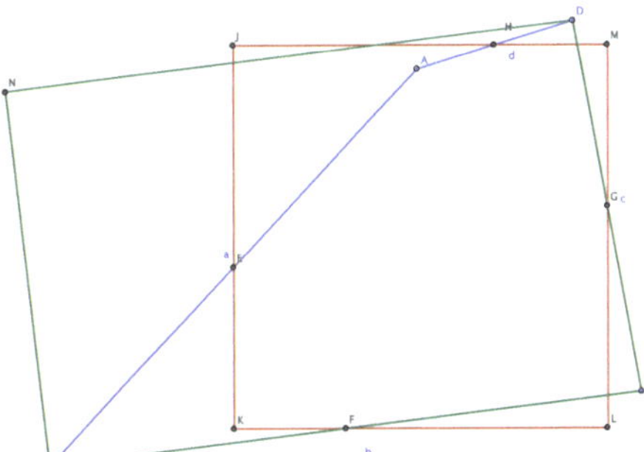

Figure 3. *NBCD* is the rectangle of minimum area for quadrilateral *ABCD* and *JKLM* is the rectangle associated by Robinson for the Example 2.

As we have discussed, some measures to characterize rectangles are based on the inner angles. Another way to achieve this is to ask Q to be a parallelogram and one of its inner angles to be a right one. As it is, we use a measure that imposes a particular condition on a rectangle instead of one on the form of Q.

Our interest is to characterize the rectangles geometrically. A well-known result in the literature is as follows:

Theorem 1. *Let Q be a quadrilateral of vertices A, B, C and D whose sides are a, b, c and d. The quadrilateral Q is a rectangle if and only if the area of the quadrilateral is written as*

$$a_R = \frac{1}{2}\sqrt{(a^2 + c^2)(b^2 + d^2)}. \tag{13}$$

The proof of this result can be found in Josefsson [13]. The interesting fact about this theorem is that it provides of an analytical expression of the area of a hypothetical rectangle formed by the sum of the squares of the opposite sides of Q and compares the square of the area of Q to identify how far it is from being a rectangle.

On the other hand, following the proof of the theorem, it is easy to see that the area a_c of any convex quadrilateral satisfies

$$2a_c \leq \sqrt{(a^2 + c^2)(b^2 + d^2)}. \tag{14}$$

Using this idea, we propose the measure

$$\mu_R(Q) = \frac{2a_-}{\sqrt{(a^2 + c^2)(b^2 + d^2)}} \tag{15}$$

where a_- is defined in Equation (12).

The measure $\mu_R(Q)$ is a good quality measure in the Field-Oddy sense, since it is continuous, bounded and capable of indentifying degenerate quadrilaterals (to triangles), as well as to identify if a quadrilateral is non-convex. This measure reaches its optimal value of 1 for rectangles.

An acceptability interval to consider that the quadrilateral is a rectangle under this measure is $[0.95, 1]$.

3.3. Quality Measure of Parallelograms

Let Q be an oriented quadrilateral of vertices $P_1 P_2 P_3 P_4$. The latter defines four oriented triangles $T_1 = T(P_4, P_1, P_2)$, $T_2 = T(P_1, P_2, P_3)$, $T_3 = T(P_2, P_3, P_4)$ and $T_4 = T(P_3, P_4, P_1)$, see the Figure 4.

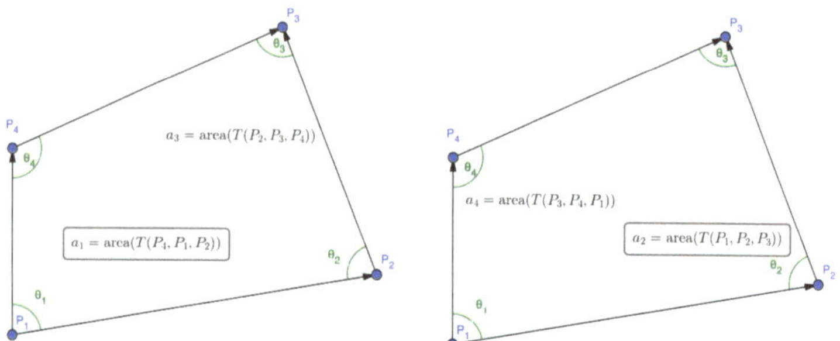

Figure 4. The four oriented triangles defined by a quadrilateral grid cell.

Let a_i be the area of the four oriented triangle and a_c be the area of grid cell Q, we have

$$g_1 = \frac{a_1}{a_c}, \quad g_2 = \frac{a_2}{a_c}, \quad g_3 = \frac{a_3}{a_c}, \quad g_4 = \frac{a_4}{a_c} \tag{16}$$

It is easy to see that a quadrilateral is a parallelogram if and only if

$$g_1 g_2 = g_3 g_4 = \frac{1}{4}, \tag{17}$$

The proof is based on follow: on one side

$$g_1 + g_3 = g_2 + g_4 = 1. \tag{18}$$

On the other hand

$$g_1 = g_2 \tag{19}$$

if and only if corresponding sides are parallel. But if we impose that g_1 and g_2 are equal $1/2$, $g_3 = g_4$ so that the other sides are parallel, see [4]. A quality measure to characterize parallelograms is

$$\mu_{p1}(Q) = 4 \min\{g_1 g_2, g_3 g_4\}. \tag{20}$$

This measure is invariant under rigid and scaling transformations.

In the Figure 5 it is shown differents level curves for $\mu_{p1}(Q)$.

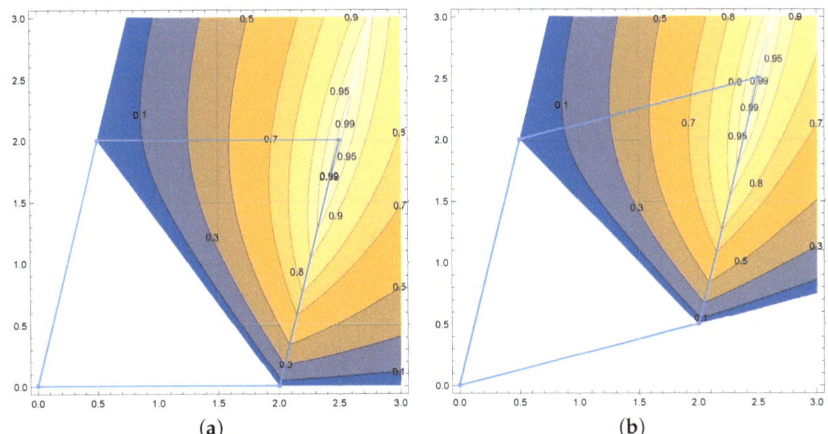

Figure 5. Level curves for (20) where (**a**) Q has 3 fixed vertices $(0,0), (2,0)$ and $(0.5,2)$ and (**b**) Q has 3 fixed vertices $(0,0), (2,0.5)$ and $(0.5,2)$.

Now, we reorder these quantities in such a way that

$$g_1 \leq g_2 \leq g_3 \leq g_4, \tag{21}$$

and use

$$\mu_{p2}(Q) = \frac{g_1 g_2}{g_3 g_4}, \tag{22}$$

as a quality measure. Again, the maximum value is 1 and it is obtained for parallelograms. This is a quality measure because it is continuous, bounded and identifies degenerate and even non-convex quadrilaterals, see [4].

In the Figure 6 it is shown different level curves for $\mu_{p2}(Q)$.

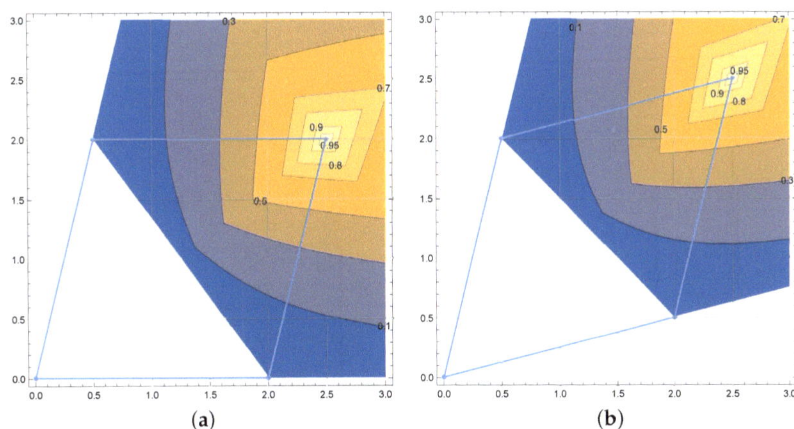

Figure 6. Level curves for (22) where (**a**) Q has 3 fixed vertices $(0,0), (2,0)$ and $(0.5,2)$ and (**b**) Q has 3 fixed vertices $(0,0), (2,0.5)$ and $(0.5,2)$.

3.4. Quality Measure of Squares

An ideal mesh is one in which its cells are close to being squares. If $\mu(T)$ is a good quality measure for triangles, the harmonic mean of the four triangles T_i is

$$\mu_s(Q) = \sigma \frac{4}{\sum_{i=1}^{4} \frac{1}{\mu(T_i)}}, \tag{23}$$

If rewriting $\mu_s(Q)$ in the form

$$\mu_s(Q) = \frac{4\sigma\mu(T_1)\mu(T_2)\mu(T_3)\mu(T_4)}{\mu(T_2)\mu(T_3)\mu(T_4) + \mu(T_1)\mu(T_3)\mu(T_4) + \mu(T_1)\mu(T_2)\mu(T_4) + \mu(T_1)\mu(T_2)\mu(T_3)}, \tag{24}$$

we obtain a good quality measure for quadrilaterals, because it is continuous, bounded, invariant under the rigid and scaling transformations and identifies degenerate and even non-convex quadrilaterals, since it inherits those properties from $\mu(T)$. Here, σ is a normalization parameter.

To characterize squares, we require a property $\mu(T)$ as seen from

Theorem 2. *If $\mu(T)$ is a good quality measure for triangles according to Field-Oddy, in which for isosceles right triangle the highest energy among all right triangles is achieved, $\mu_s(Q)$ defined in (24) is a quality measure in the Field-Oddy sense and characterizes squares at their maximum value.*

Proof. The proof is very simple; it is based on the fact that the four triangles must be congruent to have the same energy. From this it follows that the triangles must be right triangle or they do not form a quadrilateral. Now, if the lowest energy contained for right triangle only occurs when they are isosceles right triangle then $\mu_s(Q)$ defined in (24) only detects squares at its maximum value.

Some measures $\mu(T)$ for triangles with those properties are

$$\mu_1(T) = 4\sqrt{3}\frac{A}{l_1^2 + l_2^2 + l_3^2}, \quad \mu_2(T) = 2\frac{r}{R}, \quad \mu_3(T) = \frac{4\sqrt{3}}{9}\frac{A}{R^2}, \quad \mu_4(T) = \frac{4}{\sqrt{3}}\frac{A}{l_{\max}^2}$$

where $\mu_1(T)$ was proposed by Joe, [14], $\mu_2(T)$ is the *radius ratio* measure, $\mu_3(T)$ is described by Shewchuk and $\mu_4(T)$ is Cavendish's measure, see [15].

Using the quadrilateral Q with 3 fixed vertices $(0,0)$, $(2,0)$ and $(0,2)$, $\mu(Q)$ is a function of (x,y). In the Figure 7 it is shown the level curves for $\mu_s(Q)$ using $\mu_1(T)$ and $\mu_2(T)$. $\square$

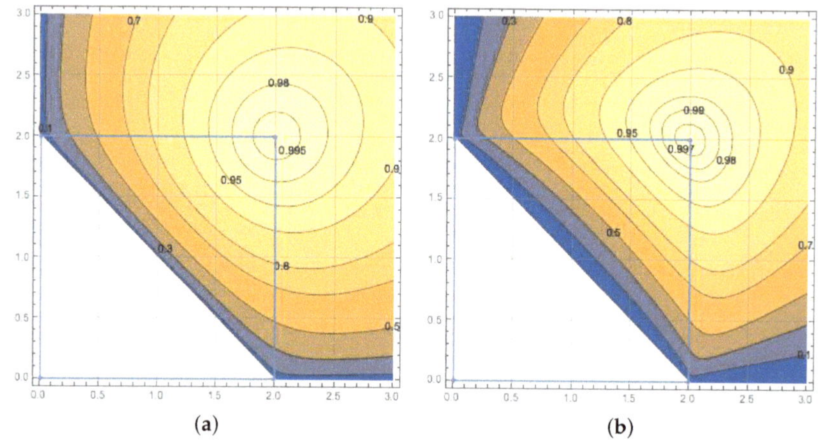

(a)   (b)

Figure 7. Level curves for (24) using (**a**) $\mu_1(T)$ and (**b**) *radius ratio* $\mu_2(T)$.

4. Some Global Quality Metrics

In many applications, it is necessary to identify the utility of a mesh through the assessment of the quality of all the mesh elements.

The assessment of the quality of a mesh can be achieved through

(1)   A visual or exploratory or inspection;
(2)   Qualitative evaluation or shape parameters;
(3)   Statistical analysis.

A first visual evaluation can involve analyzing the distribution in a histogram of values of a chosen good-quality measure $\mu(Q)$ or metric $\delta(Q)$. A second visual assessment can be performed by looking at a color map on a quality-dependent color scale. In this section, we will attempt to describe some methods or techniques with which to carry out a global qualitative assessment of the mesh by assigning a value to the mesh.

Allievi and Casal [16] proposed two measures for qualitative evaluation of orthogonality of the mesh. The first criteria is the maximum deviation of orthogonality given by

$$\text{MDO} = \max_{i,j}\{|90° - \theta_{i,j}|\} \tag{25}$$

and the second is the mean deviation of orthogonality given by

$$\text{ADO} = \frac{1}{(n-1)(m-1)} \sum_{i=2}^{m-1} \sum_{j=2}^{n-1} |90° - \theta_{i,j}|, \tag{26}$$

where $\theta_{i,j}$ are the internal angles of the mesh.

Now, let $\mu(Q)$ be a measure of quality. Other global criteria used frecuently is the average quality of a mesh G, or the mean quality, defined as

$$MQ = \bar{\mu}(G) = \frac{1}{N} \sum_{i=1}^{N} \mu(Q_i), \tag{27}$$

where N is the number of elements in G. Another global measure well known is the standard deviation or the mean square error:

$$MSE = \sigma^2 = \sqrt{\frac{1}{N} \sum_{i=1}^{N} (\mu(Q_i) - \bar{\mu}(G))^2}, \tag{28}$$

which is a value that represents the averages of all the individual differences of the observations with respect to a common reference point, which is the arithmetic mean.

As it is well known, a greater value of MSE corresponds to a greater dispersion of the values, in this case $\mu(Q)$ with respect to its mean MQ. For this reason, researchers are using the geometric mean as a measure of global quality of the mesh for a measure $\mu(Q)$ given by

$$SP = \beta = \sqrt[N]{\prod_{i=1}^{N} \mu(Q_i)}. \tag{29}$$

This quantity is well known in the literature as the mesh shape parameter or simply shape parameter, see [2].

The natural approach to evaluate the quality of an mesh from that of its elements consists of considering the best and worst element qualities, the arithmetic mean, the mean square error and the shape parameter. For the mesh of the Figure 8 with ADO = 13.21, MDO = 76.17, the corresponding results for differents quality measures are given in Table 1.

Table 1. Summary of quadrangle quality measures for the mesh in Figure 8.

Shape	Name $\mu(Q)$	Min	Max	MQ	MSE	SP
Parallelogram	AreaI	0.0381	0.9998	0.9169	0.1081	0.9068
	AreaII	0.0117	0.9973	0.8615	0.1510	0.8284
Rectangle	Lo1989	0.0160	0.9941	0.7597	0.1901	0.7241
	ScaledJacobian	0.2391	1.0000	0.9384	0.0975	0.9318
	ScaledJacobianM	0.0457	0.9965	0.8639	0.1361	0.8478
	MinRect2015	0.0555	0.9942	0.7929	0.1837	0.7607
	Rectangles2015	0.0719	0.9983	0.8999	0.1170	0.8890
Square	Lo1985	0.0339	0.9097	0.4792	0.2362	0.4162
	Hua1995	0.0830	0.9996	0.5302	0.2573	0.4621
	Knupp2000	0.0365	0.9993	0.5113	0.2588	0.4402
	Pebay2002	0.0591	0.9814	0.6119	0.2208	0.5670
	Hmean2017E	0.0797	0.9996	0.5292	0.2578	0.4607
	Hmean2017R	0.0788	0.9996	0.5292	0.2578	0.4606
	Hmean2017r	0.0297	0.9996	0.5351	0.2485	0.4713

We believe that a statistical approach should be used to qualify and quantify the geometric quality of a mesh. The shape parameter for squares or rectangles can be combined with a statistical analysis of all the elements of the mesh in the following order:

(1) How many elements are squares?
(2) How many elements are rectangles with aspect ratio less than 4?
(3) How many elements are parallelograms with aspect ratio less than 4?

We will say that the remainer cells are distorted. This can be performed as follows: consider three measures of quality $\mu_s(Q)$, $\mu_r(Q)$ and $\mu_p(Q)$ for squares, rectangles and parallelograms, and exclude rectangles from squares and parallelograms from rectangles. Let us exclude, respectively, large-aspect-ratio rectangles and large skew elements. Now let us represent those elements in a colormap, see Figure 8.

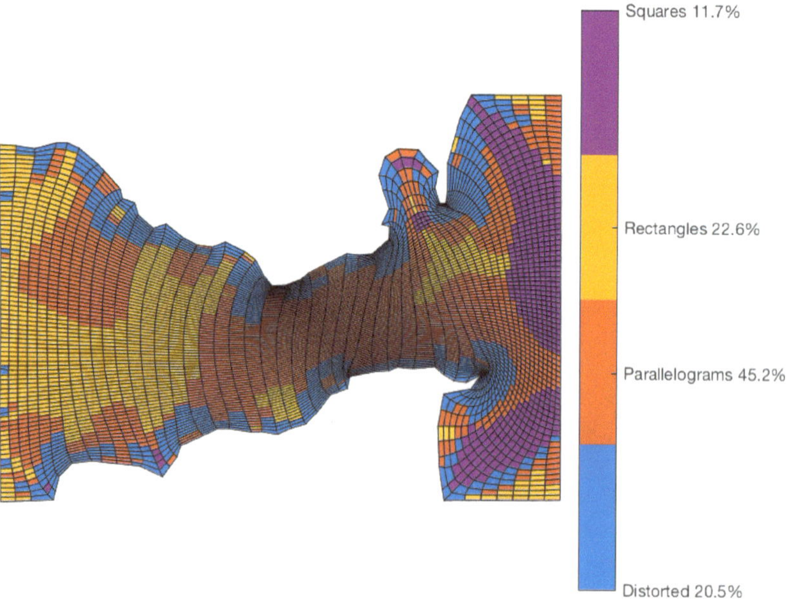

Figure 8. Characterized mesh elements with different shapes.

This technique can be very useful for meshes over irregular regions such as lakes or reservoirs.

In this paper the quality measure for curved elements are not discussed, but a good reference for curvilinear finite elements is [17].

5. Grid Quality Improvement

On Distortion of the Mesh

In general, if $\mu(Q)$ is a good quality measure for quadrilaterals, a way of measuring the distortion of a quadrilateral Q with respect to $\mu(Q)$ is using

$$f(Q) = \frac{1}{\mu(Q)}, \tag{30}$$

because if $f(Q)$ is much greater than 1, the cell will be far from the value for which $\mu(Q)$ characterizes the geometric shape of the cell Q (square, rectangle, parallelogram, etc.) and we can say that Q is a distorted quadrilateral with respect to that measure. Usually, the maximum value of a quality measure corresponds to the minimum value of the energy density over the grid, see Ivanenko [18].

Under this idea, the distortion of the mesh G can be measured as the average of the distortions of all the cells

$$F(G) = \frac{1}{Ne} \sum_{k=1}^{Ne} \frac{1}{\mu(Q_k)}, \tag{31}$$

where Ne its the number of the cells. Using this concept, we have the following definition:

Definition 2. *A grid $\hat{G}$ has better quality than the mesh $\bar{G}$ if*

$$F(\hat{G}) < F(\bar{G}), \tag{32}$$

where $F(G)$ is a distorsion measure.

As an optimization problem, improving the quality of a G mesh can be considered as the problem

$$G^* = \arg\min_G F(G) = \frac{1}{Ne} \sum_{k=1}^{Ne} \frac{1}{\mu(Q_k)}, \tag{33}$$

where the inner nodes of G are the unknowns. In this context, the discrete grid generation problem can be posed, in general, as a large scale optimization problem. The optimization problem is a large-scale one when the mesh dimensions $m \times n$ are very large. It is important to note that the initial mesh G_0 must be convex and remain so in each step of the optimization process. We use for this a Newton-like methods with bound constraints L-BFGS-B [19].

Usually, the quality measures for quadrilaterals are non-differentiable functions; in an optimization process, it is better to build a convex function with similar characteristics to the quality measure.

6. New Quality Discrete Functionals

From the proof of Theorem 1, it is easy to see that

$$2a_c \leq \sqrt{(a^2 + c^2)(b^2 + d^2)}, \tag{34}$$

for any convex quadrilateral, then

$$f_R(Q) = \frac{(a^2 + c^2)(b^2 + d^2)}{4a_c^2}, \tag{35}$$

is a positive convex function whose critical points are rectangles. With this function, we can define a discrete functional $F_R(G)$ over all the grid cells

$$F_R(G) = \frac{1}{Ne} \sum_{k=1}^{Ne} f_R(Q_k).$$

(36)

In Figure 9, the shape of the surface of F_R is sketched.

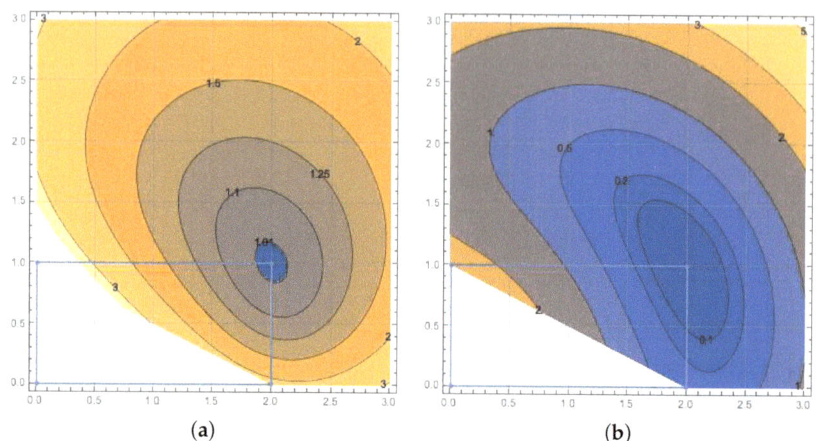

(a)   (b)

Figure 9. Level curves for (**a**) $f_R(Q)$ and (**b**) $f_r(Q)$ where Q has 3 fixed vertices $(0,0)$, $(2,0)$ and $(0,1)$.

We propose to combine this functional with a convex area functional $S_w(G)$ defined in [3]. This functional has an infinite barrier at the boundary of the set of unfolded grids.

$$F(G) = (1 - \sigma)S_w(G) + \sigma F_R(G),$$

(37)

where $\sigma > 0$. In addition, the function

$$f_R(Q) = \frac{(a^2 + c^2)(b^2 + d^2)}{4a_c^2},$$

(38)

can be interpreted (by cells) as a normalization (with respect to the Jacobian) of Knupp's area-orthogonality functional defined in [20]

$$f_{ao}(Q) = (a^2 + c^2)(b^2 + d^2).$$

(39)

As we have discussed, since the quality measures are usually non-differentiable functions, it is difficult to use them as objective functions; it is advisable to design the convex and differentiable functions $f(Q)$, whose optimal values also satisfy $\mu(Q) \approx 1$ for a specific quality measure $\mu(Q)$.

As it is known, a rectangle is a parallelogram, so its opposite sides are equal. The diagonals of a rectangle are equal and bisect each other. In Figure 10, we can see those elements.

With this idea, we propose to use a convex functional of the form

$$F_p(G) = \sum_{i,j} \left\| \frac{\mathbf{p}_{i,j} + \mathbf{p}_{i+1,j+1}}{2} - \frac{\mathbf{p}_{i+1,j} + \mathbf{p}_{i,j+1}}{2} \right\|^2,$$

(40)

over all elements of the grid G. Locally, this functional has a critical point in the cells formed by parallelograms, see Khattri [21].

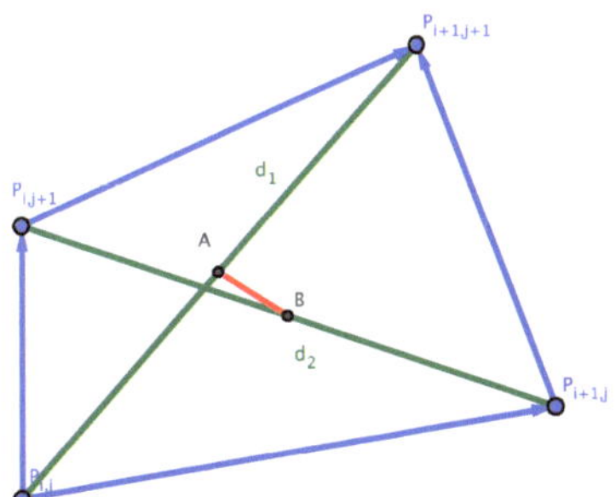

Figure 10. Diagonals of a quadrilateral and the segment joining the midpoints between them.

Optimizing $F_p(G)$ is an attempt to produce parallelograms. Now, we define a discrete functional to obtain rectangles. For each cell of G let us measure the square of the difference of the square of diagonals

$$F_d(G) = \sum_{i,j}(\|\mathbf{p}_{i,j} - \mathbf{p}_{i+1,j+1}\|^2 - \|\mathbf{p}_{i+1,j} - \mathbf{p}_{i,j+1}\|^2)^2; \tag{41}$$

combining both functionals, we obtain

$$F_r(G) = (1 - \alpha)F_p(G) + \alpha F_d(G). \tag{42}$$

If where $\alpha \geq 0$ is chosen to allow that shape of the cells can be flexible. In the practice we use $\alpha = 0.5$. $F_r(G)$ is a positive and convex functional, which has a critical point in a mesh formed by rectangles (including squares). This can always be achieved if we guarantee that in each optimization step the mesh is convex.

Therefore, we use $S_w(G)$ to guarantee and preserve the convexity of the mesh, and combine it with the latter functional in the form

$$F(G) = (1 - \sigma)S_\omega(G) + \sigma F_r(G). \tag{43}$$

Thus, with a linear convex combination between $S_\omega(G)$ and $F_r(G)$, one can generate both convex grid and close to being rectangles.

7. Examples

For both functionals, we can obtain optimal grids whose cells have a very large aspect ratio, see Figure 11.

Now, for control of the aspect ratio, we propose to use an area (volume) distortion measure functional

$$F_A(G) = \sum_{q=1}^{N} \frac{1}{\alpha(\triangle_q)^2} + \delta\alpha(\triangle_q)^2, \tag{44}$$

over all the N signed areas of all the grid cell triangles. Here, $\delta > 0$ is an adequate value, see [4]. This distortion measure has a barrier on the boundary of the set of grids consisting of convex quadrilateral cells and it is very similar to the one proposed by Garanzha [22], but now we have a better control of the global distribution of the area.

For the mesh optimized using the functional (37) with area control, we show in Figure 12 the color maps for two quality measures.

For the mesh optimized using the functional (43) with area control, we show in Figure 13 the color maps for two quality measures.

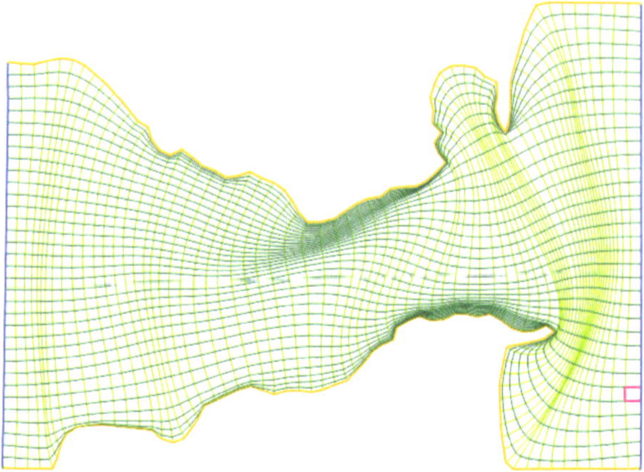

Figure 11. A mesh over the Strait of Gibraltar.

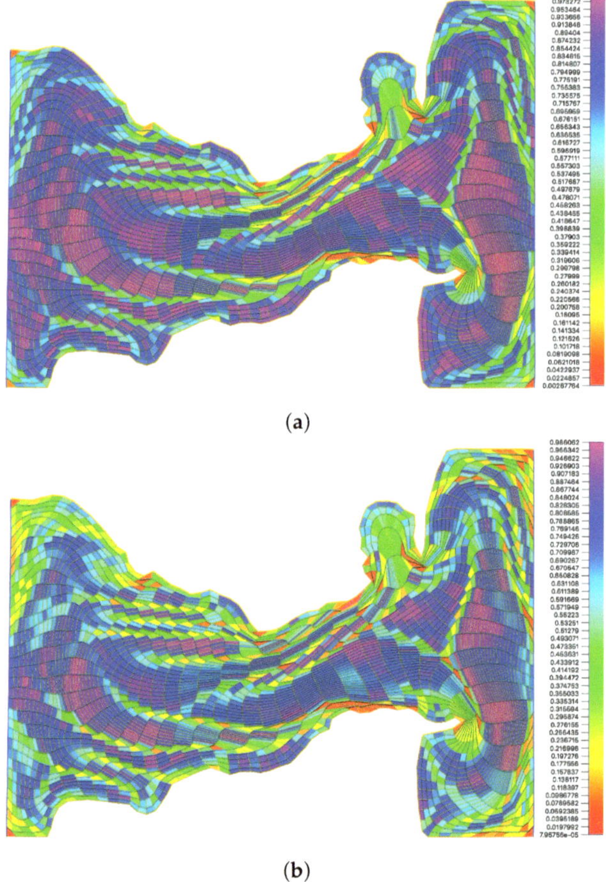

(a)

(b)

Figure 12. Color map of (**a**) rectangles quality measure and (**b**) rectangle of minimum area quality measure.

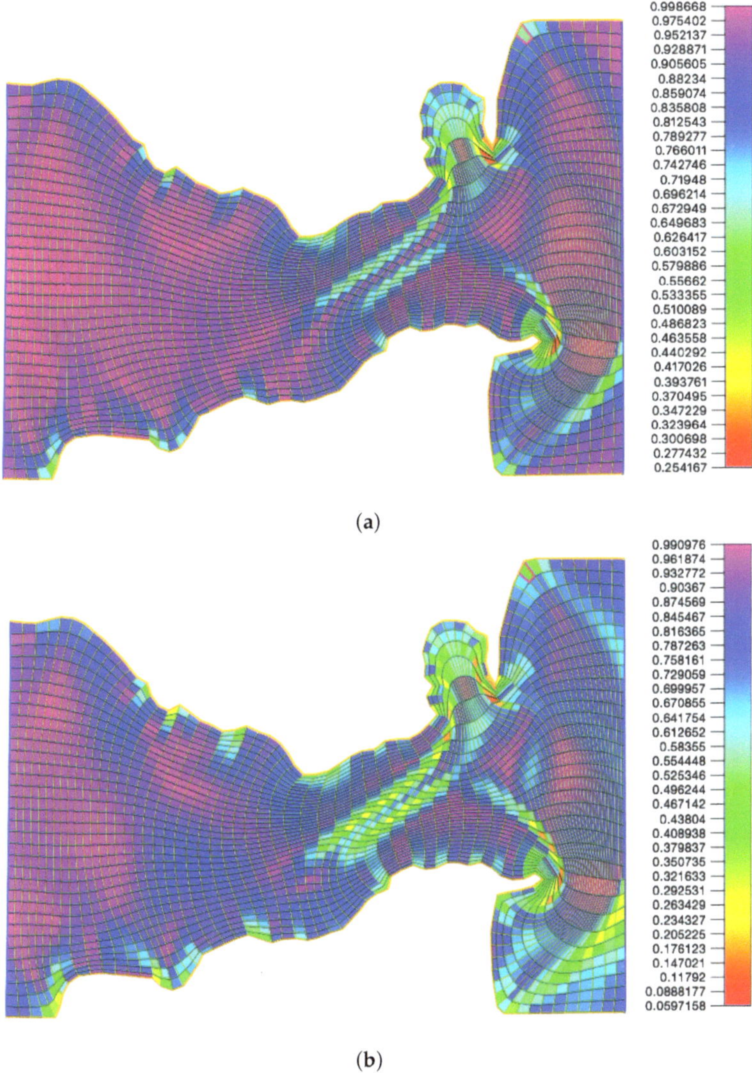

(a)

(b)

Figure 13. Color map of (**a**) rectangles quality measure and (**b**) rectangle of minimum area quality measure.

As it is well known, for irregular regions, the distorted cells accumulate near the border.

8. Conclusions

In conclusion, this paper presented an overview of classical quality measures and introduced new quality measures for quadrilaterals that help to improve meshes and the aspect ratio. We show that our aspect ratio is better than the one proposed by Robinson. We propose that a statistical approach should be used to qualify and quantify the geometric quality of a mesh. In addition, we have proposed new functionals for grid generation as alternatives for area-orthogonal grid generation. These functionals are based on the fact that the maximum value of a quality measure corresponds to the minimum value of the energy density over the grid.

Future work in this area could include extending some of these ideas to 3D, as well as exploring the potential of functionals for volume grid generation, which need to be further investigated.

Author Contributions: Conceptualization, G.F.G.F. and P.B.S.; methodology, G.F.G.F. and P.B.S.; writing—original draft preparation, G.F.G.F. and P.B.S.; writing—review and editing, G.F.G.F. and P.B.S. All authors have read and agreed to the published version of the manuscript.

Funding: This research received no external funding.

Acknowledgments: The authors would like to thank the Posgrado en Ciencias Matemáticas of UNAM for continued support.

Conflicts of Interest: The authors declare no conflict of interest.

References

1. Thompson, J.F.; Soni, B.K.; Weatherill, N.P. (Eds.) *Handbook of Grid Generation*, 1st ed.; CRC Press: Boca Raton, FL, USA, 1998; ISBN 9781420050349. [CrossRef]
2. Lo, S.H. *Finite Element Mesh Generation*; CRC Press: Boca Raton, FL, USA, 2015; ISBN 041569048X.
3. Barrera-Sánchez, P.; Cortés, J.J.; Domínguez-Mota, F.J.; González-Flores, G.F.; Tinoco-Ruiz, J.G. Smoothness and Convex Area Functional—Revisited. *SIAM J. Sci. Comput.* **2010**, *32*, 1913–1928. [CrossRef]
4. González Flores, G.F. Numerical Structured Grid Generation Adaptive and Quality Control. Ph.D. Thesis, Universidad Nacional Autónoma de México, Ciudad de México, Mexico, 2018.
5. Robinson, J. Some New Distortion Measures for Quadrilaterals. *Finite Elem. Anal. Des.* **1987**, *3*, 183–197. [CrossRef]
6. Field, D. Qualitative Measures for Initial Meshes. *Int. J. Numer. Methods Eng.* **2000**, *47*, 887–906. [CrossRef]
7. Lo, S.H. Generating Quadrilateral Elements on Plane and over Curved Surfaces. *Comput. Struct.* **1989**, *31*, 421–426. [CrossRef]
8. Knupp, P. Algebraic Mesh Quality Metrics. *SIAM J. Sci. Comput.* **2001**, *23*, 193–218. [CrossRef]
9. Van Rens, B.J.E.; Brokken, D.; Brekelmans, W.A.M.; Baaijens, F.P.T. A Two-dimensional Paving Mesh Generator for Triangles with Controllable Aspect Ratio and Quadrilaterals with High Quality. *Eng. Comput.* **1998**, *14*, 248–259. [CrossRef]
10. Remacle, J.-F.; Lambrechts, J.; Seny, B.; Marchandise, E.; Johnen, A.; Geuzaine, C. Blossom-Quad: A Non-uniform Quadrilateral Mesh Generator using a Minimum Cost Perfect Matching Algorithm. *Int. J. Numer. Methods Eng.* **2012**, *89*, 1102–1119. [CrossRef]
11. Wu, L. Realization of Quadrilateral Mesh Partition and Optimization Algorithm Based on Cloud Data. *J. Comput.* **2011**, *6*, 2519–2525. [CrossRef]
12. Lassak, M. Approximation of Convex Bodies by Rectangles. *Geom. Dedicata* **1993**, *47*, 111–117. [CrossRef]
13. Josefsson, M. Five Proofs of an Area Characterization of Rectangle. *Forum Geom.* **2013**, *13*, 17–21.
14. Joe, B. Shape Measures for Quadrilaterals, Pyramids, Wedges and Hexahedra. Technical Report ZCS2008-03. Available online: https://citeseerx.ist.psu.edu/viewdoc/download;jsessionid=CC3EA480D0EF447EE2FBC91084EC784F?doi=10.1.1.577.5216&rep=rep1&type=pdf (accessed on 31 May 2023).
15. Shewchuk, J.R. What Is a Good Linear Finite Element? Interpolation, Conditioning, Anisotropy, and Quality Measures. Available online: https://people.eecs.berkeley.edu/~jrs/papers/elemj.pdf (accessed on 31 May 2023).
16. Allievi, A.; Calisal, S.M. Application of Bubnov-Galerkin formulation to orthogonal grid generation. *J. Comput. Phys.* **1992**, *98*, 163–173. [CrossRef]
17. Johnen, A.; Geuzaine, C.; Toulorge, T.; Remacle, J.F. Quality Measures for Curvilinear Finite Elements. In *TILDA: Towards Industrial LES/DNS in Aeronautics*; Hirsch, C., Hillewaert, K., Hartmann, R., Couaillier, V., Boussuge, J.-F., Chalot, F., Bosniakov, S., Haase, W., Eds.; Notes on Numerical Fluid Mechanics and Multidisciplinary Design; Springer: Cham, Switzerland, 2021; p. 148. [CrossRef]
18. Ivanenko, S.A. Control of Cell Shapes in the Course of Grid Generation. *Comput. Math. Math. Phys.* **2000**, *40*, 1596–1684.
19. Zhu, C.; Byrd, R.H.; Nocedal, J. Algorithm 778: L-BFGS-B: Fortran Routines for Largescale Bound-constrained Optimization. *ACM Trans. Math. Softw.* **1997**, *23*, 550–560. [CrossRef]
20. Knupp, P. A Robust Elliptic Grid Generator. *J. Comput. Phys.* **1992**, *100*, 409–418. [CrossRef]
21. Khattri, S.K. A New Smoothing Algorithm for Quadrilateral and Hexahedral Meshes. In *Lecture Notes in Computer Science*; Springer: Berlin/Heidelberg, Germany, 2006; Volume 3992, pp. 239–246.
22. Garanzha, V.A. Barrier Variational Generation of Quasi-isometric Grids. *Numer. Linear Algebra Appl.* **2001**, *8*, 329–353. [CrossRef]

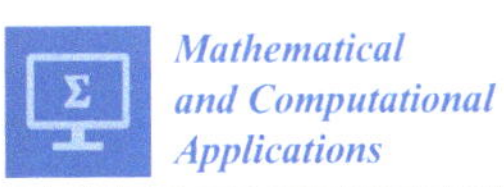

Mathematical and Computational Applications

MDPI

Article

Predictive Modeling and Control Strategies for the Transmission of Middle East Respiratory Syndrome Coronavirus

Bibi Fatima [1], Mehmet Yavuz [2,*], Mati ur Rahman [3], Ali Althobaiti [4] and Saad Althobaiti [5]

[1] Department of Mathematics, University of Malakand, Chakadara 18800, Pakistan; fatma.uom@gmail.com
[2] Department of Mathematics and Computer Sciences, Faculty of Science, Necmettin Erbakan University, 42090 Konya, Türkiye
[3] Department of Computer Science and Mathematics, Lebanese American University, Beirut 11022801, Lebanon; matimaths374@gmail.com
[4] Department of Mathematics, College of Science, Taif University, P.O. Box 11099, Taif 21944, Saudi Arabia; aa.althobaiti@tu.edu.sa
[5] Department of Sciences and Technology, Ranyah University Collage, Taif University, P.O. Box 11099, Taif 21944, Saudi Arabia; snthobaiti@tu.edu.sa
* Correspondence: mehmetyavuz@erbakan.edu.tr

Abstract: The Middle East respiratory syndrome coronavirus (MERS-CoV) is a highly infectious respiratory illness that poses a significant threat to public health. Understanding the transmission dynamics of MERS-CoV is crucial for effective control and prevention strategies. In this study, we develop a precise mathematical model to capture the transmission dynamics of MERS-CoV. We incorporate some novel parameters related to birth and mortality rates, which are essential factors influencing the spread of the virus. We obtain epidemiological data from reliable sources to estimate the model parameters. We compute its basic reproduction number (R0). Stability theory is employed to analyze the local and global properties of the model, providing insights into the system's equilibrium states and their stability. Sensitivity analysis is conducted to identify the most critical parameter affecting the transmission dynamics. Our findings revealed important insights into the transmission dynamics of MERS-CoV. The stability analysis demonstrated the existence of stable equilibrium points, indicating the long-term behavior of the epidemic. Through the evaluation of optimal control strategies, we identify effective intervention measures to mitigate the spread of MERS-CoV. Our simulations demonstrate the impact of time-dependent control variables, such as supportive care and treatment, in reducing the number of infected individuals and controlling the epidemic. The model can serve as a valuable tool for public health authorities in designing effective control and prevention strategies, ultimately reducing the burden of MERS-CoV on global health.

Keywords: MERS-CoV model; basic reproductive number; analysis of stability; equilibria points; optimality control; numerical analysis

Citation: Fatima, B.; Yavuz, M.; ur Rahman, M.; Althobaiti, A.; Althobaiti, S. Predictive Modeling and Control Strategies for the Transmission of Middle East Respiratory Syndrome Coronavirus. *Math. Comput. Appl.* **2023**, *28*, 98. https://doi.org/10.3390/mca28050098

Academic Editor: Cristiana João Soares da Silva

Received: 26 May 2023
Revised: 8 September 2023
Accepted: 26 September 2023
Published: 30 September 2023

1. Introduction

The first identification of the Middle East respiratory syndrome coronavirus (MERS-CoV), a viral respiratory illness, took place in Saudi Arabia in 2012, as reported by multiple studies, including those conducted by [1–3]. MERS-CoV is believed to have originated from an animal source and has been identified in both humans and animals. The transmission of the disease occurs through close contact with an infected individual, in any form. The World Health Organization (2019) has reported a global total of 2519 laboratory-confirmed cases of MERS-CoV infection, with 866 associated deaths. One of the largest outbreaks of MERS-CoV occurred in South Korea in 2015.

Coronaviruses constitute a diverse family of viruses that are known to infect humans, causing respiratory illnesses that can vary in severity from mild cold-like symptoms to severe respiratory syndromes, such as severe acute respiratory syndrome (SARS). MERS-CoV can cause zoonotic infections in humans through direct or indirect contact with camels or camel-related products. Additionally, human-to-human transmission has been reported, particularly in healthcare settings [4–6]. Since 2002, three novel coronaviruses have emerged and caused deadly zoonotic diseases in humans. The first was SARS, which emerged in November 2002. The second was MERS, which emerged in April 2012. The most recent and ongoing pandemic is COVID-19, which emerged in December 2019 and has affected millions of people worldwide [7–10]. The authors in [11,12] modified a mathematical model of COVID-19 by including the quarantine class and measured the disease transmission. Shen et al. studied a mathematical model of COVID-19 by presenting the vaccinated class with an optimal control analysis [13]. The authors in [14] used the data of Saudi Arabia and investigated the transmission dynamics of COVID-19. Tsay et al. analyzed the state estimation and optimal control for the COVID-19 outbreak model in the US [15]. Libotte et al. used the optimal strategy for vaccines in COVID-19 treatment considered for both mono- and multi-objective optimization [16]. Using the optimal control models also gives information about the impact of individual vaccination during an epidemic, together with the key considerations for political and economic decision making.

MERS typically presents with symptoms such as fever, cough, and shortness of breath. It is believed to spread through respiratory secretions, such as through coughing, from an infected person, similar to other coronaviruses. Several studies have investigated the potential role of camel handlers in the transmission of the virus to determine its source of infection. To understand and predict the dynamics of infectious diseases, researchers have developed various models based on biologically feasible parameters [17–19]. These models are essential tools for analyzing and forecasting the spread of diseases [20]. Although several case studies have explored the transmission of MERS-CoV, the literature on its transmission dynamics is limited. Using available data, Cauchemez et al. [21] estimated the incubation period and generation time of MERS-CoV, and calculated the reproductive numbers for both animal-to-human and human-to-human transmission. Chowell et al. [22] took a different approach and compared the reproductive numbers of SARS and MERS. Assiri et al. [23] reported one of the largest outbreaks of MERS-CoV, describing the virus as transmissible from human to human. The virus has spread globally through travel-associated cases, with reported incidences in countries including Algeria, Austria, China, Egypt, Italy, Netherlands, Philippines, South Korea, Thailand, the UK, and the US. Ground-breaking research has been published by numerous esteemed researchers, delving into the exploration of various model types, such as: SIR epidemic models [24,25], the discrete-time prey–predator model [26], and the memristor system [27]. Several infectious disease models have been investigated by researchers by using different approaches, which are available in the literature, such as [28–32].

Members of the coronavirus family, MERS-CoV and COVID-19 (caused by SARS-CoV-2) have certain similarities. It is crucial to remember that these viruses are diverse from one another and have distinctive traits and effects on human health. The following examples demonstrate how MERS-CoV can be used to treat various illnesses, including COVID-19. Both COVID-19 and MERS-CoV can cause serious respiratory infections in people. However, their overall effects differ considerably in a number of ways. MERS-CoV and COVID-19 can also be compared to influenza viruses, particularly those that cause severe respiratory infections. Different influenza viruses (A, B, and C) are what cause the illness. Despite the fact that some symptoms and the means of transmission are similar, influenza viruses have unique genetic traits and often create seasonal outbreaks. However, MERS-CoV and SARS-CoV-2 can cause pandemics or sporadic epidemics with ongoing human-to-human transmission. Several researchers have suggested mathematical models by applying different approaches to an infectious disease and studying its dynamics from different angles [33–37].

As far as the novelty is concerned, we study the model presented in [38], by incorporating the natural birth rate and death rate due to MERS-CoV. We modified the mathematical model for MERS-CoV transmission dynamics. This model consists of six groups: susceptible class $\mathcal{S}$, exposed class (or high risk latent) $\mathcal{E}$, symptomatic and infectious class $\mathcal{I}$, infectious but asymptotic class $\mathcal{A}$, hospitalized class $\mathcal{H}$, and recovery class $\mathcal{R}$. After constructing the model, the basic reproductive number is calculated by using the next generation method, and the local and global stability of the equilibrium points are determined. Lyapunov function theory is then utilized to analyze the global behavior of the model. Furthermore, the principles of optimal control theory are employed to reduce the number of infected persons and maximize the recovery rate within a given population.

2. Model Formulation

Here, we study the mathematical formulation of the deterministic model for MERS-CoV, using a set of differential equations. Specifically, the model describes the dynamics of the host population using the following system of equations:

$$
\begin{aligned}
\dot{\mathcal{S}}(t) &= b\mathrm{N} - \frac{\varphi \mathcal{I} \mathcal{S}}{\mathrm{N}} - \frac{\varphi q \mathcal{H} \mathcal{S}}{\mathrm{N}} - \eta_0 \mathcal{S}, \\
\dot{\mathcal{E}}(t) &= \frac{\varphi \mathcal{I} \mathcal{S}}{\mathrm{N}} + \frac{\varphi q \mathcal{H} \mathcal{S}}{\mathrm{N}} - (\chi + \eta_0)\mathcal{E}, \\
\dot{\mathcal{I}}(t) &= \chi \xi \mathcal{E} - (\vartheta_a + \vartheta_1)\mathcal{I} - (\eta_0 + \eta_1)\mathcal{I}, \\
\dot{\mathcal{A}}(t) &= \chi(1 - \xi)\mathcal{E} - (\eta_0 + \eta_2)\mathcal{A}, \\
\dot{\mathcal{H}}(t) &= \vartheta_a \mathcal{I} - \vartheta_\vartheta \mathcal{H} - \eta_0 \mathcal{H}, \\
\dot{\mathcal{R}}(t) &= \vartheta_1 \mathcal{I} + \vartheta_\vartheta \mathcal{H} - \eta_0 \mathcal{R},
\end{aligned}
\tag{1}
$$

with initial conditions

$$
\mathcal{S}(0) \geq 0, \; \mathcal{E}(0) \geq 0, \; \mathcal{I}(0) \geq 0, \; \mathcal{A}(0) \geq 0, \; \mathcal{H}(0) \geq, \; \mathcal{R} \geq 0,
$$

where the used parameters in the above system are: $b\mathrm{N}$ represents the rate of birth for the host populace, while the transmission rate from human to human per unit time is represented by φ. The parameter q determines the relative transmissibility of hospitalized individuals. χ represents the rate at which individuals transition from the exposed compartment $\mathcal{E}$ to the infectious compartment $\mathcal{I}$. The proportion of individuals who progress from $\mathcal{E}$ to $\mathcal{I}$ is given by ξ, while the remaining $(1 - \xi)$ progress to class $\mathcal{A}$. The average rate at which symptomatic persons are hospitalized is denoted by ϑ_a, while ϑ_1 represents the rate of recovery without hospitalization, and ϑ_ϑ represents the rate of recovery of hospitalized patients. The rate of natural death is represented by η_0, while η_1 and η_2 represent deaths due to MERS-CoV.

Assume that the total populace is represented by $\mathrm{N}(t)$ at time t, and satisfies $\mathrm{N}(t) = \mathcal{S} + \mathcal{E} + \mathcal{I} + \mathcal{A} + \mathcal{H} + \mathcal{R}$.

Adding all the equations of system (1), we have

$$
\frac{d\mathrm{N}}{dt} = b\mathrm{N} - \eta_0 \mathcal{S} - \eta_0 \mathcal{E} - (\eta_0 + \eta_1)\mathcal{I} - (\eta_0 + \eta_2)\mathcal{A} - \eta_0 \mathcal{H} - \eta_0 \mathcal{R}.
$$

Therefore, from the above relation for biological applications, the considered system (1) occurred in the closed set as

$$
\mathbb{F} = \left\{ (\mathcal{S}, \mathcal{E}, \mathcal{I}, \mathcal{A}, \mathcal{H}, \mathcal{R}) \in \mathcal{R}_+^6, 0 <, \mathcal{S} + \mathcal{E} + \mathcal{I} + \mathcal{A} + \mathcal{H} + \mathcal{R} \leq \frac{b\mathrm{N}}{\eta_0} \right\}.
$$

3. Basic Reproduction Number

The basic reproduction number determines whether an epidemic will appear or the infection will die out. It represents the expected average number of new infections that will be generated by a single infective person, both directly and indirectly, when introduced into a fully susceptible populace. In this study, we use the approach of Driessche and

Watmough [39,40] to calculate the basic reproduction number for the aforementioned system (1).

$$F = \begin{bmatrix} 0 & \frac{\varphi S_0}{N} & \frac{\varphi q S_0}{N} \\ 0 & 0 & 0 \\ 0 & 0 & 0 \end{bmatrix}, V = \begin{bmatrix} \chi + \eta_0 & 0 & 0 \\ -\chi\xi & (\vartheta_a + \vartheta_1) + (\eta_0 + \eta_1) & 0 \\ 0 & -\vartheta_a & \vartheta_\vartheta + \eta_0 \end{bmatrix} \tag{2}$$

to find

$$FV^{-1} = \begin{pmatrix} \frac{\chi\varphi\xi S_0(\vartheta_\vartheta+\eta_0+q\vartheta_a)\mathcal{I}}{N(\chi+\eta_0)(\vartheta_\vartheta+\eta_0)[(\vartheta_a+\vartheta_1)+(\eta_0+\eta_1)]} & \frac{\varphi S_0(\vartheta_\vartheta+\eta_0+q\vartheta_a)}{N(\vartheta_\vartheta+\eta_0)[(\vartheta_a+\vartheta_1)+(\eta_0+\eta_1))]} & \frac{\varphi S_0 q}{N[(\vartheta_\vartheta+\eta_0)(\vartheta_a+\vartheta_1)+(\eta_0+\eta_1)]} \\ 0 & 0 & 0 \\ 0 & 0 & 0 \end{pmatrix}.$$

Thus, the required basic reproduction number $\mathbf{R}_0$ is followed by

$$\mathbf{R}_0 = \frac{\chi\varphi\xi S_0 Q_1}{N(\chi + \eta_0)Q_2 Q_3}.$$

The terms Q_1, Q_2, and Q_3 are defined as follows:
$Q_1 = (\vartheta_\vartheta + \eta_0 + q\vartheta_a)$;
$Q_2 = (\vartheta_\vartheta + \eta_0)$;
$Q_3 = (\vartheta_a + \vartheta_1) + (\eta_0 + \eta_1)$.
These terms correspond to the susceptible individuals at the disease-free equilibrium (DFE).

Analysis of Sensitivity

Here, we conduct a sensitivity analysis of some of the parameters used in the model. This technique helps us to identify the parameters that have a significant effect on the basic reproduction number (See Table 1 and Figure 1). We use the approach as described by Chintis [41] to calculate the sensitivity index of $\mathbf{R}_0$. Specifically, the sensitivity index $\Delta_h^{\mathbf{R}_0}$ of a parameter h is presented by the formula $\Delta_h^{\mathbf{R}_0} = \frac{\partial \mathbf{R}_0}{\partial k} \frac{h}{\mathbf{R}_0}$.

Table 1. Sensitivity indices of different parameters.

Notation	Sensitivity Values	Notation	Sensitivity Values
χ	0.0384651	ϑ_1	-0.058565
ϑ_ϑ	-0.00000453	ϑ_a	-0.93704
η_0	-0.001464128	η_1	-0.041392
φ	0.99999	q	0.0000476
bN	0.99999	ξ	0.000432

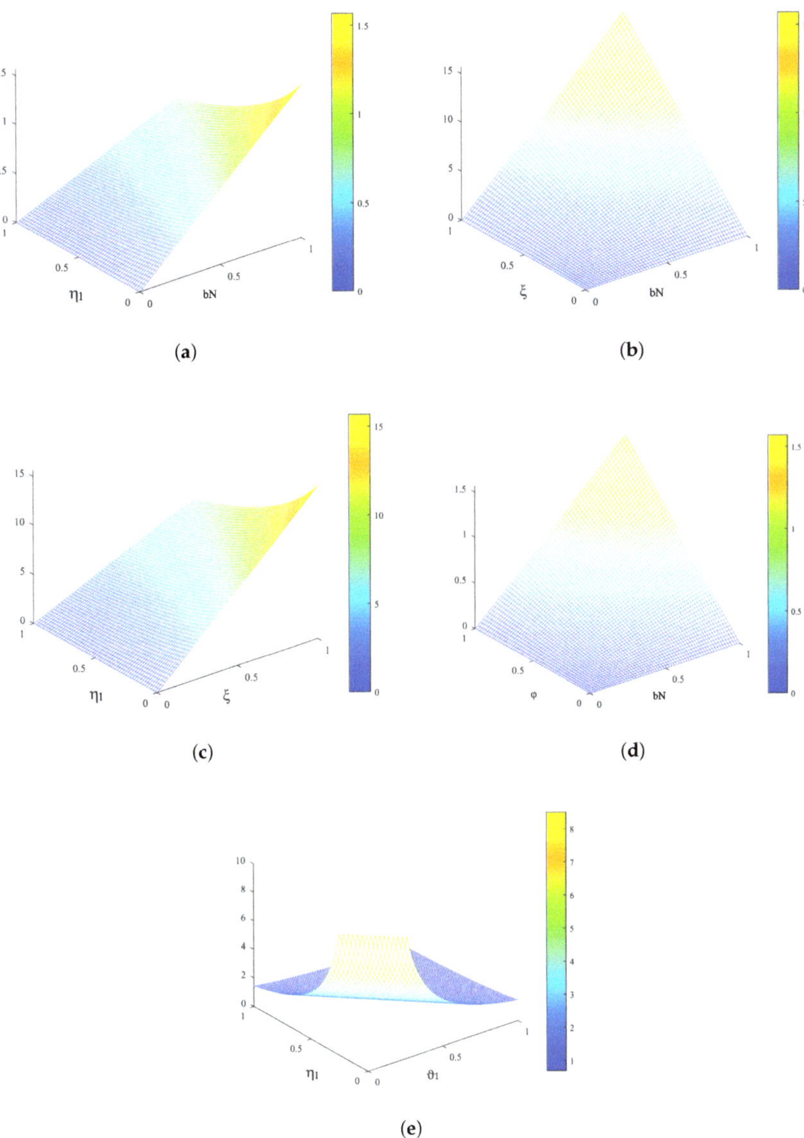

(**a**)

(**b**)

(**c**)

(**d**)

(**e**)

Figure 1. *Cont.*

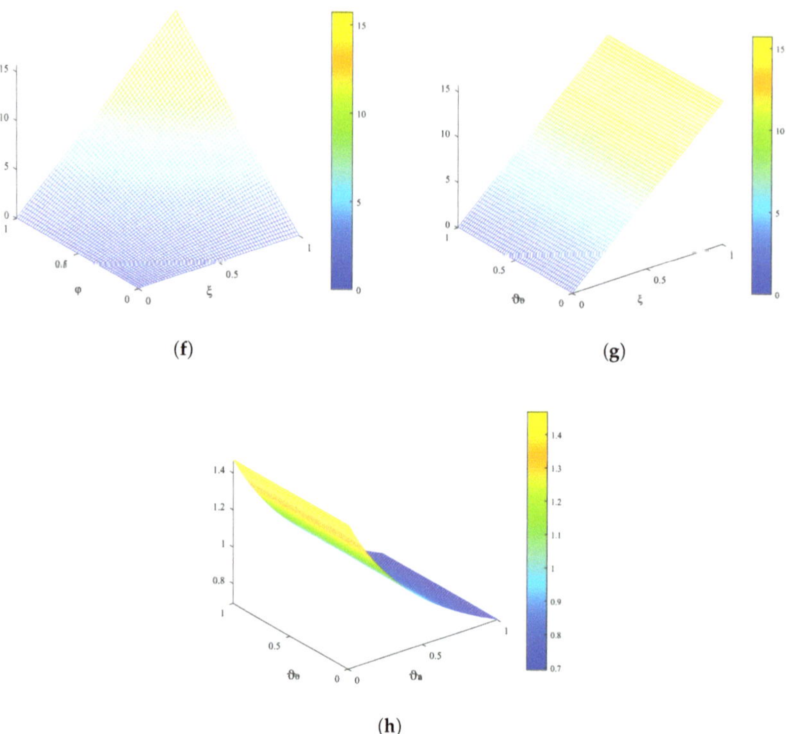

Figure 1. The graphs display the results of a sensitivity analysis on the basic reproductive number $\mathbf{R}_0$. (**a**) $\mathbf{R}_0$ with η_1, bN; (**b**) $\mathbf{R}_0$ with ξ, bN; (**c**) $\mathbf{R}_0$ with η_1, ξ; (**d**) $\mathbf{R}_0$ with φ, bN; (**e**) $\mathbf{R}_0$ with η_1, ϑ_1; (**f**) $\mathbf{R}_0$ with φ, ξ; (**g**) $\mathbf{R}_0$ with ϑ_ϑ, ξ; (**h**) $\mathbf{R}_0$ with ϑ_ϑ, ϑ_a.

4. Equilibria Points

The aforementioned system (1) has two possible equilibria: one is the disease-free equilibrium (DFE) and second one is the endemic equilibrium (EE). The DFE, denoted by $\mathscr{F}_0$, is given by $\mathscr{F}_0 = \left(\frac{bN}{\eta_0}, 0, 0, 0, 0, 0 \right)$. The EE, denoted by $\mathscr{F}_1$, is found by setting "$\mathcal{S} = \mathcal{S}_*, \mathcal{E} = \mathcal{E}_*, \mathcal{I} = \mathcal{I}_*, \mathcal{A} = \mathcal{A}_*, \mathcal{H} = \mathcal{H}_*$, and $\mathcal{R} = \mathcal{R}_*$, and the LHS of the resulting system to zero". We obtain the following expression after simplification, $\mathcal{S}_*, \mathcal{E}_*, \mathcal{I}_*, \mathcal{A}_*, \mathcal{H}_*$, and $\mathcal{R}_*$ at the EE.

$$
\begin{aligned}
\mathcal{S}_* &= \frac{NbNQ_2}{\varphi Q_2 + \varphi q \vartheta_a + \eta_0 NQ_2}, \\
\mathcal{E}_* &= \frac{NbNQ_2 + \varphi q \vartheta_a + NbNQ_2}{\varphi Q_2 + \varphi q \vartheta_a + \eta_0 NQ_2}, \\
\mathcal{I}_* &= \frac{(\chi\xi)(NbNQ_2 + (\mathbf{R}_0 - 1) + NbNQ_2)}{Q_3\varphi + \varphi q \vartheta_a + \eta_0 NQ_2}, \\
\mathcal{A}_* &= \frac{\chi(1-\xi)NbNQ_2^2 + \varphi q \vartheta_a + NbNQ_2}{\varphi Q_2 + \varphi q \vartheta_a + \eta_0 NQ_3}, \\
\mathcal{H}_* &= \frac{\vartheta_a \mathcal{I}_*}{Q_2}, \\
\mathcal{R}_* &= \frac{Q_2 \mathcal{I}_* + Q_1}{Q_2}.
\end{aligned}
\tag{3}
$$

4.1. Local Stability

We show the local asymptotic stability (LAS) of the DFE as well as the EE of the system (1) with the help of the following theorem.

Theorem 1. *If the basic reproductive number* $\mathbf{R}_0$ *is less than 1, the DFE point is LAS.*

Proof. To show the local stability of the system, about the point DFE, the Jacobian matrix for the said system (1) is

$$
J_0 = \begin{pmatrix}
-\eta_0 & 0 & -\frac{\varphi \mathcal{S}_0}{N} & 0 & \frac{\varphi q \mathcal{S}_0}{N} & 0 \\
0 & -(\chi + \eta_0) & \frac{\varphi \mathcal{S}_0}{N} & 0 & \frac{\varphi q \mathcal{S}_0}{N} & 0 \\
0 & \chi \xi & -Q_3 & 0 & 0 & 0 \\
0 & \chi(1 - \xi) & 0 & -\eta_2 & 0 & 0 \\
0 & 0 & \vartheta_a & 0 & -Q_2 & 0 \\
0 & 0 & \vartheta_1 & 0 & \vartheta_\vartheta & -\eta_0
\end{pmatrix}. \tag{4}
$$

By conducting a row operation, reducing the matrix to echelon form, the following Jacobian matrix is obtained

$$
\begin{pmatrix}
-\eta_0 & 0 & -\frac{\varphi \mathcal{S}_0}{N} & 0 & \frac{\varphi q \mathcal{S}_0}{N} & 0 \\
0 & -(\chi + \eta_0) & \frac{\varphi \mathcal{S}_0}{N} & 0 & \frac{\varphi q \mathcal{S}_0}{N} & 0 \\
0 & 0 & \mathbf{A} & \frac{\varphi q \mathcal{S}_0 \chi \xi}{N} & 0 & \frac{\varphi q \mathcal{S}_0 \chi \xi}{N} \\
0 & 0 & 0 & \mathbf{B} & \frac{\varphi q \mathcal{S}_0 \chi}{N} & 0 \\
0 & 0 & 0 & 0 & \mathbf{C} & 0 \\
0 & 0 & 0 & 0 & 0 & \mathbf{D}
\end{pmatrix}, \tag{5}
$$

$$
\begin{aligned}
\mathbf{A} &= -Q_3 Q_2 (\chi + \eta_0) - \frac{\varphi \mathcal{S}_0 \chi \xi}{N}, \\
\mathbf{B} &= -(\eta_0 + \eta_2)(\chi + \eta_0) Q_1 - \frac{\varphi \mathcal{S}_0 \chi \xi}{N}, \\
\mathbf{C} &= -Q_1 Q_2 Q_3 (\mathcal{S}_0 + \eta_0), \\
\mathbf{D} &= -\eta_0 Q_3 (\kappa + \eta_0) - \frac{(\varphi \mathcal{S}_0 \chi \xi)}{N} - [(1 - \mathbf{R}_0)(N(\chi + \eta_0) Q_2 Q_3) \varphi q \mathcal{S}_0 \chi \xi].
\end{aligned}
$$

According to [42], when $\mathbf{R}_0 < 1$, the matrices $\mathbf{A}$, $\mathbf{B}$, $\mathbf{C}$, and $\mathbf{D}$ are all negative, and the eigenvalues have negative real parts. As a result, the DFE is LAS. $\square$

Theorem 2. *If* $\mathbf{R}_0$ *is greater than 1, the EE point is LAS.*

Proof. Consider the Jacobian of the considered problem (1) at $\mathscr{F}_1$ is,

$$
J_0 = \begin{pmatrix}
-\frac{\varphi \mathcal{I}_*}{N} - \frac{\varphi \mathcal{H}_* q}{N} - \eta_0 & 0 & -\frac{\varphi \mathcal{S}_*}{N} & 0 & -\frac{\varphi q \mathcal{S}_*}{N} & 0 \\
\frac{\varphi \mathcal{I}_*}{N} + \frac{\varphi \mathcal{H}_* q}{N} & -(\chi + \eta_0) & \frac{\varphi \mathcal{S}_*}{N} & 0 & \frac{\varphi q \mathcal{S}_*}{N} & 0 \\
0 & \chi \xi & Q_3 & 0 & 0 & 0 \\
0 & \chi(1 - \xi) & 0 & -(\eta_0 + \eta_1) & 0 & 0 \\
0 & 0 & \vartheta_a & 0 & -(\vartheta_\vartheta + \eta_0) & 0 \\
0 & 0 & \vartheta_1 & 0 & \vartheta_\vartheta & -\eta_0
\end{pmatrix}. \tag{6}
$$

After performing a row operation and simplifying the resulting expressions, we obtain the following Jacobian matrix:

$$
\begin{pmatrix}
-\frac{\varphi \mathcal{I}_*}{N} - \frac{\varphi \mathcal{H}_* q}{N} - \eta_0 & 0 & -\frac{\varphi \mathcal{S}_*}{N} & 0 & -\frac{\varphi q \mathcal{S}_*}{N} & 0 \\
0 & -(\chi + \eta_0)\left(\frac{\varphi \mathcal{I}_*}{N} + \frac{\varphi \mathcal{H}_* q \mathcal{S}_*}{N} + \eta_0\right) & -\frac{\varphi \mathcal{S}_*}{N} & 0 & -\frac{\varphi q \mathcal{S}_*}{N} & 0 \\
0 & 0 & \mathbb{Z}_1 & \eta_2 \xi & 0 & 0 \\
0 & 0 & 0 & \mathbb{Z}_2 & \mathbb{Z}_3 & \mathbb{Z}_4 \\
0 & 0 & 0 & 0 & \mathbb{Z}_5 & \eta_0 \vartheta_\vartheta \\
0 & 0 & 0 & 0 & 0 & \mathbb{Z}_6
\end{pmatrix}, \tag{7}
$$

where

$$\mathbb{Z}_1 = -\xi Q_3,$$

$$\mathbb{Z}_2 = -\eta_2(\kappa + \eta_0)(\varphi \mathcal{I}_* + \varphi q \mathcal{H}_* + \eta_0)\vartheta_1,$$

$$\mathbb{Z}_3 = -\vartheta_\theta \varphi \mathcal{S}_* \eta_0 \chi \xi,$$

$$\mathbb{Z}_4 = -\eta_0 \varphi \mathcal{S}_* \chi (1 - \xi),$$

$$\mathbb{Z}_5 = -Q_2 \vartheta_1 - \vartheta_a - \vartheta_\theta,$$

$$\mathbb{Z}_6 = -\eta_0 Q_3 (\eta_2 (\chi + \eta_0)(\varphi \mathcal{I}_*$$

$$+ \varphi q \mathcal{H}_* + \eta_0)\vartheta_1 (\mathbf{R}_0 - 1) \times [(\chi + \eta_0) Q_3](\varphi \mathcal{S}_* \eta_0 \varphi (\eta_2 \xi \vartheta_1)(\vartheta_\theta + \eta_0)\vartheta_1 + \vartheta_a$$

$$+ Q_3 (\eta_0 \xi Q_3 (\eta_2 (\chi + \eta_0)))(\varphi \mathcal{I}_* + \varphi q \mathcal{H}_* + \eta_0)\vartheta_1$$

$$+ (\vartheta_\theta \varphi \mathcal{S}_* \eta_0 \chi \xi)\eta_2 \xi \vartheta_1.$$

The eigenvalues are given by

$$\zeta_1 = -\frac{\varphi \mathcal{I}_*}{N} - \frac{\varphi \mathcal{H}_* q}{N} - \eta_0 < 0,$$

$$\zeta_2 = -(\chi + \eta_0)\left(\frac{\varphi \mathcal{I}_*}{N} + \frac{\varphi \mathcal{H}_* q \mathcal{S}_*}{N} + \eta_0\right) < 0,$$

$$\zeta_3 = -\xi(\vartheta_a + \vartheta_1 + \eta_0 + \eta_1) = \mathbb{Z}_1 < 0,$$

$$\zeta_4 = -\eta_2(\kappa + \eta_0)(\varphi \mathcal{I}_* + \varphi q \mathcal{H}_* + \eta_0)\vartheta_1 = \mathbb{Z}_2, < 0,$$

$$\zeta_5 = -(\vartheta_\theta + \eta_0)\vartheta_1 - \vartheta_a - \vartheta_\theta = \mathbb{Z}_5 < 0,$$

$$\zeta_6 = \mathbb{Z}_6 < 0.$$

(8)

As per the findings reported in [43], when $\mathbf{R}_0 > 1$, all of the eigenvalues have nonpositive real parts, which indicates that the EE point is LAS. $\square$

4.2. Analysis of Global Stability

The next theorem presents that the said system is globally asymptotically stable (GAS) for the DFE and EE point.

Theorem 3. *The DFE of the system is GAS for* $\mathbf{R}_0 < 1$, *otherwise unstable.*

Proof. We define the Lyapunov function as follows:

$$U(t) = k_1(\mathcal{S} - \mathcal{S}_0) + k_2 \mathcal{E} + k_3 \mathcal{I} + k_4 \mathcal{A} + k_5 \mathcal{H}.$$

(9)

We differentiate Equation (9) and obtain:

$$U'(t) = k_1 \mathcal{S}' + k_2 E' + k_3 \mathcal{I}' + k_4 \mathcal{A}' + k_5 \mathcal{H}'.$$

(10)

Using model (1), we obtain

$$\begin{aligned}
U'(t) &= k_1\Big[b\mathrm{N} - \frac{\varphi\mathcal{I}\mathcal{S}}{\mathrm{N}} - \frac{\varphi q\mathcal{H}\mathcal{S}}{\mathrm{N}} - \eta_0 S\Big] + k_2\Big[\frac{\varphi\mathcal{I}\mathcal{S}}{\mathrm{N}} + \frac{\varphi q\mathcal{H}\mathcal{S}}{\mathrm{N}} - (\chi + \eta_0)\mathcal{E}\Big] \\
&\quad + k_3[\chi\xi\mathcal{E} - (\vartheta_a + \vartheta_1)\mathcal{I} - (\eta_0 + \eta_1)\mathcal{I}] + k_4[\chi(1 - \xi)\mathcal{E} - (\eta_0 + \eta_2)\mathcal{A}] \\
&\quad + k_5[\vartheta_a\mathcal{I} - \vartheta_\vartheta\mathcal{H} - \eta_0\mathcal{H}].
\end{aligned}$$

If we choose the positive parameter values $k_1 = k_2 = k_4 = \xi$, $k_3 = 1$, and $k_5 = q\varphi\rho$, and simplify, we obtain:

$$\begin{aligned}
U'(t) &= -\xi\eta_0(\mathcal{S} - \mathcal{S}_0) - 2\frac{\varphi q\mathcal{H}\mathcal{S}}{\mathrm{N}} - Q_1\xi\eta_0\mathcal{E} - (\eta_0 + \eta_2)\mathcal{A} \\
&\quad - Q_3[1 - \mathbf{R}_0] - Q_2 Q_3(\kappa + \eta_0) - Q_2^2\mathcal{H}.
\end{aligned}$$

where

$$\mathcal{S}_0 = \frac{b\mathrm{N}}{\eta_0},$$

Let $U'(t)$ be a function of time t, and let S and $\mathbf{R}_0$ be constants. If $\mathcal{S} > \mathcal{S}_0$ and $\mathbf{R}_0 < 1$, then $U'(t)$ is negative. If $\mathcal{S} = \mathcal{S}_0$, then $U'(t) = 0$. According to the LaSalle invariance principle [44,45], if $\mathcal{E} = \mathcal{I} = \mathcal{A} = \mathcal{H} = 0$, then the set of initial conditions for which $U'(t)$ approaches zero as t approaches infinity is an invariant set.

Therefore, the DFE $\mathscr{F}_0$ is GAS. $\square$

Theorem 4. *When* $\mathbf{R}_0 > 1$, *the EE point is GAS at* $\mathscr{F}_1$, *and unstable when* $\mathbf{R}_0 < 1$.

Proof. For the GAS of the EE point, we define the Lyapunov function as:

$$U(t) = \frac{1}{2}[p_1(\mathcal{S} - \mathcal{S}_*) + p_2(\mathcal{E} - \mathcal{E}_*) + p_3(\mathcal{I} - \mathcal{I}_*) + p_4(\mathcal{A} - \mathcal{A}_*) + p_5(\mathcal{H} - \mathcal{H}_*)]^2, \quad (11)$$

and we introduce the constants p_1, p_2, p_3, p_4, and p_5, which will be chosen later. Upon differentiating Equation (11), we obtain:

$$\begin{aligned}
U'(t) &= [p_1(\mathcal{S} - \mathcal{S}_*) + p_2(\mathcal{E} - \mathcal{E}_*) + p_3(\mathcal{I} - \mathcal{I}_*) + p_4(\mathcal{A} - \mathcal{A}_*) \\
&\quad + p_5(\mathcal{H} - \mathcal{H}_*)][p_1(\frac{ds}{dt}) + p_2(\frac{dE}{dt}) + p_3(\frac{dI}{dt}) + p_4(\frac{dA}{dt}) + p_5(\frac{dH}{dt})],
\end{aligned}$$

$$\begin{aligned}
U'(t) &= p_1(\mathcal{S} - \mathcal{S}_*) + p_2(\mathcal{E} - \mathcal{E}_*) + p_3(\mathcal{I} - \mathcal{I}_*) + p_4(\mathcal{A} - \mathcal{A}_*) + p_5(\mathcal{H} - \mathcal{H}_*)(p_1(b\mathrm{N} - \frac{\varphi\mathcal{I}\mathcal{S}}{\mathrm{N}} - \frac{\varphi q\mathcal{H}\mathcal{S}}{\mathrm{N}} - \eta_0 S) \\
&\quad + p_2(\frac{\varphi\mathcal{I}\mathcal{S}}{\mathrm{N}} + \frac{\varphi q\mathcal{H}\mathcal{S}}{\mathrm{N}} - (\chi + \eta_0)\mathcal{E} + p_3(\chi\xi\mathcal{E} - (\vartheta_a + \vartheta_1)\mathcal{I} - (\eta_0 + \eta_1)\mathcal{I}) + p_4(\chi(1 - \xi)\mathcal{E} - (\eta_0 + \eta_2)\mathcal{A} \\
&\quad + p_5(\vartheta_a\mathcal{I} - \vartheta_\vartheta\mathcal{H} - \eta_0\mathcal{H}).
\end{aligned}$$

After some calculation, we obtain, and utilizing the values of $p_1, \ldots p_5$, we obtain

$$U'(t) = -\frac{(\varphi\mathcal{S}_*)}{\mathrm{N}}(\mathcal{S} - \mathcal{S}_*) - (\mathbf{R}_0 - 1)Q_1 Q_2 - \frac{\varphi q\mathcal{H}_*\mathcal{S}_*}{\mathrm{N}}[\mathcal{E} + \mathcal{I} + \mathcal{A}] - (\vartheta_a\mathcal{I} - \vartheta_\vartheta)Q_2 Q_3^2\mathcal{H}_*$$

For $\mathcal{S} = \mathcal{S}_*$ and $(\vartheta_a > \vartheta_\vartheta)$ for $\mathbf{R}_0$ is greater than 1; thus, the proof is finished. $\square$

5. Results and Discussion

In this context, we substantiate our analytical discoveries through the application of the fourth-order Runge–Kutta method [46]. We select certain parameters for illustrative purposes, while obtaining others from published data sources [38]. The parameters employed in the simulation are chosen with careful consideration of their biological plausibility. The ensuing set of parameters is utilized for the subsequent analysis. $\varphi = 0.007; q = 0.003; \chi = 0.005;$ $\xi = 0.0001; \eta_0 = 0.0003; \eta_1 = 0.0001; \zeta = 0.002; \vartheta_1 = 0.001; \vartheta_a = 0.000001; \vartheta_\vartheta = 0.0007;$ and $bN = 0.00004$. To validate the analytical findings of the proposed model concerning the DFE, we employed the aforementioned parameter values. Subsequently, we computed the DFE point's coordinates and the threshold parameter $\mathbf{R}_0$ as $(7.98337196, 0, 0, 0, 0)$ and (0.043732), respectively. The simulation outcomes utilizing the aforementioned parameters are depicted in Figures 2 and 3, thereby substantiating the analytical conclusions outlined in the theorem. To corroborate this, we employed the linear stability analysis technique and introduced perturbations to the initial compartmental population values. Remarkably, these perturbed values consistently converged to the DFE, underscoring its robustness against varying initial conditions, $\mathcal{S}(0) = 1000, \mathcal{E}(0) = 800, \mathcal{I}(0) = 600, \mathcal{A}(0) = 500, \mathcal{H}(0) = 400$, and $\mathcal{R}(0) = 300$. Drawing from the theoretical interpretation of the data, a definitive conclusion can be drawn: when the value of $\mathbf{R}_0$ is below 1, the disease transmission will inevitably diminish over time. This is evidenced by the convergence of every solution curve to a stable position, as depicted in the corresponding plots.

$$
\begin{aligned}
\frac{\mathcal{S}^{i+1} - \mathcal{S}^i}{l} &= bN - \frac{\varphi \mathcal{I}^i \mathcal{S}^{i+1}}{N} - \frac{\varphi q \mathcal{H}^i \mathcal{S}^{i+1}}{N} - \eta_0 \mathcal{S}^{i+1}, \\
\frac{\mathcal{E}^{i+1} - \mathcal{E}^i}{l} &= \frac{\varphi \mathcal{I}^i \mathcal{S}^{i+1}}{N} + \frac{\varphi q \mathcal{H}^i \mathcal{S}^{i+1}}{N} - (\chi + \eta_0) \mathcal{E}^{i+1}, \\
\frac{\mathcal{I}^{i+1} - \mathcal{I}^i}{l} &= \chi \xi \mathcal{E}^{i+1} - (\vartheta_a + \vartheta_1) \mathcal{I}^{i+1} - (\eta_0 + \eta_1) \mathcal{I}^{i+1}, \\
\frac{\mathcal{A}^{i+1} - \mathcal{A}^i}{l} &= \chi(1 - \xi) \mathcal{E}^{i+1} - (\eta_0 + \eta_2) \mathcal{A}^{i+1}, \\
\frac{\mathcal{H}^{i+1} - \mathcal{H}^i}{l} &= \vartheta_a \mathcal{I}^{i+1} - \vartheta_\vartheta \mathcal{H}^{i+1} - \eta_0 \mathcal{H}^{i+1}, \\
\frac{\mathcal{R}^{i+1} - \mathcal{R}^i}{l} &= \vartheta_1 \mathcal{I}^{i+1} + \vartheta_\vartheta \mathcal{H}^{i+1} - \eta_0 \mathcal{R}^{i+1}.
\end{aligned}
$$

Used Algorithm

Step 1: $(\mathcal{S}_0, \mathcal{E}_0, \mathcal{I}_0, \mathcal{A}_0, \mathcal{H}_0, \mathcal{R}_0 = 0)$.

Step 2: Let $i = 1, 2 \ldots n - 1$.

$$
\begin{aligned}
\mathcal{S}^{i+1} &= \frac{Nlb}{\varphi \mathcal{I}^i \mathcal{S}^{i+1} + \varphi q \mathcal{H}^i \mathcal{S}^{i+1}l + \eta_0 lN} + \frac{\mathcal{S}^{i+1}}{\varphi \mathcal{I}^i \mathcal{S}^{i+1} + \varphi q \mathcal{H}^i \mathcal{S}^{i+1}l + \eta_0 lN}, \\
\mathcal{E}^{i+1} &= \frac{l \varphi \mathcal{I}^i \mathcal{S}^{i+1}}{N(1 + l(\chi + \eta_0))} + \frac{l \varphi q \mathcal{H}^i \mathcal{S}^{i+1}}{N(1 + l(\chi + \eta_0))} + \frac{\mathcal{E}^{i+1}}{(1 + l(\chi + \eta_0))}, \\
\mathcal{I}^{i+1} &= \frac{l \chi \xi \mathcal{E}^{i+1}}{(1 + l(\vartheta_a + \vartheta_1) + (\eta_0 + \eta_1)l)} + \frac{\mathcal{I}^{i+1}}{1 + l(\vartheta_a + \vartheta_1) + (\eta_0 + \eta_1)l}, \\
\mathcal{A}^{i+1} &= \frac{l \chi(1 - \xi) \mathcal{E}^{i+1}}{1 + (\eta_0 + \eta_2)l} + \frac{\mathcal{A}^{i+1}}{1 + (\eta_0 + \eta_2)l}, \\
\mathcal{H}^{i+1} &= \frac{l \vartheta_a \mathcal{I}^{i+1}}{1 + \eta_0 l + \vartheta_\vartheta l} + \frac{\mathcal{H}^{i+1}}{1 + \vartheta_\vartheta + \eta_0 l}, \\
\mathcal{R}^{i+1} &= \frac{l \vartheta_1 \mathcal{I}^{i+1}}{1 + \eta_0 l} + \frac{l \vartheta_\vartheta \mathcal{H}^{i+1}}{1 + \eta_0 l} + \frac{\mathcal{R}^{i+1}}{1 + \eta_0 l}.
\end{aligned}
$$

Step 3: Let $i = 1, 2, 3, \ldots, n - 1$, by letting "$\mathcal{S}_*(t_i) = \mathcal{S}_*, \mathcal{E}_*(t_i) = \mathcal{E}_*, \mathcal{I}_*(t_i) = \mathcal{I}_*,$ $\mathcal{A}_*(t_i) = \mathcal{A}_*, \mathcal{H}_*(t_i) = \mathcal{H}_*, \mathcal{R}_*(t_i) = \mathcal{R}_*.$"

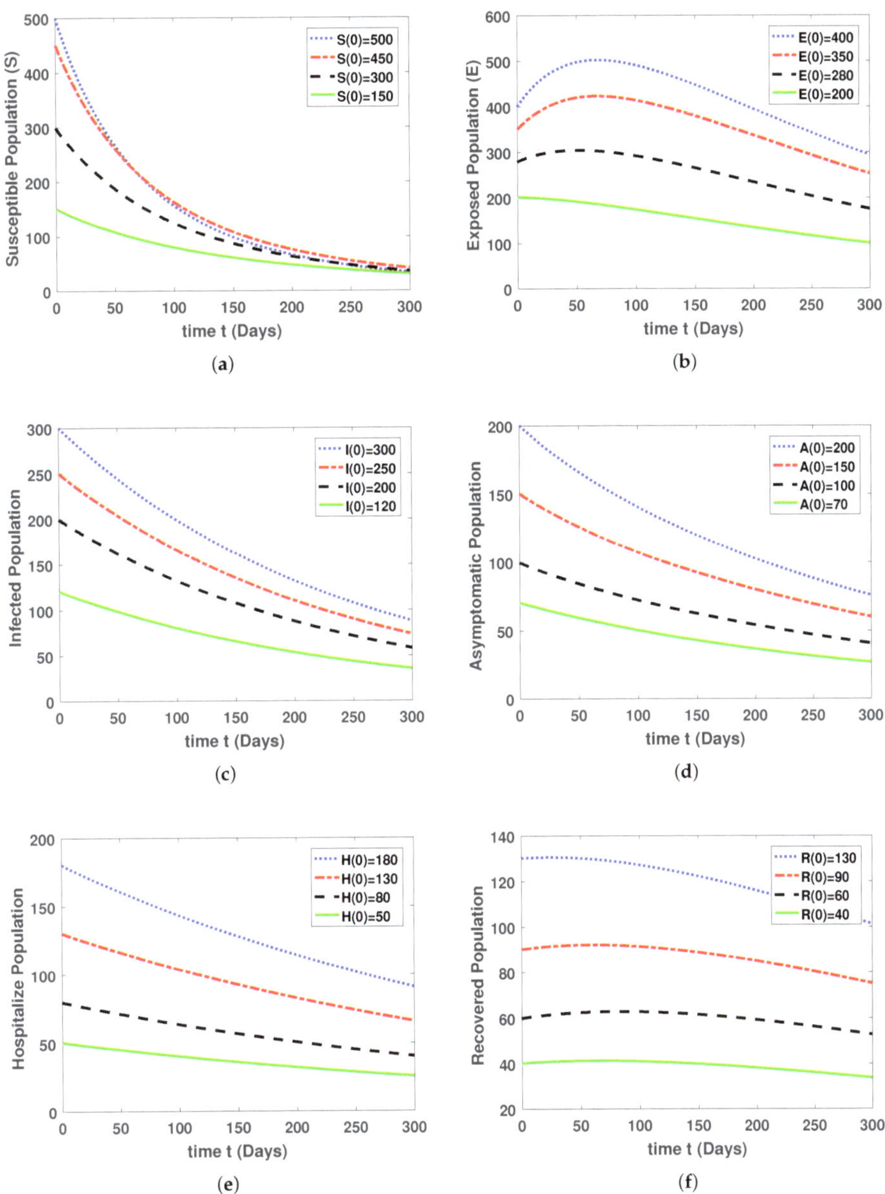

Figure 2. The time dynamics of the compartmental populations in model (1) are shown graphically for an initial population value. (**a**) Class of $\mathcal{S}$; (**b**) class of $\mathcal{E}$; (**c**) class of $\mathcal{I}$; (**d**) class of $\mathcal{A}$; (**e**) class of $\mathcal{H}$; (**f**) class of $\mathcal{R}$.

Subsequently, we proceed to explore the system's dynamics around the EE by assuming an alternate set of parameters: $\varphi = 0.17$, $q = 0.03$, $\chi = 0.05$, $\xi = 0.01$, $\eta_0 = 0.03$, $\eta_1 = 0.031$, $\zeta = 0.052$, $\vartheta_1 = 0.041$, $\vartheta_a = 0.000001$, $\vartheta_\theta = 0.0007$, and $bN = 0.004$. Using these parameter values acquired earlier, we calculate the endemic equilibrium points and the associated $\mathbf{R}_0$ for the model (1). When $\mathbf{R}_0 > 1$, the endemic equilibrium point is determined to be $(40.76549, 110.908700, 70.45321, 85.934214, 85.7659321, 120.7659321)$, with a calculated value of $\mathbf{R}_0 = 7.13587$. Assuming the same initial population sizes for the compartments as in the previous analysis, the graphical results indicate that the populations of susceptible, exposed, infected asymptomatic, hospitalized, and recovered individuals initially undergo fluctuations before eventually stabilizing at their respective equilibrium values. For the parameter values employed in this study, the equilibrium point is $(24.76549, 99.908700, 22.45321, 90.934214, 85.7659321, 120.7659321)$.

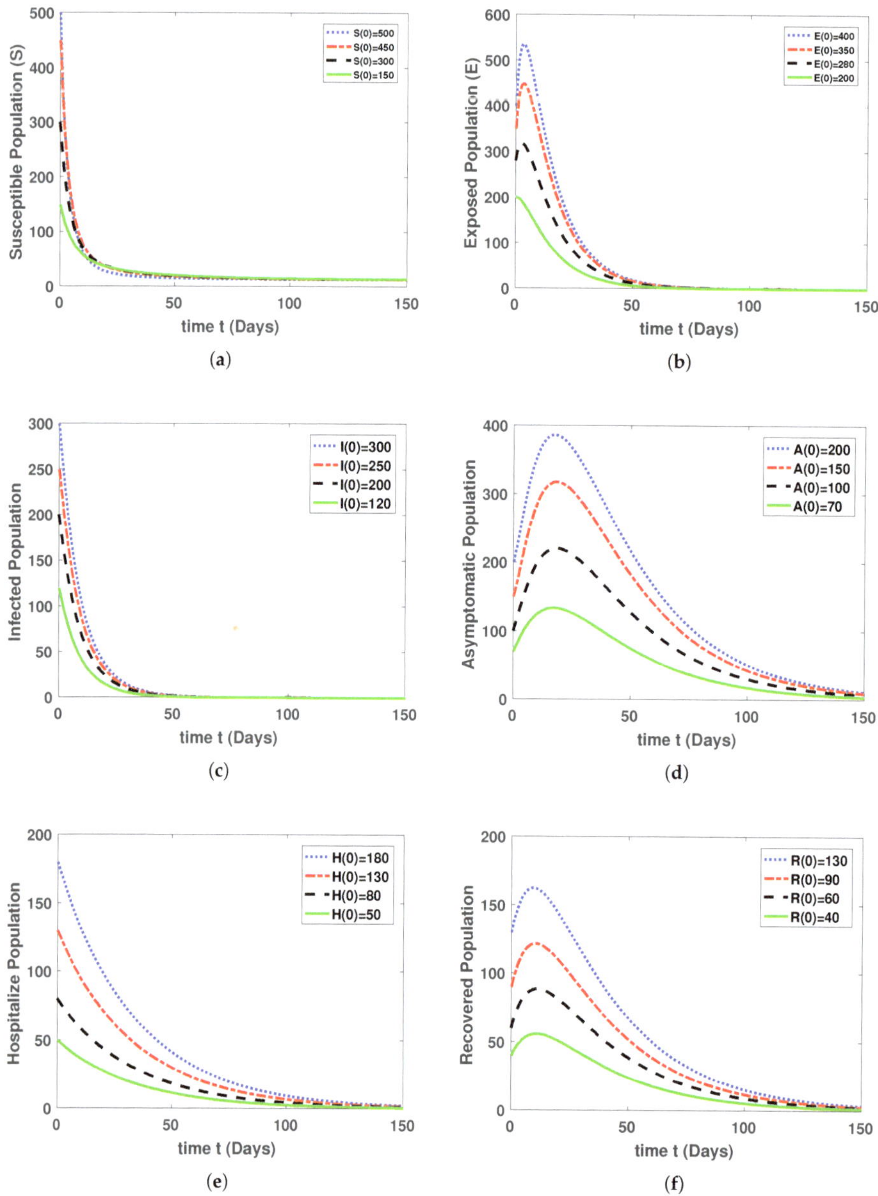

Figure 3. The time dynamics of the compartmental populations in model (1) are shown graphically for initial population values. (**a**) Class of $\mathcal{S}$; (**b**) class of $\mathcal{E}$; (**c**) class of $\mathcal{I}$; (**d**) class of $\mathcal{A}$; (**e**) class of $\mathcal{H}$; (**f**) class of $\mathcal{R}$.

6. Analysis of Optimal Control

Here, we aim to establish an effective control strategy to prevent the spread of MERS in the population. Optimal control theory is a powerful mathematical technique that can be applied to design control schemes for a variety of infectious diseases. To achieve this, we apply optimal control theory, as described in previous works [46–49], to establish an appropriate control strategy. Our objective in this study is to reduce the prevalence of MERS in the populace by increasing the number of persons who recover from the disease, denoted as $\mathcal{R}$, and decreasing the number of individuals who are infectious, denoted as $\mathcal{I}$, and hospitalized, denoted as $\mathcal{H}$, by implementing time-dependent control variables such as treatment $v_1(t)$ and care $v_2(t)$. In model (1), we take six state variables $\mathcal{S}, \mathcal{E}, \mathcal{I}, \mathcal{A}, \mathcal{H}$, and $\mathcal{R}$. Now, for the control problem, we take the two control variables, that is treatment $v_1(t)$ and care $v_2(t)$. Hence, we have the successive optimal control problem to reduce the objective functional

$$J(v_1, v_2) = \int_0^T [c_1 \mathcal{I}(t) + c_2 \mathcal{H}(t) + \frac{1}{2}(c_3 v_1^2(t) + c_4 v_2^2(t)]dt \tag{12}$$

subject to

$$\begin{aligned}
\dot{\mathcal{S}}(t) &= b\mathrm{N} - \frac{\varphi \mathcal{I} \mathcal{S}}{\mathrm{N}} - \frac{\varphi q \mathcal{H} \mathcal{S}}{\mathrm{N}} - \eta_0 \mathcal{S}, \\
\dot{\mathcal{E}}(t) &= \frac{\varphi \mathcal{I} \mathcal{S}}{\mathrm{N}} + \frac{\varphi q \mathcal{H} \mathcal{S}}{\mathrm{N}} - (\chi + \eta_0)\mathcal{E}, \\
\dot{\mathcal{I}}(t) &= \chi \xi \mathcal{E} - (\vartheta_a + \vartheta_1)\mathcal{I} - (\eta_0 + \eta_1) - v_1 \mathcal{I}, \\
\dot{\mathcal{A}}(t) &= \chi(1 - \xi)\mathcal{E} - (\eta_0 + \eta_2)\mathcal{A}, \\
\dot{\mathcal{H}}(t) &= \vartheta_a \mathcal{I} - \vartheta_\vartheta \mathcal{H} - \eta_0 \mathcal{H} - v_2 \mathcal{H}, \\
\dot{\mathcal{R}}(t) &= \vartheta_1 \mathcal{I} + \vartheta_\vartheta \mathcal{H} - \eta_0 \mathcal{R} + v_1 \mathcal{I} + v_2 \mathcal{H},
\end{aligned} \tag{13}$$

with initial conditions

$$\text{``}\mathcal{S}(0) \geq 0, \mathcal{E}(0) \geq 0, \mathcal{I}(0) \geq 0, \mathcal{A}(0) \geq 0, \mathcal{H}(0) \geq 0, \mathcal{R}(0) \geq 0\text{''}.$$

Equation (12) includes weight constants c_1, c_2, c_3, and c_4 that correspond to the relative importance of infected people I and hospitalized individual H in the objective function. The parameters $\frac{1}{2}c_3 v_1^2$ and $\frac{1}{2}c_4 v_2^2$ represent the costs associated with self-care and treatment. The primary objective is to evaluate the control function to achieve a specific goal.

$$J(v_1^*, v_2^*) = min\{J(v_1, v_2), v_1, v_2 \in U\} \tag{14}$$

dependent on control system (13), where U in Equation (14) is known as the control set and is presented as,

$$\text{``}U = \{(v_1, v_2)/v_i(t) \text{ is Lebesgue measurable on } [0,1], 0 \leq v_i(t) \leq 1, i = 1, 2\}.\text{''} \tag{15}$$

Before proceeding, it is important to establish the existence of control variables. According to Kamien and Aldila's study [47], a solution for a state system can be found when the controls are bounded and Lebesgue measurable, in addition to satisfying the initial conditions. Thus, we can suppose that the considered control model can be formulated in the manner presented below.

$$\frac{d\phi}{dt} = \mathscr{A}\phi + \mathscr{B}\phi.$$

From the above equation $\phi = (\mathcal{S}, \mathcal{E}, \mathcal{I}, \mathcal{A}, \mathcal{H}, \mathcal{R})$, where $\mathscr{A}(\phi)$ and $\mathscr{B}(\phi)$ denote the linear and nonlinear bounded coefficient $\ni$

$$J_0 = \begin{pmatrix} -\eta_0 & 0 & 0 & 0 & 0 & 0 \\ 0 & -(\chi + \eta_0) & 0 & 0 & 0 & 0 \\ 0 & \chi\xi & -Q_3 & 0 & 0 & 0 \\ 0 & \chi(1-\xi) & 0 & -\eta_2 & 0 & 0 \\ 0 & 0 & \vartheta_a & 0 & -Q_2 & 0 \\ 0 & 0 & \vartheta_1 & 0 & \vartheta_\theta & -\eta_0 \end{pmatrix}. \tag{16}$$

$$B(\phi) = \begin{pmatrix} bN - \frac{\varphi \mathcal{I} \mathcal{S}}{N} - \frac{\varphi q \mathcal{H} \mathcal{S}}{N} \\ \frac{\varphi \mathcal{I} \mathcal{S}}{N} + \frac{\varphi q \mathcal{H} \mathcal{S}}{N} \\ 0 \\ 0 \\ 0 \\ 0 \end{pmatrix}. \tag{17}$$

Letting $L(\phi) = \mathcal{A}\phi + F\mathcal{E}$,

$$\begin{aligned}
|F(\phi_1) - F(\phi_2)| &\leq m_1|\mathcal{S}_1 - \mathcal{S}_2| + m_2|\mathcal{E}_1 - \mathcal{E}_2| + m_3|\mathcal{I}_1 - \mathcal{I}_2| + m_4|\mathcal{A}_1 - \mathcal{A}_2| + m_5|\mathcal{H}_1 - \mathcal{H}_2| + m_6|\mathcal{R}_1 - \mathcal{R}_2| \\
&\leq N|\mathcal{S}_1 - \mathcal{S}_2| + |\mathcal{E}_1 - \mathcal{E}_2| + |\mathcal{I}_1 - \mathcal{I}_2| + |\mathcal{A}_1 - \mathcal{A}_2| + |\mathcal{H}_1 - \mathcal{H}_2| + |\mathcal{R}_1 - \mathcal{R}_2|.
\end{aligned}$$

Here, $N = \max(m_1, m_2, m_3, m_4, m_5, m_6)$ is a constant that is independent of the state variables in the aforementioned system. We also express

$$|L(\phi_1) - L(\phi_2)| \leq M|(\phi_1) - (\phi_2)|.$$

The solution for (13) exists due to the nonnegativity of the model state variables $\mathcal{S}, \mathcal{E}, \mathcal{I}, \mathcal{A}, \mathcal{H}$, and $\mathcal{R}$. Furthermore, it has been shown that the function L is Lipschitz uniformly continuous, and where $M = (N, \|\mathcal{K}\|) < \infty$. Based on the properties mentioned earlier, we present the following theorem to establish the existence of a solution for model (1), which we then proceed to prove.

Theorem 5. *For the control problem in Equations (12) and (13) there exists an optimal control as* $v^* = (v_1^*, v_2^*) \in U.$

Proof. It is evident that the control and state variables in system (1) are positive. Additionally, the control variables set U is a closed and convex set, as mentioned in the problem statement. Furthermore, the control system is bounded, implying the compactness of the system. The integral in the objective function of the optimization problem, given by $c_1\mathcal{I} + c_2\mathcal{H} + \frac{1}{2}(c_3 v_1^2(t) + c_4 v_2^2(t))$, is also convex w.r.t the control set U. This convexity guarantees the existence results for optimal control for the optimal control variables (v_1^*, v_2^*). $\square$

6.1. Methods

Next to show the optimal solution to the control model (12) and (13), we can apply the Lagrangian and Hamiltonian methods, described in the equation below

$$L(\mathcal{I}, \mathcal{H}, v_1, v_2) = c_1\mathcal{I} + c_2\mathcal{H} + \frac{1}{2}(c_3 v_1^2(t) + c_4 v_2^2(t)).$$

To describe the Hamiltonian $(\mathcal{H})$, by utilizing the notation $\vartheta = (\vartheta_1, \vartheta_2, \vartheta_3, \vartheta_4, \vartheta_5, \vartheta_6)$ and $y = (y_1, y_2, y_3, y_4, y_5, y_6)$, thus

$$\mathcal{H}(x, v, \vartheta) = L(x, v) + \vartheta Z(x, v),$$

where

$$Z_1 = b\mathrm{N} - \frac{\varphi \mathcal{IS}}{\mathrm{N}} - \frac{\varphi q \mathcal{HS}}{\mathrm{N}} - \eta_0 \mathcal{S},$$

$$Z_2 = \frac{\varphi \mathcal{IS}}{\mathrm{N}} + \frac{\varphi q \mathcal{HS}}{\mathrm{N}} - (\chi + \eta_0)\mathcal{E},$$

$$Z_3 = \chi \zeta \mathcal{E} - (\vartheta_a + \vartheta_1)\mathcal{I} - (\eta_0 + \eta_1) - v_1(t)\mathcal{I}(t),$$

$$Z_4 = \chi(1 - \zeta)\mathcal{E} - (\eta_0 + \eta_2)\mathcal{A}, \tag{18}$$

$$Z_5 = \vartheta_a \mathcal{I} - \vartheta_\theta \mathcal{H} - \eta_0 \mathcal{H} - v_2(t)\mathcal{H}(t),$$

$$Z_6 = \vartheta_1 \mathcal{I} + \vartheta_\theta \mathcal{H} - \eta_0 \mathcal{R} + v_1(t)\mathcal{I}(t) + v_2(t)\mathcal{H}(t),$$

Thus we apply the Pontryagin Maximum Principle [50,51] to the Hamiltonian in order to determine the optimal solution. According to this principle, if (x^*, v^*) is an optimal solution, then there must exist a function $\vartheta \ni$:

$$\frac{dx}{dt} = \frac{\partial \mathcal{H}}{\partial \vartheta}, 0 = \frac{\partial \mathcal{H}}{\partial u},$$

$$\vartheta(t)' = -\frac{\partial \mathcal{H}}{\partial x}.$$

$$\mathcal{H}(t, x^*, v^*, \vartheta)\partial x = max_{v_1, v_2, v_3, v_4 \in [0,1]} \mathcal{H}(x^*(t), v_1, v_2 \vartheta(t)); \tag{19}$$

with

$$\vartheta(t_f) = 0, \tag{20}$$

The principles outlined in Equation (19) are utilized to determine the adjoint system (adjoint variables) and optimal control variables. Based on these principles, the following result can be obtained.

Theorem 6. *Suppose* $\mathcal{S}_*$, $\mathcal{E}_*$, $\mathcal{I}_*$, $\mathcal{A}_*$, $\mathcal{H}_*$, *and* $\mathcal{R}_*$ *represent the optimal state solutions for the system, obtained using the combined optimal control variables* (v_1^*, v_2^*) *that were derived through the numerical solution of the optimality system. The optimal control problem is defined by the objective function* (12) *and the control system* (13). *Then* $\exists$ *adjoint variables* $\vartheta_1(t)$, $\vartheta_2(t)$, $\vartheta_3(t)$, *and* $\vartheta_4(t)$, $\vartheta_5(t)$, $\vartheta_6(t)$ *satisfy*

$$\begin{aligned}
\vartheta_1'(t) &= -\mathcal{A}_1 + (\vartheta_2 - \vartheta_1)\varphi \mathcal{I}_* + (\vartheta_2 - \vartheta_1)\varphi q \mathcal{H}_* - \eta_0 \vartheta_1, \\
\vartheta_2'(t) &= -\mathcal{A}_2 + (\vartheta_4 - \vartheta_2)\varphi \mathrm{N}^* + (\chi + \eta_0)\vartheta_2 - \vartheta_1 v_1^* - \vartheta_3 \zeta, \\
\vartheta_3'(t) &= -\mathcal{A}_3 + (\vartheta_2 - \vartheta_1)\varphi \mathcal{S}_* + (\vartheta_5 - \vartheta_3)\vartheta_a + (\vartheta_6 - \vartheta_3)\vartheta_1 + (\vartheta_6 - \vartheta_3)v_1(t) - v_1 \vartheta_3, \\
\vartheta_4'(t) &= -\mathcal{A}_4 + (v_2 - u_0)\vartheta_4, \\
\vartheta_5'(t) &= -\mathcal{A}_5 - (\vartheta_2 - \vartheta_1)\varphi q \mathcal{S}_* + (\vartheta_6 - \vartheta_5)\vartheta_\theta - (u_0 + v_2)\vartheta_5, \\
\vartheta_6'(t) &= -\mathcal{A}_6 + u_0 \vartheta_6,
\end{aligned} \tag{21}$$

with boundary conditions.

Additionally, the optimal control parameters $v_1(t)$ *and* $v_2(t)$ *are obtained through numerical solutions of the optimality problem and are presented below.*

$$v_1^*(t) = max\{min\{\frac{(\vartheta_6 - \vartheta_3)\mathcal{I}_*}{B_1}, 1\}, 0\}, \tag{22}$$

$$v_2^*(t) = max\{min\{\frac{(\vartheta_6 - \vartheta_5)\mathcal{H}_*}{B_2}, 1\}, 0\}. \tag{23}$$

Proof. The adjoint problem described by Equation (21) is obtained through the uses of the Pontryagin Maximum Principle given by Equation (19), while the transversal conditions arise from $\vartheta(T) = 0$. The set of optimal functions v_1^*, v_2' is obtained using $\frac{\partial H}{\partial u}$. In the following section, we present numerical solutions to the optimality system in order to provide a clearer understanding for the reader, as opposed to relying solely on analytical results. The optimality problem is expressed by several components, including the control problem (13), the adjoint model (21), the boundary (terminal) conditions, and the optimal control functions. By solving these components numerically, we can gain valuable insights into the behavior of the system and assess its performance. $\square$

6.2. Results and Discussion for Optimal Control

We utilize the Runge–Kutta method of order four to solve the optimal control system (13), in order to investigate the effects of self-care and treatment. To find the solution of the state system (12) with initial conditions in the time interval [0, 50], we employ the forward Runge–Kutta procedure. Similarly, the backward Runge–Kutta technique is used to solve the adjoint system (21) in the same interval with the assistance of the transversality condition. Below are the parameters that we used for the simulation: $bN = 0.0071$; $\varphi = 0.00041$; $q = 0.0000123$; $\chi = 0.0000123$; $\xi = 0.0000123$; $\vartheta_1 = 0.003907997$; $\vartheta_a = 0.98$; $\vartheta_\theta = 0.0000404720925$; $q = 0.017816$; $\rho = 0.00007$; and $\eta_0 = 0.00997$. The weight constants c_1, c_2, c_3, and c_4 were chosen based on biological feasibility. Specifically, we set $c_1 = 0.6610000$, $c_2 = 0.54450$, $c_3 = 0.0090030$, $c_4 = 0.44440$. The results obtained from the simulations are presented in Figures 4 and 5.

Figures 4 and 5 show the variations in the number of all compartments with and without control measures implemented.

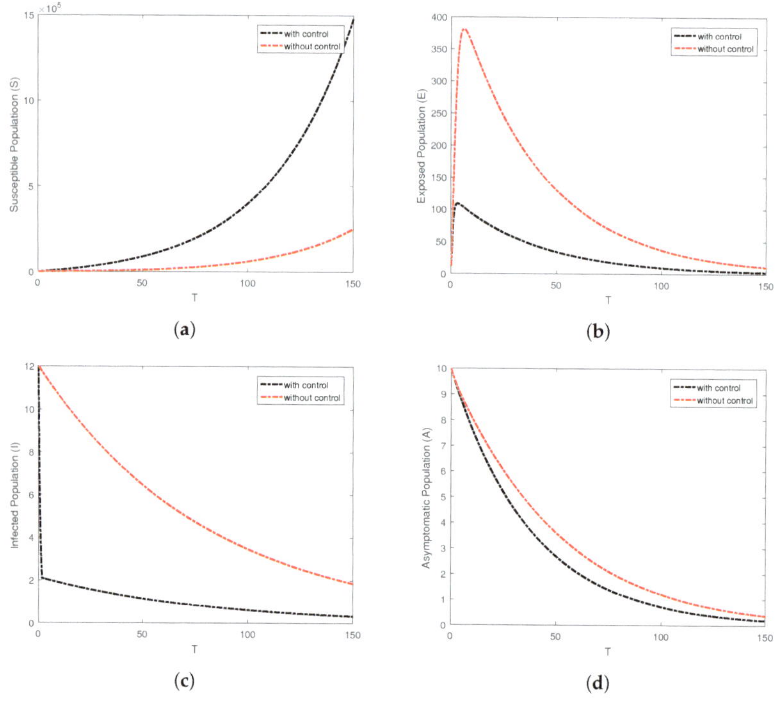

Figure 4. The visual representations demonstrate the changes in the compartmental population over time, comparing the scenarios with and without control measures implemented. (**a**) Susceptible populace; (**b**) exposed populace; (**c**) infected population; (**d**) asymptomatic population.

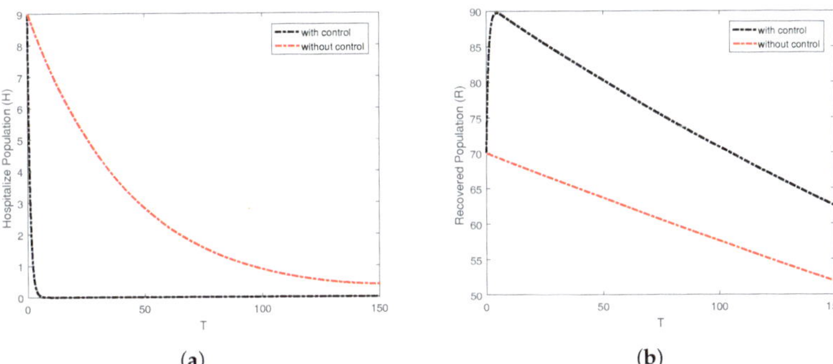

Figure 5. The visual representations demonstrate the changes in the compartmental population over time, comparing the scenarios with and without control measures implemented. (**a**) Hospitalized population; (**b**) recovered population.

7. Conclusions

The objective of this study is to developed a more realistic mathematical model that captures the transmission dynamics of the MERS-CoV. This is accomplished by introducing new parameters for the birth and death rates in the host populace. The threshold number $\mathbf{R}_0$ is a measure used to estimate the potential spread of a disease within a populace, and it can be calculated from a model to quantify the transmissibility of MERS-CoV. The model is analyzed using stability theory to identify conditions for local and global stability, and the most sensitive parameter is determined through a sensitivity analysis of $\mathbf{R}_0$. An optimal control problem is formulated with the goal of minimizing the number of infected persons and maximizing the number of recoveries in the population. The effectiveness of the approach is verified through numerical simulations, which demonstrate the stability of the results.

Author Contributions: Conceptualization, B.F., M.Y. and M.u.R.; methodology, A.A. and S.A.; software, B.F.; validation, M.Y. and M.u.R.; formal analysis, M.Y.; investigation, A.A.; resources, S.A.; data curation, B.F.; writing—original draft preparation, B.F., M.Y. and M.u.R.; writing—review and editing, M.Y. and A.A.; supervision, S.A. All authors have read and agreed to the published version of the manuscript.

Funding: This research received no external funding.

Data Availability Statement: Not applicable.

Acknowledgments: The researchers would like to acknowledge Deanship of Scientific Research, Taif University for funding this work.

Conflicts of Interest: The authors declare no conflict of interest.

References

1. AlTawfiq, J.A.; Hinedi, K.; Ghandour, J. Middle East respiratory syndrome coronavirus: A case-control study of hospitalized patients. *Clin. Infect. Dis.* **2014**, *59*, 160–165. [CrossRef] [PubMed]
2. Azhar, E.I.; Kafrawy, E.I.; Hassan, S.A.; Al-Saeed, A.M.; Hashem, M.S.; Madani, A.M. Evidence for camel to human transmission of MERS coronavirus. *N. Engl. J. Med.* **2014**, *370*, 2499–2505. [CrossRef] [PubMed]
3. Arabi, Y.M.; Arifi, A.A.; Balkhy, H.H. Clinical course and outcomes of critically ill patients with Middle East respiratory syndrome coronavirus infection. *Ann. Intern. Med.* **2014**, *160*, 389–397. [CrossRef] [PubMed]
4. Madani, T.A.; Azhar, E.I.; Hashem, A.M. Evidence for camel-to-human transmission of MERS coronavirus. *N. Engl. J. Med.* **2014**, *371*, 1360. [PubMed]
5. Breban, R.; Riou, J.; Fontanet, A. Interhuman transmissibility of Middle East respiratory coronavirus: Estimation of pandemic risk. *Lancet* **2013**, *382*, 694–699. [CrossRef]
6. Fisman, D.N.; Ashleigh, R.T. The epidemiology of MERS-CoV. *Lancet Infect. Dis.* **2014**, *14*, 6–7. [CrossRef]

7. Zhong, N.S.; Zheng, B.J.; Li, Y.M.; Poon, L.L.M.; Xie, Z.H.; Chan, K.H.; Li, P.H.; Tan, S.Y.; Chang, Q.; Xie, J.P.; et al. Epidemiology and cause of severe acute respiratory syndrome (SARS) in Guangdong, People's Republic of China, in February, 2003. *Lancet* **2003**, *362*, 1353–1358. [CrossRef]
8. Ksiazek, T.G.; Erdman, D.; Goldsmith, C.S.; Zaki, S.R.; Peret, T.; Emery, S.; Tong, S.; Urbani, C.; Comer, J.A.; Lim, W.; et al. A novel coronavirus associated with severe acute respiratory syndrome. *N. Engl. J. Med.* **2003**, *348*, 1953–1966. [CrossRef]
9. Zaki, A.M.; Van Boheemen, S.; Bestebroer, T.M.; Osterhaus, A.D.; Fouchier, R.A. Isolation of a novel coronavirus from a man with pneumonia in Saudi Arabia. *N. Engl. J. Med.* **2012**, *367*, 1814–1820. [CrossRef]
10. Ciotti, M.; Ciccozzi, M.; Terrinoni, A.; Jiang, W.C.; Wang, C.B.; Bernardini, S. The COVID-19 pandemic. *Crit. Rev. Clin. Lab. Sci.* **2020**, *57*, 365–388. [CrossRef]
11. Mandal, M.; Jana, S.; Nandi, S.K.; Khatua, A.; Adak, S.; Kar, T.K. A model based study on the dynamics of COVID-19: Prediction and control. *Chaos Solitons Fractals* **2020**, *136*, 109889. [CrossRef] [PubMed]
12. Atede, A.O.; Omame, A.; Inyama, S.C. A fractional order vaccination model for COVID-19 incorporating environmental transmission: A case study using Nigerian data. *Bull. Biomath.* **2023**, *1*, 78–110. [CrossRef]
13. Shen, Z.H.; Chu, Y.M.; Khan, M.A.; Muhammad, S.; Al-Hartomy, O.A.; Higazy, M. Mathematical modeling and optimal control of the COVID-19 dynamics. *Results Phys.* **2021**, *31*, 105028. [CrossRef] [PubMed]
14. Alqarni, M.S.; Alghamdi, M.; Muhammad, T.; Alshomrani, A.S.; Khan, M.A. Mathematical modeling for novel coronavirus (COVID-19) and control. *Numer. Methods Partial. Differ. Equ.* **2022**, *38*, 760–776. [CrossRef]
15. Tsay, C.; Lejarza, F.; Stadtherr, M.A.; Baldea, M. Modeling, state estimation, and optimal control for the US COVID-19 outbreak. *Sci. Rep.* **2020**, *10*, 10711. [CrossRef]
16. Libotte, G.B.; Lobato, F.S.; Platt, G.M.; Neto, A.J.S. Determination of an optimal control strategy for vaccine administration in COVID-19 pandemic treatment. *Comput. Methods Programs Biomed.* **2020**, *196*, 105664. [CrossRef]
17. Ullah, I.; Ahmad, S.; Al-Mdallal, Q.; Khan, Z.A.; Khan, H.; Khan, A. Stability analysis of a dynamical model of tuberculosis with incomplete treatment. *Adv. Differ. Equ.* **2020**, *2020*, 499. [CrossRef]
18. Hajji Mohamed, A.; Al-Mdallal, Q. Numerical simulations of a delay model for immune system-tumor interaction. *Sultan Qaboos Univ. J. Sci. (SQUJS)* **2018**, *23*, 19–31.
19. Asif, M.; Haider, N.; Al-Mdallal, Q.; Khan, I. A Haar wavelet collocation approach for solving one and two-dimensional second-order linear and nonlinear hyperbolic telegraph equations. *Numer. Methods Partial. Differ. Equ.* **2020**, *36*, 1962–1981. [CrossRef]
20. Kermack, W.O.; McKendrick, A.G. Contributions to the mathematical theory of epidemics, part 1. *Proc. R. Soc. Edinburgh. Sect. A Math.* **1927**, *115*, 700–721.
21. Cauchemez, S.; Fraser, C.; Van Kerkhove, M.D.; Donnelly, C.A.; Riley, S.; Rambaut, A.; Enouf, V.; van der Werf, S.; Ferguson, N.M. Middle East respiratory syndrome coronavirus: Quantification of the extent of the epidemic, surveillance biases, and transmissibility. *Lancet Infect. Dis.* **2013**, *14*, 50–56. [CrossRef]
22. Chowell, G.; Blumberg, S.; Simonsen, L.; Miller, M.A.; Viboud, C. Synthesizing data and models for the spread of MERS-CoV, 2013: Key role of index cases and hospital transmission. *Epidemics* **2014**, *9*, 40–51. [CrossRef] [PubMed]
23. Assiri, A.; Al-Tawfiq, J.A.; Al-Rabeeah, A.A.; Al-Rabiah, F.A.; Al-Hajjar, S.; Al-Barrak, A. Epidemiological, demographic, and clinical characteristics of 47 cases of Middle East respiratory syndrome coronavirus disease from Saudi Arabia: A descriptive study. *Lancet Infect. Dis.* **2013**, *13*, 752–761. [CrossRef] [PubMed]
24. Li, B.; Eskandari, Z.; Avazzadeh, Z. Dynamical behaviors of an SIR epidemic model with discrete time. *Fractal Fract.* **2022**, *6*, 659. [CrossRef]
25. Li, B.; Eskandari, Z. Dynamical analysis of a discrete-time SIR epidemic model. *J. Frankl. Inst.* **2023**, *360*, 7989–8007. [CrossRef]
26. Li, B.; Eskandari, Z.; Avazzadeh, Z. Strong resonance bifurcations for a discrete-time prey–predator model. *J. Appl. Math. Comput.* **2023**, *69*, 2421–2438. [CrossRef]
27. Jiang, X.; Li, J.; Li, B.; Yin, W.; Sun, L.; Chen, X. Bifurcation, chaos, and circuit realisation of a new four-dimensional memristor system. *Int. J. Nonlinear Sci. Numer. Simul.* **2022**. [CrossRef]
28. Sabbar, Y. Asymptotic extinction and persistence of a perturbed epidemic model with different intervention measures and standard lévy jumps. *Bull. Biomath.* **2023**, *1*, 58–77. [CrossRef]
29. Adoum, A.H.; Haggar, M.S.D.; Ntaganda, J.M. Mathematical modelling of a glucose-insulin system for type 2 diabetic patients in Chad. *Math. Model. Numer. Simul. Appl.* **2022**, *2*, 244–251. [CrossRef]
30. Ahmed, I.; Akgül, A.; Jarad, F.; Kumam, P.; Nonlaopon, K. A Caputo-Fabrizio fractional-order cholera model and its sensitivity analysis. *Math. Model. Numer. Simul. Appl.* **2023**, *3*, 170–187. [CrossRef]
31. Singh, T.; Vaishali; Adlakha, N. Numerical investigations and simulation of calcium distribution in the alpha-cell. *Bull. Biomath.* **2023**, *1*, 40–57. [CrossRef]
32. Odionyenma, U.B.; Ikenna, N.; Bolaji, B. Analysis of a model to control the co-dynamics of Chlamydia and Gonorrhea using Caputo fractional derivative. *Math. Model. Numer. Simul. Appl.* **2023**, *3*, 111–140. [CrossRef]
33. Yao, T.-T.; Qian, J.-D.; Zhu, W.-Y.; Wang, Y.; Wang, G.-Q. A systematic review of lopinavir therapy for SARS coronavirus and MERS coronavirus—A possible reference for coronavirus disease-19 treatment option. *J. Med. Virol.* **2020**, *92*, 556–563. [CrossRef] [PubMed]

34. Fatima, B.; Yavuz, M.; Rahman, M.u.; Al-Duais, F.S. Modeling the epidemic trend of middle eastern respiratory syndrome coronavirus with optimal control. *Math. Biosci. Eng.* **2023**, *20*, 11847–11874. [CrossRef]
35. Afshar, Z.M.; Ebrahimpour, S.; Javanian, M.; Koppolu, V.; Vasigala, V.K.R.; Hasanpour, A.H.; Babazadeh, A. Coronavirus disease 2019 (COVID-19), MERS and SARS: Similarity and difference. *J. Acute Dis.* **2020**, *9*, 194–199.
36. Ahmad, S.; Dong, Q.; Rahman, M.u. Dynamics of a fractional-order COVID-19 model under the nonsingular kernel of Caputo-Fabrizio operator. *Math. Model. Numer. Simul. Appl.* **2022**, *2*, 228–243. [CrossRef]
37. ur Rahman, M.; Arfan, M.; Baleanu, D. Piecewise fractional analysis of the migration effect in plant-pathogen-herbivore interactions. *Bull. Biomath.* **2023**, *1*, 1–23. [CrossRef]
38. Yunhwan, K.; Sunmi, L.; Chaeshin, C.; Seoyun, C.; Saeme, H.; Youngseo, S. The Characteristics of Middle Eastern Respiratory Syndrome Coronavirus Transmission Dynamics in South Korea. *Osong Public Health Res. Perspect.* **2016**, *7*, 49–55.
39. Driessch, V.; Watmough, J. Reproduction numbers and sub threshold endemic equilibria for compartmental models of disease transmission. *Math. Biosci.* **2002**, *180*, 29–48. [CrossRef]
40. Van Den Driessche, P.; Watmough, J. *Mathematical Epidemiology*; Springer: Berlin/Heidelberg, Germany, 2008.
41. Chitnis, N.; Cushing, J.M.; Hyman, J.M. Bifurcation analysis of a mathematical model for malaria transmission. *SIAM J. Appl. Math.* **2006**, *67*, 24–45. [CrossRef]
42. Khan, T.; Zaman, G.; Chohan, M.I. The transmission dynamic and optimal control of acute and chronic hepatitis B. *J. Biol. Dyn.* **2017**, *11*, 172–189. [CrossRef]
43. Zaman, G.; Han Kang, Y.; Jung, I.H. Stability analysis and optimal vaccination of an SIR epidemic model. *Biosystems* **2008**, *93*, 240–249. [CrossRef] [PubMed]
44. LaSalle, J.P. *The Stability of Dynamical System*; SIAM: Philadelphia, PA, USA, 1976.
45. LaSalle, J.P. Stability of nonautonomous system. *Nonlinear Anal.* **1976**, *1*, 83–90. [CrossRef]
46. Khan, T.; Ullah, Z.; Ali, N.; Zaman, G. Modeling and control of the hepatitis b virus spreading using an epidemic model. *Chaos Solitons Fractals* **2019**, *124*, 1–9. [CrossRef]
47. Kamien, M.; Schwartz, N. *Dynamic Optimization*, 2nd ed.; North-Holland: Amsterdam, The Netherlands, 1991.
48. Hattaf, K.; Yousfi, N. Optimal control of a delayed HIV infection model with immune response using an efficient numerical method. *Int. Sch. Res. Not.* **2012**, *2012*, 215124. [CrossRef]
49. Evirgen F.; Ozköse, F.; Yavuz, M.; Ozdemir, N. Real data-based optimal control strategies for assessing the impact of the Omicron variant on heart attacks. *AIMS Bioeng.* **2023**, *10*, 218–239. [CrossRef]
50. Lashari, A.A.; Zaman, G. Global dynamics of vector-borne diseases with horizontal transmission in host population. *Comput. Math. Appl.* **2001**, *61*, 745–754. [CrossRef]
51. Aseev, S.M.; Kryazhimskii, A.V. The Pontryagin maximum principle and optimal economic growth problems. *Proc. Steklov Inst. Math.* **2007**, *257*, 1–255. [CrossRef]

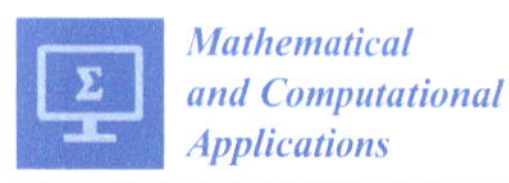
Mathematical and Computational Applications

MDPI

Article

On Generalized Dominance Structures for Multi-Objective Optimization

Kalyanmoy Deb [1,*] and Matthias Ehrgott [2]

[1] Computational Optimization and Innovation (COIN) Laboratory, Department of Electrical and Computer Engineering, Michigan State University, East Lansing, MI 48824, USA

[2] Department of Management Science, Lancaster University, Lancaster LA1 4YW, UK; m.ehrgott@lancaster.ac.uk

* Correspondence: kdeb@egr.msu.edu

Abstract: Various dominance structures have been proposed in the multi-objective optimization literature. However, a systematic procedure to understand their effect in determining the resulting optimal set for generic domination principles, besides the standard Pareto-dominance principle, is lacking. In this paper, we analyze and lay out properties of generalized dominance structures which help provide insights for resulting optimal solutions. We introduce the concept of the anti-dominance structure, derived from the chosen dominance structure, to explain how the resulting non-dominated or optimal set can be identified easily compared to using the dominance structure directly. The concept allows a unified explanation of optimal solutions for both single- and multi-objective optimization problems. The anti-dominance structure is applied to analyze respective optimal solutions for most popularly used static and spatially changing dominance structures. The theoretical and deductive results of this study can be utilized to create more meaningful dominance structures for practical problems, understand and identify resulting optimal solutions, and help develop better test problems and algorithms for multi-objective optimization.

Keywords: dominance principles; multi-objective optimization; evolutionary algorithms

Citation: Deb, K.; Ehrgott, M. On Generalized Dominance Structures for Multi-Objective Optimization. *Math. Comput. Appl.* **2023**, *28*, 100. https://doi.org/10.3390/mca28050100

Academic Editor: Leonardo Trujillo

Received: 6 November 2022
Revised: 1 September 2023
Accepted: 30 September 2023
Published: 7 October 2023

1. Introduction

In a practical multi-objective optimization study, users should have the flexibility in choosing a dominance structure which would involve objective preferences and priorities of users. In most evolutionary multi-objective optimization (EMO) studies, a set of Pareto-optimal solutions are attempted to be found by an evolutionary population-based algorithm and the choice of a single preferred solution from the Pareto-optimal set is deferred as a post-optimality decision-making task [1,2]. In the recent past, some exceptions to this principle have shown that a generic dominance structure can be created with desired preference information for the resulting EMO to converge to a single preferred Pareto-optimal solution or to focus to a preferred Pareto-optimal region at the end of the optimization task [3–7]. The practicalities and merits of both approaches from computational and decision-making points of view can be debated, but if the preference information is easier to obtain, the latter approach can be appealing from a computational viewpoint.

The EMO and classical optimization literature has proposed a number of alternate dominance structures for this purpose [8–12], not always motivated by the decision-making preferences used in the real world, rather from considering interesting geometric constructions aided by their practical significance. However, the literature lacks a systematic study outlining what properties a generalized dominance structure must have such that an EMO will end up finding a non-empty optimal solution set with certain desired properties. This study attempts to fill this gap and answer the following questions. Does any arbitrarily chosen dominance structure generate a non-empty optimal solution set? For a given dominance structure, how does one identify the resulting optimal solution set for a problem?

If a dominance structure fails to produce an optimal solution, are there ways to modify it to find the desired non-empty optimal solution set? To have confidence in the derived principles, do they reveal known properties of the existing dominance structures?

To achieve our goal, we have formalized the concept of an *anti-dominance* structure, which is intricately dependent on the chosen dominance structure; however, we argue that it is more useful than the dominance structure in answering the above questions and providing a better understanding of the outcome of the chosen dominance structure. In addition to the conditions for a non-empty optimal solution set, we discover a number of interesting and useful properties of these structures to determine if the respective optimal solution set has a single optimum or multiple optima. Contrary to the general belief, it has been clearly demonstrated here that dominance structures satisfying semi-transitive properties can lead to an optimal solution set and that a satisfaction of the transitive property need not be a hard restriction. Although demonstrated using two- and three-objective problems in this paper, the concepts of this paper are applicable to many-objective problems as well. Overall, the results of this paper should enable researchers to obtain a better insight into a direct understanding of generalized dominance structures and the resulting optimal solution set, which may be useful in achieving various EMO activities, such as multi-objective test problem generation, efficient algorithm development with a knowledge of sources of algorithmic inefficiencies in finding the true optimal solution set, and the design of meaningful generalized dominance structures for effective application.

In the rest of the paper, we discuss the optimality conditions for single-objective optimization problems in a generalized manner through an inferior structure concept in Section 2. Based on uni-modal and multi-modal single-objective optimization, the concept of an anti-inferior structure is introduced so the concept of inferior and anti-inferior (or anti-dominance) structures can be carried over to multi-objective optimization in Section 3. Section 4 applies the concept of an anti-dominance structure to explain the working of a number of existing generalized dominance principles proposed in the literature. Then, in Section 5, we extend the use of dominance and anti-dominance structures for spatially changing dominance relationships. Finally, in Section 6, we present the conclusions of this study.

2. Optimality Principles for Single-Objective Optimization

For a single-objective minimization problem:

$$\begin{aligned} \text{Minimize} \quad & f(\mathbf{x}), \\ \text{Subject to} \quad & g_j(\mathbf{x}) \leq 0, \quad j = 1, 2, \ldots, J, \end{aligned} \tag{1}$$

in which $\mathbf{x} \in \mathbb{R}^n$ is the variable vector, $f : \mathbb{R}^n \to \mathbb{R}$ is the objective function, and $g_j : \mathbb{R}^n \to \mathbb{R}$ is the j-th inequality constraint. A solution is called *feasible* if all J constraints are satisfied by the solution. Let us denote the feasible solution set $\mathbf{X} = \{\mathbf{x} | g_j(\mathbf{x}) \leq 0, \forall j\}$ as the set of all feasible solutions in the search space and the feasible objective set $\mathbf{Z} = \{f(\mathbf{x}) | \mathbf{x} \in \mathbf{X}\}$. If a solution is not feasible, it is called an *infeasible* solution.

To arrive at the definition of the optimal solution for the problem stated above, we first define the concept of the *inferiority solution set* $\omega(\mathbf{x})$ of a feasible solution $\mathbf{x}$. Every member of $\omega(\mathbf{x})$ is worse than $\mathbf{x}$ in terms of the given objective function. To understand the inferiority set, we first define the inferiority condition between two feasible solutions.

Definition 1 (Single-objective inferiority condition). *For a pair of feasible solutions $\mathbf{x} \in \mathbf{X}$ and $\mathbf{y} \in \mathbf{X}$, $\mathbf{y}$ is inferior to $\mathbf{x}$ (or mathematically, $\mathbf{x} \prec \mathbf{y}$) in a single-objective sense if $f(\mathbf{x}) < f(\mathbf{y})$.*

From this condition, we derive an inferior solution set $\omega(\mathbf{x})$, as follows:

Definition 2 (Single-objective inferior set of $\mathbf{x}$). *The set of all feasible solutions $\mathbf{y} \in \mathbf{X}$ for which $\mathbf{x} \prec \mathbf{y}$ is defined as the inferior set $\omega(\mathbf{x})$ of $\mathbf{x} \in \mathbf{X}$, and the set of respective objective values is defined as the inferior objective set $\Omega(\mathbf{x}) = \{f(\mathbf{y}) | \mathbf{x} \prec \mathbf{y}, \forall \mathbf{y} \in \mathbf{X}\}$.*

For minimization problems, the inferior objective set of $\mathbf{x} \in \mathbf{X}$ can also be written as follows:

$$\Omega(\mathbf{x}) = f(\mathbf{x}) + \{\delta | \delta = f(\mathbf{y}) - f(\mathbf{x}) > 0, \forall \mathbf{y} \in \mathbf{X}\}. \tag{2}$$

The summation is in the Minkowski sense. For a problem, the user can provide a generalized *null inferior* structure Ω_0 at the origin to define the desired inferiority condition ($\Omega_0 = \{\delta | \delta > 0\}$). Note that the definition of Ω_0 may not depend on the knowledge of the feasible objective space $\mathbf{Z}$ and can be defined purely based on the desired condition for a generalized "optimal" solution. The following relationships between $\Omega(\mathbf{x})$ and Ω_0 can be written:

$$\Omega(\mathbf{x}) \quad \subset \quad f(\mathbf{x}) + \Omega_0, \tag{3}$$
$$\Omega(\mathbf{x}) \quad = \quad (f(\mathbf{x}) + \Omega_0) \cap \mathbf{Z}. \tag{4}$$

The above inferiority condition respects the following properties:

- **Irreflexive property:** A solution $\mathbf{x}$ is not inferior to itself, that is, $\mathbf{x} \not\prec \mathbf{x}$.
- **Asymmetric property:** If $\mathbf{x} \prec \mathbf{y}$, then $\mathbf{y} \not\prec \mathbf{x}$.
- **Transitive property:** If $\mathbf{x} \prec \mathbf{y}$ and $\mathbf{y} \prec \mathbf{z}$, then $\mathbf{x} \prec \mathbf{z}$.

Graphically, the above inferiority condition can be demonstrated with a sketch of the objective function on a real line ($\mathbb{R}$), as shown in Figure 1a. In the sketch, $\Omega(\mathbf{x})$ represents the entire green line on the right of $f(\mathbf{x})$, excluding the value of $\mathbf{f}(\mathbf{x})$.

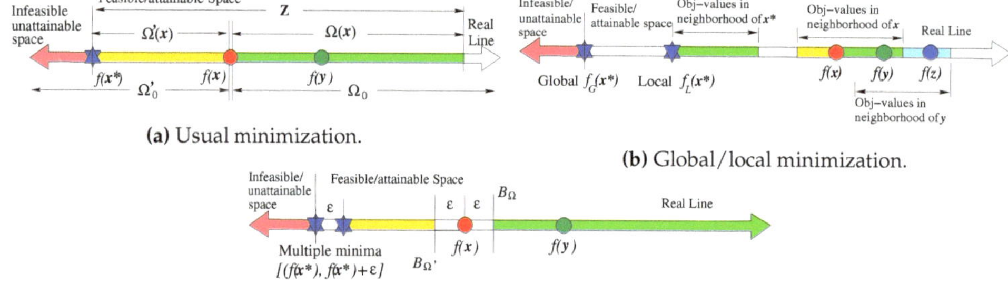

(**a**) Usual minimization.

(**b**) Global/local minimization.

(**c**) Epsilon-minimization.

Figure 1. Inferiority and optimality definitions are illustrated for a single-objective problem.

Note that a simple check on $\Omega(\mathbf{x})$ is not adequate for defining an optimal solution appropriately, particularly if there exist multiple optimal solutions to the problem. To define the optimal solution, we need to define an *anti-inferior* solution set $\omega'(\mathbf{x})$ and its corresponding anti-inferior objective set $\Omega'(\mathbf{x})$ as follows:

Definition 3 (Anti-inferiority condition). *For a pair of feasible solutions $\mathbf{x} \in \mathbf{X}$ and $\mathbf{y} \in \mathbf{X}$, $\mathbf{y}$ is anti-inferior to $\mathbf{x}$ in a single-objective sense if $f(\mathbf{x}) > f(\mathbf{y})$.*

A little thought will lead us to define the *anti-inferior* solution set $\omega'(\mathbf{x}) = \{\mathbf{y} | \mathbf{y} \prec \mathbf{x}, \mathbf{y} \in \mathbf{X}\}$ and the *anti-inferior objective set* $\Omega'(\mathbf{x}) = f(\mathbf{x}) + \{\delta | \delta = f(\mathbf{y}) - f(\mathbf{x}) < 0, \forall \mathbf{y} \in \mathbf{X}\}$. For the sketch in Figure 1a, the golden line left of $f(\mathbf{x})$ (excluding $f(\mathbf{x})$) denotes the anti-inferior objective set. The set $\Omega'(\mathbf{x})$ is important in determining the optimal solution, as there cannot exist any better solution than the optimal solution:

Definition 4 (Single-objective optimality condition). *A solution $\mathbf{x}^* \in \mathbf{X}$ is an (global) optimal solution in the single-objective sense if its anti-inferior objective set is empty or $\Omega'(\mathbf{x}^*) = \emptyset$.*

For the solution $\mathbf{x}^*$ in Figure 1a, $\Omega(\mathbf{x}^*) = \varnothing$ (there is no feasible objective value left of $f(\mathbf{x}^*)$). Hence, it is an optimal solution to the problem. This is true for every global optimal solution. The opposite is also true.

Theorem 1 (Empty anti-inferior set). *If $\mathbf{x}^*$ is an (global) optimal solution, its anti-inferior set is empty, or, $\Omega'(\mathbf{x}^*) = \varnothing$.*

The proof is intuitive. The definition and theorem also suggest that at the optimal solution $\mathbf{x}^*$, $\Omega(\mathbf{x}^*) \cap \Omega'(\mathbf{x}^*) = \varnothing$. From the asymmetric property, it is also true that $\Omega(\mathbf{x}) \cap \Omega'(\mathbf{x}) = \varnothing$ for every feasible point $\mathbf{x}$. However, this may not be true at every feasible point in the search space and still have an existence of an optimal solution $\mathbf{x}^*$ by satisfying $\Omega(\mathbf{x}^*) \cap \Omega'(\mathbf{x}^*) = \varnothing$. However, if this condition is not met at every feasible point, an algorithm to find optimal solutions may have difficulties in progressing towards the optimal region.

Let us now define a *null anti-inferior* structure at the origin as $\Omega'_0 = \{\delta | \delta < 0\}$, so that $\Omega'(\mathbf{x}) \subset f(\mathbf{x}) + \Omega'_0$ and $\Omega'(\mathbf{x}) = (f(\mathbf{x}) + \Omega'_0) \cap \mathbf{Z}$. It is interesting to note from their mathematical constructs that $\Omega'_0 = -\Omega_0$. Because the user has the liberty to choose any null inferior structure Ω_0 for defining certain desired properties of the resulting optimal solution(s), it is advisable to meet the condition $\Omega_0 \cap \Omega'_0 = \varnothing$ to guarantee asymmetric property everywhere in the search space, irrespective of the nature of $\mathbf{Z}$. This may allow the smooth operation of an optimization algorithm. However, in some esoteric cases, an optimal solution still may exist without satisfying this condition with Ω_0. This happens when there exists no feasible solution at the overlapping region of inferior and anti-inferior structures applied at the optimal solution. This discussion brings us to the following theorem.

Theorem 2 (Overlapping inferior and anti-inferior sets). *For a null inferior structure Ω_0 with non-empty $\Omega_0 \cap \Omega'_0$, an optimal solution $\mathbf{x}^*$ exists only if $(f(\mathbf{x}^*) + (\Omega_0 \cap \Omega'_0)) \cap \mathbf{Z} = f(\mathbf{x}^*)$.*

Note that the Definition 4 is also valid for each of the multiple global optimal solutions if they exist in a problem, as all such optimal solutions have an identical $f(\mathbf{x}^*)$ value. However, for a local optimal solution, the condition $\mathbf{x} \in \mathbf{X}$ needs to be appended with a local neighborhood restriction of $\mathbf{y}$, $\mathcal{B}_\gamma(\mathbf{x}) = \{\mathbf{y} | (\|\mathbf{y} - \mathbf{x}\|_2 \leq \gamma) \wedge (\mathbf{y} \in \mathbf{X})\}$, around $\mathbf{x}$.

Definition 5 (Single-objective local inferiority condition). *For a pair of feasible solutions $\mathbf{x} \in \mathbf{X}$ and $\mathbf{y} \in \mathcal{B}_\gamma(\mathbf{x})$, $\mathbf{y}$ is local inferior to $\mathbf{x}$ (say, $\mathbf{x} \prec_L \mathbf{y}$) in a single-objective sense if $f(\mathbf{x}) < f(\mathbf{y})$.*

Figure 1b marks the respective objective values in the neighborhood of $\mathbf{x}$. It is understood that $\mathbf{y}$ is within a radius of γ around $\mathbf{x}$ and has a worse objective value than $\mathbf{x}$, hence $\mathbf{x} \prec \mathbf{y}$ in the local sense. Note that the anti-inferior solution and objective sets for a local optimal solution can also be defined by restricting the members within the neighborhood. For brevity, we do not define them here. The irreflexive and asymmetric properties are still valid for the local inferior structure; however, the transitivity property may not be satisfied among three feasible solutions ($\mathbf{x} \in \mathbf{X}$, $\mathbf{y} \in \mathbf{X}$, and $\mathbf{z} \in \mathbf{X}$). To have $\mathbf{x} \prec_L \mathbf{y}$, the solution $\mathbf{y}$ must be in the neighborhood of $\mathbf{x}$, and to have $\mathbf{y} \prec_L \mathbf{z}$, the solution $\mathbf{z}$ must be in the neighborhood of $\mathbf{y}$; however, these conditions do not require that $\mathbf{z}$ must be in the neighborhood of $\mathbf{x}$. Hence, $\mathbf{z}$ may not lie in $\mathcal{B}_\gamma(\mathbf{x})$, and vice versa. Thus, the transitivity property may not be satisfied for the local inferiority structure. This brings us to the semi-transitive property for a generic inferiority structure:

- **Semi-transitive Property:** If $\mathbf{x} \prec \mathbf{y}$ and $\mathbf{y} \prec \mathbf{z}$, then $\mathbf{z} \not\prec \mathbf{x}$.

It is not as strong as the transitive property. This property does not require that the solution $\mathbf{x}$ be better than $\mathbf{z}$, but enforces that $\mathbf{z}$ must not be better than $\mathbf{x}$. Clearly, if an inferiority structure does not satisfy the transitivity property, it must *necessarily* satisfy the semi-transitive property for it to break the inferiority cycle and finally result in an optimal solution.

Theorem 3 (Nonexistence of optimum). *If both transitive and semi-transitive properties are violated by an inferiority structure, there cannot exist an optimal solution.*

Proof. This can be proven by contradiction. Assume that there exists an optimal solution $\mathbf{x}^*$ and also consider two other feasible solutions $\mathbf{y}$ and $\mathbf{z}$, in which $\mathbf{y}$ satisfies $\mathbf{x}^* \prec \mathbf{y}$ and $\mathbf{z}$ satisfies $\mathbf{y} \prec \mathbf{z}$. Because neither the transitivity nor semi-transitivity condition is satisfied, it is possible that $\mathbf{x}^* \not\prec \mathbf{z}$ and $\mathbf{z} \prec \mathbf{x}^*$, thereby violating the optimality condition stated in Definition 4. $\square$

2.1. Epsilon-Inferiority Conditions

Let us reconsider Equation (2) and generalize it with a non-negative parameter ϵ (Equation (2) was defined with $\epsilon = 0$):

$$\Omega(\mathbf{x}, \epsilon) = f(\mathbf{x}) + \{\delta | \delta = f(\mathbf{y}) - f(\mathbf{x}) > \epsilon, \forall \mathbf{x} \in \mathbf{X}, \mathbf{y} \in \mathbf{X}\}. \tag{5}$$

Equation (5) allows a generalization of the inferior objective set with an epsilon-optimality condition in which a solution $\mathbf{y}$ is considered epsilon-inferior to $\mathbf{x}$, if $f(\mathbf{x}) < f(\mathbf{y}) - \epsilon$, allowing a more practice-oriented inferiority definition, as shown in Figure 1c. Writing Ω in terms of the difference in objective values ($\delta = f(\mathbf{y}) - f(\mathbf{x})$), we can define a *null* inferior set, $\Omega_0(\epsilon) = \{\delta | \delta > \epsilon\}$. Thus, $\Omega(\mathbf{x}, \epsilon) \subset f(\mathbf{x}) + \Omega_0(\epsilon)$ in the Minkowski sense. The defining boundary of the inferior set $\Omega_0(\epsilon)$ ($B_\Omega = \epsilon$ in this case) is an important feature for a user to define the desired inferiority of a solution in the feasible objective set $\mathbf{Z}$. The boundary of the inferior set can be inclusive (B_Ω^I) to the set or exclusive (B_Ω^E). Note that the epsilon-inferiority condition defined in Equation (5) is valid for B_Ω^E.

It is interesting to note that the respective anti-inferior objective set is $\Omega'(\mathbf{x}, \epsilon) \subset f(\mathbf{x}) + \Omega'_0(\epsilon)$, where $\Omega'_0 = \{\delta | \delta < -\epsilon\}$ is the null anti-inferior set for the epsilon-inferior condition. In the same manner as before, we notice that $\Omega'_0(\epsilon) = -\Omega_0(\epsilon)$. All solutions $\mathbf{y}$ which have an objective value smaller than ϵ from $f(\mathbf{x})$ are now epsilon-inferior to $\mathbf{x}$. The defining boundary of the anti-inferior set is $B_{\Omega'}^E = -\epsilon$ for the epsilon-inferiority definition. Figure 1c shows that solutions with objective values left of $B_{\Omega'^E}$ are inferior to $\mathbf{x}$. By definition, irreflexive, asymmetric, and transitive properties are satisfied by the epsilon-inferiority structure.

The epsilon-inferiority condition extends the definition of optimality conditions for problems having multiple optimal solutions. All solutions $\mathbf{x}$ which are within ϵ from the optimal solution's objective value $f(\mathbf{x}^*)$ are epsilon-optimal. This discussion helps to define optimality conditions for multi-objective optimization.

3. Optimality Principles for Multi-Objective Optimization

A constrained multi-objective optimization problem having M conflicting objectives, J inequality constraints, and no equality constraints is formulated as follows:

$$\begin{aligned} \text{Minimize} \quad & \mathbf{f}(\mathbf{x}) = \{f_1(\mathbf{x}), f_2(\mathbf{x}), \dots, f_M(\mathbf{x})\}, \\ \text{Subject to} \quad & g_j(\mathbf{x}) \leq 0, \quad j = 1, 2, \dots, J. \end{aligned} \tag{6}$$

The set of feasible solutions satisfying all constraints is denoted with $\mathbf{X}$ and the respective objective vectors constitute the feasible objective set $\mathbf{Z}$. First, we extend and define the inferiority of a solution over another as follows:

Definition 6 (Multi-objective inferiority condition). *For a pair of solutions* $\mathbf{x} \in \mathbf{X}$ *and* $\mathbf{y} \in \mathbf{X}$, $\mathbf{y}$ *is inferior to* $\mathbf{x}$ *in a multi-objective sense if* $\mathbf{x}$ *dominates* $\mathbf{y}$ ($\mathbf{x} \prec \mathbf{y}$).

The above requires a suitable definition of dominance. The popularly used Pareto-dominance condition ($\prec_P$) is defined as follows:

Definition 7 (Pareto-dominance ($\prec_P$) condition). *A solution* $\mathbf{x} \in \mathbf{X}$ *Pareto-dominates another solution* $\mathbf{y} \in \mathbf{X}$, *if the following two conditions are true: (i)* $f_i(\mathbf{x}) \leq f_i(\mathbf{y})$ *for all* $i = 1, 2, \ldots, M$, *and (ii)* $f_j(\mathbf{x}) < f_j(\mathbf{y})$ *for at least one* $j = 1, 2, \ldots, M$.

Note that for the single-objective case, $M = 1$ and the above dominance condition becomes equivalent to the inferiority condition defined in Definition 1. As with the single-objective case, the inferior objective set is defined as follows: $\Omega(\mathbf{x}) = \{\mathbf{f}(\mathbf{y}) | \mathbf{x} \prec \mathbf{y}, \forall \mathbf{y} \in \mathbf{X}\}$. To define the optimality condition (Pareto-optimality condition with $\prec_P$), we need to have the definition of anti-inferior objective set $\Omega'(\mathbf{x}) = \{\mathbf{f}(\mathbf{y}) | \mathbf{y} \prec \mathbf{x}, \forall \mathbf{y} \in \mathbf{X}\}$, derived from a generalized dominance structure $\Omega(\mathbf{x})$.

It is clear from the above discussion that the dominance concept between two solutions prevalent in multi-objective optimization literature is equivalent to the inferiority concept introduced here for single-objective optimization. Hence, we call $\Omega(\mathbf{x})$ and $\Omega'(\mathbf{x})$ sets the *dominance* and *anti-dominance* sets of $\mathbf{x}$, respectively, in the context of multi-objective optimization. Like in the single-objective case, a user can define the null dominance structure Ω_0, a bounded M-dimensional set for which origin is not a member for multi-objective optimization. The resulting dominance set at a point $\mathbf{x}$ is then defined as $\Omega(\mathbf{x}) = (\mathbf{f}(\mathbf{x}) + \Omega_0) \cap \mathbf{Z}$. Clearly, $\Omega(\mathbf{x}) \subset \mathbf{f}(\mathbf{x}) + \Omega_0$. Figure 2 shows a specific dominance structure and its null structure. The boundary B_Ω of Ω_0 in the M-dimensional objective space can be obtained from Ω_0. Then, the resulting null anti-dominance structure Ω_0' can be determined from Ω_0 with a procedure described in Section 3.2 in order to define the respective generalized optimal solutions for the supplied dominance structure Ω_0. Extending the concept from single-objective optimization, the defined generalized dominance structure must satisfy irreflexive and asymmetry properties mentioned for the single-objective case and also the transitivity or semi-transitivity property stated in Theorem 3.

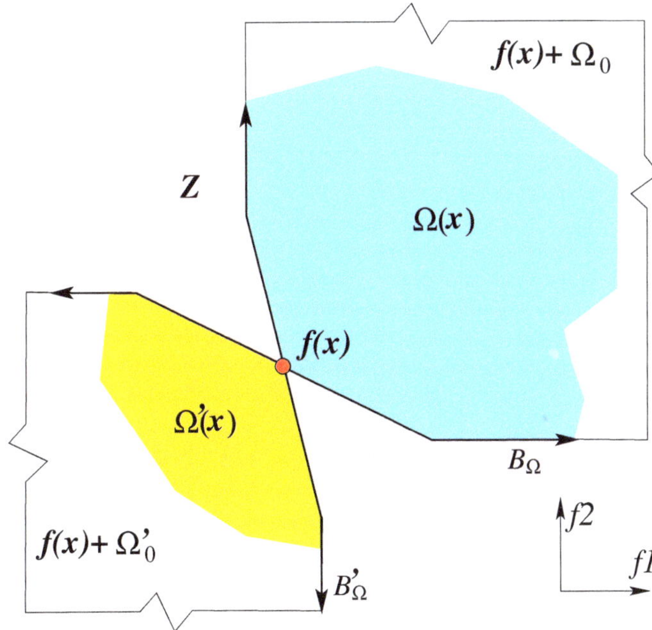

Figure 2. Relationship between $\Omega(\mathbf{x})$ and Ω_0. Here, Ω_0 and Ω_0' are non-overlapping.

Knowing the null anti-dominance structure $\Omega_0' = \{\delta | \delta < 0\}$ and defining anti-dominance objective set $\Omega'(\mathbf{x}) = (\mathbf{f}(\mathbf{x}) + \Omega_0') \cap \mathbf{Z}$, an optimal solution for a multi-objective optimization problem can be defined as follows:

Definition 8 (Multi-objective optimality condition). *A solution* $\mathbf{x}^* \in \mathbf{X}$ *is optimal in the multi-objective sense with a generalized dominance structure* Ω_0 *at the origin if there does not exist any feasible solution in the anti-inferior objective set, or* $\Omega'(\mathbf{x}^*) = \varnothing$.

The resulting objective vector $\mathbf{f}^*$ lies in the optimal set $\mathbf{Z}^*$. Note that the above definition can result in multiple optimal solutions. In fact, if the optimal solution $\mathbf{x}^{*,i}$ of the single-objective constrained optimization problem with the i-th objective function is different from $(\mathbf{x}^{*,j})$ that of j-th objective function, then $\mathbf{x}^{*,i} \not\prec \mathbf{x}^{*,j}$ and $\mathbf{x}^{*,j} \not\prec \mathbf{x}^{*,i}$ can both occur. Both solutions then become optimal solutions to the multi-objective optimization problem. Besides these individual optimal solutions, many other *compromise* solutions, trading off the objectives, may exist in the feasible solution set. For the Pareto-dominance structure ($\prec_P$) with $\Omega_0 = \mathbb{R}_+^M \setminus \{\mathbf{0}\}$, the respective optimal solutions are called *Pareto-optimal* solutions ($\mathbf{x}^P$) and the set is referred to as the Pareto-optimal set. The respective set of objective vectors (set $\mathbf{Z}^{*,P}$) is called to constitute a *Pareto-optimal front* (PF), in the parlance of the EMO literature.

As with Theorem 1 for the single-objective case, the following is also true:

Theorem 4 (Empty anti-dominance set for multi-objective optimization). *If* $\mathbf{x}^*$ *is an optimal solution, its anti-inferior objective set is empty, or,* $\Omega'(\mathbf{x}^*) = \varnothing$.

Although it is advisable to construct a dominance structure Ω_0 such that $\Omega_0 \cap \Omega_0' = \varnothing$ to achieve a smooth progress of an optimization algorithm towards the optimal region, for some scenarios, an overlapping Ω_0 and Ω_0' structure can also cause an optimal solution $\mathbf{x}^*$ to exist, particularly when $(\mathbf{f}(\mathbf{x}^*) + (\Omega_0 \cap \Omega_0')) \cap \mathbf{Z} = \mathbf{f}(\mathbf{x}^*)$, meaning that there does not exist any other feasible objective vector other than the optimal objective vector at the intersection of dominance and anti-dominance objectives sets constructed at the optimal solution.

3.1. Defining a Generalized Dominance Structure for Multi-Objective Optimization

The defining boundary for the dominance structure is useful to define a generalized dominance condition for the multi-objective case. Extending the boundary of the dominance structure $\Omega_0(\epsilon)$ discussed for epsilon-inferiority condition in Section 2.1, one can define a boundary $\Omega_0(\epsilon)$ at the origin to declare the part of the M-dimensional objective space (with ϵ-vector) that are dominated (or worse) than the origin, where $\epsilon = (\epsilon_1, \ldots, \epsilon_M) \in \mathbb{R}_+^M$ is the vector of changes in objective values. Figure 3 shows one such generalized dominance structure (Ω_0) at the origin. It implies that, up to a limit of $-\epsilon_1$ change on f_1 from the origin, the origin is inferior to any point with a trade-off (loss/gain) larger than T_1. A similar trade-off of T_2 exists for a limit of $-\epsilon_2$ change in f_2. Note that the generic $\Omega_0(\epsilon)$ structure allows any arbitrary definition of the dominance structure as a function of changes in objectives, compared to the objective-wise settings in the level sets [13] and epsilon-dominance principles [14]. A redefinition of the dominance structure will create a different set of optimal solutions than the Pareto-optimal solutions. Hence, such an approach will allow users to find respective optimal solution set for any chosen dominance structure. However, before we find the generalized optimal solution set, let us find the relationship between Ω_0 and Ω_0'.

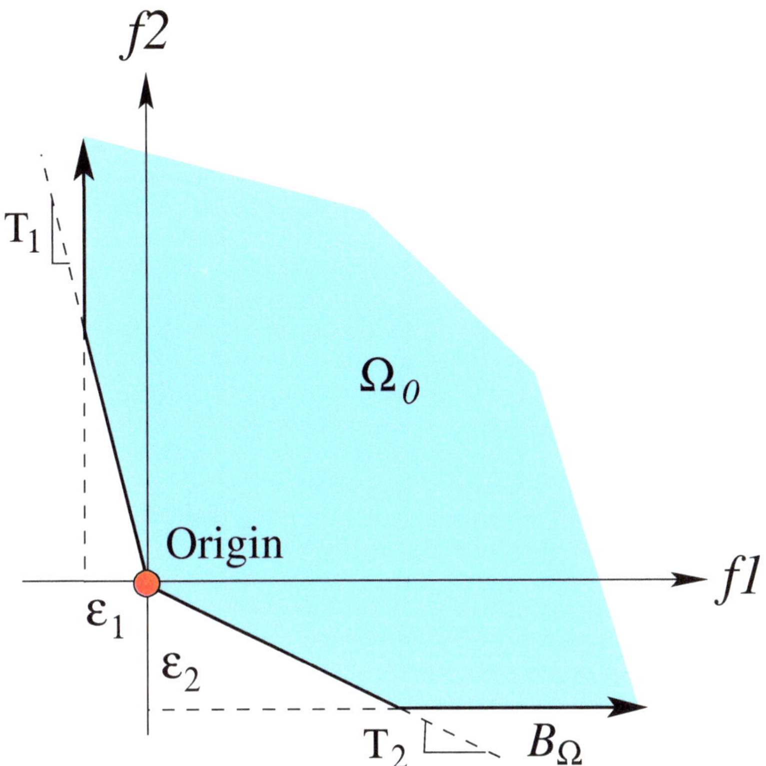

Figure 3. A generalized dominance structure Ω_0 and its boundary B_Ω.

3.2. Relationship Between Dominance and Anti-Dominance Structures

Every member of the dominance structure Ω_0 is dominated by (i.e., worse than) the origin and every member of anti-dominance structure Ω'_0 dominates (i.e., is better than) the origin. As indicated for the single-objective case, the two sets are related by a simple relation: $\Omega'_0 = -\Omega_0$. The following theorem states that this is a universal property, even for the multi-objective case.

Theorem 5 (Relationship between dominance and anti-dominance structures). *For any null dominance set Ω_0, its null anti-dominance structures $\Omega'_0 = -\Omega_0$.*

Proof. Let us consider Figure 4 for a proof with a generic dominance structure (Ω_0) defined at the origin (point O). Let us consider a generic point A at $\mathbf{d} \in \Omega_0$ from O. Let us now construct a point B at $\mathbf{d}' = -\mathbf{d}$ from O. Now, we construct the dominance structure Ω_0 at B (shown in shaded region), as if B is the new origin. Then, the original origin (point O) is now at $\mathbf{d}$ location from the new origin (B). Because $\mathbf{d}$ is inside the set Ω_0, the new origin B dominates the original origin O. This is true for every $\mathbf{d}$, and thus Ω'_0 can be constructed with negative vectors of every member of Ω_0. $\square$

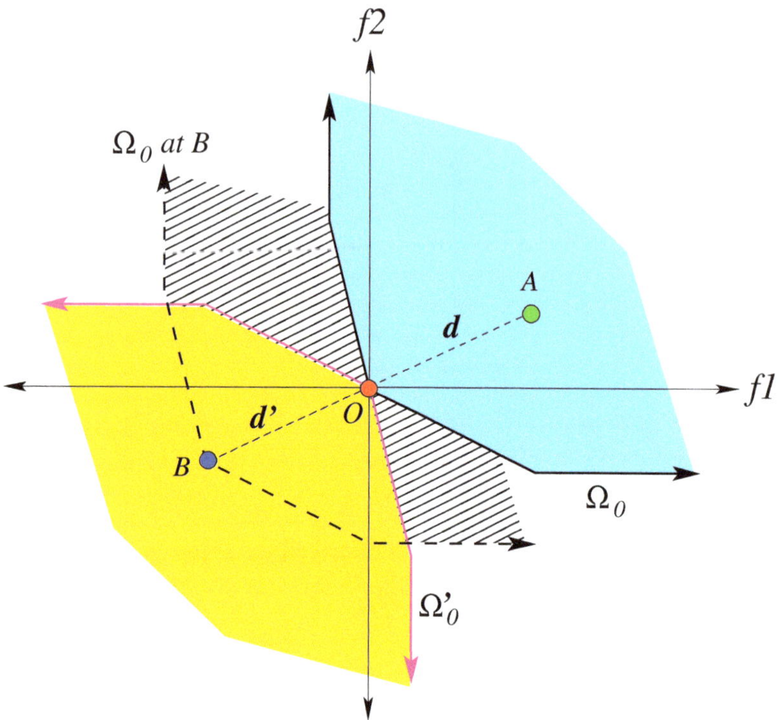

Figure 4. Illustration of anti-dominance structure Ω'_0.

Corollary 1. *For any dominance structure Ω_0 with a defining boundary B_Ω, the defining boundary of the anti-dominance structure $B_{\Omega'} = -B_\Omega$.*

As with the single-objective case, we now discuss the properties that a generalized dominance (GD) structure (Ω_0) must have:

- **Irreflexive property:** Ω_0 (and its boundary B_Ω) must exclude the **0**-vector (origin) from its set.
- **Asymmetric property:** $\Omega_0 \cap \Omega'_0 = \varnothing$ (recommended, as discussed in the paragraph before Theorem 2).
- **Transitive property:** This requires a chain of Ω_0 consideration and requires further discussion (provided in Section 3.2.1).

The first two properties indicate that a GD structure can have an inclusive boundary with points on the boundary B_Ω^I being inferior to the origin or an exclusive boundary on which the points are not inferior to the origin.

The above also indicates the following corollary is true, as in multi-objective objective space, overlapping solutions between Ω_0 and Ω'_0 can arise from the boundary (B_Ω^I) of the chosen dominance structure:

Corollary 2. *For $M > 1$, if $\Omega_0 \cup \Omega'_0 = \mathbb{R}^M \backslash \{\mathbf{0}\}$, no optimal solution exists.*

The clause indicates that boundary B_Ω must be included in both Ω_0 and Ω'_0. Because this contradicts $\Omega_0 \cap \Omega'_0 = \varnothing$, the corollary is true. For $M = 1$, the clause is true only when $B_\Omega^I = B_{\Omega'}^I = 0$ (origin). Because the definition of Ω_0 excludes the origin, the corollary may not be true for $M = 1$.

Corollary 3. *If $\Omega_0 \cup \Omega'_0 = \mathbb{R}^M \backslash (B_\Omega \cup B_{\Omega'} \cup \{\mathbf{0}\})$, the exclusive boundaries must be equal (that is, $B_\Omega^E = B_{\Omega'}^E$) and all optimal solutions must lie on B_Ω passing through one of the optimal solutions.*

The above theorem is valid for the exclusive boundary GDs. The weighted-sum dominance structure (Figure 9d) satisfies the above condition and forces all optimal solutions to lie on the boundary plane $B_{\Omega(\mathbf{x}^*)}^E$ having a normal vector ($\mathbf{w}$) and passing through any of the optimal solutions $\mathbf{x}^*$. Another GD structure is shown in Figure 5.

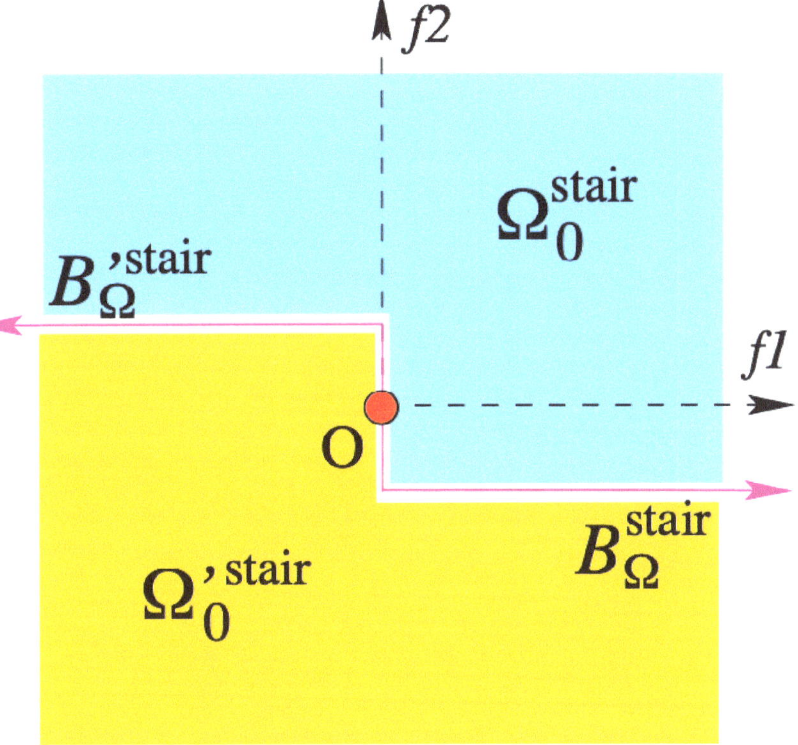

Figure 5. A dominance structure satisfying Corollary 3.

3.2.1. Transitive and Semi-Transitive Properties

It has been previously mentioned that the generalized dominance structure Ω_0 having irreflexive, asymmetric, and semi-transitive properties can produce a non-empty optimal solution set, contrary to the general belief that a transitive property is a must. Let us first demonstrate this fact graphically with the generalized dominance structure considered in Figure 3. We observe from Figure 6 that when a dominated or an inferior point $\mathbf{y}$ is chosen from Ω_0^{GD} at $\mathbf{x}$ and another point $\mathbf{z}$ is chosen from Ω_0^{GD} at $\mathbf{y}$, $\mathbf{x}$ may not dominate $\mathbf{z}$, in general. This violates the transitive property, but we observe from the figure that this specific dominance structure satisfies the semi-transitive property in that $\mathbf{z}$ does not dominate $\mathbf{x}$, as $\mathbf{z}$ does not lie inside the $\Omega'(\mathbf{x})$ set. Thus, an important task is to determine the true nature of the *effective dominance* structure when intermediate points such as $\mathbf{y}$ are allowed in the optimization process.

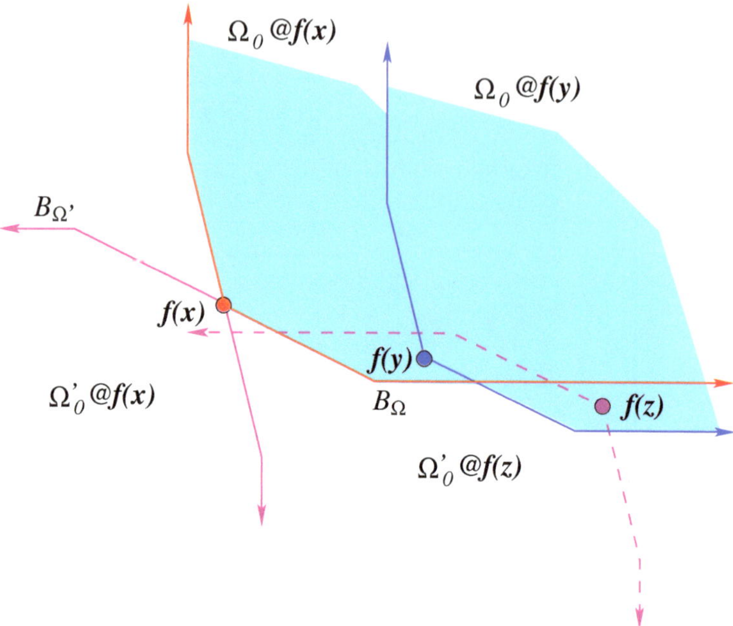

Figure 6. Ω_0^{GD} is not transitive but follows the semi-transitive property.

For a population-based optimization algorithm, such as evolutionary multi-objective optimization (EMO) [1,2], the set of non-dominated population members (which are not dominated by any other population member) is determined by comparing every population member with every other for dominance. Thus, in an EMO population, if all three solutions (**x**, **y**, and **z**) exist, **z** will not be in the same non-dominated set with **x** due to the presence of **y** as a *catalyst* in the population. This may not be possible for a point-based multi-objective optimization approach, which works mostly by comparing two competing solutions at every iteration. If we continue the chain of dominance structure on the specific problem in Figure 3, we observe that with the presence of catalyst solutions, the effective dominant structure of **x** becomes a cone, shown in Figure 7, which has the transitive property [13].

This example illustrates how a semi-transitive dominance (STD) structure can exhibit an effective transitive dominance (ETD) structure in the presence of a population of catalyst solutions, such as solution **y**. This is an important distinction between population-based and point-based multi-objective optimization algorithms, allowing EMO researchers and practitioners to consider a more relaxed dominance structures to suit their practical needs. This concept leads to the following theorem.

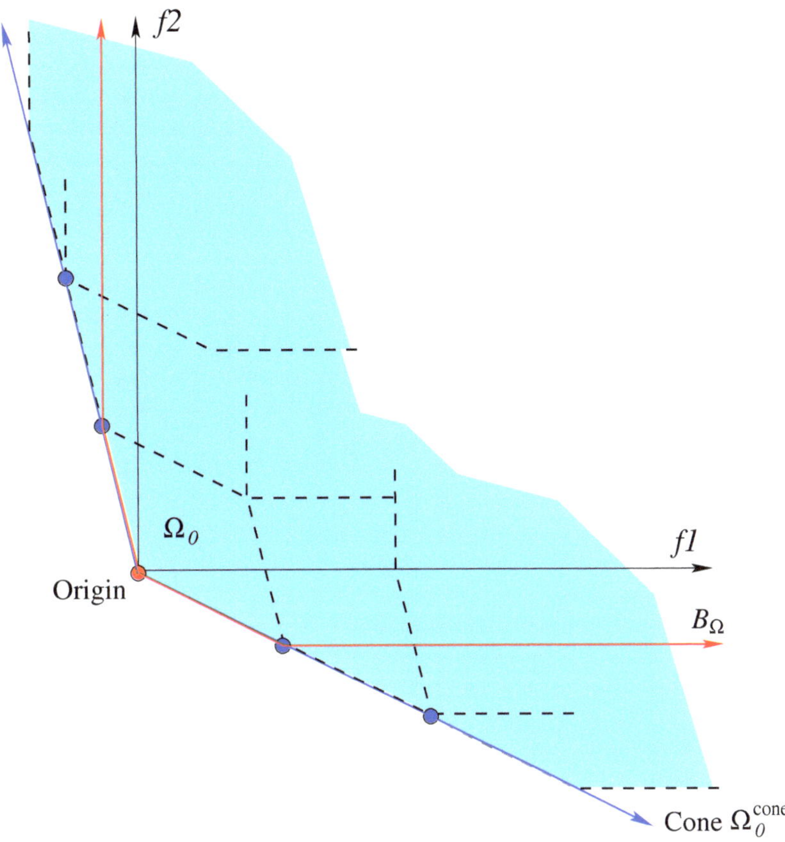

Figure 7. Repeated application of semi-transitive Ω_0 creates a transitive dominance condition Ω_0^{cone}.

Theorem 6 (Semi-transitive and effective transitive dominance structure relation). *For a semi-transitive dominance structure having $\Omega_0 \cap \Omega'_0 = \varnothing$, its effective transitive dominance structure also follows the same principle: $\Omega_0^{\text{ETD}} \cap \Omega_0'^{\text{ETD}} = \varnothing$.*

Proof. Consider three points $\mathbf{x}$, $\mathbf{y}$, and $\mathbf{z}$ in Figure 6. The semi-transitive property indicates that $\mathbf{x} \prec_{\text{GD}} \mathbf{y}$, and $\mathbf{y} \prec_{\text{GD}} \mathbf{z}$, but $\mathbf{z} \not\prec_{\text{GD}} \mathbf{x}$. The final property indicates that $\mathbf{z}$ cannot lie in $\Omega_0'^{\text{GD}}$ of $\mathbf{x}$. Clearly, $\mathbf{y} \in \Omega_0^{\text{ETD}}$ and $\mathbf{z} \in \Omega_0^{\text{ETD}}$ of $\mathbf{x}$, as the ETD structure is the collection of all such dominated points, thereby yielding $\mathbf{x} \prec_{\text{ETD}} \mathbf{z}$. If $\mathbf{z}$ must lie in $\Omega_0'^{\text{ETD}}$ as well, it means that there exists a chain of points starting with $\mathbf{z}$, making $\mathbf{z} \prec_{\text{GD}} \mathbf{y}'$ and $\mathbf{y}' \prec_{\text{GD}} \mathbf{x}$. Because each and every point $\mathbf{y}'$ which is dominated by $\mathbf{z}$ is also a member of Ω_0^{ETD} (by semi-transitive property of GD structure), $\mathbf{y}' \not\prec_{\text{GD}} \mathbf{x}$. Hence, the chain breaks, and $\mathbf{z}$ cannot be a member of both Ω_0^{ETD} and $\Omega_0'^{\text{ETD}}$. $\square$

3.2.2. Further Illustration of Semi-Transitive Property

Let us define a circle-dominance structure Ω_0^{circle} at the origin indicating the region inside (and not on) the blue circle, which is dominated by the origin, as shown in Figure 8. The origin is at $5\pi/4$ radian from the positive f_1-axis set at the center of the circle. Its anti-dominance structure Ω'_0 is shown in golden color. Clearly, $\Omega_0^{\text{circle}} \cap \Omega_0'^{\text{circle}} = \varnothing$. Thus, this circle dominance structure is expected to produce a non-empty optimal solution set. However, to understand the exact nature of the optimal solution set, we notice that the above circle dominance structure is semi-transitive. An analysis reveals that the effective transitive dominance (ETD) structure is a region above the $-45^{\deg}$ line at the origin: $\Omega_0^{\text{ETD,circle}} = \{\mathbf{f}(\boldsymbol{\epsilon})| f_1(\boldsymbol{\epsilon}) + f_2(\boldsymbol{\epsilon}) > 0\}$, shown in the figure. The respective anti-dominance structure is $\Omega'^{\text{ETD,circle}}_0 = \{\mathbf{f}(\boldsymbol{\epsilon})| f_1(\boldsymbol{\epsilon}) + f_2(\boldsymbol{\epsilon}) < 0\}$. A careful thought reveals that the ETD structure is identical to the weighted-sum dominance structure with equal weight to each objective and according to Theorem 6, the respective ETD and anti-ETD structures are also non-overlapping. Thus, although we wanted to establish a circle-dominance concept to find respective optimal points, an EMO will effectively establish a weighted-sum dominance structure with equal weight to find the optimal points. Thus, $\Omega_0^{\text{ETD,circle}} = \Omega_0^{\text{cone}}$ with equal weights.

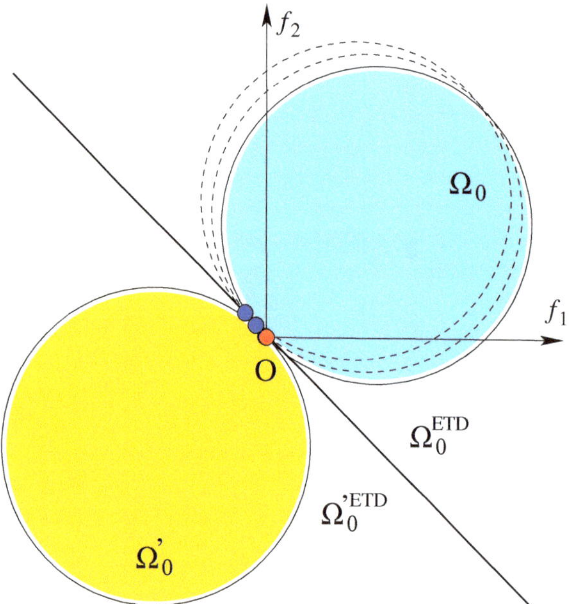

Figure 8. A user-specified circle dominance Ω_0^{circle} results in weighted-sum dominance as an effective transitive dominance structure.

For semi-transitive dominance structures, the results (both theoretical and experimental) of this paper extend to their effective transitive dominance structures as well.

3.3. Commonly Used Dominance Structures

Figure 9 shows Ω_0 for a number of commonly used dominance structures in the literature and presents their respective Ω'_0 set for two-objective problems. They are extendable to higher dimensions as well.

First, the Pareto-dominance structure ($\Omega_0^P = \{\boldsymbol{\epsilon}| \epsilon_i \geq 0, \forall i \wedge \boldsymbol{\epsilon} \neq \mathbf{0}\}$), in which the origin dominates all points in its first quadrant (for two objectives) without itself, is shown in Figure 9a. Its respective $\Omega'^P_0 = -\Omega_0^P = \{\boldsymbol{\epsilon}| \epsilon_i \leq 0, \forall i \wedge \boldsymbol{\epsilon} \neq \mathbf{0}\}$ is, by definition, the third quadrant without the origin. Because $\Omega_0^P \cup \Omega'^P_0 \subset \mathbb{R}^M \backslash \{\mathbf{0}\}$ (second and fourth quadrants are not in the set $\Omega_0^P \cup \Omega'^P_0$, hence the subset symbol $\subset$), the optimal solution set is likely to have multiple optimal points.

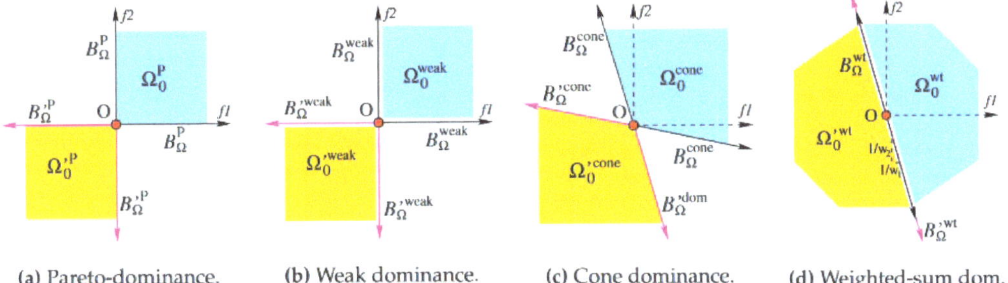

(a) Pareto-dominance. (b) Weak dominance. (c) Cone dominance. (d) Weighted-sum dom.

Figure 9. Commonly used dominance structures. In each case, $\Omega_0 \cap \Omega'_0 = \varnothing$. For Figure 9b,d, the boundary is exclusive, whereas in the other two the boundary is inclusive.

Second, the weak dominance structure ($\Omega_0^{\text{weak}} = \{\boldsymbol{\epsilon}|\ \epsilon_i > 0, \forall i\}$), in which a point dominates all points in the interior of its first quadrant (for two objectives), is shown in Figure 9b. Its respective $\Omega_0'^{\text{weak}} = -\Omega_0^{\text{weak}} = \{\boldsymbol{\epsilon}|\ \epsilon_i < 0, \forall i\}$ is the interior of the third quadrant.

Next, Ω_0^{cone} and its respective $\Omega_0'^{\text{cone}}$ of the commonly used cone dominance structure [13,15] are shown in Figure 9c. Note that for a wider cone structure (cone angle more than 180 degrees for two objectives), the asymmetric property is violated, meaning $\Omega_0^{\text{cone}} \cap \Omega_0'^{\text{cone}} \neq \varnothing$. Such a structure will result in an empty optimal solution set.

For the weighted-sum approach, $\Omega_0^{\text{wt}} = \{\boldsymbol{\epsilon}|\ \sum_{i=1}^{M} w_i \epsilon_i > 0\}$, the resulting $\Omega_0'^{\text{wt}} = \{\boldsymbol{\epsilon}|\ \sum_{i=1}^{M} w_i \epsilon_i < 0\}$ and is shown Figure 9d. Here, $B_{\Omega'}^E = B_\Omega^E$. According to Corollary 3, the optimal point(s) must lie on the hyperplane boundary B_Ω^E passing through one of the optimal solutions.

We discuss a few more existing generalized dominance structures in Section 4, but next we discuss an important matter of identifying the generalized optimal solution set for a given generalized dominance structure.

3.4. Identifying Generalized Non-Dominated Set

For a given Ω_0 structure, a member of the generalized non-dominated (GND) set in a finite population of objective vectors P is defined as follows:

Definition 9 (Generalized non-dominated set). *A feasible objective vector* **f** *in a finite set P is a member of the generalized non-dominated set* $P^{\text{GND}} \subset P$ *with a null GD structure* Ω_0 *if no other member of P lies in the anti-dominance set at* **f**, *or* $\mathbf{z} \notin (\mathbf{f} + \Omega'_0)$, $\forall \mathbf{z} \in P$.

Let us reconsider Figure 10 with a GD structure with four members in set P: points O, A, B, and C. The above definition can be used to identify if the point O is a non-dominated point in P. There can be two approaches for checking it. First, the Ω_0 set can be translated to every other feasible point in P, such as A, B and C, and check if O lies on the respective Ω_0 set. With three other points in P, three translations of Ω_0 are needed. As is clear from Figure 10, point O does not lie in any of the Ω_0 sets of three population members. Only at the end of three checks, we know that O is a non-dominated point in P. The second approach can be executed with the anti-dominance structure and the set Ω'_0 can be put only on O and check whether any of the other population members lies on the respective Ω'_0 set. It is observed that none of the points (A, B, or C) lie in Ω'_0, meaning that point O is a non-dominated point in P. Because the latter involves a single translation of the Ω'_0 set, rather than translating Ω_0 multiple times, the second approach is a computationally faster non-domination check approach [16]. In this regard, the creation of Ω'_0 from a given Ω_0 becomes an important task for non-domination check. On the same account, it is also useful for non-dominated sorting-based algorithms, such as the NSGA series [16–18]. The above discussion also results in the following theorem.

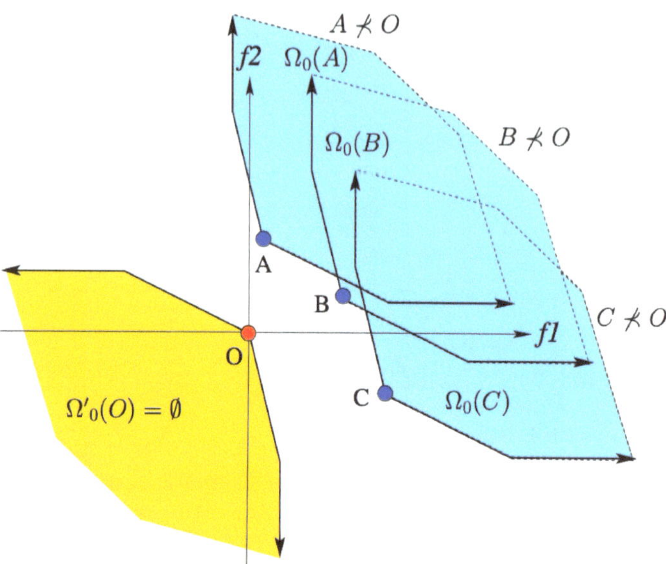

Figure 10. The set Ω_0 needs to be applied many times to identify a GND point, but Ω_0' needs to be applied once at point O to identify if it is on the GND set.

Theorem 7 (Identical optimal solution sets). *For two identical Ω_0 (or, two identical Ω'_0 sets), the respective optimal solution sets are identical.*

Proof. Because for a given Ω_0, Ω'_0 is unique and the non-domination check is performed with Ω'_0, if the resulting Ω'_0 for two dominance principles are identical, the resulting optimal solution sets will also be identical. □

3.5. Identifying Generalized Optimal Solution Set

If Ω_0 satisfies all three properties (irreflexive, asymmetric and transitive or semi-transitive), then there will exist a non-empty optimal solution set. Ω'_0 can be used directly with the following theorem to identify an optimal point:

Theorem 8 (Generalized optimal solution set). *A feasible point $\mathbf{x}$ is optimal with respect to a generalized dominance structure Ω_0 if $(\mathbf{f}(\mathbf{x}) + \Omega'_0) \cap \mathbf{Z} = \emptyset$.*

Proof. Because there does not exist any feasible objective vector in the anti-dominance set of $\mathbf{x}$ in the entire feasible search space $\mathbf{Z}$, there is no solution to dominate (in the sense of Ω_0) it. Hence, $\mathbf{x}$ is an optimal point. □

The above theorem can be used to achieve the following tasks computationally or theoretically.

1. First, it can be used to test if an objective vector $\mathbf{f}$ is a potential optimal point, as discussed above, but instead of restricting the check in a finite sampled set P, every feasible point from the search space must be considered. Although it is a computationally challenging task, the concept can be used theoretically or in a geometry-based checking procedure.
2. Second, Ω'_0 can be used to identify the entire optimal solution set for a given feasible structure $\mathbf{Z}$. This task will be useful for studies involving test problems and requires

an identification of the exact optimal objective set ($\mathbf{Z}^*$) from $\mathbf{Z}$ for a given dominance structure. The theoretical procedure is to identify Ω'_0 set for every point in $\mathbf{Z}$ systematically and by repeating the test only to Ω'_0 members in a nested manner. This will allow a faster computational procedure to identify the optimal solution set.

3. Third, knowing one or more optimal points, Ω_0 can help identify further optimal points quickly by eliminating the dominated solutions from its Ω_0 set and narrowing down the search to find further optimal points. However, in such a task, often the relevant boundary points of $\mathbf{Z}$ can be tested for their optimality. Starting with extreme boundary points of $\mathbf{Z}$, Ω'_0 can immediately verify if the point is a member of the optimal solution set. If yes, the test can continue to the neighboring extreme boundary point, and so on. If no, the test will identify the points in the Ω'_0 set that dominate the extreme point and a new test can be executed on members of the Ω'_0 set.

Clearly, Ω'_0 enables a faster way to identify the optimal solution set than Ω_0, simply because of the former's ability to identify points that dominate the current point under consideration, thereby not only allowing to determine if the current point is an optimal point but also narrowing down the search for further optimal points. For the generalized dominance structure (Ω_0^{cone}) shown in the inset of Figure 11, the resulting $\mathbf{Z}^{*,\text{cone}}$ is determined for a hypothetical $\mathbf{Z}$, shown in the figure. The $\mathbf{Z}^{*,P}$ according to the Pareto-dominance principle is the entire line 1–10, whereas the $\mathbf{Z}^{*,\text{cone}}$ consists of line segments 2–4, 6–7, and 9–10. For the extreme boundary point 1, Ω'_0 is not empty, as shown by the overlap of $\mathbf{Z}$ and Ω'_0 at point 1. Thus, point 1 cannot be an optimal point.

Ω'_0 also helps to identify the boundary region (line segments between 1 and 10) which must be tested next. Because the relevant boundary in this problem comes from piece-wise linear segments and a cone-dominance structure is used, we restrict our testing only to extreme points of the line segments. By testing point 2 with Ω'_0, it is clear that $\Omega'_0 = \varnothing$. Thus, point 2 is a member of the optimal solution set. This can continue systematically to identify the entire optimal solution set (line segments 2–4, 6–7, and 9–10). Cone-dominance brings in useful properties, such as, finding *proper* Pareto-optimal solutions [15], finding the partial preferred Pareto set [1], and helping to eliminate dominant resistant solutions in a population, thereby enabling NSGA-II-like algorithms to solve many-objective problems well [19].

The set Ω_0 at any feasible point $\mathbf{f}$ can also be used to identify a special scenario:

Theorem 9 (Singleton optimal objective vector). *If* $\mathbf{Z} - (\mathbf{f}(\mathbf{x}^*) + \Omega_0) = \{\mathbf{f}(\mathbf{x}^*)\}$, *then the dominance structure* Ω_0 *produces a single optimal objective vector* $(\mathbf{f}(\mathbf{x}^*))$ *for the problem.*

Proof. The condition signifies that $\mathbf{x}^*$ dominates every point in $\mathbf{Z}$ and thus, no member of $\mathbf{Z}$ dominates $\mathbf{x}^*$. Hence, it is the only optimal point. $\square$

Corollary 4 (Singleton solution). *If the condition in the above theorem is satisfied by a unique solution* $\mathbf{x}^*$, *then* $\mathbf{x}^*$ *is the only strictly optimal solution in the search space.*

The single-objective optimality condition satisfies the above condition and results in a singleton global optimal objective value. If the weighted-sum dominance structure, demonstrated in Figure 9d, satisfies the above condition for a unique solution $\mathbf{x}$, it is the singleton optimal point for the problem [15]. Figure 12 illustrates one such example.

The following generic properties of dominance structures are also useful for identifying optimal solution sets [13].

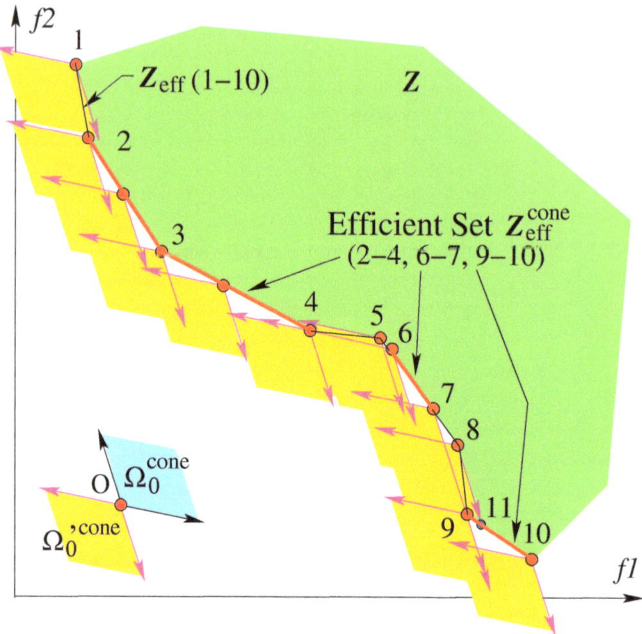

Figure 11. Identification process of $\mathbf{Z}_{\mathrm{eff}}^{\mathrm{cone}}$ using a generalized dominance structure Ω_0^{cone}.

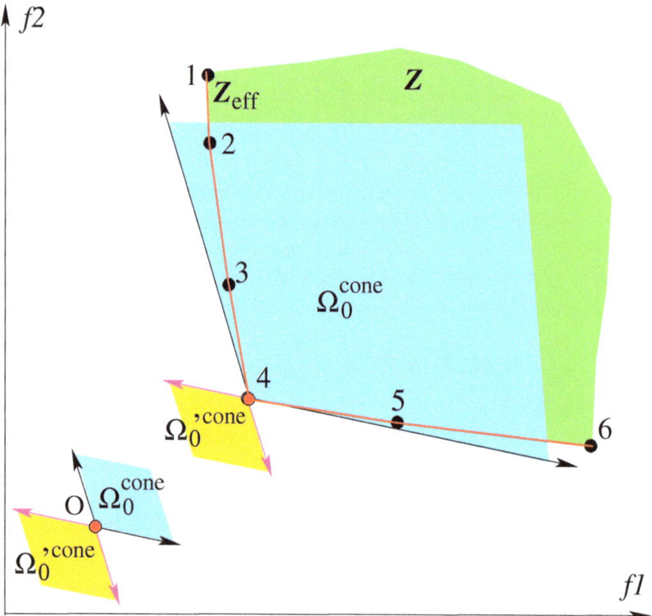

Figure 12. An illustration of cone-dominance structure supporting Theorem 9.

Theorem 10. *If $\Omega_0^{(1)}$ is a subset of $\Omega_0^{(2)}$, then the resulting optimal solution set $\mathbf{Z}^{*,(2)}$ is a subset of $\mathbf{Z}^{*,(1)}$.*

Proof. Because $\Omega_0^{(1)} \subset \Omega_0^{(2)}$, $\Omega'^{(1)}_0 \subset \Omega'^{(2)}_0$. Thus, every member of $\Omega'^{(1)}_0$ that dominates $\mathbf{x}$ also exists in $\Omega'^{(2)}_0$ and dominates $\mathbf{x}$. Moreover, there can be additional members of $\Omega'^{(2)}_0$ that dominate $\mathbf{x}$. Hence, the optimal solution set $\mathbf{Z}^{*,(2)}$ is a subset of $\mathbf{Z}^{*,(1)}$. □

Two corollaries follow from this theorem.

Corollary 5. *For any $\Omega_0^{GD} \subset \Omega_0^P$, $\mathbf{Z}^{*,P} \subset \mathbf{Z}^{*,GD}$.*

The above corollary means that any dominance structure that is weaker than the Pareto-dominance will include originally dominated points as optimal points, thereby causing a larger optimal solution set. An example is when $\Omega_0 = \Omega_0^{\text{weak}}$. The Pareto-optimal set is a subset of the *weakly* Pareto-optimal set.

Corollary 6. *For any $\Omega_0^P \subset \Omega_0^{GD}$, $\mathbf{Z}^{*,GD} \subset \mathbf{Z}^{*,P}$.*

The above indicates that any dominance structure that is stronger than the Pareto-dominance structure may not indicate some Pareto-optimal points as optimal, thereby having a reduced number of optimal solutions. An example is when $\Omega_0 = \Omega_0^{\text{cone}}$ (with an obtuse angle). The cone-optimal set $\mathbf{Z}^{*,\text{cone}}$ is a subset of Pareto-optimal set $\mathbf{Z}^{*,P}$.

3.6. Theoretical and Practical Optimal Sets

In addition to identifying the theoretical optimal solution set by the above procedure, there is a practical aspect which we discuss next. For practical reasons, one can define a generalized dominance structure that has an overlapping dominated and anti-dominance sets having $\Omega_0 \cap \Omega'_0 \neq \varnothing$. For such a structure, no theoretical optimal point exists. However, the use of such a dominance structure can still produce artificial optimal points with a multi-objective optimization algorithm due to certain algorithmic inaccuracies and an often-used implementational adjustment. We illustrate these aspects with the epsilon-dominance structure [20].

Epsilon-dominance was proposed to obtain Pareto-optimal solutions with a certain pre-specified (ϵ_i) difference in the i-th objective values, even in continuous search space problems. The epsilon-dominance structure is defined as follows:

Definition 10 (Epsilon-Dominance). *A feasible solution $\mathbf{x} \in \mathbf{X}$ epsilon-dominates (The original study [20] defined with a product term). Furthermore, notice that this definition is different from epsilon-inferior structure defined in Section 2.1. another feasible solution $\mathbf{y} \in \mathbf{X}$, if $f_i(\mathbf{x}) \leq f_i(\mathbf{y}) + \epsilon_i$ for each $i = 1, \ldots, M$.*

The respective Ω_0 and Ω'_0 sets are shown in Figure 13a in blue and golden shaded regions, respectively. It is clear that $\Omega_0 \cap \Omega'_0 \neq \varnothing$, having a small overlapping rectangular region around the point O. Thus, there does not exist a theoretical optimal solution set for this dominance structure. However, an EMO algorithm may still find optimized solutions using this dominance structure but, due to search inefficiencies associated in the algorithm, run for a finite number of solution evaluations. In other occasions, additional implementational changes are forcibly introduced to find a discrete set of optimized solutions with at least ϵ_i difference in the i-th objective between any two optimized solutions.

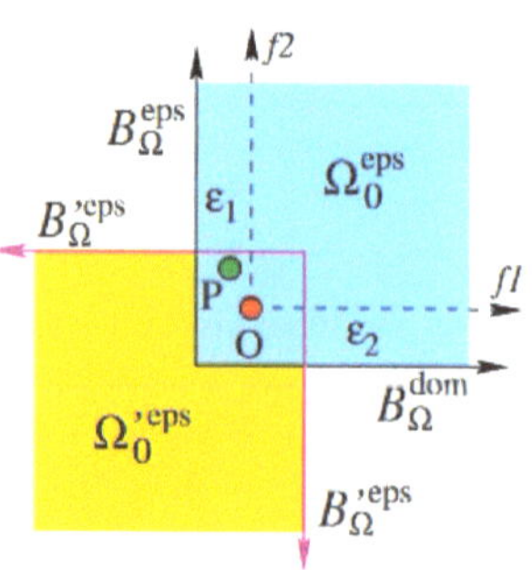

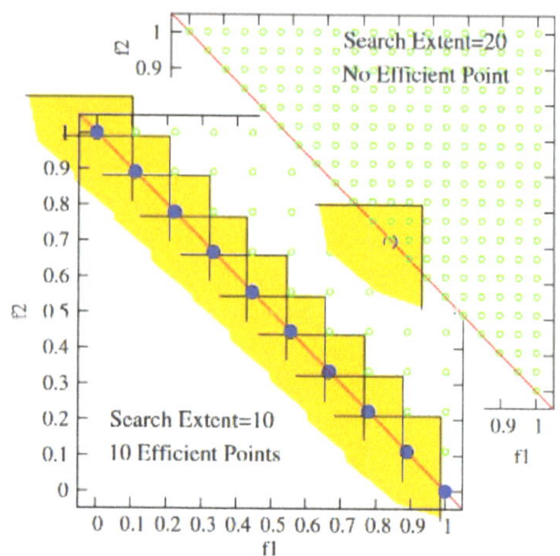

(a) Epsilon dominance . Ω'^{eps}_0 also includes blue rectangle around O.

(b) Epsilon optimal solution set due to discreteness in search space.

Figure 13. Ω^{eps}_0 does not produce any theoretical optimal point due to the overlap between Ω_0 and Ω'_0, although in practice it may produce an artificial optimized solution set.

We discuss these ideas, as they can be generically used with some other GD structures.

3.6.1. Search Inefficiency

For a point (say, $\mathbf{x}$) close to the optimal set, Ω'_0 at $\mathbf{x}$ indicates the set of points that GD-dominates $\mathbf{x}$. However, if this objective space is relatively small or is difficult to discover by an algorithm's search operators, the point $\mathbf{x}$ may still be declared as an optimal point. Consider point 8 in Figure 11. It is not an optimal point based on Ω^{cone}_0, due to the existence of a tiny objective space (region 8–9–11). If a multi-objective optimization algorithm fails to find any point in this tiny objective region which would dominate point 8, point 8 will be wrongly declared as an optimal point.

3.6.2. Compatibility of Dominance Structure with Discreteness in Search Space

Artificial optimal points may also emerge if the search space is discrete, causing certain critical points to stay non-dominated due to unavailability of other points in the feasible search space to dominate them. Figure 13b shows two scenarios with epsilon-dominance structure with $\epsilon_i = 0.1$ for $i = 1, 2$ to find optimal points for a discrete search space problem on a linear Pareto-optimal front ($f_i \in [0, 1]$ for $i = 1, 2$). In the first scenario (the main part of the figure), each objective value comes at an interval of $1/9$. The epsilon-dominance structure finds 10 optimal points, each of whose Ω'_0 is empty, as shown in the figure. Each discrete point clears the Ω'_0 region for other discrete points to constitute 10 optimal points. The second scenario (shown in the inlet figure) has a finer discreteness (f_i values are available at an interval of $1/19$). The inlet figure shows that now no optimal point is discovered, as for every discrete point on the boundary of $\mathbf{Z}$, there are a few points in its Ω'_0 set, making it dominated. This example illustrates that despite the non-existence of any optimal solution, a correct combination of the discreteness of the search space and the chosen dominance structure may allow a non-empty optimized solution set to be found.

3.6.3. Implementation of Adjusted Dominance Principle

The overlapping dominant structure can be modified before using it in an EMO algorithm so that non-overlapping Ω_0 and Ω'_0 sets are achieved to produce a non-empty optimal solution set in a problem. For example, the epsilon-dominance structure can be modified by removing the overlapping region between Ω_0 and Ω'_0, as follows and as illustrated in Figure 14a:

$$\Omega_0 \;\leftarrow\; \Omega_0 \backslash (\Omega_0 \cap \Omega'_0), \tag{7}$$

$$\Omega'_0 \;\leftarrow\; \Omega'_0 \backslash (\Omega_0 \cap \Omega'_0). \tag{8}$$

The above operations guarantee that adjusted sets are non-overlapping: $\Omega_0^{\text{Adj}} \cap \Omega'^{\text{Adj}}_0 = \varnothing$. Figure 14a identifies the same 10 optimal solutions as in Figure 13b with the above adjusted dominance structure for the coarse search space (interval of 1/9). However, when a finer search space (interval of 1/19) is used, all 20 discrete optimal points are found with the same epsilon vector ($\epsilon_i = 0.1$). The non-overlapping property of the adjusted dominance structure discovers a non-empty optimal set, but it seems that the number of optimal solutions cannot be controlled directly by setting the ϵ-vector. The following implementational change fixes this issue.

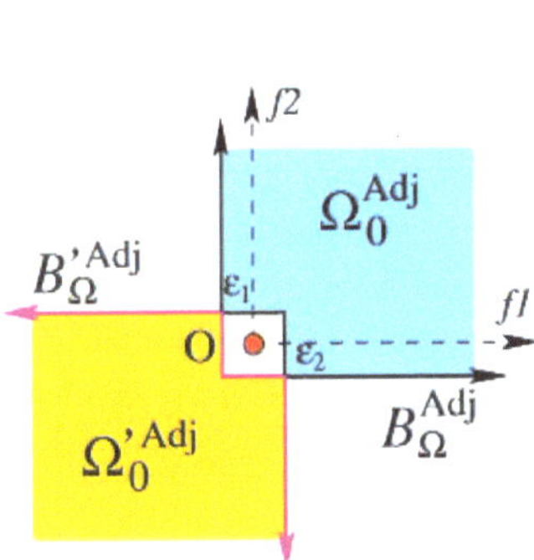

(a) Adjusted epsilon-dominance.

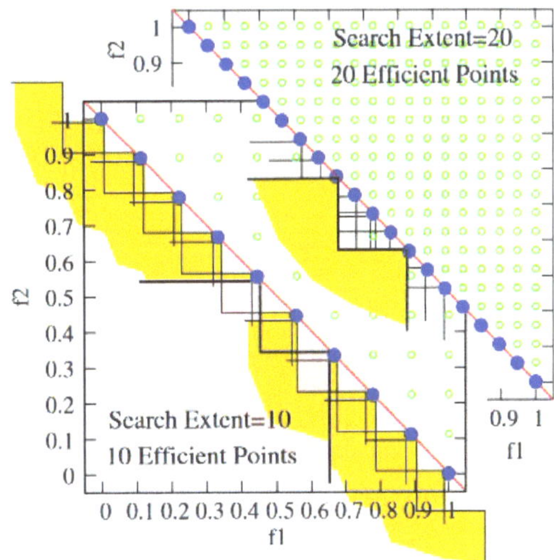

(b) Adjusted epsilon-optimal set.

Figure 14. Adjusted epsilon-dominance structure can produce an optimal solution set.

3.6.4. Implementation of Grid Dominance Principle

Another approach adopted by EMO researchers is the use of a fixed grid structure in the objective space with size ϵ_i along the i-th objective axis [21–23]. Every objective vector $\mathbf{f}$ is now replaced with its grid vector (the lower left corner point of the grid in which $\mathbf{f}$ lies). In the grid-dominance structure, a Pareto-dominance structure is applied with grid vector of points, and not with the points themselves. If two points in the same grid result in the same grid vector, then the one closer (in the Euclidean sense) to the grid vector dominates the other, as shown in Figure 15a. Note that the resulting dominance and anti-dominance sets are non-overlapping, or $\Omega_0^{\text{grid}} \cap \Omega'^{\text{grid}}_0 = \varnothing$. This change in the dominance structure produces a non-empty optimal solution set with well-distributed points. Figure 15b shows that irrespective of the discreteness in the search space, the same number of optimal points are obtained for both levels of discreteness in the search space (interval of 1/9 and 1/19). Because the same epsilon-vector ($\epsilon_i = 0.1$) is used in the dominance structure, the final outcomes are identical for both scenarios. Despite the existence of more discrete Pareto-optimal solutions in the search space, the ϵ-vector keeps the cardinality of the optimal set checked.

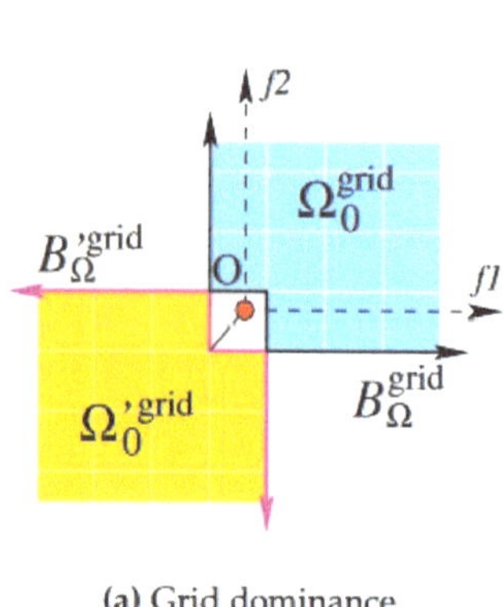

(a) Grid dominance.

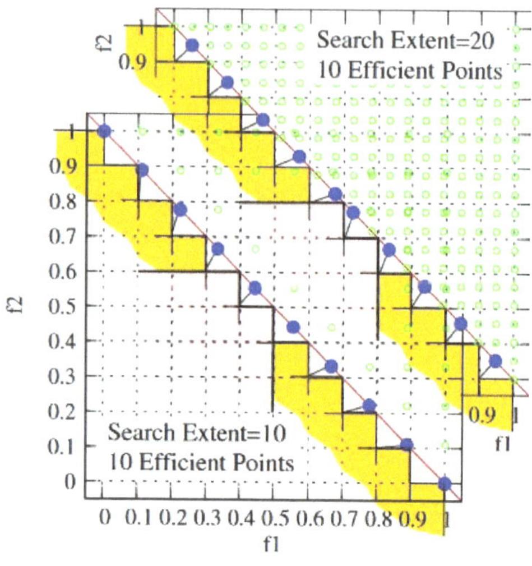

(b) Grid optimal solution set.

Figure 15. Grids are fixed in objective space. Ω_0^{grid} can produce a well-distributed set of optimal points.

These algorithmic and implementational adjustments of a generalized dominance structure may produce a different outcome than their Pareto-dominance structure. A proper analysis of the derived Ω'_0 for the chosen Ω_0 structure may reveal the expected outcome from an EMO run. Moreover, such a thought process is expected to reveal a better understanding of relationships between Ω_0 and Ω'_0 and allow new and innovative dominance structures and new algorithmic implementations to be used for different types of problems.

4. Other Existing Dominance Structures

Next, we consider a few other existing dominance structures from the EMO literature and attempt to reveal their properties based on the above fundamental principles of generalized dominance structures. Visual descriptions of the boundary for the dominated objective space (Ω_0) of some of these dominance structures were presented in [9].

4.1. α-Dominance Structure

The α-dominance structure [12] was proposed in 2001 and is identical to the cone-dominance structure, described in Figure 9c.

4.2. Cone-Epsilon Dominance Structure

The cone-epsilon dominance structure [24] uses the grid-based epsilon-dominance concept, discussed in Section 3.6.4, but instead of distance-based dominance for points within the occupying grid, it uses an acute cone dominance principle, as shown in Figure 16a. Clearly, $\Omega_0^{\text{cone-e}} \cap \Omega'^{\text{cone-e}}_0 \neq \varnothing$, thereby producing no theoretical optimal solution. However, an adjusted dominance principle (shown in Figure 16b), discussed in Section 3.6.3, can be implemented to find a set of optimal solutions.

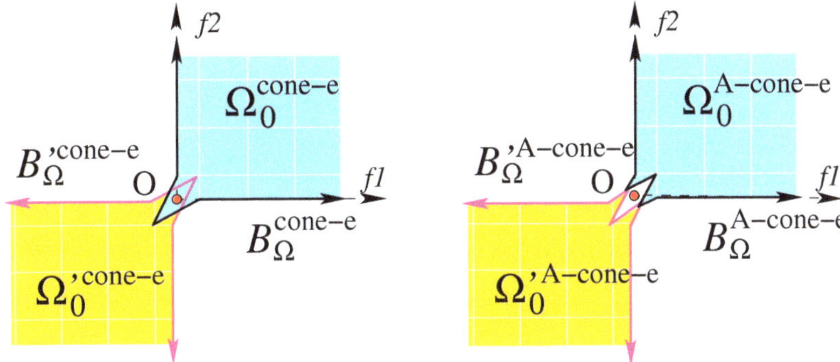

(a) Cone–epsilon domination   (b) Adj. cone–epsilon domination

Figure 16. Cone-epsilon dominance.

4.3. CN and CN-Alpha Dominance Structures

The CN-dominance structure is shown in Figure 17a. Clearly, $\Omega_0^{CN} \cap \Omega_0'^{CN} \neq \varnothing$, making no theoretical optimal solution. The adjusted $\Omega_0'^{CN} \leftarrow \Omega_0'^{CN} \backslash (\Omega_0^{CN} \cap \Omega_0'^{CN})$ structure will produce a set of CN-optimal solutions and is equivalent to the adjusted epsilon-dominance structure (shown in Figure 17b), discussed in Section 3.6.3.

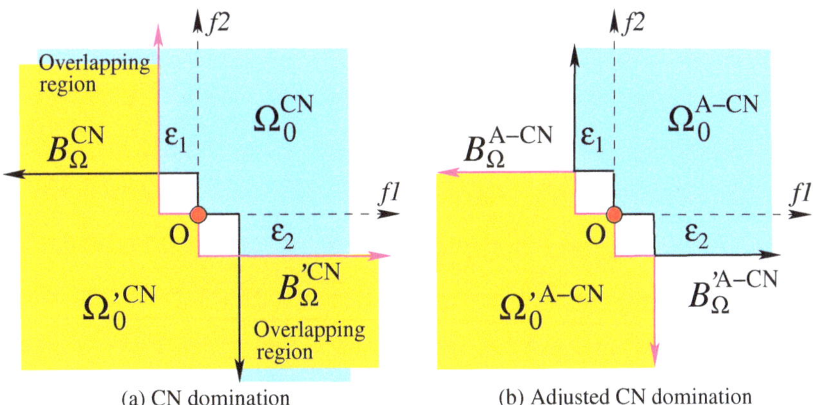

(a) CN domination   (b) Adjusted CN domination

Figure 17. CN domination.

4.4. CN-α Domination Structure

The CN-α dominance structure uses the cone dominance structure with a predefined cone in the vicinity (within ϵ-vector), as shown in Figure 18a. The overlap between $\Omega_0^{CN\text{-}\alpha}$ and $\Omega_0'^{CN\text{-}\alpha}$ exists. To find an optimal solution set, the adjusted anti-dominance structure (Figure 18b) can be used.

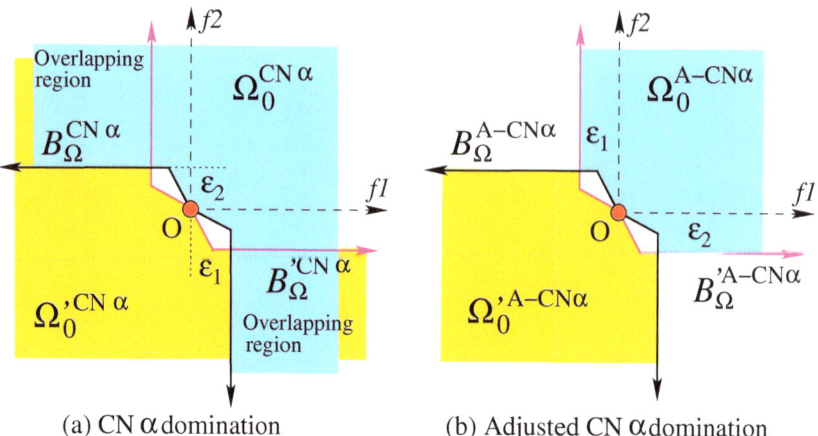

(a) CN α domination (b) Adjusted CN α domination

Figure 18. CN-alpha domination.

4.5. Nonlinear Dominance Structure (NLAD)

A nonlinear dominance structure was proposed in [25]. It uses a cubic dominance boundary to define the dominated region (Figure 19). For certain cubic parameters, a non-overlapping $\Omega_0'^{NLAD}$ can be formulated and used to find the respective optimal solution set.

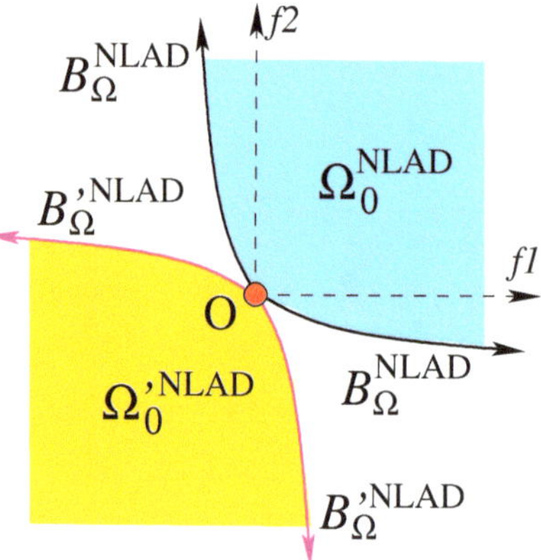

Figure 19. NLAD dominance.

Other nonlinear dominance definitions exist that combine objectives in a nonlinear manner $\phi(\mathbf{f})$. They impose a dominance structure which prefers solutions having a small value of ϕ [26,27] and other structures that simply map every objective into a nonlinear function $\phi_i(f_i)$ and check Pareto-dominance based on ϕ_i, such as CDAS-dominance structure [28].

4.6. D-Dominance Structure

The D-dominance structure was defined in [29] and was based on the PBI metric, shown in Figure 20:

Definition 11 (D-dominance). *A solution* $\mathbf{x} \in \mathbf{X}$ *D-dominates another solution* $\mathbf{y} \in \mathbf{X}$*, if* $d_1(\mathbf{x}) + d_2(\mathbf{f}(\mathbf{y}) - \mathbf{f}(\mathbf{x}))\cot\beta < d_1(\mathbf{y})$.

In the figure, the clause in the definition refers to OA < OB. The distances d_1 and d_2 are along and orthogonal directions of a weight vector $\mathbf{w}$ from a reference point (usually, the ideal point), as shown in Figure 20. It is similar to the edge-rotated cone-dominance structures [30,31], except in more than two-objective problems, the dominated region is conical, rather than prismatic. For a pair of solutions $\mathbf{x}$ and $\mathbf{y}$, the authors defined a condition for which $\mathbf{x}$ is better and also a condition for which $\mathbf{y}$ is better and if both conditions are not satisfied, then both solutions are non-dominated according to D-dominance structure. In some sense, the dominated and anti-dominance sets can be defined from the definition, making this study one of the precursors of the anti-dominance concept introduced in this study.

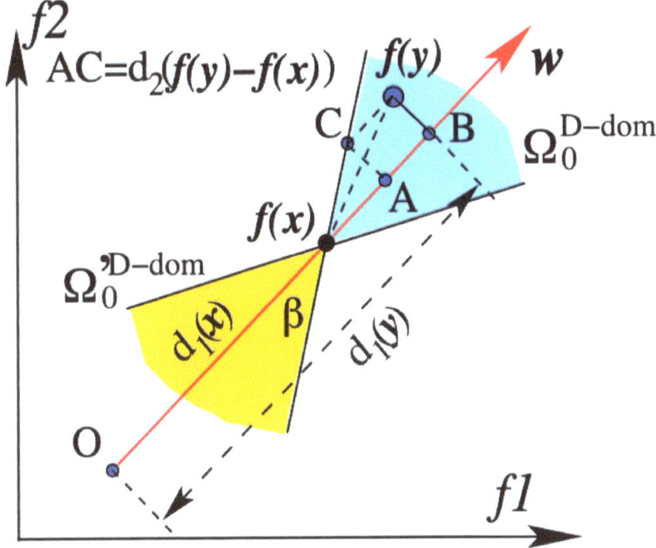

Figure 20. D-dominance.

4.7. Strength Dominance Relationship (SDR) Structure

For the reference vector-based optimization, [9] proposed a new dominance structure in which points associated with the reference vector are compared with the sum of objective functions:

Definition 12 (SDR-dominance). *A solution* $\mathbf{x} \in \mathbf{X}$ *SDR-dominates another solution* $\mathbf{y} \in \mathbf{X}$ *if*

$$\begin{cases} \sum_{i=1}^{M} f_i(\mathbf{x}) < \sum_{i=1}^{M} f_i(\mathbf{y}), & \text{for } \angle(\mathbf{f}(\mathbf{x}), \mathbf{f}(\mathbf{y})) \leq \theta; \\ \sum_{i=1}^{M} f_i(\mathbf{x}) \dfrac{\angle(\mathbf{f}(\mathbf{x}), \mathbf{f}(\mathbf{y}))}{\theta} < \sum_{i=1}^{M} f_i(\mathbf{y}), & \text{otherwise.} \end{cases}$$

However, when an associated point is compared with an unassociated point, angles between the points are considered, resulting in a dominance structure Ω_0^{SDR} shown in Figure 21. If the entire objective space is checked for SDR-dominance structure, the null anti-dominance structure Ω'_0 can be constructed from Ω_0, as shown in the figure. Clearly, they are non-overlapping sets with no common points on boundaries of these sets and are expected to produce a set of optimal solutions.

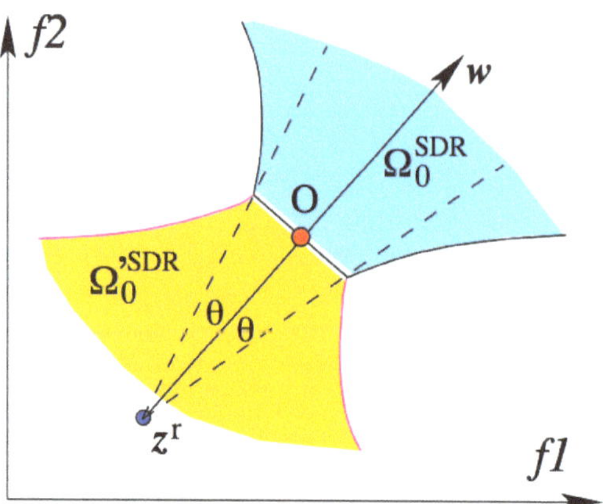

Figure 21. SDR domination.

4.8. $(1-k)$-Dominance Structure

For two solutions **x** and **y**, three quantities are first computed [8]: (i) $n_b(\mathbf{x},\mathbf{y})$, count of number of objectives in which **x** is better than **y**, (ii) $n_e(\mathbf{x},\mathbf{y})$, count of number of objectives in which **x** is identical to **y**, and (iii) $n_w(\mathbf{x},\mathbf{y})$, count of number of objectives in which **x** is worse than **y**, which is not directly used in the definition, but $n_w = M - n_b - n_e$.

Definition 13. *A solution $\mathbf{x} \in \mathbf{X}$ $(1-k)$-dominates another solution $\mathbf{y} \in \mathbf{X}$ if (i) $n_e(\mathbf{x},\mathbf{y}) < M$, and (ii) $n_b(\mathbf{x},\mathbf{y}) \geq \frac{M-n_e(\mathbf{x},\mathbf{y})}{k+1}$.*

For $k = 0$, $(1-k)$-dominance becomes Pareto-dominance, but for $k > 1$, it results in a reduced optimal solution set. Figures 22 and 23 show Ω_0 and Ω'_0 for a three-objective case having $k = 1$ with $n_e = 1$ and $n_e = 0$, respectively. Both sets are non-overlapping in both cases.

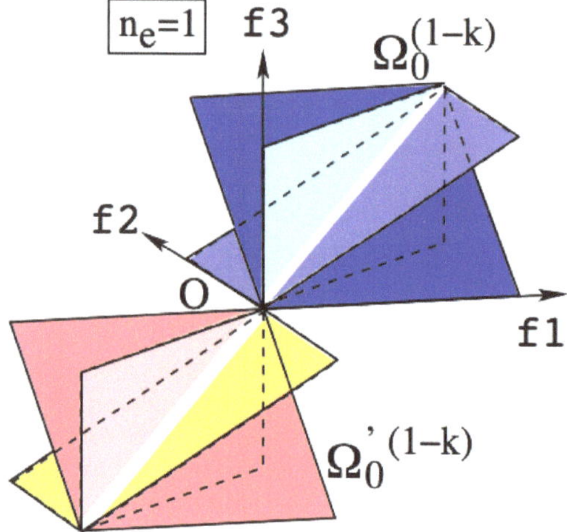

Figure 22. $(1-k)$-dominance structure with $M = 3$, $k = 1$, and $n_e = 1$.

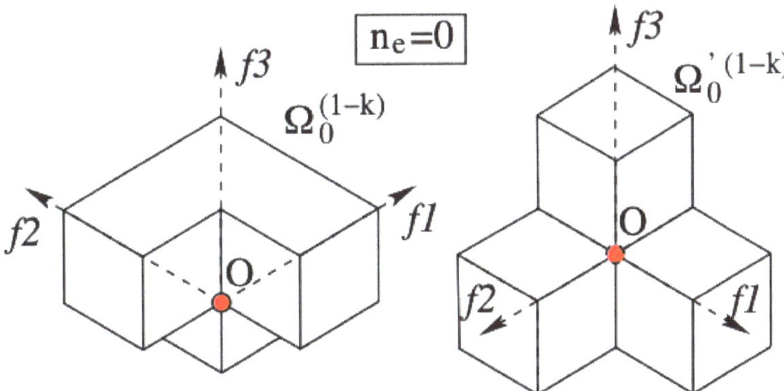

Figure 23. $(1 - k)$-dominance structure with $M = 3, k = 1$, and $n_e = 0$.

This dominance is also known as the fuzzy-dominance structure. Its effect becomes prominent for more than two objectives.

4.9. L-Dominance Structure

Borrowing the idea from the above $(1 - k)$-dominance structure, ref. [32] proposed the L-dominance structure:

Definition 14 (L-Dominance). *A solution* $\mathbf{x}$ *L-dominates another solution* $\mathbf{y}$ *if (i)* $n_b(\mathbf{x}, \mathbf{y}) > n_w(\mathbf{x}, \mathbf{y})$ *and (ii)* $\|\mathbf{f}(\mathbf{x})\| < \|\mathbf{f}(\mathbf{y})\|$.

Normalization of objectives is performed before computing the above checks. L-dominance causes more solutions to be dominated than the Pareto-dominance mainly due to the first condition, hence L-dominance is likely to cause a reduced L-optimal set compared to the Pareto-optimal set. The contribution of the second condition in defining the dominance structure is to emphasize optimal solutions closer to the ideal point.

4.10. $(M - 1)$-Generalized Pareto-Dominance Structure

Following the concept of cone-dominance structure, ref. [33] proposed a cone which does not extend it along one of the chosen objective axes (say, the k-th one). Thereafter, they suggested to use M different GPD cones, each with a different k, to determine the overall non-dominated set. The concept of $(M - 1)$-GPD is similar to the cone-dominance described before.

5. Spatially Dependent Ω_0 Structure

The above discussions are applicable to static dominance structures in which the same dominance structure is applicable at every point in the feasible objective space. However, multi-objective optimization researchers have developed other dominance structures, which change with the location of $\mathbf{f}$ in the objective space [34,35]. A few such dominance structures are reference vector (RV)-based dominance, such as achievement scalarization function (ASF)-based dominance structure [36], PBI metric-based dominance [37], LHiFD [10], w-dominance [38], and location-based dominance, such as angle-dominance structures [39,40]. Some of these structures are applied to a number of pre-defined RVs in the objective space and each RV is expected to produce a single optimal solution. We attempt to explain here how our proposed anti-dominance structure concept can be used to identify respective optimal solution(s).

5.1. ASF-Based Dominance Structure

Consider the ASF approach [36] first, in which the minimization of the following scalarizing function produces a weakly Pareto-optimal point:

$$\text{Minimize} \quad \text{ASF}(\mathbf{x}) = \max_{i=1}^{M} \frac{f_i(\mathbf{x}) - z_i^r}{w_i},$$
$$\text{subject to} \quad \mathbf{x} \in \mathbf{X}. \tag{9}$$

The reference vector $\mathbf{w}$ and reference point $\mathbf{z}^r$ are fixed in the objective space, as shown in Figure 24. Because points above or below the reference vector $\mathbf{w}$ have different dominance principles, theoretically, Ω_0 and Ω'_0 sets vary from point to point and also depend on the chosen RV. However, considering the contour of ASF function, we can perform a mapping of **f**-vector onto the RV line, marked as the $\mathbf{w}$ line in Figure 24. Thereafter, we can define a Ω_0 (resulting in a Ω'_0) on $\mathbf{w}$-vector, as follows. For a point A, its mapped point a on $\mathbf{w}$-vector is first found. All objective vectors that are mapped above a (away from $\mathbf{z}^r$) on $\mathbf{w}$-line belong to point A's Ω_0 set. Similarly, all points that are mapped below (near $\mathbf{z}^r$) belong to Ω'_0 at A, thereby applying the proposed dominated and anti-dominance structure definitions along the RV line. A little thought will reveal that the respective objective vectors of Ω'_0 set will come from the region marked in golden color in the objective space, meaning any point in the golden region ASF-dominates point A. These adjusted and mapped dominated and anti-dominance sets are non-overlapping to each other and when applied to the entire feasible $\mathbf{Z}$, will result in a single optimal point, marked with a red point, according to Theorem 9.

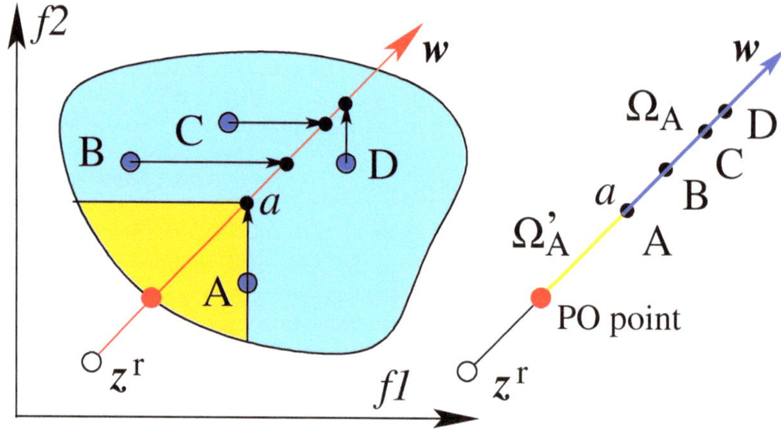

Figure 24. ASF dominance.

Other reference point-based dominance structures, such as g-dominance [41], r-dominance [42], p-dominance [43], and ar-dominance [44] can also be analyzed using the above concept.

5.2. θ-Dominance or PBI-Dominance Structure

A similar adjusted and mapped dominance structure can also be applied to the PBI metric-based dominance structure [11], as shown in Figure 25. The PBI metric uses a user-specified parameter θ, which is equal to $\tan(\gamma)$ (γ is half of the cone angle) to create a penalty function to combine perpendicular and parallel distances along the $\mathbf{w}$-line. It is clear that point A PBI-dominates B, which again PBI-dominates C. The region that dominates point A is marked in golden color. The Ω_A created at mapped point A contains mapped points B and C and the respective Ω'_A represents all points from the golden region, thereby following the principle of PBI-based dominance structure.

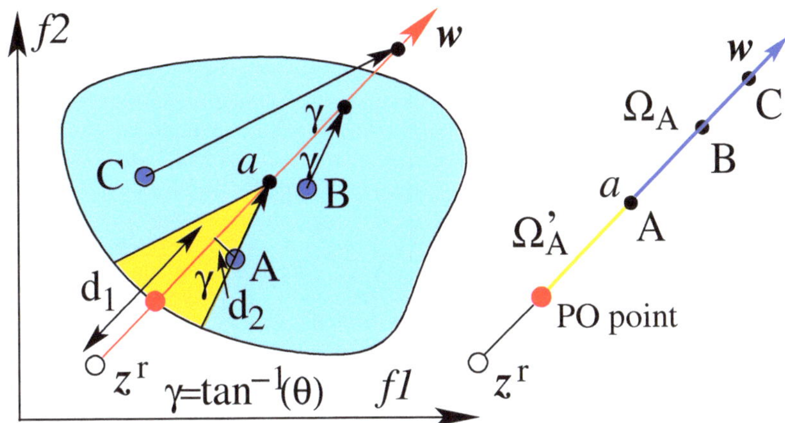

Figure 25. PBI dominance.

5.3. Angle-Dominance Structure

Instead of the objective values, angles (α_i, $i = 1, \ldots, M$) from M anchor points on objective axes can be computed for any **f**-vector by extending the nadir point by k times (>1) and locating its coordinates on the objective axes. The angle-dominance [39] is then defined as follows:

Definition 15 (Angle-dominance). *A solution* $\mathbf{x} \in \mathbf{X}$ *angle-dominates another solution* $\mathbf{y} \in \mathbf{X}$ *if* $\alpha_i(\mathbf{x}) \leq \alpha_i(\mathbf{y})$ *for all* $i = 1, 2, \ldots, M$ *and* $\alpha_j(\mathbf{x}) < \alpha_j(\mathbf{y})$ *for at least one* $j = 1, 2, \ldots, M$.

As can be seen from Figure 26, it is similar to the cone-dominance principle, except that the cone depends on the point **x**. To determine the optimal set, **f**-vectors can be converted to **α**-vectors, as shown in Figure 26 and a Pareto-dominance structure can be applied on the angle space.

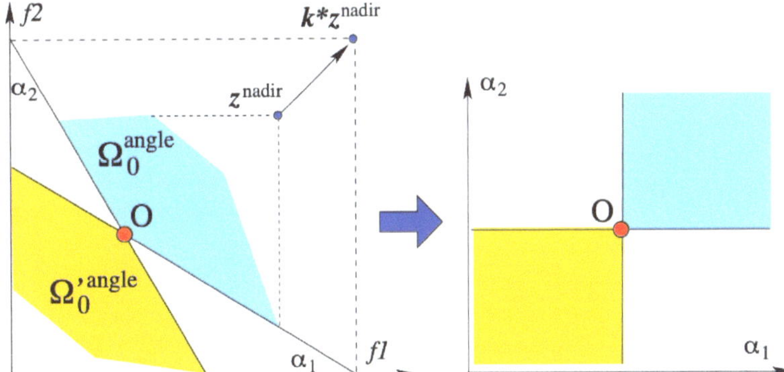

Figure 26. Angle-dominance structure can be converted to angle-space, on which Pareto-dominance can be applied.

Besides the above positionally defined dominance structures in the objective space, dominance structures that depend on a given sample of points, such as, ranking dominance [45,46], and dominance structures that do not use objective vectors and are restricted to variable space relationships, are difficult to analyze to have a comprehensive insight on resulting optimal solution set; however, certain innovative mapping techniques, following the ones proposed above, may be derived in such cases.

6. Conclusions

This study has introduced a concept of anti-dominance structure for a chosen static dominance structure by extending the optimality conditions in a single-objective problem to multi-modal problems and then to multi-objective problems. It has been shown that the anti-dominance structure can be useful for identifying the respective non-dominated set in a finite population or perceiving the true optimal solution set in two and three-objective problems. Importantly, the study has brought out certain fundamental principles which every practical dominance structure should have for it to generate a non-empty optimal solution set. However, suitable adjustments to generic dominance structures have been proposed to make them more practically applicable.

To demonstrate their use, dominated and anti-dominance structures have been identified for most popular dominance structures in EMO and classical multi-objective optimization literature. In some situations, an adjustment with mapped objective vectors has been proposed to analyze the resulting optimal solutions. The reasons for their necessary adjustment have been presented in the light of the anti-dominance structures.

It has been shown that with a population-based multi-objective optimization algorithm (such as an EMO algorithm), transitivity of the dominance structure need not be a strict requirement. Semi-transitivity property of dominance structures breaks the cycle of domination among three solutions and allows a transitive relationship to be implicitly formed from the existence of intermediate population members. In this sense, population-based evolutionary multi-objective optimization algorithms have an edge for using more flexible dominated structures possessing semi-transitive properties.

Although not used in this paper, the anti-inferiority concept can be extended to find special and practically relevant (but non-optimal) solutions by developing generalized inferiority conditions for single-objective optimization. An extension of the anti-dominance structure for constrained dominance principles [47,48] will motivate further development of optimal constrained multi-objective optimization algorithms. There is a practical need for developing a GUI-based system in which the user can provide any desired dominance structure and the system creates the respective adjusted anti-dominance structure automatically and runs an EMO algorithm with it. This paper has also revealed that for semi-transitive dominance structures, optimization algorithms must find appropriate catalyst intermediate solutions to establish the non-dominated structure. Whether EMO algorithms will be slower in converging to the resulting optimal solution set compared to algorithms which use transitive dominance structures will be an interesting future study. In this regard, it will be interesting to find if there can ever exist a theoretical dominance structure that may cause a domination cycle with four or more solutions. Finally, providing preference elicitation through a number of pair-wise comparisons of objective vectors and creating a meaningful generalized dominance structure from them through the use of developed principles of this paper, would make another interesting practically significant study.

Author Contributions: Conceptualization, K.D.; methodology, K.D.; validation, K.D. and M.E.; formal analysis, K.D. and M.E.; investigation, K.D. and M.E.; resources, K.D.; data curation, K.D.; writing—original draft preparation, K.D.; writing—review and editing, K.D. and M.E.; visualization, K.D. and M.E.; supervision, K.D. and M.E.; project administration, K.D. All authors have read and agreed to the published version of the manuscript.

Funding: This research received no external funding.

Data Availability Statement: This paper does not use any existing data. The concepts are theoretical and can be coded for their use in any optimization algorithm.

Conflicts of Interest: The authors declare no conflict of interest.

References

1. Deb, K. *Multi-Objective Optimization Using Evolutionary Algorithms*; Wiley: Chichester, UK, 2001.
2. Coello, C.A.C.; VanVeldhuizen, D.A.; Lamont, G. *Evolutionary Algorithms for Solving Multi-Objective Problems*; Kluwer: Boston, MA, USA, 2002.
3. Deb, K.; Sinha, A.; Korhonen, P.; Wallenius, J. An Interactive Evolutionary Multi-Objective Optimization Method Based on Progressively Approximated Value Functions. *IEEE Trans. Evol. Comput.* **2010**, *14*, 723–739. [CrossRef]
4. Fowler, J.W.; Gel, E.S.; Koksalan, M.; Korhonen, P.; Marquis, J.L.; Wallenius, J. Interactive Evolutionary Multi-Objective Optimization for Quasi-Concave Preference Functions. *Submitt. Eur. J. Oper. Res.* **2010**, *206*, 417–425. [CrossRef]
5. Branke, J.; Greco, S.; Slowinski, R.; Zielniewicz, P. Interactive evolutionary multiobjective optimization using robust ordinal regression. In Proceedings of the Fifth International Conference on Evolutionary Multi-Criterion Optimization (EMO-09), Nantes, France, 7–10 April 2009; Springer: Berlin, Germany, 2009; pp. 554–568.
6. Kadziński, M.; Tomczyk, M.K.; Słowiński, R. Preference-based cone contraction algorithms for interactive evolutionary multiple objective optimization. *Swarm Evol. Comput.* **2020**, *52*, 100602. [CrossRef]
7. Tomczyk, M.K.; Kadziński, M. Robust indicator-based algorithm for interactive evolutionary multiple objective optimization. In Proceedings of the Genetic and Evolutionary Computation Conference, Lisbon, Portugal, 13–17 July 2019; pp. 629–637.
8. Farina, M.; Amato, P. A fuzzy definition of "optimality" for many-criteria optimization problems. *IEEE Trans. Syst. Man-Cybern.-Part Syst. Humans* **2004**, *34*, 315–326. [CrossRef]
9. Tian, Y.; Cheng, R.; Zhang, X.; Su, Y.; Jin, Y. A strengthened dominance relation considering convergence and diversity for evolutionary many-objective optimization. *IEEE Trans. Evol. Comput.* **2018**, *23*, 331–345. [CrossRef]
10. Saxena, D.K.; Mittal, S.; Kapoor, S.; Deb, K. A Localized High-Fidelity-Dominance based Many-Objective Evolutionary Algorithm. *IEEE Trans. Evol. Comput.* **2022**. [CrossRef]
11. Yuan, Y.; Xu, H.; Wang, B.; Yao, X. A new dominance relation-based evolutionary algorithm for many-objective optimization. *IEEE Trans. Evol. Comput.* **2015**, *20*, 16–37. [CrossRef]
12. Ikeda, K.; Kita, H.; Kobayashi, S. Failure of Pareto-based MOEAs: Does non-dominated really mean near to optimal? In Proceedings of the IEEE Congress on Evolutionary Computation, Seoul, Korea, 27–30 May 2001; pp. 957–962.
13. Ehrgott, M. *Multicriteria Optimization*; Springer: Berlin, Germany, 2000.
14. Steuer, R.E. *Multiple Criteria Optimization: Theory, Computation and Application*; Wiley: New York, NY, USA, 1986.
15. Miettinen, K. *Nonlinear Multiobjective Optimization*; Kluwer: Boston, MA, USA, 1999.
16. Deb, K.; Agrawal, S.; Pratap, A.; Meyarivan, T. A fast and Elitist multi-objective Genetic Algorithm: NSGA-II. *IEEE Trans. Evol. Comput.* **2002**, *6*, 182–197. [CrossRef]
17. Srinivas, N.; Deb, K. Multi-Objective function optimization using non-dominated sorting genetic algorithms. *Evol. Comput. J.* **1994**, *2*, 221–248. [CrossRef]
18. Deb, K.; Jain, H. An Evolutionary Many-Objective Optimization Algorithm Using Reference-point Based Non-dominated Sorting Approach, Part I: Solving Problems with Box Constraints. *IEEE Trans. Evol. Comput.* **2014**, *18*, 577–601. [CrossRef]
19. Pang, L.M.; Ishibuchi, H.; Shang, K. NSGA-II with simple modification works well on a wide variety of many-objective problems. *IEEE Access* **2020**, *8*, 190240–190250. [CrossRef]
20. Laumanns, M.; Rudolph, G.; Schwefel, H.P. A spatial predator-prey approach to multi-objective optimization: A preliminary study. In Proceedings of the Parallel Problem Solving from Nature, V, Amsterdam, The Netherlands, 27–30 September 1998; pp. 241–249.
21. Laumanns, M.; Thiele, L.; Deb, K.; Zitzler, E. Combining Convergence and Diversity in Evolutionary Multi-objective Optimization. *Evol. Comput.* **2002**, *10*, 263–282. [CrossRef] [PubMed]
22. Deb, K.; Mohan, M.; Mishra, S. Evaluating the $\in$-domination based multi-objective evolutionary algorithm for a quick computation of Pareto-optimal solutions. *Evol. Comput.* **2005**, *13*, 501–525. [CrossRef]
23. Hernández-Díaz, A.G.; Santana-Quintero, L.V.; Coello, C.A.C.; Molina, J. Pareto adaptive ϵ-dominance. *Evol. Comput. J.* **2007**, *15*, 493–517. [CrossRef] [PubMed]
24. Batista, L.S.; Campelo, F.; Guimaraes, F.G.; Ramírez, J.A. Pareto cone ε-dominance: Improving convergence and diversity in multiobjective evolutionary algorithms. In Proceedings of the International Conference on Evolutionary Multi-Criterion Optimization, Ouro Preto, Brazil, 5–8 April 2011; Springer: Berlin/Heidelberg, Germany, 2011, pp. 76–90.
25. Liu, J.; Wang, Y.; Wei, S.; Wu, X.; Tong, W. A parameterless penalty rule-based fitness estimation for decomposition-based many-objective optimization evolutionary algorithm. *IEEE Access* **2019**, *7*, 81701–81716. [CrossRef]
26. Le, K.; Landa-Silva, D. Obtaining better non-dominated sets using volume dominance. In Proceedings of the 2007 IEEE Congress on Evolutionary Computation, Singapore, 25–28 September 2007; pp. 3119–3126.
27. Dai, C.; Wang, Y.; Ye, M. A new evolutionary algorithm based on contraction method for many-objective optimization problems. *Appl. Math. Comput.* **2014**, *245*, 191–205. [CrossRef]
28. Sato, H.; Aguirre, H.E.; Tanaka, K. Controlling dominance area of solutions and its impact on the performance of MOEAs. In Proceedings of the International Conference on Evolutionary Multi-Criterion Optimization, Sendai, Japan, 5–8 March 2007; Springer: Berlin/Heidelberg, Germany, 2007; pp. 5–20.

29. Chen, L.; Liu, H.L.; Tan, K.C. Decomposition based dominance relationship for evolutionary many-objective algorithm. In Proceedings of the 2017 IEEE Symposium Series on Computational Intelligence (SSCI), Honolulu, HI, USA, 27 November–1 December 2017; pp. 1–6.
30. Wang, Y.; Deutz, A.; Bäck, T.; Emmerich, M. Edge-rotated cone orders in multi-objective evolutionary algorithms for improved convergence and preference articulation. In Proceedings of the 2020 IEEE Symposium Series on Computational Intelligence (SSCI), Canberra, Australia, 1 December–4 December 2020; pp. 165–172.
31. Shukla, P.K.; Emmerich, M.; Deutz, A. A theoretical analysis of curvature based preference models. In Proceedings of the International Conference on Evolutionary Multi-Criterion Optimization, Sheffield, UK, 19–22 March 2013; pp. 367–382.
32. Zou, X.; Chen, Y.; Liu, M.; Kang, L. A new evolutionary algorithm for solving many-objective optimization problems. *IEEE Trans. Syst. Man, Cybern. Part B* **2008**, *38*, 1402–1412.
33. Zhu, S.; Xu, L.; Goodman, E.D.; Lu, Z. A new many objective evolutionary algorithm based on generalized Pareto dominance. *IEEE Trans. Cybern.* **2021**, *52*, 7776–7790. [CrossRef]
34. Eichfelder, G. *Variable Ordering Structures in Vector Optimization*; Springer Science & Business Media: Berlin/Heidelberg, Germany, 2014.
35. Shukla, P.K.; Hirsch, C.; Schmeck, H. In search of equitable solutions using multi-objective evolutionary algorithms. In Proceedings of the International Conference on Parallel Problem Solving from Nature, Krakov, Poland, 11–15 September 2010; pp. 687–696.
36. Wierzbicki, A.P. The use of reference objectives in multiobjective optimization. In *Multiple Criteria Decision Making Theory and Applications*; Fandel, G.; Gal, T., Eds.; Springer: Berlin, Germany, 1980; pp. 468–486.
37. Zhang, Q.; Li, H. MOEA/D: A multiobjective evolutionary algorithm based on decomposition. *IEEE Trans. Evol. Comput.* **2007**, *11*, 712–731. [CrossRef]
38. Szlapczynski, R.; Szlapczynska, J. W-dominance: Tradeoff-inspired dominance relation for preference-based evolutionary multi-objective optimization. *Swarm Evol. Comput.* **2021**, *63*, 100866. [CrossRef]
39. Liu, Y.; Zhu, N.; Li, K.; Li, M.; Zheng, J.; Li, K. An angle dominance criterion for evolutionary many-objective optimization. *Inf. Sci.* **2020**, *509*, 376–399. [CrossRef]
40. Braun, M.; Shukla, P.; Schmeck, H. Angle-based preference models in multi-objective optimization. In Proceedings of the International Conference on Evolutionary Multi-Criterion Optimization, Münster, Germany, 19–22 March 2017; pp. 88–102.
41. Molina, J.; Santana, L.V.; Hernández-Díaz, A.G.; Coello, C.A.C.; Caballero, R. g-dominance: Reference point based dominance for multiobjective metaheuristics. *Eur. J. Oper. Res.* **2009**, *197*, 685–692. [CrossRef]
42. Said, L.B.; Bechikh, S.; Ghédira, K. The r-dominance: A new dominance relation for interactive evolutionary multicriteria decision making. *IEEE Trans. Evol. Comput.* **2010**, *14*, 801–818. [CrossRef]
43. Hu, J.; Yu, G.; Zheng, J.; Zou, J. A preference-based multi-objective evolutionary algorithm using preference selection radius. *Soft Comput.* **2017**, *21*, 5025–5051. [CrossRef]
44. Yi, J.; Bai, J.; He, H.; Peng, J.; Tang, D. ar-MOEA: A novel preference-based dominance relation for evolutionary multiobjective optimization. *IEEE Trans. Evol. Comput.* **2018**, *23*, 788–802. [CrossRef]
45. Kukkonen, S.; Lampinen, J. Ranking-dominance and many-objective optimization. In Proceedings of the 2007 IEEE Congress on Evolutionary Computation, Singapore, 25–28 September 2007; pp. 3983–3990.
46. Qasim, S.Z.; Ismail, M.A. Rode: Ranking-dominance-based algorithm for many-objective optimization with opposition-based differential evolution. *Arab. J. Sci. Eng.* **2020**, *45*, 10079–10096. [CrossRef]
47. Fan, Z.; Fang, Y.; Li, W.; Cai, X.; Wei, C.; Goodman, E.D. MOEA/D with angle-based constrained dominance principle for constrained multi-objective optimization problems. *Appl. Soft Comput.* **2019**, *74*, 621–633. [CrossRef]
48. Deb, K.; Pratap, A.; Meyarivan, T. Constrained test problems for multi-objective evolutionary optimization. In Proceedings of the International Conference on Evolutionary Multi-Criterion Optimization, Zurich, Switzerland, 7–9 March 2001; pp. 284–298.

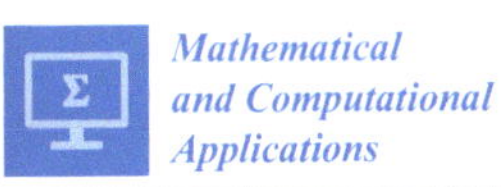
Mathematical and Computational Applications

Article

Observer-Based State Estimation for Recurrent Neural Networks: An Output-Predicting and LPV-Based Approach

Wanlin Wang, Jinxiong Chen and Zhenkun Huang *

School of Science, Jimei University, Xiamen 361021, China; wanlin19981025@outlook.com (W.W.);
chenjinxiong@wuyiu.edu.cn (J.C.)
* Correspondence: hzk974226@jmu.edu.cn

Abstract: An innovative cascade predictor is presented in this study to forecast the state of recurrent neural networks (RNNs) with delayed output. This cascade predictor is a chain-structured observer, as opposed to the conventional single observer, and is made up of several sub-observers that individually estimate the state of the neurons at various periods. This new cascade predictor is more useful than the conventional single observer in predicting neural network states when the output delay is arbitrarily large but known. In contrast to examining the stability of error systems solely employing the Lyapunov–Krasovskii functional (LKF), several new global asymptotic stability standards are obtained by combining the application of the Linear Parameter Varying (LPV) approach, LKF and convex principle. Finally, a series of numerical simulations verify the efficacy of the obtained results.

Keywords: cascade predictor; recurrent neural networks; delayed output; linear parameter varying approach

1. Introduction

Over the past decades, delayed recurrent neural networks were successfully applied in many fields, including pattern recognition, image processing, and combinatorial optimization [1–5], and the dynamic behaviors of RNNs have quickly become a research hotspot. At present, many stability results about the dynamic behavior of RNNs have been obtained [6–12]. Meanwhile, the state information of neurons is very important, because it may participate in the design process of control law, such as feedback control. Therefore, neural networks' state estimation research is of significant importance in practical applications.

The issue of state estimation for RNNs is currently of great interest to many scholars, and many significant results have been made [13–20]. In [13], the authors discussed the state estimation problem for delayed RNNs and obtained delay-independent results using LMI technique. The state estimation problem for Markov jump RNNs with distributed delays was discussed in [14]; the authors proposed an effective LMI technique to solve the problem of neuron states' estimation. The state estimation problem of uncertain RNNs was addressed via a robust state estimator in [15], and it was shown that the suggested robust estimator can be ensured by the feasibility of solving a set of LMIs. An interesting delay partition method was proposed in [17]; the authors used this method to investigate the state estimation problem for delayed static neural networks. In [20], the authors solved the memristive neural networks' (MNNs) state estimation problem by using a novel full-order state observer. It is well known that the effectiveness of the designed observer is usually related to system parameters and the size of various time delays. For example, in most of the works mentioned above, the output states do not have a time delay or the size of the delay is limited to a small range, and a full-order observer is designed based on the measured output. In [21], during the identification of RNN models, a subspace encoder is co-estimated to reconstruct the state of the model from past input and output data.

Citation: Wang, W.; Chen, J.; Huang, Z. Observer-Based State Estimation for Recurrent Neural Networks: An Output-Predicting and LPV-Based Approach. *Math. Comput. Appl.* **2023**, *28*, 104. https://doi.org/10.3390/mca28060104

Academic Editor: Guillermo Valencia-Palomo

Received: 5 August 2023
Revised: 12 October 2023
Accepted: 13 October 2023
Published: 25 October 2023

However, such an explicit form of an observer might run into difficulties if the state delay is not known, and needs an excessively large number of past input–output samples.

For arbitrary large known output delays, it is still an open problem to construct an effective observer to predict the current accurate states of the neuron. In fact, the proposal of the cascade predictor in the field of nonlinear systems has attracted much attention from researchers in recent years. In [22], the authors first proposed a cascade predictor for a class of triangular nonlinear systems that have only output delay; the cascade predictor is made up of a series of subsystems, and each subsystem has a similar structure. However, due to the complexity of the structure of the cascade predictor in [22], the observation estimation is not easily implemented by computer simulation. Since then, the cascade predictor has been studied and a series of results have been achieved on such triangular nonlinear systems with only output delay [23–25]. It can be observed that the cascade predictor has a promising application value in the state estimation of delay systems. It will be a challenge to incorporate cascade predictors into the state estimation of RNNs, and new state estimation approaches may arise.

Inspired by the arguments mentioned above, we focus on the state estimation of delayed RNNs based on cascade predictors. The designed cascade predictor is composed of a limited amount of subsystems; each subsystem estimates the neuron states at different delays, and the last subsystem estimates the current actual states of the neuron. The following are the paper's primary innovative ideas.

(1) This paper theoretically describes the reason why a single observer cannot observe the neuron state information when the output delay is large enough. Then, inspired by [22,23], we design a new cascade predictor for estimating the state of RNNs with state and output delays. To our best knowledge, this is the first time that a cascade predictor has been applied to state estimation in neural networks.

(2) For the activation function, most papers usually use the traditional Lipschitz condition hypothesis; however, for the activation with the large Lipschitz constant, it may indirectly lead to the conservatism of the design process and theoretical results. To overcome these difficulties, a new reformulated Lipschitz property of the activation function, which is the outcome of applying the LPV approach to the Lipschitz condition, is provided. This property is motivated by [26,27] and can lessen conservatism in the observer design process.

(3) In contrast to [14,16,20], the case when the output states have an arbitrarily large delay is explored, and the state prediction problem of delayed RNNs is resolved based on the measured output. A set of LMIs may be used to calculate the observer gain, and new adequate requirements for the global asymptotic stability of each error system are obtained based on the LKF, the LMI technique, and the convex principle.

The structure of this paper is as follows. The RNNs model and its associated assumptions are introduced in Section 2 of this article. The main results of this paper are presented in Section 3. The efficiency of the results obtained is demonstrated in Section 4 by numerical simulations. Section 5 closes with a general conclusion.

Notations: $\mathscr{R}$, $\mathscr{Z}$ denote, respectively, the set of real numbers and the positive integer set. $\mathscr{R}^n$ represents the n dimensional Euclidean space with the Euclidean norm $\| \cdot \|$. $\mathscr{R}^{n \times m}$ denotes the set of all $n \times m$ real matrices. The superscript "T" and "-1" represent the transpose and inverse of a matrix. $X > Y$ ($X < Y$) means that $X - Y$ is a positive (negative) matrix. $\|A\|$ denotes the operator norm of matrix A, i.e., $A = (a_{ij})$, $\|A\| = \sqrt{\lambda_{max}(A^T A)}$, where $\lambda_{max}(A)$ is the largest eigenvalue of A. $diag\{\cdots\}$ represents a block diagonal matrix. The symbol "$*$" denotes the symmetric term of the matrix. Let $\tau > 0$, $\mathscr{C}\left([-\tau, 0]; \mathscr{R}^n\right)$ denote the family of continuous functions ψ from $[-\tau, 0]$ to $\mathscr{R}^n$. I stands for an identity matrix with the proper dimensions.

2. Problem Formulation

Consider the following RNNs with delayed output [28,29]:

$$\begin{cases} \dot{x}(t) = -Ax(t) + W_0 f\big(x(t)\big) + W_1 f\big(x(t - h_x)\big), \\ y(t) = Cx(t - h_y), \\ x(s) = \phi(s), s \in [-\tau, 0], \end{cases} \tag{1}$$

where $x(t) = [x_1(t), \cdots, x_n(t)]^T \in \mathscr{R}^n$ denotes the state vector, $A = diag\{a_1, \cdots, a_n\}$ is a diagonal matrix with $a_i > 0$, W_0 and W_1 represent the connection weight matrices, $f(x(t)) = [f_1(x_1(t)), \cdots, f_n(x_n(t))]^T \in \mathscr{R}^n$ denotes the activation function, C is a output matrix, $y(t)$ represents the measured output. h_x, h_y denote the known discrete delays, and $\phi(s) \in \mathscr{C}([-\tau, 0]; \mathscr{R}^n)$ is an initial condition, $\tau = max\{h_x, h_y\}$.

The primary objective of this research is to construct an effective observer that can accurately predict the neuron states when the output delay h_y is arbitrarily large yet known. Next, for a subsequent analysis, the following corresponding lemmas are given.

Assumption 1. *The activation function $f_i(\cdot)$ is bounded and satisfies*

$$|f_i(u) - f_i(v)| \leq l_i|u - v|, \forall u, v \in \mathscr{R}, \tag{2}$$

where $f_i(0) = 0$ and $l_i > 0$ is a Lipschitz constant.

Some conservative conditions in the observer design may result from the Lipschitz condition of the activation function in (2). However, it is generally known that LPV approach can reduce the Lipschitz condition's conservatism, making it useful for designing observers for nonlinear systems with a large Lipschitz constant [26,27]. Here, we will extend this method to RNNs (1) and derive the subsequent lemma.

Lemma 1. *The activation function $f(\cdot)$ has the following two properties that are equal:*

(1) Lipschitz property: $f_i(\cdot)$ is l_i -Lipschitz, i.e.,

$$|f_i(x_i) - f_i(y_i)| \leq l_i|x_i - y_i|, \ \forall x_i, y_i \in \mathscr{R}. \tag{3}$$

(2) Lipschitz property reformulated: for all $i = 1, \cdots, n$, there exist functions $\psi_{ii}(t) : \mathscr{R} \to \mathscr{R}$ and constants $\overline{\gamma_{ii}}$, $\underline{\gamma_{ii}}$, such that

$$f(x) - f(y) = \sum_{i=1}^{n} \psi_{ii}(t) H_{ii}(x - y), \ \forall x, y \in \mathscr{R}^n \tag{4}$$

with $\underline{\gamma_{ii}} \leq \psi_{ii}(t) \leq \overline{\gamma_{ii}}$, where $H_{ii} = e_n(i)e_n^T(i)$ and $e_n(i) = [0, \ldots, \overset{i-th}{1}, \ldots, 0]^T \in \mathscr{R}^n$.

The proofs of Lemma 1 are similar to Lemma 6 and Lemma 7 in [26]; we omit it here. Note that $\psi_{ii}(t)$ in (4) is expressed as follows

$$\psi_{ii}(t) = \begin{cases} \dfrac{f_i(x_i) - f_i(y_i)}{x_i - y_i}, & x_i \neq y_i, \\ 0, & x_i = y_i. \end{cases} \tag{5}$$

Remark 1. *Compared with the traditional global Lipschitz condition hypothesis of activation functions in [13–16,30], the reformulation (4) in Lemma 1 offers a best less conservative Lipschitz condition and deals with the activation functions $f(x)$ with the best accuracy. For instance, for $f(x) = [tanh(x_1) + \frac{1}{2}sin(x_1), \frac{1}{2}cos(x_2) + \frac{1}{2}(|x_2 + 1| - |x_2 - 1|)]^T$, we have $\underline{\gamma_{11}} = -\frac{1}{2}$, $\overline{\gamma_{11}} = \frac{3}{2}$, $\underline{\gamma_{22}} = -\frac{1}{2}$, $\overline{\gamma_{22}} = \frac{3}{2}$. For $g(x) = [\frac{3}{4}(|x_1 + 1| - |x_1 - 1|), \frac{3}{2}tanh(x_2)]^T$, we have $\underline{\gamma_{11}} = 0$, $\overline{\gamma_{11}} = \frac{3}{2}$, $\underline{\gamma_{22}} = 0$, $\overline{\gamma_{22}} = \frac{3}{2}$. If we use the global Lipschitz condition in Assumption 1,* *we can only obtain $|f_i(x_i) - f_i(y_i)| \leq \frac{3}{2}|x_i - y_i|$ and $|g_i(x_i) - g_i(y_i)| \leq \frac{3}{2}|x_i - y_i|$, $i = 1, 2$, we cannot accurately distinguish between $f(x)$ and $g(x)$, and the related properties of $f(x)$ and $g(x)$ cannot be effectively utilized.*

Lemma 2 (Finsler's Lemma [31]). *For $x \in \mathscr{R}^n$, $M \in \mathscr{R}^{n \times n}$ is a symmetric matrix, and the matrix $\mathscr{G} \in \mathscr{R}^{m \times n}$, such that $rank(\mathscr{G}) < n$. The subsequent properties are equivalent:*

$$(1)\ x^T M x < 0,\ \forall x \in \mathscr{R}^n / \mathscr{G} x = 0,\ x \neq 0,$$

$$(2)\ \mathscr{G}^{\perp^T} M \mathscr{G}^\perp < 0,$$

where $\mathscr{G}^\perp$ is a right orthogonal complement of $\mathscr{G}$.

Lemma 3 (Moon's Inequality [32]). *Assume that $x(s) \in \mathscr{R}^{n_a}, y(s) \in \mathscr{R}^{n_b}$ are defined on the interval Ω and $Y \in \mathscr{R}^{n_a \times n_b}$. Then, for matrices $D \in \mathscr{R}^{n_a \times n_a}, T \in \mathscr{R}^{n_a \times n_b}$ and $Z \in \mathscr{R}^{n_b \times n_b}$, the following holds:*

$$-2 \int_\Omega x^T(s) Y y(s) ds \leq \int_\Omega \begin{bmatrix} x(s) \\ y(s) \end{bmatrix}^T \cdot \begin{bmatrix} D & T - Y \\ * & Z \end{bmatrix} \cdot \begin{bmatrix} x(s) \\ y(s) \end{bmatrix} ds,$$

where

$$\begin{bmatrix} D & T \\ * & Z \end{bmatrix} > 0.$$

3. Results

3.1. Single Observer

In this section, we will employ a full-order observer to handle the matter of RNNs' state estimation. First, in order to accurately estimate the RNNs' state information, we will design the full-order observer based on the measured delayed output as

$$\dot{\hat{x}}(t) = -A\hat{x}(t) + W_0 f(\hat{x}(t)) + W_1 f(\hat{x}(t - h_x)) + L(C\hat{x}(t - h_y) - y(t)), \tag{6}$$

where $\hat{x}(t)$ is an estimation of the state $x(t)$ of (1). Then, by defining the estimation error $e(t) = \hat{x}(t) - x(t)$, we can obtain the error system given, as follows

$$\dot{e}(t) = -Ae(t) + W_0 \Delta f(t) + W_1 \Delta f(t - h_x) + LCe(t - h_y), \tag{7}$$

where $\Delta f(t) = f(\hat{x}(t)) - f(x(t))$. Due to Lemma 1, there are functions $\psi_{ii}(t)$ and $\psi_{ii}^{h_x}(t)$, such that

$$\begin{cases} \Delta f(t) = \sum_{i=1}^n \psi_{ii}(t) H_{ii} e(t), \\ \Delta f(t - h_x) = \sum_{i=1}^n \psi_{ii}^{h_x}(t) H_{ii} e(t - h_x), \end{cases} \tag{8}$$

where $\psi_{ii}^{h_x}(t) = \psi_{ii}(t - h_x)$.

Define the time-varying matrices $\Psi(t) = diag\{\psi_{11}(t), \cdots, \psi_{nn}(t)\}$, $\Psi_{h_x}(t) = diag\{\psi_{11}^{h_x}(t), \cdots, \psi_{nn}^{h_x}(t)\}$, and bounded convex set $\mathcal{H}_n$, where the vertex set of $\mathcal{H}_n$ is defined as

$$V_{\mathcal{H}_n} = [\phi = diag[\phi_{11}, \phi_{22}, \cdots, \phi_{nn}] \in \mathscr{R}^{n \times n} | \phi_{ii} \in \{\underline{\gamma_{ii}}, \overline{\gamma_{ii}}\}]. \tag{9}$$

It is obvious that time-varying matrix parameters $\Psi(t)$ and $\Psi_{h_x}(t)$ belong to the bounded convex set $\mathcal{H}_n$. Now, we define the following matrices:

$$\begin{cases} \mathcal{A}(\Psi(t)) = -A + W_0 \sum_{i=1}^{n} \psi_{ii}(t) H_{ii} = -A + W_0 \Psi(t), \\[2mm] \mathcal{B}(\Psi_{h_x}(t)) = W_1 \sum_{i=1}^{n} \psi_{ii}^{h_x}(t) H_{ii} = W_1 \Psi_{h_x}(t), \end{cases} \tag{10}$$

then, by using (10), the LPV error system (7) can be reconstructed as

$$\dot{e}(t) = \mathcal{A}(\Psi(t))e(t) + \mathcal{B}(\Psi_{h_x}(t))e(t - h_x) + LCe(t - h_y). \tag{11}$$

The sufficient condition for the global asymptotic stability of error system (11) is presented in the following theorem.

Theorem 1. *The error system (11) is globally asymptotically stable for all $h_y \in [0, h^*]$, if there exist matrices $P > 0$, $Q > 0$, $M > 0$, $Z > 0$, $S > 0$, a matrix R, and positive scalars $\rho_i > 0, i = 1, 2,$ such that for $\forall \Psi, \Psi_{h_x} \in V_{\mathcal{H}_n}$, the following LMIs hold with observer gain $L = P^{-1}R$:*

$$\begin{bmatrix} \Omega_1 & P\mathcal{B}(\Psi_{h_x}) + \frac{Z}{h_x} & RC + \frac{S}{h^*} & h_x \mathcal{A}^T(\Psi)P & h^* \mathcal{A}^T(\Psi)P \\ * & \Omega_2 & 0 & h_x \mathcal{B}^T(\Psi_{h_x})P & h^* \mathcal{B}^T(\Psi_{h_x})P \\ * & * & \Omega_3 & h_x C^T R^T & h^* C^T R^T \\ * & * & * & \Omega_4 & 0 \\ * & * & * & * & \Omega_5 \end{bmatrix} < 0, \tag{12}$$

where $\Omega_1 = P\mathcal{A}(\Psi) + \mathcal{A}^T(\Psi)P + Q + M - \frac{Z}{h_x} - \frac{S}{h^*}$, $\Omega_2 = -Q - \frac{Z}{h_x}$, $\Omega_3 = -M - \frac{S}{h^*}$, $\Omega_4 = -h_x(2\rho_1 P - \rho_1^2 Z)$ *and* $\Omega_5 = -h^*(2\rho_2 P - \rho_2^2 S)$.

Proof. Consider the following Lyapunov–Krasovskii functions as

$$\begin{aligned} V(t) =\, & e^T(t)Pe(t) + \int_{t-h_x}^{t} e^T(\tau)Qe(\tau)d\tau + \int_{t-h_y}^{t} e^T(\tau)Me(\tau)d\tau \\ & + \int_{-h_x}^{0}\int_{t+\beta}^{t} \dot{e}^T(\tau)Z\dot{e}(\tau)d\tau d\beta + \int_{-h_y}^{0}\int_{t+\beta}^{t} \dot{e}^T(\tau)S\dot{e}(\tau)d\tau d\beta, \end{aligned} \tag{13}$$

and the time derivative of $V(t)$ can be evaluated as

$$\begin{aligned} \dot{V}(t) =\, & 2e^T(t)P\dot{e}(t) + e^T(t)Qe(t) - e^T(t - h_x)Qe(t - h_x) + e^T(t)Me(t) \\ & - e^T(t - h_y)Me(t - h_y) + h_x \dot{e}^T(t)Z\dot{e}(t) - \int_{t-h_x}^{t} \dot{e}^T(\tau)Z\dot{e}(\tau)d\tau \\ & + h_y \dot{e}^T(t)S\dot{e}(t) - \int_{t-h_y}^{t} \dot{e}^T(\tau)S\dot{e}(\tau)d\tau. \end{aligned} \tag{14}$$

Applying the Jensen's Inequality [33], we can obtain

$$-\int_{t-h_x}^{t} \dot{e}^T(\tau)Z\dot{e}(\tau)d\tau \leq -\frac{1}{h_x}\Delta e_{h_x}^T(t)Z\Delta e_{h_x}(t), \tag{15}$$

$$-\int_{t-h_y}^{t} \dot{e}^T(\tau)S\dot{e}(\tau)d\tau \leq -\frac{1}{h_y}\Delta e_{h_y}^T(t)S\Delta e_{h_y}(t), \tag{16}$$

where $\Delta e_{h_x}(t) = e(t) - e(t - h_x)$, $\Delta e_{h_y}(t) = e(t) - e(t - h_y)$.

By using (14)–(16), $\dot{V}(t)$ satisfies

$$\dot{V}(t) \leq \chi^T(t)Y(h_x, h_y)\chi(t), \tag{17}$$

where

$$\chi(t) = [\dot{e}^T(t), e^T(t), e^T(t - h_x), e^T(t - h_y), \Delta e_{h_x}^T(t), \Delta e_{h_y}^T(t)]^T,$$

$$Y(h_x, h_y) = \begin{bmatrix} h_x Z + h_y S & P & 0 & 0 & 0 & 0 \\ * & Q + M & 0 & 0 & 0 & 0 \\ * & * & -Q & 0 & 0 & 0 \\ * & * & * & -M & 0 & 0 \\ * & * & * & * & -\frac{Z}{h_x} & 0 \\ * & * & * & * & * & -\frac{S}{h_y} \end{bmatrix}.$$

Moreover, it follows from the error system (11) and the definition of $y_{h_x}(t)$ and $y_{h_y}(t)$ that $\Gamma\big(\Psi(t), \Psi_{h_x}(t)\big)\chi(t) = 0$ with

$$\Gamma\big(\Psi(t), \Psi_{h_x}(t)\big) = \begin{bmatrix} I & -\mathcal{A}\big(\Psi(t)\big) & -\mathcal{B}\big(\Psi_{h_x}(t)\big) & -P^{-1}RC & 0 & 0 \\ 0 & -I & I & 0 & I & 0 \\ 0 & -I & 0 & I & 0 & 0 \end{bmatrix}.$$

Therefore, the error system (11) is globally asymptotically stable if, for all $\Gamma\big(\Psi(t), \Psi_{h_x}(t)\big)\chi(t) = 0$ with $\chi(t) \neq 0$, there holds $\chi^T(t)Y(h_x, h_y)\chi(t) < 0$. Then, according to Lemma 2 and the convexity principle [34], $\chi^T(t)Y(h_x, h_y)\chi(t) < 0$ is equivalent to

$$\big(\Gamma^{\perp}(\Psi, \Psi_{h_x})\big)^T Y(h_x, h_y)\Gamma^{\perp}(\Psi, \Psi_{h_x}) < 0, \ \forall \Psi, \Psi_{h_x} \in \mathcal{V}_{\mathcal{H}_n}, \tag{18}$$

where $\Gamma^{\perp}(\Psi, \Psi_{h_x})$ is a right orthogonal complement of $\Gamma(\Psi, \Psi_{h_x})$ and

$$\Gamma^{\perp}(\Psi, \Psi_{h_x}) = \begin{bmatrix} \mathcal{A}(\Psi) & \mathcal{B}(\Psi_{h_x}) & P^{-1}RC \\ I & 0 & 0 \\ 0 & I & 0 \\ 0 & 0 & I \\ I & -I & 0 \\ I & 0 & -I \end{bmatrix}. \tag{19}$$

Further, (18) can be rewritten as

$$\begin{bmatrix} P\mathcal{A}(\Psi) + \mathcal{A}^T(\Psi)P + Q + M & P\mathcal{B}(\Psi_{h_x}) & RC \\ * & -Q & 0 \\ * & * & -M \end{bmatrix} + h_x \Pi_1 Z \Pi_1^T$$
$$- \frac{1}{h_x}\Pi_2 Z \Pi_2^T + h_y \Pi_1 S \Pi_1^T - \frac{1}{h_y}\Pi_3 S \Pi_3^T < 0, \ \forall \Psi, \Psi_{h_x} \in \mathcal{V}_{\mathcal{H}_n}, \tag{20}$$

where $\Pi_1 = [\mathcal{A}(\Psi), \mathcal{B}(\Psi_{h_x}), P^{-1}RC]^T$, $\Pi_2 = [I, -I, 0]^T$ and $\Pi_3 = [I, 0, -I]^T$. Since $Z > 0$ and $S > 0$, (20) cannot hold when h_y is large enough. Therefore, there exists an upper bound h^* for h_y such that (20) holds when $h_y \in [0, h^*]$, and $\big(\Gamma^{\perp}(\Psi, \Psi_{h_x})\big)^T Y(h_x, h_y)\Gamma^{\perp}(\Psi, \Psi_{h_x}) < 0$, $\forall \Psi, \Psi_{h_x} \in \mathcal{V}_{\mathcal{H}_n}$ is a sufficient condition for (18).

By employing a Schur complement [34], $\big(\Gamma^{\perp}(\Psi, \Psi_{h_x})\big)^T Y(h_x, h_y)\Gamma^{\perp}(\Psi, \Psi_{h_x}) < 0$ with $\forall \Psi, \Psi_{h_x} \in \mathcal{V}_{\mathcal{H}_n}$ is equivalent to

$$\begin{bmatrix} \Omega_1 & P\mathcal{B}(\Psi_{h_x}) + \frac{Z}{h_x} & RC + \frac{S}{h^*} & h_x \mathcal{A}^T(\Psi)Z & h^* \mathcal{A}^T(\Psi)S \\ * & \Omega_2 & 0 & h_x \mathcal{B}^T(\Psi_{h_x})Z & h^* \mathcal{B}^T(\Psi_{h_x})S \\ * & * & \Omega_3 & h_x C^T R^T P^{-1}Z & h^* C^T R^T P^{-1}S \\ * & * & * & -h_x Z & 0 \\ * & * & * & * & -h^* S \end{bmatrix} < 0, \tag{21}$$

with $\forall \Psi, \Psi_{h_x} \in V_{\mathcal{H}_n}$, then, multiplying both sides of (21) on the left and on the right by $diag\{I, I, I, PZ^{-1}, PS^{-1}\}$ and its transpose, respectively, and using the inequalities $-PZ^{-1}P \leq -2\rho_1 P + \rho_1^2 Z$, $-PS^{-1}P \leq -2\rho_2 P + \rho_2^2 S$, we can deduce that (12) is a sufficient condition for (21). We finish the proof. $\square$

Remark 2. *Different from the stability analysis of the nonlinear error system in [12,13,15,16,20], we provide an LPV formulation of the error system for RNNs with Lipschitz activation functions, which leads us to study the stability of the linear error system (11) by using the convexity principle. Obviously, our LPV-based approach is a useful tool for the state estimation of RNNs.*

Remark 3. *It follows from Theorem 1, which depends on the output delay, that the designed full-order observer (6) cannot predict the current state of RNNs (1) if $h_y \gg h*$. This is due to our inability to choose the appropriate observer gain L to stabilize the error system (11). From the later numerical simulations, it is clear that this is a drawback of the full-order observer. In addition, the conditions shown in (12) can be checked against a set of fixed values by standard LMI routines and an estimate of h^* is obtained. The algorithm for finding a feasible solution to (12) is summarized as follows:*
Step 1: *Fix the value of h^* to constant $\bar{h}$ and make an initial guess for $\bar{h}$.*
Step 2: *Fix the value of ρ_1, ρ_2 to some constants $\bar{\rho}_1, \bar{\rho}_2$ and make an initial guess for the values of $\bar{\rho}_1, \bar{\rho}_2$.*
Step 3: *Solve the LMI (12) for L with the fixed values $\bar{\rho}_1, \bar{\rho}_2$ and $\bar{h}$; if a feasible value of L cannot be computed, return to step 2 to reset the initial values of $\bar{\rho}_1$ and $\bar{\rho}_2$; if a feasible value of L can be computed, return to step 1 and increase the value of $\bar{h}$ until L cannot be solved.*

3.2. Cascade Predictor

If $h \gg h^*$, the full-order observer (6) will fail. In this case, cascade predictor design can be used to solve this problem. Let $h = \frac{h_y}{m}$, $m \in \mathcal{Z}$ and define

$$x_i(t): \begin{cases} \dot{x}_i(t) = -Ax_i(t) + W_0 f(x_i(t)) + W_1 f(x_i(t - h_x)), t \in [(m-i)h, +\infty), \\[2mm] x_i(s) = \phi(s - (m-i)h), \; s \in [-\tau + (m-i)h, (m-i)h], \end{cases} \tag{22}$$

where $i = 1, 2, \cdots, m$. Through mathematical analysis, we obtain $x_i(t) = x(t - h_y + i \cdot h) = x_{i+1}(t - h)$, $i = 1, 2, \cdots, m-1$ and $x_m(t) = x(t)$. Then, the cascade predictor can be constructed as

$$\begin{cases} \dot{\hat{x}}_1(t) = -A\hat{x}_1(t) + W_0 f(\hat{x}_1(t)) + W_1 f(\hat{x}_1(t - h_x)) + L_1(C\hat{x}_1(t - h) - y(t)), \\ \dot{\hat{x}}_2(t) = -A\hat{x}_2(t) + W_0 f(\hat{x}_2(t)) + W_1 f(\hat{x}_2(t - h_x)) + L_2(C\hat{x}_2(t - h) - \hat{y}_1(t)), \\ \qquad \vdots \\ \dot{\hat{x}}_m(t) = -A\hat{x}_m(t) + W_0 f(\hat{x}_m(t)) + W_1 f(\hat{x}_m(t - h_x)) + L_m(C\hat{x}_m(t - h) - \hat{y}_{m-1}(t)), \end{cases} \tag{23}$$

where $\phi_i(s) \in \mathscr{C}([-\tau_1, 0]; \mathscr{R}^n)$ is an initial condition for the subsystem $\hat{x}_i$, $i = 1, 2, \cdots, m$, $\tau_1 = max\{h_x, h\}$, and $\hat{y}_i(t) = Cx_i(t)$, $i = 1, 2, \cdots, m-1$. In the cascade predictor (23), the subsystem $\hat{x}_i(t)$ estimates the state $x_i(t)$, $i = 1, 2, \ldots, m-1$, and the subsystem $\hat{x}_m(t)$ estimates the state $x(t)$.

Remark 4. *The idea behind the cascade predictor (23) is that regardless of how long the output delay h_y is, we can split it into m small time periods. Then, each sub-observer $\hat{x}_i(t)$ in the cascade predictor estimates the delayed state $x(t - h_y + i\frac{h_y}{m})$, and the last sub-observer $\hat{x}_m(t)$ estimates the current state $x(t)$. Compared with [14–20], which discuss the state estimation of neural networks with only a small output delay or even no output delay using a full-order observer, the output delay h_y in this paper is arbitrarily large yet known, and, in this sense, this is an advancement in the study of neural networks' state estimation.*

Remark 5. *Moreover, the idea of this novel predictor was first proposed in [22], and the authors used this predictor with a chain structure for a class of triangular nonlinear systems with only the output delay. In this paper, we will discuss the state estimation problem of RNNs with both the state delay and output delay using this novel cascade predictor.*

Next, define the estimation error $e_i(t) = \hat{x}_i(t) - x_i(t)$, $i = 1, 2, \cdots, m$; then, similar to (10)–(11), which use the LPV approach, we can obtain the following error systems

$$\begin{cases} \dot{e}_1(t) = \mathcal{A}(\Psi(t))e_1(t) + \mathcal{B}(\Psi_{h_x}(t))e_1(t - h_x) + L_1 C e_1(t - h), \\ \dot{e}_j(t) = \mathcal{A}(\Psi(t))e_j(t) + \mathcal{B}(\Psi_{h_x}(t))e_j(t - h_x) + L_j C e_j(t - h) - L_j e_{j-1}(t), j = 2, 3, \cdots, m. \end{cases} \tag{24}$$

Theorem 2. *For given output delay h_y and scalar $m \in \mathscr{Z}$, the error systems (24) are globally asymptotically stable if there exist matrices $P_j > 0$, $Q_j > 0$, $M_j > 0$, $Z_j > 0$, $S_j > 0$, $X_i > 0$, $D_i > 0$, $i = 1, 2, \cdots, m$, matrices R_i, Y_i, T_i, $i = 1, 2, \cdots, m$ and a positive scalar $\gamma > 0$, such that $L_i = P_i^{-1} R_i$, $i = 1, 2, \cdots, m$ and the following LMIs are satisfied:*

$$\begin{bmatrix} \Xi_i & -T_i + P_i \mathcal{B}(\Psi_{h_x}) & -Y_i + R_i C & \gamma \mathcal{A}(\Psi) P_i \\ * & -Q_i & 0 & \gamma \mathcal{B}^T(\Psi_{h_x}) P_i \\ * & * & -M_i & \gamma C^T R_i^T \\ * & * & * & -2\gamma P_i + h_x Z_i + h S_i \end{bmatrix} < 0, \tag{25}$$

with $\forall \Psi \in \mathcal{V}_{\mathcal{H}_n}$, $\forall \Psi_{h_x} \in \mathcal{V}_{\mathcal{H}_n}$ and

$$\begin{bmatrix} X_i & Y_i \\ * & S_i \end{bmatrix} < 0, \tag{26}$$

$$\begin{bmatrix} D_i & T_i \\ * & Z_i \end{bmatrix} < 0, \tag{27}$$

where $\Xi_i = P_i \mathcal{A}(\Psi) + \mathcal{A}^T(\Psi) P_i + T_i + T_i^T + Y_i + Y_i^T + h_x D_i + h X_i + Q_i + M_i$.

Proof. The stability of the error systems (24) will be proved gradually:
    Step 1: we consider the first error system $e_1(t)$ in (24):

$$\dot{e}_1(t) = \mathcal{A}(\Psi(t))e_1(t) + \mathcal{B}(\Psi_{h_x}(t))e_1(t - h_x) + L_1 C e_1(t - h). \tag{28}$$

Using the Newton–Leibniz formula, we have

$$\begin{cases} e_1(t - h_x) = e_1(t) - \int_{t-h_x}^{t} \dot{e}_1(\tau) d\tau, \\ e_1(t - h) = e_1(t) - \int_{t-h}^{t} \dot{e}_1(\tau) d\tau, \end{cases} \tag{29}$$

then, $\dot{e}_1(t)$ in (28) can be reconstructed as

$$\dot{e}_1(t) = \left[\mathcal{A}(\Psi(t)) + \mathcal{B}(\Psi_{h_x}(t)) + L_1 C \right] e_1(t) - \mathcal{B}(\Psi_{h_x}(t)) \int_{t-h_x}^{t} \dot{e}_1(\tau) d\tau - L_1 C \int_{t-h}^{t} \dot{e}_1(\tau) d\tau. \tag{30}$$

Constructing the Lyapunov–Krasovskii functions as follows

$$V_1(t) = \overbrace{e_1^T(t)P_1e_1(t)}^{V_{11}(t)} + \overbrace{\int_{t-h_x}^t e_1^T(\tau)Q_1e_1(\tau)d\tau}^{V_{12}(t)} + \overbrace{\int_{t-h}^t e_1^T(\tau)M_1e_1(\tau)d\tau}^{V_{13}(t)}$$

$$+ \overbrace{\int_{-h_x}^0 \int_{t+\eta}^t \dot{e}_1^T(\tau)Z_1\dot{e}_1(\tau)d\tau d\eta}^{V_{14}(t)} + \overbrace{\int_{-h}^0 \int_{t+\eta}^t \dot{e}_1^T(\tau)S_1\dot{e}_1(\tau)d\tau d\eta}^{V_{15}(t)}. \tag{31}$$

then, the derivative of $V_{11}(t)$ along (30) is as follows

$$\dot{V}_{11}(t) = e_1^T(t)\left[P_1\mathcal{A}(\Psi(t)) + \mathcal{A}^T(\Psi(t))P_1 + P_1\mathcal{B}(\Psi_{h_x}(t)) + \mathcal{B}^T(\Psi_{h_x}(t))P_1 + R_1C + C^TR_1\right]$$

$$\times e_1(t) - 2e_1^T(t)P_1\mathcal{B}(\Psi_{h_x}(t))\int_{t-h_x}^t \dot{e}_1(\tau)d\tau - 2e_1^T(t)R_1C\int_{t-h}^t \dot{e}_1(\tau)d\tau. \tag{32}$$

According to Lemma 2 and (26)–(27), we have

$$-2e_1^T(t)P_1\mathcal{B}(\Psi_{h_x}(t))\int_{t-h_x}^t \dot{e}_1(\tau)d\tau \leq h_x e_1^T(t)D_1e_1(t) + 2e_1^T(t)\left[T_1 - P_1\mathcal{B}(\Psi_{h_x}(t))\right]$$

$$\times \left[e_1(t) - e_1(t-h_x)\right] + \int_{t-h_x}^t \dot{e}_1^T(\tau)Z_1\dot{e}_1(\tau)d\tau, \tag{33}$$

$$-2e_1^T(t)R_1C\int_{t-h}^t \dot{e}_1(\tau)d\tau \leq h e_1^T(t)X_1e_1(t) + 2e_1^T(t)\left[Y_1 - R_1C\right]\left[e_1(t) - e_1(t-h)\right]$$

$$+ \int_{t-h}^t \dot{e}_1^T(\tau)S_1\dot{e}_1(\tau)d\tau. \tag{34}$$

By (32)–(34), $\dot{V}_{11}(t)$ satisfies

$$\dot{V}_{11} \leq e_1^T(t)\left[P_1\mathcal{A}(\Psi(t)) + \mathcal{A}^T(\Psi(t))P_1 + T_1^T + T_1 + Y_1^T + Y_1 + h_xD_1 + hX_1\right]e_1(t)$$

$$- 2e_1^T(t)\left[T_1 - P_1\mathcal{B}(\Psi_{h_x}(t))\right]e_1(t-h_x) - 2e_1^T(t)\left[Y_1 - R_1C\right]e_1(t-h)$$

$$+ \int_{t-h_x}^t \dot{e}_1^T(\tau)Z_1\dot{e}_1(\tau)d\tau + \int_{t-h}^t \dot{e}_1^T(\tau)S_1\dot{e}_1(\tau)d\tau. \tag{35}$$

The derivatives of $V_{12}(t)$, $V_{13}(t)$, $V_{14}(t)$ and $V_{15}(t)$ are as follows

$$\dot{V}_{12}(t) = e_1^T(t)Q_1e_1(t) - e_1^T(t-h_x)Q_1e_1(t-h_x), \tag{36}$$

$$\dot{V}_{13}(t) = e_1^T(t)M_1e_1(t) - e_1^T(t-h)M_1e_1(t-h), \tag{37}$$

$$\dot{V}_{14}(t) = h_x\dot{e}_1^T(t)Z_1\dot{e}_1(t) - \int_{t-h_x}^t \dot{e}_1^T(\tau)Z_1\dot{e}_1(\tau)d\tau, \tag{38}$$

$$\dot{V}_{15}(t) = h\dot{e}_1^T(t)S_1\dot{e}_1(t) - \int_{t-h}^t \dot{e}_1^T(\tau)S_1\dot{e}_1(\tau)d\tau, \tag{39}$$

and for arbitrary constant $\gamma > 0$, we have

$$-2\gamma\dot{e}_1^T(t)P_1\left[\dot{e}_1(t) - \mathcal{A}(\Psi(t))e_1(t) - \mathcal{B}(\Psi_{h_x}(t))e_1(t-h_x) - P_1^{-1}R_1Ce_1(t-h)\right] = 0. \tag{40}$$

Combining (35)–(40), we obtain

$$\dot{V}_1(t) \leq \xi_1^T(t)\Omega_1\xi_1(t), \tag{41}$$

where $\xi_1(t) = [e_1^T(t), e_1^T(t-h_x), e_1^T(t-h), \dot{e}_1^T(t)]^T$ and

$$
\Omega_1 = \begin{bmatrix} \Xi_1 & -T_1 + P_1\mathcal{B}(\Psi_{h_x}(t)) & -Y_1 + R_1 C & \gamma\mathcal{A}(\Psi(t))P_1 \\ * & -Q_1 & 0 & \gamma\mathcal{B}^T(\Psi_{h_x}(t))P_1 \\ * & * & -M_1 & \gamma C^T R_1^T \\ * & * & * & -2\gamma P_1 + h_x Z_1 + hS_1 \end{bmatrix}.
$$

Due to $\forall\Psi \in \mathcal{V}_{\mathcal{H}_n}$, $\forall\Psi_{h_x} \in \mathcal{V}_{\mathcal{H}_n}$, according to the convex principle, one obtains

$$
\Omega_1 < 0, \ \forall\Psi, \Psi_{h_x} \in \mathcal{V}_{\mathcal{H}_n}, \tag{42}
$$

which leads to $\dot{V}_1(t) < 0$. Then, it follows from (41) that

$$
\dot{V}_1(t) \le -\lambda_{min}(-\Omega_1)\xi_1^T(t)\xi_1(t) \le -\lambda_{min}(-\Omega_1)e_1^T(t)e_1(t), \tag{43}
$$

this indicates that the error system $e_1(t)$ is globally asymptotically stable. Obviously, conditions (25) ensure that $\Omega_1 < 0$ holds.

Step j: To recursively prove the stability of the error system $e_j(t)$, we assume that $e_{j-1}(t)$ is globally asymptotically stable. Similar to (28) and using the Newton–Leibniz formula, $e_j(t)$ $(j = 2, 3, \cdots, m)$ can be rewritten, as follows

$$
\begin{aligned}
\dot{e}_j(t) = &\left[\mathcal{A}(\Psi(t)) + \mathcal{B}(\Psi_{h_x}(t)) + L_j C\right]e_j(t) - \mathcal{B}(\Psi_{h_x}(t))\int_{t-h_x}^t \dot{e}_j(\tau)d\tau - L_j C \int_{t-h}^t \dot{e}_j(\tau)d\tau \\
&- L_j C e_{j-1}(t).
\end{aligned} \tag{44}
$$

Then, we construct the following Lyapunov–Krasovskill functions

$$
V_j(t) = \overbrace{e_j^T(t)P_j e_j(t)}^{V_{j1}(t)} + \overbrace{\int_{t-h_x}^t e_j^T(\tau)Q_j e_j(\tau)d\tau}^{V_{j2}(t)} + \overbrace{\int_{t-h}^t e_j^T(\tau)M_j e_j(\tau)d\tau}^{V_{j3}(t)}
$$
$$
+ \overbrace{\int_{-h_x}^0 \int_{t+\eta}^t \dot{e}_j^T(\tau)Z_j\dot{e}_j(\tau)d\tau d\eta}^{V_{j4}(t)} + \overbrace{\int_{-h}^0 \int_{t+\eta}^t \dot{e}_j^T(\tau)S_j\dot{e}_j(\tau)d\tau d\eta}^{V_{j5}(t)}. \tag{45}
$$

Taking the derivative of $V_{j1}(t)$ along with (44), we have

$$
\begin{aligned}
\dot{V}_{j1}(t) = &e_j^T(t)\left[P_j\mathcal{A}(\Psi(t)) + \mathcal{A}^T(\Psi(t))P_j + P_j\mathcal{B}(\Psi_{h_x}(t)) + \mathcal{B}^T(\Psi_{h_x}(t))P_j + R_j C + C^T R_j\right] \times \\
&e_j(t) - 2e_j^T(t)P_j\mathcal{B}(\Psi_{h_x}(t))\int_{t-h_x}^t \dot{e}_j(\tau)d\tau - 2e_j^T(t)R_j C\int_{t-h}^t \dot{e}_j(\tau)d\tau \\
&- e_j^T(t)R_j C e_{j-1}(t).
\end{aligned} \tag{46}
$$

Now, by using Young's inequality [27], we obtain

$$
-e_j^T(t)R_j C e_{j-1}(t) \le \varepsilon_1 e_j^T(t)e_j(t) + \frac{1}{\varepsilon_1}\|R_j C\|^2\|e_{j-1}\|^2, \tag{47}
$$

with $\varepsilon_1 > 0$. Then, similar to (33)–(34) in step 1 and using (46)–(47), we have

$$
\begin{aligned}
\dot{V}_{j1} \le &e_j^T(t)\left[P_j\mathcal{A}(\Psi(t)) + \mathcal{A}^T(\Psi(t))P_j + T_j^T + T_j + Y_j^T + Y_j + h_x D_j + hX_j + \varepsilon_1 I\right]e_j(t) \\
&- 2e_j^T(t)\left[T_j - P_j\mathcal{B}(\Psi_{h_x}(t))\right]e_j(t-h_x) - 2e_j^T(t)\left[Y_j - R_j C\right]e_j(t-h) \\
&+ \int_{t-h_x}^t \dot{e}_j^T(\tau)Z_j\dot{e}_j(\tau)d\tau + \int_{t-h}^t \dot{e}_j^T(\tau)S_j\dot{e}_j(\tau)d\tau + \frac{1}{\varepsilon_1}\|R_j C\|^2\|e_{j-1}\|^2.
\end{aligned} \tag{48}
$$

For arbitrary constants $\gamma > 0$ and $\varepsilon_2 > 0$, we obtain

$$\begin{cases} -2\gamma \dot{e}_j^T(t)P_j[\dot{e}_j(t) - \mathcal{A}(\Psi(t))e_j(t) - \mathcal{B}(\Psi_{h_x}(t))e_j(t - h_x) - P_j^{-1}R_jCe_j(t - h) \\ \quad + P_j^{-1}R_jCe_{j-1}(t)] = 0, \\ -2\gamma e_j^T(t)R_jCe_{j-1}(t) \le \varepsilon_2 e_j^T(t)e_j(t) + \dfrac{\gamma^2}{\varepsilon_2}\|R_jC\|^2\|e_{j-1}\|^2. \end{cases} \tag{49}$$

Then, combining the derivative of $\sum_{i=2}^{5} V_{ji}(t)$ and using (48)–(49), we have

$$\dot{V}_j(t) \le \xi_j^T(t)\hat{\Omega}_j\xi_j(t) + \left(\frac{1}{\varepsilon_1} + \frac{\gamma^2}{\varepsilon_2}\right)\|R_jC\|^2\|e_{j-1}\|^2, \tag{50}$$

where $\xi_j(t) = [e_j^T(t), e_j^T(t - h_x), e_j^T(t - h), \dot{e}_j^T(t)]^T$,

$$\hat{\Omega}_j = \begin{bmatrix} \hat{\Xi}_j & -T_j + P_j\mathcal{B}(\Psi_{h_x}(t)) & -Y_j + R_jC & \gamma\mathcal{A}(\Psi(t))P_j \\ * & -Q_j & 0 & \gamma\mathcal{B}^T(\Psi_{h_x}(t))P_j \\ * & * & -M_j & \gamma C^T R_j^T \\ * & * & * & -2\gamma P_j + h_x Z_j + hS_j + \varepsilon_2 I \end{bmatrix}$$

and $\hat{\Xi}_j = P_j\mathcal{A}(\Psi(t)) + \mathcal{A}^T(\Psi(t))P_j + T_j + T_j^T + Y_j + Y_j^T + h_x D_j + hX_j + Q_j + M_j + \varepsilon_1 I$. Similar to Ω_1 in step 1, if

$$\hat{\Omega}_j < 0, \ \forall \Psi, \Psi_{h_x} \in \mathcal{V}_{\mathcal{H}_n} \tag{51}$$

is true, we have

$$\dot{V}_j(t) \le -\lambda_{min}(-\hat{\Omega}_j)e_j^T(t)e_j(t) + \left(\frac{1}{\varepsilon_1} + \frac{\gamma^2}{\varepsilon_2}\right)\|R_jC\|^2\|e_{j-1}\|^2. \tag{52}$$

Then, employing the comparison Lemma [35], we can conclude that if $e_{j-1}(t)$ is globally asymptotically stable, then $e_j(t)$ is also globally asymptotically stable.

Furthermore, it not difficult to observe that $\hat{\Omega}_j = \Omega_j + \varepsilon_1\Pi_1^T\Pi_1 + \varepsilon_2\Pi_2^T\Pi_2$, where $\Pi_1 = [I, 0, 0, 0]^T$, $\Pi_2 = [0, 0, 0, I]^T$, and

$$\Omega_j = \begin{bmatrix} \Xi_j & -T_j + P_j\mathcal{B}(\Psi_{h_x}) & -Y_j + R_jC & \gamma\mathcal{A}(\Psi)P_j \\ * & -Q_j & 0 & \gamma\mathcal{B}^T(\Psi_{h_x})P_j \\ * & * & -M_j & \gamma C^T R_j^T \\ * & * & * & -2\gamma P_j + h_x Z_j + hS_j \end{bmatrix} < 0.$$

Since ε_1 and ε_2 are selected arbitrarily, $\Omega_j < 0$ ensures that $\hat{\Omega}_j < 0$ holds when ε_1 and ε_2 are sufficiently small. Finally, it observes that conditions (25) ensure that $\Omega_j < 0$ holds. We finish the proof. $\square$

Remark 6. *It follows from Theorem 2 that for a given output delay h_y, whether the LMI set (25), (26) and (27) have feasible solutions depends on the size of the parameter $h = \frac{h_y}{m}$. Obviously, for a large output delay of h_y, a sufficiently large m can ensure that the LMI sets (25), (26) and (27) have feasible solutions. However, the complexity of the cascade predictor is proportional to m; in other words, a larger m will reduce the observational performance of the cascade predictor. Therefore, we should choose a suitable value of m to balance the stability requirement and the complexity limitation of the cascade predictor.*

4. Numerical Simulation

This section provides a series of numerical simulations to demonstrate the efficacy of our results.

Let $x(t) = [x_1(t), x_2(t)]^T$. Consider the RNNs (1) with the following parameters:

$$A = \begin{bmatrix} 1 & 0 \\ 0 & 1 \end{bmatrix}, \; W_0 = \begin{bmatrix} 2.0 & -0.1 \\ -5.0 & 3.0 \end{bmatrix}, \; W_1 = \begin{bmatrix} -1.5 & -0.1 \\ -0.2 & -2.5 \end{bmatrix}, \; C = \begin{bmatrix} 3 & 0.3 \\ 0.3 & 3 \end{bmatrix},$$

$$h_x = 1, \; \phi(s) = [2,2]^T, \; f(x(t)) = [tanh(x_1(t)), tanh(x_2(t))]^T,$$

where h_y exists if we take the value later. It can be observed from Figure 1 that these RNNs have complex chaotic behaviors. Then, from Theorem 1, it follows that there is a nonfeasible solution to the LMI set (12) when $h^* > 0.24$. Therefore, when $h_y \leq 0.24$, we only use the full-order observer (6). When $h_y > 0.24$, we use the cascade predictor (23).

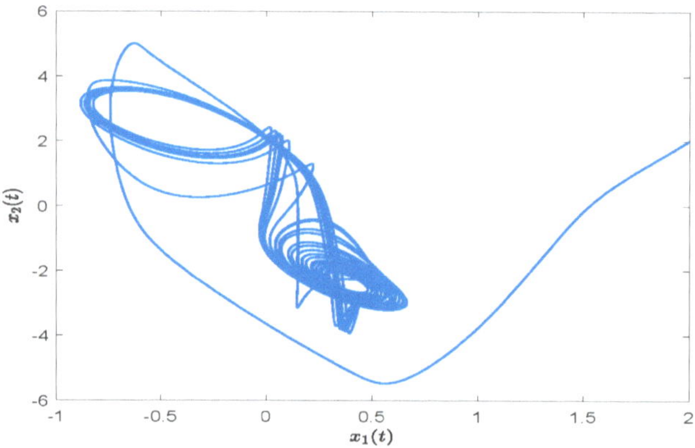

Figure 1. The phase space trajectory of RNNs (1).

Example 1. *For $h_y = 0.23$, assume that the initial conditions of the full-order observer $\hat{x}$ are $\hat{x}(s) = [-2, -2]^T, s \in [-1, 0)$. Then, from Theorem 1, we can obtain feasible solutions:*

$$P = \begin{bmatrix} 15.9540 & 0.3889 \\ 0.3889 & 2.3080 \end{bmatrix}, L = \begin{bmatrix} -1.4728 & 0.1894 \\ 1.1641 & -2.3357 \end{bmatrix}.$$

The simulation results are shown in Figures 2 and 3. Figure 2 represents the trajectory profiles of $\hat{x}(t)$ and $x(t)$, which implies the validity of our designed full-order observer (6). Figure 3 represents the converged trajectory of $\|e(t)\| = \|\hat{x}(t) - x(t)\|$, which implies the convergence performance of the full-order observer (6). However, when $h = 0.25$, it can be observed from Figures 4 and 5 that the full-order observer cannot accurately estimate the state of the original system and the estimation error $\|e(t)\|$ becomes larger, since $h = 0.25$ is greater than $h^* = 0.24$.

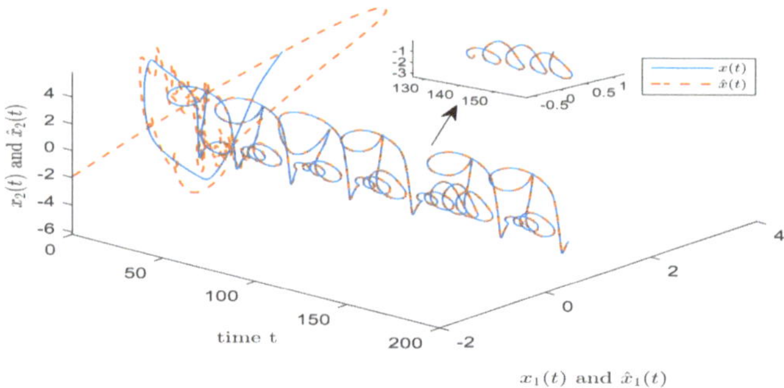

Figure 2. Profile on trajectories of $x(t)$ and $\hat{x}(t)$.

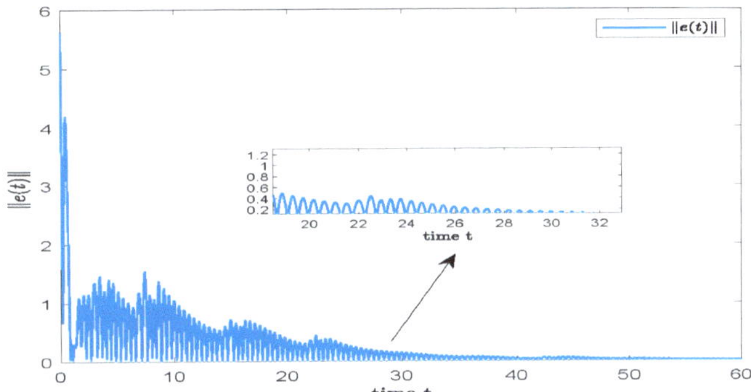

Figure 3. The estimation error $\|e(t)\|$.

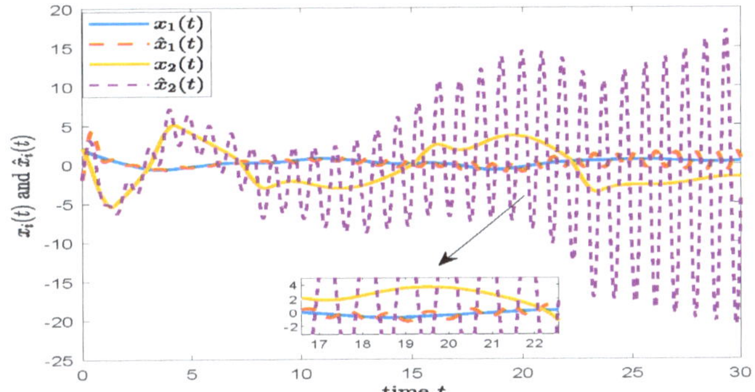

Figure 4. The state $x_i(t)$ and $\hat{x}_i(t)$, $i = 1, 2$.

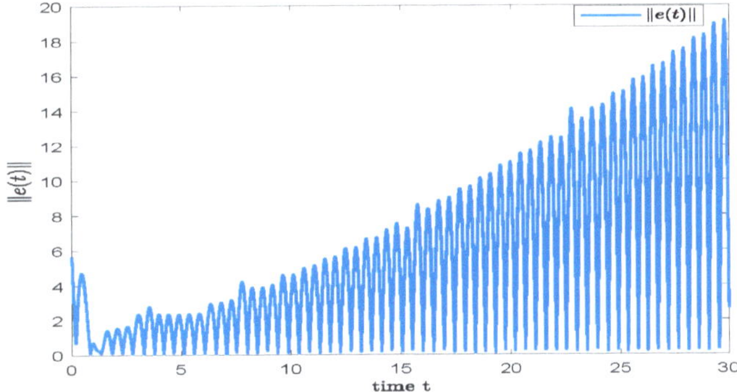

Figure 5. The estimation error $\|e(t)\|$.

Example 2. *This example considers the case $h_y > 0.24$; thus, we only use the cascade predictor (23).*
(i) For $h_y = 0.5$, we select $m = 5$ and $h = 0.1$ and assume that the initial conditions of $\hat{x}_i$, $i = 1, \cdots, 5$ are $\hat{x}_i(s) = [-2, -2]^T$, $s \in [-1, 0)$. Then, from Theorem 2, we obtain feasible solutions:

$$P_i = \begin{bmatrix} 142.1455 & -3.8554 \\ -3.8554 & 16.5776 \end{bmatrix}, L_i = \begin{bmatrix} -1.5129 & 0.1483 \\ 1.5729 & -2.8071 \end{bmatrix} (i = 1, \cdots 5).$$

As illustrated in Figures 6 and 7, the cascade predictor is valid and the estimation error $\|e_5(t)\| = \|\hat{x}_5(t) - x(t)\|$ finally converges to 0.

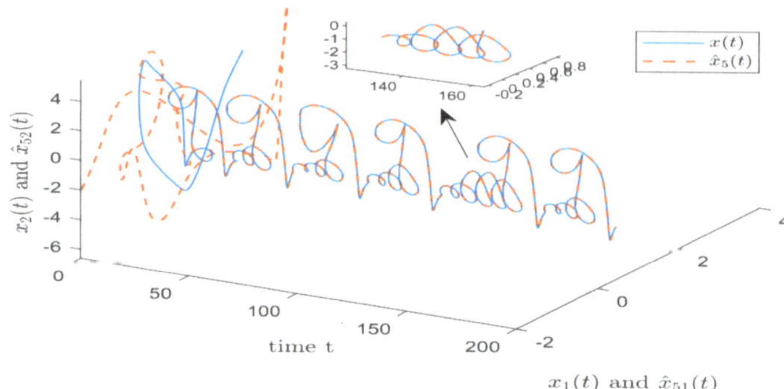

Figure 6. Profile on trajectories of $x(t)$ and $\hat{x}_5(t)$.

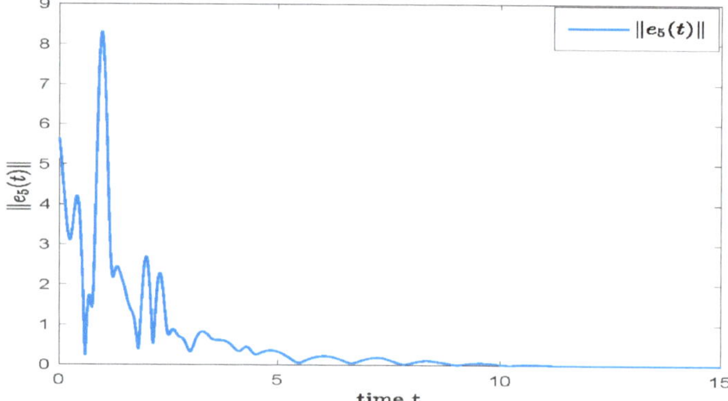

Figure 7. The estimation error $\|e_5(t)\|$.

(ii) For $h_y = 1$, we select $m = 10$ and $h = 0.1$ and assume that the initial conditions of $\hat{x}_i$, $i = 1, \cdots, 10$ are $\hat{x}_i(s) = [-1, -1]^T$, $s \in [-1, 0)$. Since the value of h is equal to the value of h in i), the observer gain L_i $(i = 1, \cdots, 10)$ can be equal to the observer gain in (i). Then, from Figures 8 and 9, it is clear that the cascade predictor is valid, and the observer error $\|e_{10}(t)\| = \|\hat{x}_{10}(t) - x(t)\|$ finally converges to 0.

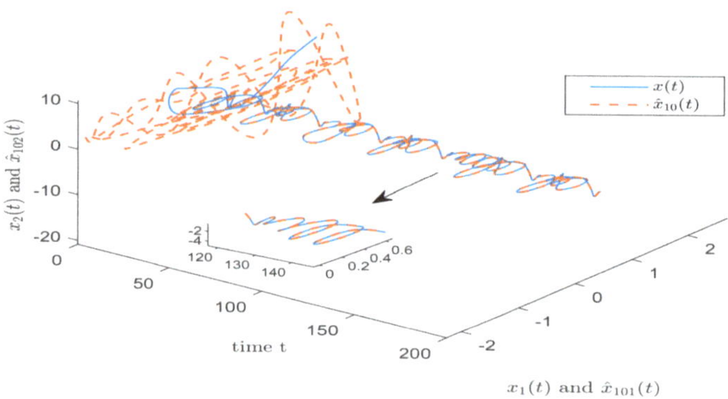

Figure 8. Profile on trajectories of $x(t)$ and $\hat{x}_{10}(t)$.

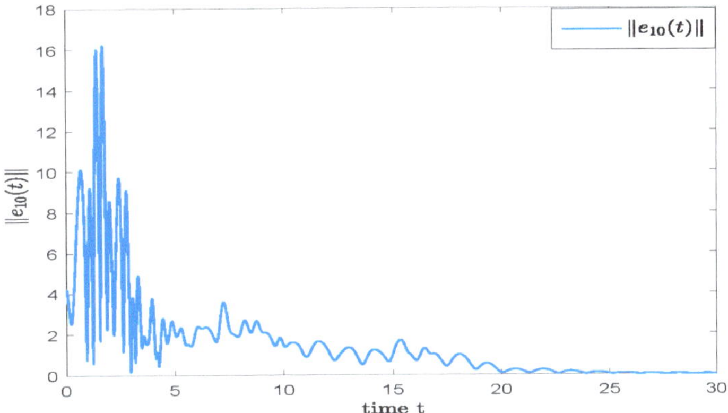

Figure 9. The estimation error $\|e_{10}(t)\|$.

Example 3. *This example will further discuss the effect of the size of the output delay on the convergence of the two predictors. The simulation results are shown in Figures 10–12, and the influence of m and h on the convergence time is given in Table 1, where "*" denoting the single observer is not valid ($h_y > h^*$).*

From Table 1, we can clearly observe that for both predictors, the output delay is directly proportional to the convergence time, and the larger the output delay, the longer the convergence time. In addition, from the experimental results, it can be concluded that, although the cascade predictor can solve the problem of an arbitrarily large output delay, as h_y increases, we will have to choose more subsystems to transmit the state information, which leads to the accumulation of error information and increases the cost of observation.

Table 1. The convergence time for two types of predictors by setting observer error $\|e\| = 0.1$.

Predictor\h_y		$h_y = 0.1$	$h_y = 0.2$	$h_y = 0.5$	$h_y = 1$	$h_y = 1.5$	$h_y = 2$
Convergence time	Simple observer	1.6	8.5	*	*	*	*
	Cascade predictor	1.8 (m = 1)	3.6 (m = 2)	9.1 (m = 5)	16.3 (m = 10)	27.4 (m = 15)	31.5 (m = 20)

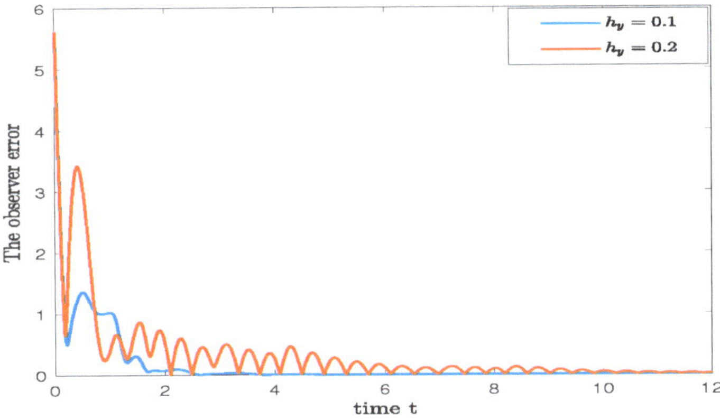

Figure 10. Convergence of single observer at different delays.

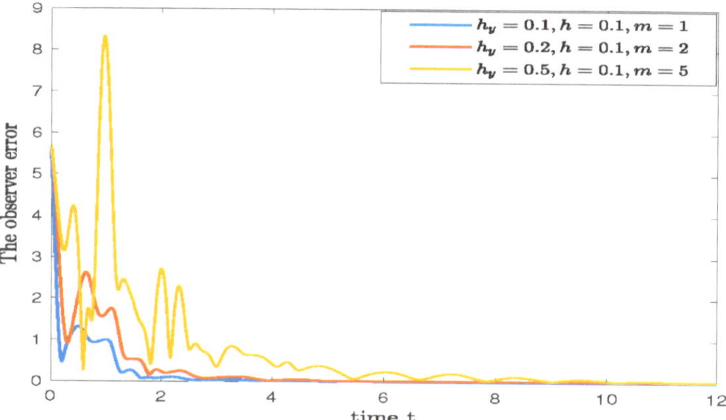

Figure 11. Convergence of cascade predictor at different delays.

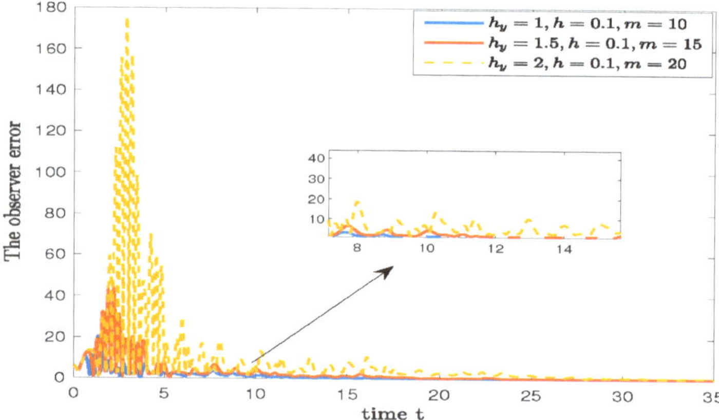

Figure 12. Convergence of cascade predictor at different delays.

5. Conclusions

In this research, we investigate the RNNs' state estimation by proposing an output-predicting and LPV approach. Due to the LPV approach, LKF and convex principle, several new conditions for the global asymptotic stability of the error system have been established. Compared with the traditional observer in [14–20], the chain-structured cascade predictor is more useful in the state estimation of neural networks. Different from [12,13,15,16,20], we use the LPV approach to convert nonlinear error dynamic systems into linear error systems, which greatly reduces the difficulty of the stability analysis. Finally, a series of numerical simulations show the effectiveness of the cascade predictor.

Author Contributions: Conceptualization, W.W. and Z.H.; methodology, W.W.; software, W.W. and J.C.; formal analysis, W.W.; writing—original draft preparation, W.W.; writing—review and editing, W.W. and Z.H. All authors have read and agreed to the published version of the manuscript.

Funding: This work was supported in part by the National Natural Science Foundation of China under Grant 61573005, and in part by the Natural Science Foundation of Fujian Province under Grant 2018J01417 and Grant 2019J01330.

Data Availability Statement: Not applicable.

Conflicts of Interest: The authors declare no conflict of interest.

References

1. Chua, L.; Yang, L. Cellular neural networks: Application. *IEEE Trans. Circuits Syst.* **1998**, *35*, 1273–1290. [CrossRef]
2. Cichocki, A.; Unbehauen, R. *Neural Networks for Optimization and Signal Processing*; Wiley: Chichester, UK, 1993.
3. Joya, G.; Atencia, M.; Sandoval, F. Hopfield neural networks for optimization: Study of the different dynamics. *Neurocomputing* **2002**, *43*, 219–237. [CrossRef]
4. Li, W.; Lee, T. Hopfield neural networks for affine invariant matching. *IEEE Trans. Neural Netw.* **2001**, *12*, 1400–1410. [CrossRef]
5. Yong, S.; Scott, P.; Nasrabadi, N. Object recognition using multilayer Hopfield neural network. *IEEE Trans. Image Process.* **1997**, *6*, 357–372. [CrossRef] [PubMed]
6. Wang, Z.; Liu, L.; Shan, Q.; Zhang, H. Stability criteria for recurrent neural networks with time-varying delay based on secondary delay partitioning method. *IEEE Trans. Neural Netw. Learn. Syst.* **2015**, *26*, 2589–2595. [CrossRef] [PubMed]
7. Zhang, C.; He, Y.; Jiang, L.; Wu, M. Delay-dependent stability criteria for generalized neural networks with two delay components. *IEEE Trans. Neural Netw. Learn. Syst.* **2014**, *25*, 1263–1276. [CrossRef]
8. Zhang, X.; Han, Q. Global asymptotic stability analysis for delayed neural networks using a matrix-based quadratic convex approach. *Neural Netw.* **2014**, *54*, 57–69. [CrossRef] [PubMed]
9. Wang, Z.; Zhang, H.; Jiang, B. LMI-based approach for global asymptotic stability analysis of recurrent neural networks with various delays and structures. *IEEE Trans. Neural Netw.* **2011**, *22*, 1032–1045. [CrossRef] [PubMed]
10. Liu, Y.J.; Lee, S.M.; Kwon, O.M; Park, J.H. New approach to stability criteria for generalized neural networks with interval time-varying delays. *Neurocomputing* **2015**, *149*, 1544–1551. [CrossRef]
11. Wu, Z.; Shi, P.; Su, H.; Chu, J. Delay-dependent stability analysis for switched neural networks with time-varying delay. *IEEE Trans. Syst. Man Cybern. Part B (Cybern.)* **2011**, *41*, 1522–1530. [CrossRef]
12. Zhu, Q.; Cao, J. Stability of Markovian jump neural networks with impulse control and time varying delays. *Nonlinear Anal. Real World Appl.* **2012**, *13*, 2259–2270. [CrossRef]
13. Wang, Z.; Ho, D.W.C.; Liu, X. State estimation for delayed neural networks. *IEEE Trans. Neural Netw.* **2005**, *16*, 279–284. [CrossRef] [PubMed]
14. Wang, Z.; Liu, R.; Liu, Y. State estimation for jumping recurrent neural networks with discrete and distributed delays. *Neural Netw.* **2009**, *22*, 41–48. [CrossRef] [PubMed]
15. Wang, Z.; Wang, J.; Wu, Y. State estimation for recurrent neural networks with unknown delays: A robust analysis approach. *Neurocomputing* **2017**, *227*, 29–36. [CrossRef]
16. Huang, H.; Feng, G.; Cao, J. Robust state estimation for uncertain neural networks with time-varying delay. *IEEE Trans. Neural Netw.* **2008**, *19*, 1329–1339. [CrossRef]
17. Huang, H.; Feng, G.; Cao, J. State estimation for static neural networks with time-varying delay. *Neural Netw.* **2010**, *23*, 1202–1207. [CrossRef]
18. Ren, J.; Zhu, H.; Zhong, S.; Ding, Y.; Shi, K. State estimation for neural networks with multiple time delays. *Neurocomputing* **2015**, *151*, 501–510. [CrossRef]
19. Liu, M.; Chen, H. H_∞ state estimation for discrete-time delayed systems of the neural network type with multiple missing measurements. *IEEE Trans. Neural Netw. Learn. Syst.* **2015**, *26*, 2987–2998. [CrossRef]
20. Guo, M.; Zhu, S.; Liu, X. Observer-based state estimation for memristive neural networks with time-varying delay. *Knowl.-Based Syst.* **2022**, *246*, 108707. [CrossRef]
21. Beintema, G.I.; Schoukens, M.; Toth, R. Deep subspace encoders for nonlinear system identification. *Automatica* **2023**, *156*, 111210. [CrossRef]
22. Germani, A.; Manes, C.; Pepe, P. A new approach to state observation of nonlinear systems with delayed output. *IEEE Trans. Autom. Control* **2002**, *47*, 96–101. [CrossRef]
23. Ahmed-Ali, T.; Cherrier, E.; Lamnabhi-Lagarrigue, F. Cascade high gain predictors for a class of nonlinear systems. *IEEE Trans. Autom. Control* **2012**, *57*, 224–229. [CrossRef]
24. Farza, M.; M'Saad, M.; Menard, T.; Fall, M.L.; Gehan, O.; Pigeon, E. Simple cascade observer for a class of nonlinear systems with long output delays. *IEEE Trans. Autom. Control* **2015**, *60*, 3338–3343. [CrossRef]
25. Farza, M.; Hernandez-Gonzalez, O.; Menard, T.; Targui, B.; M'Saad, M.; Astorga-Zaragoza, C.M. Cascade observer design for a class of uncertain nonlinear systems with delayed outputs. *Automatica* **2018**, *89*, 125–134. [CrossRef]
26. Zemouche, A.; Boutayeb, M. On LMI conditions to design observers for Lipschitz nonlinear systems. *Automatica* **2013**, *49*, 585–591. [CrossRef]
27. Adil, A.; Hamaz, A.; N'Doye, I.; Zemouche, A.; Laleg-Kirati, T.M.; Bedouhene, F. On high-gain observer design for nonlinear systems with delayed output measurements. *Automatica* **2022**, *141*, 11281. [CrossRef]
28. Huang, H.; Huang, T.; Chen, X.; Qian, C. Exponential stabilization of delayed recurrent neural networks: A state estimation based approach. *Neural Netw.* **2013**, *48*, 153–157. [CrossRef] [PubMed]
29. Zhang, Z.; He, Y.; Zhang, C.; Wu, M. Exponential stabilization of neural networks with time-varying delay by periodically intermittent control. *Neurocomputing* **2016**, *207*, 469–475. [CrossRef]
30. Huang, H.; Feng, G. Delay-dependent h_∞ and generalized h_2 filtering for delayed neural networks. *IEEE Trans. Circuits Syst. Regul. Pap.* **2009**, *56*, 846–857. [CrossRef]

31. Gonzalez, A. Improved results on stability analysis of time-varying delay systems via delay partitioning method and Finsler's lemma. *J. Frankl. Inst.* **2022**, *359*, 7632–7649. [CrossRef]
32. Moon, Y.S.; Park, P.; Kwon, W. H.; Lee, Y.S. Delay-dependent robust stabilization of uncertain state-delayed systems. *Int. J. Control* **2001**, *74*, 1447–1455. [CrossRef]
33. Gu, K.L.; Kharitonov, V.; Chen, J. *Stability of Time-Delay Systems*; Springer: Berlin, Germany, 2003.
34. Body, S.; Ghaoui, L.E.; Feron, E.; Balakrishnan, V. *Linear Matrix Inequalities in System and Control Theory*; SIAM: Philadelphia, PA, USA, 1994.
35. Khalil, H.K. *Nonlinear Systems*; Prentice-Hall: Englewood Cliffs, NJ, USA, 2002.

Mathematical and Computational Applications

Article

Modelization of Low-Cost Maneuvers for an Areostationary Preliminary Mission Design

Marta M. Sánchez-García [1], Gonzalo Barderas [1,2,*] and Pilar Romero [1,2]

1   U.D. Astronomía y Geodesia, Facultad de Matemáticas, Universidad Complutense de Madrid, Plaza de las Ciencias, 3, E-28040 Madrid, Spain; martsa08@ucm.es (M.M.S.-G.); pilar_romero@mat.ucm.es (P.R.)
2   Instituto de Matemática Interdisciplinar, Facultad de Matemáticas, Universidad Complutense de Madrid, Plaza de las Ciencias, 3, E-28040 Madrid, Spain
*   Correspondence: gonzalobm@mat.ucm.es

Abstract: The aim of this paper is to analyze the determination of interplanetary trajectories from Earth to Mars to evaluate the cost of the required impulse magnitudes for an areostationary orbiter mission design. Such analysis is first conducted by solving the Lambert orbital boundary value problem and studying the launch and arrival conditions for various date combinations. Then, genetic algorithms are applied to investigate the minimum-energy transfer orbit. Afterwards, an iterative procedure is used to determine the heliocentric elliptic transfer orbit that matches at the entry point of Mars's sphere of influence with an areocentric hyperbolic orbit imposing specific conditions on inclination and periapsis radius. Finally, the maneuvers needed to obtain an areostationary orbit are numerically computed for different objective condition values at the Mars entry point to evaluate an areostationary preliminary mission cost for further study and characterization. Results show that, for the dates of the minimum-energy Earth–Mars transfer trajectory, a low value for the maneuvers to achieve an areostationary orbit is obtained for an arrival hyperbola with the minimum possible inclination and a capture into an elliptical trajectory with a low periapsis radius and an apoapsis at the stationary orbit. For a 2026 mission with a TOF of 304 for the minimum-energy Earth–Mars transfer trajectory, for a capture with a periapsis of 300 km above the Mars surface the value achieved will be 2.083 km/s.

Keywords: areostationary mission planning; Earth–Mars transfer trajectories; hyperbolic orbit matching; Lambert problem

Citation: Sánchez-García, M.M.; Barderas, G.; Romero, P. Modelization of Low-Cost Maneuvers for an Areostationary Preliminary Mission Design. *Math. Comput. Appl.* **2023**, *28*, 105. https://doi.org/10.3390/mca28060105

Academic Editor: Gianluigi Rozza

Received: 20 September 2023
Revised: 15 October 2023
Accepted: 25 October 2023
Published: 27 October 2023

1. Introduction

Small relay satellites in areostationary orbit are considered the most efficient candidates to support the telecommunication needs in the 2020s [1–7]. Areostationary orbiters, like geostationary satellites for Earth [8,9], can provide continuous access at very high data rates to remotely supervise a significant population of probes and robotic missions on the Martian surface. The determination of transfer trajectories from Earth to Mars aimed at lowering costs in terms of impulses has become a key factor in mission planning, allowing for larger payloads to be transported at a minimum energy cost.

In this work, we analyze the design of an interplanetary Earth–Mars transfer to reach the areostationary orbit with the minimum impulsive maneuvers cost. Several authors have studied the optimization of interplanetary trajectories: in [10,11] transfer trajectories to the Moon and Jupiter, respectively, passing close to a Lagrangian point, are considered; in [12], a method is developed to obtain approximate near-optimal low-thrust interplanetary transfers using solar electric propulsion spacecraft; in [13], the optimization is performed with a cost function with variable coefficients; in [14], launch constraints are imposed for the optimization. We derive the heliocentric elliptic transfer characterizing the launch windows using an heuristic optimization method for determining an optimal time of flight

(TOF) that minimizes the characteristic energies [15,16]. We will analyze the sensibility of this parameter in the optimization of impulsive maneuvers.

The first step consists in solving the Lambert problem [17] for various combinations of departure and arrival dates. Departure characteristic energy and hyperbolic arrival velocity plots are usually examined to investigate possible transfer windows [18]. We use genetic algorithms [19] to simultaneously minimize these two key parameters within these launch windows, comparing their performance.

Then, we match this interplanetary transfer with an entry hyperbola around Mars. The classic patched conic problem has been used to achieve a continuous trajectory composed of the trajectory between two planets and the planetocentric trajectory [20–24]. We use the iterative procedure [25] with imposed conditions on the periapsis distance, the arrival hyperbolic inclination, and a fixed radius for the Mars sphere of influence (SOI), and analyze the changes in the B-plane [26] due to the variations in the arrival asymptote direction. This iterative procedure enables the evaluation of these selected parameters in order to minimize fuel consumption for planning an areostationary mission. Once the fully matched trajectory to arrive at Mars is obtained, the maneuvers necessary to capture the orbiter and to place it in the areostationary orbit are analyzed.

The paper is organized as follows. In Section 2, we describe the dynamical model of the minimum-energy launch window problem. The determination of the Earth–Mars transfer trajectory with imposed hyperbolic arrival trajectory conditions is presented in Section 3. Section 4 describes the maneuvers performed to capture the spacecraft into an areostationary orbit and the numerical simulations to evaluate these maneuvers for different conditions. Finally, in Section 5, we briefly summarize the main conclusions.

2. Minimum-Energy Launch Window for Earth–Mars Transfer Trajectories

We first analyze the determination of interplanetary trajectories from Earth to Mars by minimizing the required energy at Earth departure and Mars arrival. We assume point mass gravitational forces for Earth and Mars within their respective spheres of influence and an unperturbed Keplerian orbit around the Sun.

The key parameters commonly used [27,28] to analyze the Earth–Mars mission launch opportunities are the characteristic energy at departure from Earth, $C_3 = V_{\infty_E}^2$, the hyperbolic excess velocity to escape from Earth, V_{∞_E}, and Mars arrival hyperbolic excess velocity, V_{∞_M}. In order to obtain these two parameters, it is necessary to first solve the Lambert orbital boundary value problem for the heliocentric spacecraft position, $\mathbf{r}_{s_S}$,

$$\ddot{\mathbf{r}}_{s_S} = -\mu_S \frac{\mathbf{r}_{s_S}}{r_{s_S}^3}, \tag{1}$$

constrained by two points, P_1 and P_2, and an elapsed TOF, $t_2 = t_1 + \text{TOF}$, as illustrated in Figure 1,

$$\mathbf{r}_{s_S}(t_1) = \mathbf{r}_E, \tag{2}$$

$$\mathbf{r}_{s_S}(t_2) = \mathbf{r}_M, \tag{3}$$

where $\mathbf{r}_E$ and $\mathbf{r}_M$ are the heliocentric position vectors for the Earth at t_1 and Mars at t_2, respectively.

The solution to the Lambert problem results in an elliptic conic section connecting P_1 and P_2. We consider the shortcut solution that satisfies the boundary conditions, based on the iterative procedure, choosing the time transfer function introduced by Lancaster [29] as parameter for the iteration. The solution results in an elliptic conic section connecting P_1 and P_2, with departure and arrival velocities, $\mathbf{V}_{s_{S1}}$ and $\mathbf{V}_{s_{S2}}$, at t_1 and t_2, respectively.

For each Earth departure and Mars arrival combination of dates, $\mathbf{V}_{\infty_E}$ and $\mathbf{V}_{\infty_M}$ change as $\mathbf{V}_{s_{S1}}$ and $\mathbf{V}_{s_{S2}}$ change according to

$$\mathbf{V}_{\infty_E} = \mathbf{V}_{s_{S1}} - \mathbf{V}_{E_{S1}}, \tag{4}$$

$$\mathbf{V}_{\infty_M} = \mathbf{V}_{s_{S2}} - \mathbf{V}_{M_{S2}}, \tag{5}$$

where $\mathbf{V}_{E_{S1}}$ and $\mathbf{V}_{M_{S2}}$ are the heliocentric velocity vectors for the Earth at t_1 and Mars at t_2, respectively.

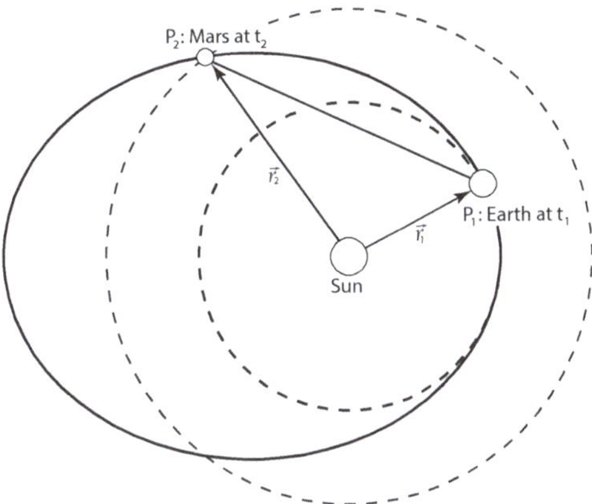

Figure 1. Lambert orbital boundary value problem and the heliocentric elliptic trajectory.

To analyze launch and arrival window opportunities, we first focus on data visualization [15,18] of the departure characteristic energy C_3 and the hyperbolic arrival velocity V_{∞_M} for various combinations of departure and arrival dates. We use Matlab software available from [30] to solve the Lambert problem, first obtaining a reduced launch window for the minimum-energy solution. The porkchop plots [31,32] shown in Figure 2 depict the contour lines of constant C_3 in km^2/s^2 and V_{∞_M} in km/s for the 2019–2029 departure and the 2020–2030 arrival time frames. It is possible to observe that the launch and arrival windows that give the minimum values approximately repeat every Mars synodic period of about 780 days. In more detail, Figure 3a shows the departure characteristic energy and hyperbolic arrival velocity contour plots for the Earth–Mars transfer covering 27 months, from 1 July 2019 to 1 November 2021, and Figure 3b presents these plots for the departure window on July 2020 and the arrival window on February 2021.

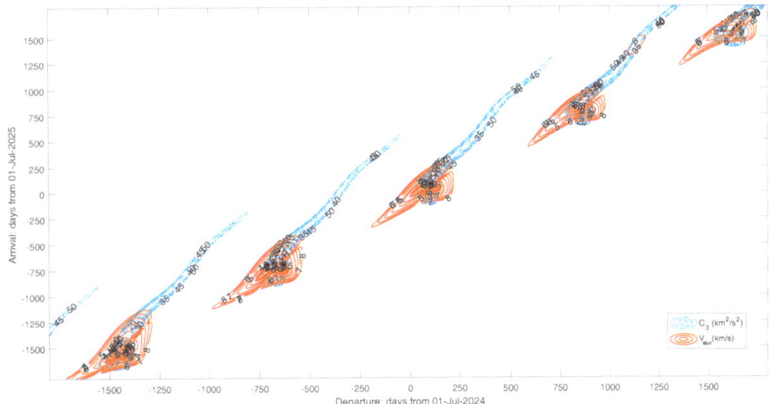

Figure 2. Departure characteristic energy and hyperbolic arrival velocity contour plots for the Earth–Mars transfer spanning 10 years.

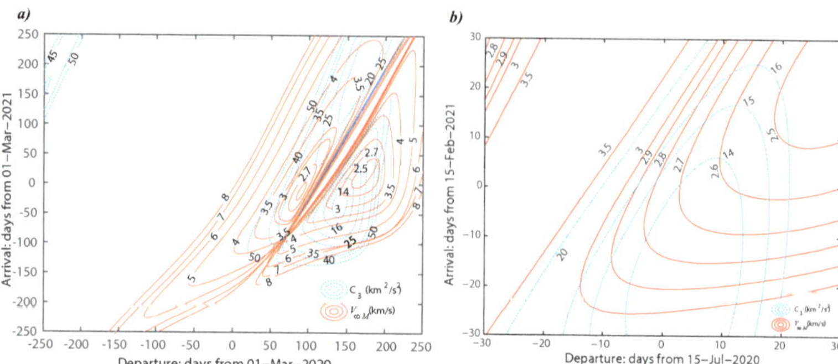

Figure 3. Departure characteristic energy and hyperbolic arrival velocity contour plots for the Earth–Mars transfer (**a**) from July 2019 to November 2021 and (**b**) from July 2020 to February 2021.

Now, we search for the solution minimizing the equation:

$$C = C_3 + V_{\infty M}. \tag{6}$$

The minimum C in Equation (6) tends to give lower values of the impulsive maneuvers required, first to obtain an Earth escape velocity, and after, at the Mars arriving hyperbolic orbit, to reduce the hyperbolic excess velocity to capture the probe.

To minimize Equation (6), applied to the reduced windows previously estimated from porkchop plots, we now use a Matlab optimizer [33] that includes a library dedicated to genetic algorithms with different implementations of selection and crossover functions (see [19]). In order to analyze the accuracy of the genetic algorithms when applied to this problem, we compare the performance of the *Remainder* and the *Stochastic Uniform* functions as selection functions to select the individuals that contribute to the population at the next generation, and the *Heuristic*, the *Scattered*, and the *Single point* rules as crossover functions to combine two individuals to form the next generation for populations of 100, 500, and 2000 individuals.

Table 1 summarizes the key results to compare the genetic algorithms performances. It can be concluded that the considered selection and crossover functions do not significantly change the results. CPU time depends on the population size, leading to equivalent results.

Moreover, the results in Table 1 are in agreement with the trajectories defined by the different missions launched in the year 2020. The *Mars 2020* mission (EEUU) [34] was launched on 30 July 2020, and its rover, the Perseverance rover, landed on 18 February 2021, with TOF = 203 days; The *Tianwen-1* mission (China) [35] was launched on 23 July 2020 and arrived in Mars on 10 February 2021, with a TOF of 202 days. The Emirates Mars Mission (UAE) [36] was launched on 19 July 2020, and arrived at the orbit around Mars on 9 February 2021, with a TOF of 205 days.

In order to analyze how a lower TOF impacts $V_{\infty E}$ and $V_{\infty M}$, we compare the heliocentric Earth–Mars optimal transfer orbit with the resulting orbit when a TOF 31 days less than the optimal minimum-energy trajectory is imposed. We choose a population of 500 individuals, the *Stochastic Uniform* function as the selection function, and the *Heuristic* as the crossover function. The resulting heliocentric elliptical orbits are illustrated in Figure 4, and their orbital elements are listed in Table 2. For the optimal minimum-energy orbit with a TOF of 197 days, departure on 20 July 2020 and arrival on 1 February 2021, we obtain values of $V_{\infty E} = 3.6361$ km/s and $V_{\infty M} = 2.7682$ km/s. When reducing the TOF to 166 days, departure and arrival dates change to 22 July 2020 and 4 January 2021, respectively, resulting in a more eccentric orbit, with $V_{\infty M} = 3.5888$ km/s significantly increased.

Table 1. Simulation scenarios in a launch window in 2020 to compare the genetic algorithm performances minimizing the cost function, $C = C_3 + V_{\infty_M}$, of the required C_3 energy and the arrival velocity, V_{∞_M}.

Pop.	Crossover	Selection	CPU Time (s)	Departure Time	TOF (Days)	C_3 (km^2/s^2)	V_{∞_M} (km/s)	C
				Departure Date 20 July 2020				
100	Heuristic	remainder	10.86	01:19:08	196.9397	13.2216	2.7681	15.9897
		stoch. unif.	10.36	01:05:27	196.9253	13.2212	2.7685	15.9897
	Scattered	remainder	10.30	01:04:01	196.9501	13.2213	2.7684	15.9897
		stoch. unif.	10.49	01:01:17	196.9281	13.2212	2.7685	15.9897
	Single pt	remainder	10.14	01:06:02	196.9657	13.2218	2.7679	15.9897
		stoch. unif.	10.35	01:14:40	196.9566	13.2217	2.7680	15.9897
500	Heuristic	remainder	48.30	01:06:27	196.9288	13.2212	2.7685	15.9897
		stoch. unif.	47.86	01:13:05	196.9420	13.2215	2.7682	15.9897
	Scattered	remainder	48.31	01:12:22	196.9420	13.2215	2.7682	15.9897
		stoch. unif.	47.60	01:21:17	196.9537	13.2218	2.7679	15.9897
	Single pt	remainder	47.32	01:24:10	196.9407	13.2217	2.7681	15.9898
		stoch. unif.	47.19	01:16:41	196.9450	13.2216	2.7681	15.9897
2000	Heuristic	remainder	188.06	01:04:09	196.9393	13.2214	2.7683	15.9897
		stoch. unif.	186.57	01:07:45	196.9332	13.2213	2.7684	15.9897
	Scattered	remainder	222.98	01:16:41	196.9450	13.2216	2.7681	15.9897
		stoch. unif.	229.03	01:10:38	196.9361	13.2214	2.7683	15.9897
	Single pt	remainder	238.38	01:08:54	196.9343	13.2214	2.7683	15.9897
		stoch. unif.	237.49	01:26:54	196.9480	13.2218	2.7679	15.9897

Table 2. Comparison of Earth–Mars optimal transfer heliocentric orbital parameters with respect to the mean ecliptic and equinox of J2000 for (a) the optimal launch and arrival dates in the window from 1 July 2019 to 1 November 2021 and (b) the launch and arrival dates in the window from 1 July 2019 to 1 November 2021 with the constraint of having 31 days of flight less than the optimal.

Parameter	(a)	(b)
Departure Date, t_1	20 July 2020 01:13:05	22 July 2020 15:20:14
Arrival Date, t_2	1 February 2021	4 January 2021
Arrival time	23:49:34 h	15:21:40 h
Semimajor axis, a_{s_S} (km)	198,312,598.97	202,972,264.04
Eccentricity, e_{s_S} (unitless)	0.23346	0.25130
Inclination, i_{s_S} (deg)	1.73626	0.72898
Ascending node long., Ω_{s_S} (deg)	297.4399	299.8067
Arg. of the perihelium, ω_{s_S} (deg)	359.6131	358.1402
True anomaly, v_{s_S} (deg)	0.4652	2.0420
V_{∞_E} (km/s)	3.6361	3.7803
V_{∞_M} (km/s)	2.7682	3.5888
Time of flight, TOF (days)	196.9420	166.0010
Total cost, C	15.9897	17.8795

For the analysis henceforth, we focus on the optimal orbit for a transference in the launch window of 2026. We consider the window with the earliest launch date on 1 March 2026 and latest arrival date on 1 November 2027. We consider the same parameters for the genetic algorithms as considered in the case of the launch window of 2020: a population of 500 individuals, the *Stochastic Uniform* function as the selection function, and the *Heuristic* as crossover function. The obtained trajectory has its launch date on 31 October 2026 and arrival date on 31 August 2027. Table 3 presents the heliocentric elliptic orbital elements for this transference. The results, particularly the higher TOF value with respect to 2020, are in good agreement with the mission analysis referenced for year 2026 in [28].

Table 3. Elliptic heliocentric orbital elements for the optimal launch and arrival dates in the window from 1 March 2026 to 1 November 2027.

Parameter	Value
Departure Date, t_1	31 October 2026 05:42:13 h
Arrival Date, t_2	31 August 2027 16:47:12 h
Semimajor axis, a_{s_S} (km)	189,961,652.992134
Eccentricity, e_{s_S} (unitless)	0.218496
Inclination, i_{s_S} (deg)	0.8695
Periapsis argument, ω_{s_S} (deg)	4.1768
Right ascension node longitude, Ω_{s_S} (deg)	37.5815
True anomaly, v_{s_S} (deg)	197.9132
V_{∞_E} (km/s)	3.0311
V_{∞_M} (km/s)	2.5913
Time of flight, TOF (days)	304.4618

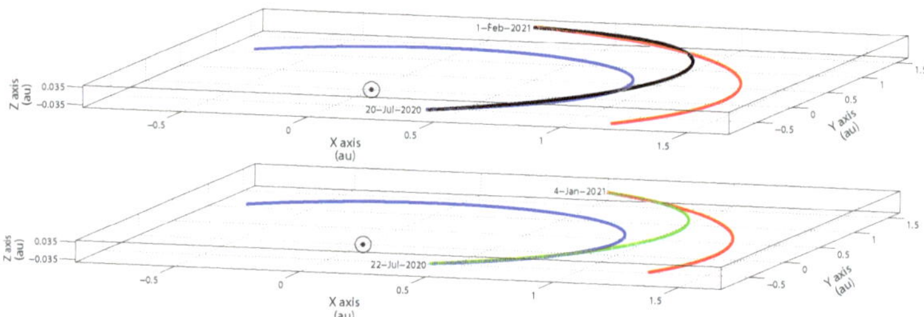

Figure 4. Minimum-energy optimal trajectory for the Earth–Mars transfer for departure date on 1 July 2019 and arrival date on 1 November 2021 (**top**) in comparison with a trajectory with 31 fewer days of flight (**bottom**).

3. Determination of Earth–Mars Trajectories with Hyperbolic Orbital Objective Values

Once the launch and arrival dates for the optimal minimum-energy solution have been determined, we move on to deal with the determination of the orbit for the entry point at the Mars SOI. To this end, we implement a procedure for matching the Earth–Mars elliptic transfer orbit and the Mars arrival hyperbolic orbit, fixing the periapsis distance, r_{ps_M}, the arrival hyperbolic inclination, i_{s_M}, and the radius for the SOI. We first present the determination of the arrival hyperbolic orbit and subsequent computation of the position and velocity at the matching point, which will be iterated with the Earth–Mars elliptic transfer afterwards. The goal is to analyze the changes in the B-plane [26] due to the variations in the arrival asymptote direction.

The iterative procedure used to compute the entry point at the SOI, in which both the heliocentric elliptic transfer orbit for the obtained TOF and the areocentric hyperbolic arrival orbit match, is formulated as follows:

$$\mathbf{r}_{s_M}^{(i+1)} = \mathbf{g}\left(\mathbf{f}\left(\mathbf{r}_{s_M}^{(i)}\right)\right). \tag{7}$$

The function $\mathbf{f}$ provides the areocentric ecliptic transfer velocity, $\mathbf{V}_{\infty_M}$, at the Mars SOI entry point, $\mathbf{r}_{s_M}$:

$$\mathbf{f}(\mathbf{r}_{s_M}^{(i)}) = \mathbf{V}_{\infty_M}^{(i)}, \tag{8}$$

obtaining first $\mathbf{V}_{s_{S_2}}$, by solving Lambert's problem (1) with a modified condition (3),

$$\mathbf{r}_{s_S}(t_2) = \mathbf{r}_M + \mathbf{r}_{s_M}, \tag{9}$$

and then using Equation (5).

The function $\mathbf{g}$ in (7) provides the position vector, $\mathbf{r}_{s_M}$, in the areocentric ecliptic reference frame:

$$\mathbf{g}(\mathbf{V}^{(i)}_{\infty_M}) = \mathbf{r}^{(i+1)}_{s_M}. \tag{10}$$

This function gathers a set of expressions to obtain $\mathbf{r}_{s_M}$ at the SOI for the objective values of i_{s_M} and r_{ps_M}, based on [22,24,25], fixing the radius of the SOI instead of the time that the spacecraft is inside the SOI, as in [22].

To this end, we first determine the areoequatorial coordinates (α, δ) of the arrival asymptote, given by $-\mathbf{V}_{s_M}$ (obtained by transforming $\mathbf{V}_{\infty_M}$ to the areocentric areoequatorial reference frame), to obtain the parameter σ as

$$\sigma = \arcsin \frac{\tan \delta}{\tan i_{s_M}}, \tag{11}$$

defining the minimum inclination as the value of $\|\delta\|$.

For a direct orbit, the right ascension of the ascending node, Ω_{s_M}, can be computed in two different ways:

$$\begin{aligned}
\Omega_{s_M} &= \alpha - \sigma & (V_{s_{Mz}} > 0), \tag{12}\\
\Omega_{s_M} &= \alpha + \sigma + \pi & (V_{s_{Mz}} < 0). \tag{13}
\end{aligned}$$

The value of V_{s_M} determines the semimajor axis, a_{s_M}. Then, the eccentricity of the hyperbola, e_{s_M}, for a periapsis radius, r_{ps_M}, is fixed as

$$e_{s_M} = 1 + \frac{r_{ps_M}}{a_{s_M}}. \tag{14}$$

The true anomaly, v_{s_M}, of the spacecraft at the SOI is determined using the standard procedure (i.e., [37,38]).

According to Figure 5, the unitary vectors defining the local reference system for the arrival asymptote, $\{\mathbf{u}_T, \mathbf{u}_B, \mathbf{u}_H\}$, are obtained as

$$\mathbf{u}_T = \frac{-\mathbf{V}_{s_M}}{\|\mathbf{V}_{s_M}\|}, \tag{15}$$

$$\mathbf{u}_H = \begin{bmatrix} \sin i_{s_M} \sin \Omega_{s_M} \\ -\sin i_{s_M} \cos \Omega_{s_M} \\ \cos i_{s_M} \end{bmatrix}, \tag{16}$$

$$\mathbf{u}_B = \mathbf{u}_H \times \mathbf{u}_T. \tag{17}$$

The plane containing $\mathbf{u}_B$, the planet center, and perpendicular to the arrival orbit is known as the B-plane, which is considered a fundamental tool in analyzing planetary arrivals [26]. With this iterative procedure, we update the B-plane according to the objective values imposed (see Figure 6).

The vectors (15)–(17) are then transformed to a local reference system in the periapsis, $\left\{\mathbf{u}_{R_p}, \mathbf{u}_{T_p}, \mathbf{u}_H\right\}$, according to

$$\begin{aligned}
\mathbf{u}_{R_p} &= -\sin \eta \, \mathbf{u}_T + \cos \eta \, \mathbf{u}_N, \tag{18}\\
\mathbf{u}_{T_p} &= -\cos \eta \, \mathbf{u}_T - \sin \eta \, \mathbf{u}_N, \tag{19}
\end{aligned}$$

where

$$\eta = \arcsin \frac{1}{e_{s_M}}. \tag{20}$$

Then, the areocentric areoequatorial position $\mathbf{R}_{s_M} = (X_{s_M}, Y_{s_M}, Z_{s_M})$ vector components at the SOI are determined in terms of the base vectors $\left\{ \mathbf{u}_{R_p}, \mathbf{u}_{T_p}, \mathbf{u}_H \right\}$, in a classical way, as

$$\mathbf{R}_{s_M} = r_{s_M} \left(\cos v_{s_M} \, \mathbf{u}_{R_p} + \sin v_{s_M} \, \mathbf{u}_{T_p} \right). \tag{21}$$

Finally, the position vector $\mathbf{R}_{s_M}$ is transformed to obtain $\mathbf{r}_{s_M}$ in the areocentric ecliptic reference frame, as a result of the function $\mathbf{g}$ in (10).

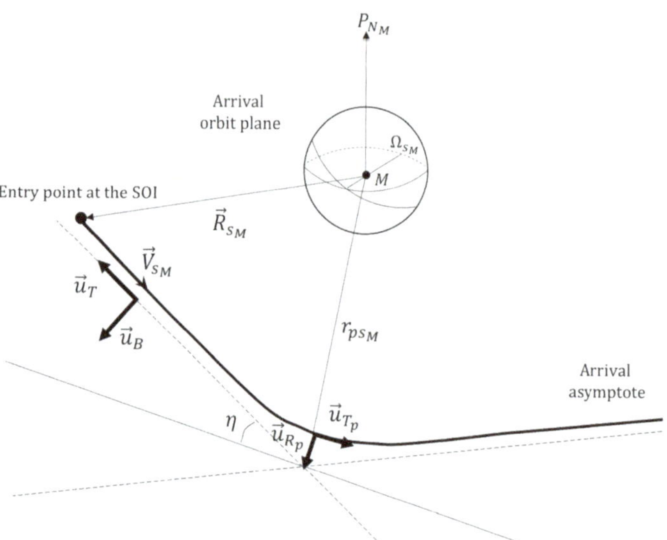

Figure 5. Local reference system vectors $\{\vec{u}_T, \vec{u}_B\}$ and $\left\{\vec{u}_{R_p}, \vec{u}_{T_p}\right\}$ in the arrival hyperbolic orbit plane defined by the normal vector $\vec{u}_H$.

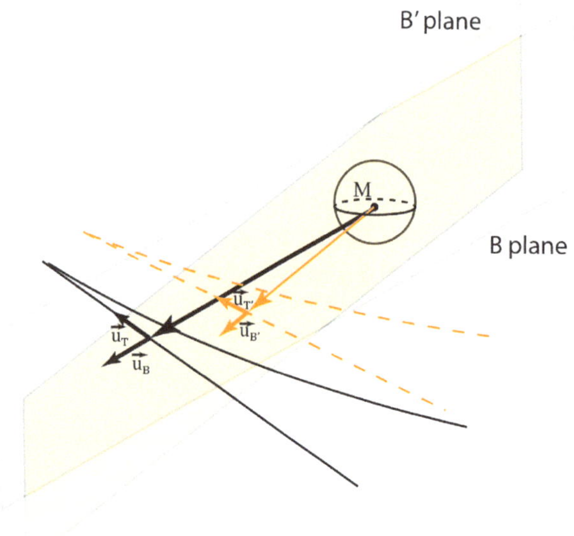

Figure 6. Changes in the B-plane orientation due to the arrival asymptotic velocity variations.

We consider that the iterative method defined by (7) converges, for a given tolerance τ, when the following condition is achieved:

$$\left\| \mathbf{r}_{s_M}^{(i+1)} - \mathbf{r}_{s_M}^{(i)} \right\| < \tau. \tag{22}$$

For comparison with Table 3, in Table 4, we present the results for the elliptic transfer orbit in the heliocentric ecliptic reference frame for the minimum-energy TOF (304.4618 days), the minimum arrival hyperbolic inclination ($i_{s_M min} = 16°.1167$), $r_{ps_M} = 20{,}428$ km, and a SOI of $r_{I_M} = 577{,}239$ km. We note that V_{∞_M} changes by 0.015 km/s approximately. Also, the resulting areocentric hyperbolic arrival trajectory orbital elements in the areoequatorial reference frame are shown in Table 4.

Table 4. Elliptic heliocentric orbital elements for (a) the Earth–Mars transfer orbit at the Mars sphere of influence entry point and final hyperbolic areocentric orbital elements and (b) launch date on 31 October 2026 and arrival date on 31 August 2027, for the objective values $i_{s_M} = i_{s_M min} = 16°.1167$ and $r_{ps_M} = 20{,}428$ km.

Orbit	Parameter	Units	Value
(a)	Departure date, t_1	(UT)	31 October 2026 05:42:13 h
	Arrival date, t_2	(UT)	31 August 2027 16:47:12 h
	Semimajor axis, a_{s_S}	(km)	189,905,238.422086
	Eccentricity, e_{s_S}	(unitless)	0.218286
	Inclination, i_{s_S}	(deg)	0.9311
	Periapsis argument, ω_{s_S}	(deg)	4.3094
	Right ascension node longitude, Ω_{s_S}	(deg)	37.5723
	True anomaly, v_{s_S}	(deg)	197.9302
	V_{∞_E}	(km/s)	3.0333
	V_{∞_M}	(km/s)	2.5763
	Time of flight, TOF	(days)	304.4618
(b)	Semimajor axis, a_{s_M}	(km)	6600.229103
	Eccentricity, e_{s_M}	(unitless)	4.0950441
	Inclination, i_{s_M}	(deg)	16.1167
	Periapsis argument, ω_{s_M}	(deg)	194.1344
	Right ascension node longitude, Ω_{s_M}	(deg)	162.7116
	True anomaly at t_2, v_{s_M}	(deg)	258.4533
	Periapsis radius, r_{ps_M}	(km)	20,467.9232
	Arrival at periapsis date, t_p	(UT)	31 August 2027 20:03:55

4. Mars Arrival Maneuvers Evaluation for an Areostationary Mission

We now conduct a preliminary evaluation of the total impulsive maneuvers ΔV_M needed to capture the probe and to place it in an areostationary orbit:

$$\Delta V_M = \Delta V_c + \Delta V_P + \Delta V_A + \Delta V_i, \tag{23}$$

where ΔV_c is the capture maneuver to avoid the probe leaving the SOI on a flyby trajectory; ΔV_P and ΔV_A are the two Hohmann transfer maneuvers, at the perigee and apogee, respectively, designed to obtain a transfer orbit from the capture orbit to the target areostationary orbit; and ΔV_i is the inclination correction maneuver to reach the final zero desired inclination.

If we consider a ΔV_c maneuver at the periapsis of the hyperbolic orbit in order to obtain a circular capture orbit, its magnitude would be

$$\Delta V_c = \sqrt{V_{\infty_M}^2 + 2\frac{\mu_M}{r_{ps_M}}} - \sqrt{\frac{\mu_M}{r_{ps_M}}}. \tag{24}$$

The hyperbolic periapsis distance, r_{ps_M}, is calculated as follows:

$$r_{ps_\text{M}} = a_{s_\text{M}}(e_{s_\text{M}} - 1), \tag{25}$$

with a_{s_M} as the hyperbolic semimajor axis and e_{s_M} as the hyperbolic eccentricity.

The two Hohmann transfer maneuvers required to achieve an orbit at the areostationary semimajor axis, $r_A = 20{,}428$ km, in the case that $r_{ps_\text{M}} < r_A$, are calculated as

$$\Delta V_P = \sqrt{\frac{\mu_M}{r_{ps_\text{M}}}}\left(\sqrt{\frac{2r_A}{r_{ps_\text{M}} + r_A}} - 1\right), \tag{26}$$

$$\Delta V_A = \sqrt{\frac{\mu_M}{r_A}}\left(1 - \sqrt{\frac{2r_{ps_\text{M}}}{r_{ps_\text{M}} + r_A}}\right). \tag{27}$$

Finally, the inclination maneuver is performed at the stationary distance in order to minimize its magnitude. For a maneuver at the node, this is obtained as

$$\Delta V_i = \sqrt{2\frac{\mu_M}{r_A} - 2\frac{\mu_M}{r_A}\cos i_{s_\text{M}}}, \tag{28}$$

where i_{s_M} is the hyperbolic inclination.

In the case that $r_{ps_\text{M}} > r_A$, the inclination maneuver at the node would be the first to be performed at the r_{ps_M} distance:

$$\Delta V_i = \sqrt{2\frac{\mu_M}{r_{ps_\text{M}}} - 2\frac{\mu_M}{r_{ps_\text{M}}}\cos i_{s_\text{M}}}. \tag{29}$$

Then, the two Hohmann maneuvers would be performed according to the following equations:

$$\Delta V_P = \sqrt{\frac{\mu_M}{r_{ps_\text{M}}}}\left(1 - \sqrt{\frac{2r_A}{r_{ps_\text{M}} + r_A}}\right), \tag{30}$$

$$\Delta V_A = \sqrt{\frac{\mu_M}{r_A}}\left(\sqrt{\frac{2r_{ps_\text{M}}}{r_{ps_\text{M}} + r_A}} - 1\right). \tag{31}$$

As can be seen from Equations (24) to (31), the periapsis distance, r_{ps_M}, and the orbital inclination of the arrival trajectory with respect to Mars, i_{s_M}, are the key design parameters in order to minimize the required impulses. Corrections to change the approach asymptotic plane are not considered, and neither are the combinations of the capture maneuver with the first Hohmann maneuver or the inclination maneuver with the second Hohmann maneuver.

Several numerical simulations are carried out to analyze the magnitude of these impulses due to the variations in the B-plane when different imposed values for TOF, i_{s_M}, and r_{ps_M} are considered.

We fix the launch window for 2026 according to the parameters given in Table 3 for the launch and arrival dates connecting Earth and Mars that minimize V_{∞_M}. In the first set of simulations, we consider different imposed objective values for the inclination, i_{s_M}, and the periapsis radius, r_{ps_M}, to determine the entry point at the SOI, using the iterative procedure defined in (7) to evaluate the total impulse ΔV_M in Equation (23). We consider different values for i_{s_M}, ranging in the interval $[i_{s_\text{M}min}, 90°)$, and for r_{ps_M}, in the interval [15,000 km, 25,000 km], with the arrival velocity, V_{∞_M}, recalculated for the different entry points.

Figure 7a,b represent the results obtained for V_{∞_M} with different values of r_{ps_M} and i_{s_M}, respectively. For the objective conditions of $i_{s_\text{M}} = 16°.1167$, which corresponds to the $i_{s_\text{M}min}$ for $r_{ps_\text{M}} = r_A = 20{,}428$ km, a value of $V_{\infty_\text{M}} = 2.5763$ km/s is obtained. As can be observed in Figure 7a for $i_{s_\text{M}min}$, V_{∞_M} decreases linearly when increasing r_{ps_M} from 2.5768 km/s to 2.5759 km/s. Figure 7b shows variations of about 2×10^{-3} km/s in V_{∞_M}, for the areostationary distance, as the objective inclination increases from $i_{s_\text{M}min}$ to $90°$. Table 5 summarizes the values for the C_3, V_{∞_M}, and total cost C of Equation (6) for the minimum and maximum values of r_{ps_M} and i_{s_M} considered.

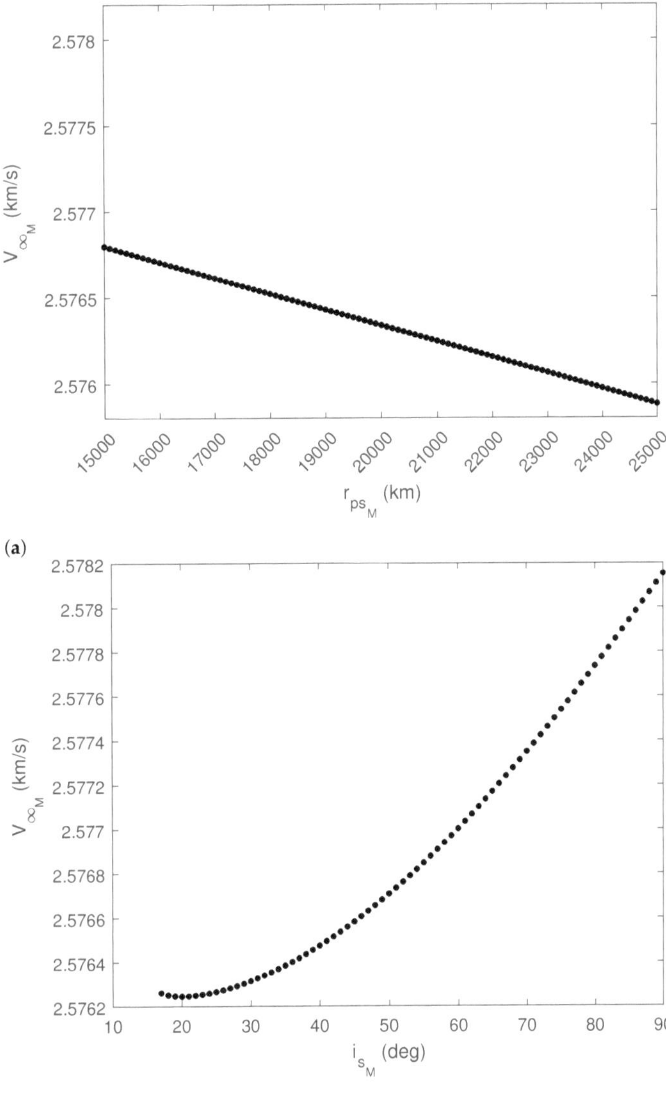

Figure 7. V_{∞_M} values for (**a**) the $i_{s_M min}$ for different periapsis radii, r_{ps_M}, ranging in the interval [15,000 km, 25,000 km] and (**b**) the objective values of $r_{ps_M} = r_A = 20{,}428$ km for different inclination values, i_{s_M}, ranging in the interval $\left[i_{s_M min} = 16°.1167, 90° \right]$.

Figure 8a,b represent the values of ΔV_c, ΔV_i, ΔV_A, ΔV_P, and the total ΔV_M for different values of r_{ps_M} and i_{s_M} to achieve the circular areostationary orbit at r_A with zero areoequatorial inclination. In the case that $r_{ps_M} < r_A$, the maneuvers are computed using (24) and (26)–(28). In the other case, the maneuvers are computed by means of (24) and (29)–(31).

Table 5. Interplanetary transfer energy variations for the considered combinations of i_{s_M} and the maximum and minimum r_{ps_M} and of $r_{ps_M} = r_A$ with the maximum and minimum i_{s_M}.

i_{s_M} (deg)	r_{ps_M} (km)	C_3 (km^2/s^2)	V_{∞_M} (km/s)	C
$i_{s_M min} = 16.1158$	15,000	9.2014	2.5768	11.7782
$i_{s_M min} = 16.1175$	25,000	9.2008	2.5759	11.7767
$i_{s_M min} = 16.1167$	$r_A = 20,428$	9.2011	2.5763	11.7774
90	$r_A = 20,428$	9.2123	2.5781	11.7904

As can be seen in Figure 8a, with the above-described strategy, the values for the capture maneuver ΔV_c change, for $i_{s_M min}$, from 1.8246 km/s for r_{ps_M} = 15,000 km to 1.8631 km/s for r_{ps_M} = 25,000 km, depending on the different objective periapsis radius considered. Obviously, the Hohmann maneuver vanishes for a capture at r_A, and its value varies from 0.2404 km/s for r_{ps_M} = 15,000 km to 0.1387 km/s for r_{ps_M} = 25,000. The inclination maneuver, ΔV_i, to achieve zero inclination, from the $i_{s_M min}$ for each r_{ps_M}, has a slight variation from 0.4062 km/s to 0.3673 km/s. The minimum for the total amount, ΔV_M, has a value of 2.2493 km/s, which corresponds to $r_{ps_M} = r_A$. This magnitude rises to 2.4712 km/s for r_{ps_M} = 15,000 km and to 2.3691 km/s for r_{ps_M} = 25,000 km.

Figure 8b shows that, for $r_{ps_M} = r_A$, with $\Delta V_H = 0$ km/s, ΔV_M increases when different values for the objective hyperbolic inclination are considered. From the same minimum of $\Delta V_M = 2.2493$ km/s for $i_{s_M min} = 16°.1167$, the total quantity of impulses rises to 3.8922 km/s, as illustrated in Figure 8a.

In a second set of simulations, we consider a different strategy to evaluate the Oberth effect that, due to the potential energy for a capture near the Mars surface, allows one to reduce the required ΔV_M. To this end, we consider a capture into an elliptical trajectory with a low periapsis radius and an apoapsis at the stationary orbit. Then, a circularization process is performed at the apoapsis. For a capture at a periapsis of $r_{ps_M} = r_M + 300$ km into an ellipse with an apoapsis of r_A, the value of the impulse required to obtain a circular orbit at the areostationary distance is decreased to 1.6764 km/s, with $\Delta V_c = 1.0294$ km/s and $\Delta V_A = 0.6470$ km/s. The total $\Delta V_M = 2.0834$ km/s reduces 0.1659 km/s with respect to the previous strategy.

Finally, we analyze the influence of the TOF in ΔV_M for a possible launch date delay. We consider up to two weeks delay from 31 October 2026 for the launch date and up to two months after 31 August 2027 for the arrival date for the minimum i_{s_M}. In the first case, a capture into a circular orbit of $r_{ps_M} = r_A$ is considered, as shown in Figure 9a. In the second case, a capture into an ellipse with $r_{ps_M} = r_M + 300$ and an apoapsis of r_A is considered, as shown in Figure 9b. It can be noticed, by comparing Figure 9a,b, that the second case is less sensitive to TOF variations. For case (a), a delay of 14 days at departure and 60 at arrival, increases the value of 2.25 km/s by 0.75 km/s, corresponding to the minimum-energy transfer TOF of 304 days, instead of an increase of 0.52 km/s from the value 2.08 km/s in case (b). In both cases, a launch delay of 14 days maintaining a TOF of 304 days would increase ΔV_M by 0.25 km/s.

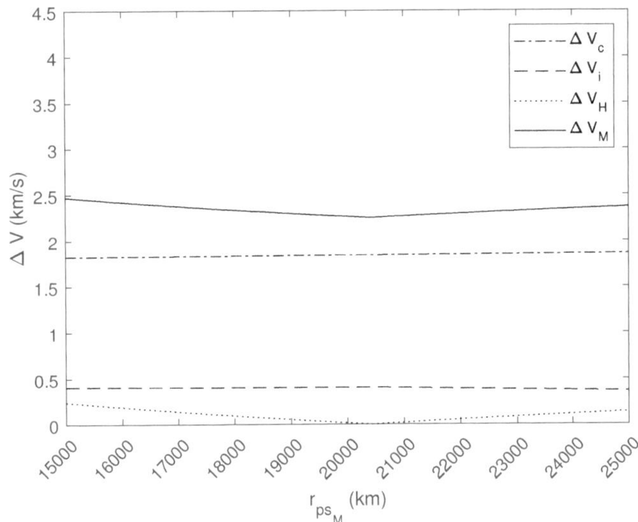

(a)

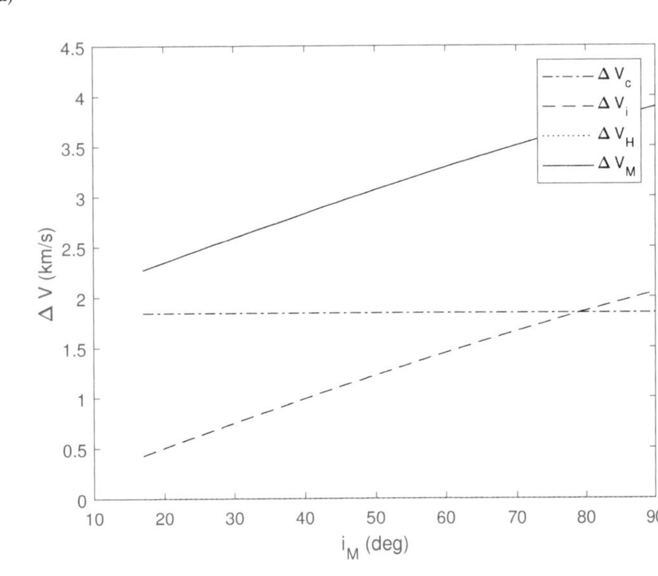

(b)

Figure 8. ΔV_c, ΔV_i, ΔV_A, ΔV_P, and the total ΔV_M values for (**a**) the $i_{s_M min}$ for different periapsis radii, r_{ps_M}, ranging in the interval [15,000 km, 25,000 km] and (**b**) the objective values of $r_{ps_M} = r_A = 20{,}428$ km for different inclinations, i_{s_M}, ranging in the interval $\left[i_{s_M min} = 16°\!.1167, 90° \right]$.

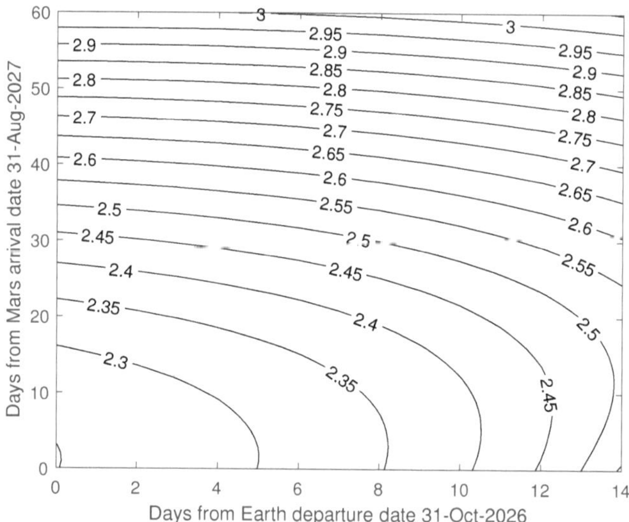

(**a**)

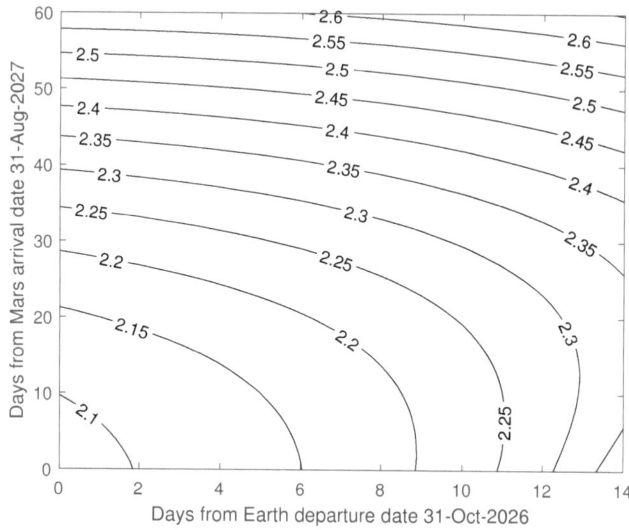

(**b**)

Figure 9. ΔV values for a combination of dates ranging from 31 October 2026 to 13 November 2026 for the launch date and from 31 August 2027 to 30 October 2027 for the arrival date, for the following objective conditions: (**a**) minimum i_{s_M} and $r_{ps_M} = r_A = 20{,}428$ km; (**b**) minimum i_{s_M} and a capture into an ellipse with a periapsis of $r_{ps_M} = r_M + 300$ km and an apoapsis of r_A.

5. Conclusions

In this paper, we conducted a preliminary analysis of the impulsive maneuvers cost to transfer a spacecraft from Earth to Mars and to position it in an areostationary orbit. We first obtained the minimum-energy interplanetary transfer from Earth to Mars by applying genetic algorithms to select launch and arrival dates. Several simulations were carried out to analyze the performance of the genetic algorithms. Results show that differences in the final energy cost and the time of flight are negligible, and the only significant change is the CPU time needed to converge, which is dependent on the population size.

With the optimized launch and arrival dates, an iterative procedure was used to match the interplanetary trajectory obtained with the genetic algorithms, with an entry hyperbola defined by imposing objective conditions for the inclination and the periapsis radius of the orbit. Two different strategies were computed to evaluate the cost of the mission, ΔV_M: The first includes a capture maneuver to a circular orbit at different periapsis radii, as well as Hohmann maneuvers and an inclination maneuver; The second includes a capture maneuver to an elliptic orbit with a low periapsis and an apoapsis at the areostationary orbit. Simulations with different imposed conditions on the entry hyperbola were conducted depending on the two key parameters: the hyperbolic inclination, i_{s_M}, and the periapsis radius, r_{ps_M}. For a circular capture at the stationary radius, results show that for a 2026 mission with a TOF of 304 days for the minimum-energy Earth–Mars transfer trajectory, the values achieved are $\Delta V_c = 1.84$, $\Delta V_H = 0$ and $\Delta V_i = 0.41$, being the total impulse $\Delta V_M = 2.25$ km/s for the minimum possible inclination $i_{s_M} = 16°1167$ and $r_{ps_M} = 20{,}428$ km corresponding to an areostationary radius. For an elliptical capture with a periapsis of $r_{ps_M} = r_M + 300$ km and an apoapsis of 20,428, the values achieved are $\Delta V_c = 1.03$, $\Delta V_A = 0.65$, and $\Delta V_i = 0.41$, the total impulse amounting to $\Delta V_M = 2.08$ km/s. A launch delay of two weeks would increase this minimum value of ΔV_M by 0.25 km/s in both cases. The capture into an ellipse with a delay of 14 days at departure and 60 at arrival is less sensitive to TOF variations, increasing the total ΔV_M by 0.52 km/s, instead of 0.75 for the direct circular capture.

Author Contributions: All authors contributed equally to this work. All authors have read and agreed to the published version of the manuscript.

Funding: This research received no external funding.

Data Availability Statement: No new data were created or analyzed in this study. Data sharing is not applicable to this article.

Conflicts of Interest: The authors declare no conflict of interest.

Abbreviations

The following abbreviations are used in this manuscript:

CPU	Central Processing Unit
SOI	Sphere Of Influence
TOF	Time Of Flight

References

1. Edwards, C.; Arnold, B.; DePaula, R.; Kazz, G.; Lee, C.; Noreen, G. Relay communications strategies for Mars exploration through 2020. *Acta Astronaut.* **2006**, *59*, 310–318. [CrossRef]
2. Edwards, C.; DePaula, R. Key telecommunications technologies for increasing data return for future Mars exploration. *Acta Astronaut.* **2007**, *61*, 131–138. [CrossRef]
3. Jentsch, C.; Rathke, A.; Wallner, O. Interplanetary communication: A review of future missions. In Proceedings of the 2009 International Workshop on Satellite and Space Communications, Siena, Italy, 9–11 September 2009; pp. 291–294. [CrossRef]
4. Podnar, G.; Dolan, J.; Elfes, A. Telesupervised robotic systems and the human exploration of Mars. *J. Cosmol.* **2010**, *12*, 4058–4067.
5. Romero, P.; Pablos, B.; Barderas, G. Analysis of orbit determination from Earth-based tracking for relay satellites in a perturbed areostationary orbit. *Acta Astronaut.* **2017**, *136*, 434–442. [CrossRef]
6. Montabone, L.; Heavens, N.; Babuscia, A.; Barba, N.; Battalio, J.; Bertrand, T.; Edwards, C.; Guzewich, S.; Kahre, M.; Kass, D.; et al. Observing Mars from Areostationary orbit: Benefits and applications. In Proceedings of the Mars Exploration Program Analysis Group (MEPAG) #38, Virtual, 15–17 April 2020.
7. Montabone, L.; Heavens, N.; Alvarellos, J.L.; Lillis, R.; Aye, M.; Liuzzi, G.; Babuscia, A.; Mischna, M.A.; Barba, N.; Newman, C.E.; et al. Observing Mars from Areostationary Orbit: Benefits and Applications. Available online: https://doi.org/10.13140/RG.2.2.21498.72643 (accessed on 20 September 2023).
8. Romero, P.; Gambi, J. Optimal control in the east/west station-keeping manoeuvres for geostationary satellites. *Aerosp. Sci. Technol.* **2004**, *8*, 729–734. [CrossRef]

9. Romero, P.; Gambi, J.; Patiño, E. Stationkeeping manoeuvres for geostationary satellites using feedback control techniques. *Aerosp. Sci. Technol.* **2007**, *11*, 229–237. [CrossRef]
10. Prado, A.; Broucke, R. Transfer orbits in the Earth-Moon system using a regularized model. *J. Guid. Control. Dyn.* **1996**, *19*, 929–933. [CrossRef]
11. Broucke, R.; Prado, A. Jupiter swing-by trajectories passing near the Earth. *Adv. Astronaut. Sci.* **1993**, *82*, 1159–1176.
12. Kluever, C. Efficient Computation of Optimal Interplanetary Trajectories Using Solar Electric Propulsion. *J. Guid. Control. Dyn.* **2014**, *38*, 5. [CrossRef]
13. Li, X.; Qiao, D.; Chen, H. Interplanetary transfer optimization using cost function with variable coefficients. *Astrodynamics* **2019**, *3*, 173–188. [CrossRef]
14. Chen, L.; Li, J. Optimization of Earth-Mars transfer trajectories with launch constraints. *Astrophys. Space Sci.* **2022**, *367*, 12. [CrossRef]
15. Woolley, R.; Whetsel, C. On the nature of Earth-Mars Porkchop plots. *Adv. Astronaut. Sci.* **2013**, *148*, 413–426.
16. Sanchez-Garcia, M.M.; Barderas, G.; Romero, P. Analysis of the optimization for an Earth to Mars areostationary mission. In Proceedings of the European Planetary Science Congress, Virtual, 21 September–9 October 2020; Volume 14, p. EPSC2020-134. [CrossRef]
17. Gooding, R.H. A procedure for the solution of Lambert's orbital boundary-value problem. *Celest. Mech. Dyn. Astron.* **1990**, *48*, 145–165. [CrossRef]
18. Conte, D. Survey of Earth-Mars trajectories using Lambert's Problem and Applications. Ph.D. Thesis, The Pennsylvania State University, State College, PA, USA, 2014.
19. Matlab. Genetic Algorithms Options, 2022. Available online: https://es.mathworks.com/help/gads/genetic-algorithm-options.html (accessed on 20 September 2023).
20. Clarke, V. C., Jr.; Bollman, W.E.; Feitis, P.H.; Roth, R.Y. *Design Parameters for Ballistic Interplanetary Trajectories, Part II: One-Way Transfers to Mercury and Jupiter*; Technical Report 32–77. JPL; US Gov.: Washington, DC, USA, 1966.
21. Cornelisse, J.W. Trajectory analysis for interplanetary missions. *ESA J.* **1978**, *2*, 131–144.
22. Parvathi, S.P.; Ramanan, R.V. Direct Transfer Trajectory Design Options for Interplanetary Orbiter Missions using an Iterative Patched Conic Method. *Adv. Space Res.* **2016**, *59*, 1763–1774. [CrossRef]
23. Parvathi, S.P.; Ramanan, R.V. Direct interplanetary trajectory design with a precise V-infinity targeting technique. In Proceedings of the 2017 First International Conference on Recent Advances in Aerospace Engineering (ICRAAE), Coimbatore, India, 3–4 March 2017; pp. 1–6. [CrossRef]
24. Iwabuchi, M.; Satoh, S.; Yamada, K. Smooth and continuous interplanetary trajectory design of spacecraft using iterative patched-conic method. *Acta Astronaut.* **2021**, *185*, 58–69. [CrossRef]
25. Bond, V.R. *Matched-Conic Solutions to Round-Trip Interplanetary Trajectory Problems That Insure State-Vector Continuity at All Boundaries*; Technical Note D-4942; NASA: Washington, DC, USA, 1969.
26. Farnocchia, D.; Eggl, S.; Chodas, P.; Giorgini, J.; Chesley, S. Planetary encounter analysis on the B-plane: A comprehensive formulation. *Celest. Mech. Dyn. Astron.* **2019**, *131*, 36. [CrossRef]
27. George, L.E.; Kos, L.D. *Interplanetary Mission Design Handbook: Earth-to-Mars Mission Opportunities and Mars-to-Earth Return Opportunities, 2009–2024*; Technical Report TM-1998-208533; NASA: Washington, DC, USA, 1998.
28. Burke, L.M.; Falck, R.D.; McGuire, M.L. *Interplanetary Mission Design Handbook: Earth-to-Mars Mission Opportunities 2026 to 2045*; Technical Report TM-2010-216764; NASA: Washington, DC, USA, 2010.
29. Lancaster, E.R.; Blanchard, R.C. *A Unified Form of Lambert's Theorem*; Technical Note D-5368; NASA: Washington, DC, USA, 1969.
30. Oldenhuis, R. Robust Solver for Lambert's Orbital-Boundary Value Problem, 2017. Available online: http://es.mathworks.com/matlabcentral/fileexchange/26348-robust-solver-for-lambert-s-orbital-boundary-value-problem?? (accessed on 20 September 2023).
31. Conte, D.; Di Carlo, M.; Ho, K.; Spencer, D.; Vasile, M. Earth-Mars transfers through Moon Distant Retrograde Orbits. *Acta Astronaut.* **2017**, *143*, 372–379. [CrossRef]
32. Conte, D.; Spencer, D. Mission Analysis for Earth to Mars-Phobos Distant Retrograde Orbits. *Acta Astronaut.* **2018**, *151*, 761–771. [CrossRef]
33. Trajectory Optimization Tool, v2.1.1. 2011. Available online: http://www.orbithangar.com/searchid.php?ID=5418 (accessed on 20 September 2023).
34. Farley, K.A.; Williford, K.H.; Stack, K.M.; Bhartia, R.; Chen, A.; de la Torre, M.; Hand, K.P.; Goreva, Y.; Herd, C.D.K.; Hueso, R.; et al. Mars 2020 Mission Overview. *Space Sci. Rev.* **2020**, *216*, 142. [CrossRef]
35. Jiang, X.; Yang, B.; Li, S. Overview of China's 2020 Mars mission design and navigation. *Astrodynamics* **2018**, *2*, 1–11. [CrossRef]
36. Sharaf, O.; Amiri, S.; AlDhafri, S.; Withnell, P.; Brain, D. Sending Hope to Mars. *Nat. Astron.* **2020**, *4*, 722. [CrossRef]
37. Capderou, M. *Satellites. Orbits and Missions*; Springer: Paris, France, 2005.
38. Montenbruck, O.; Gill, E. *Satellite Orbits: Models, Methods, and Applications*; Springer: Berlin/Heidelberg, Germany, 2000.

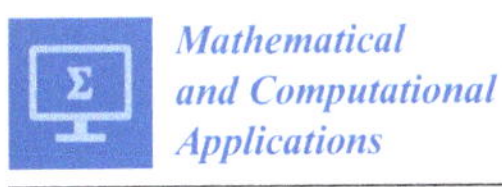
Mathematical and Computational Applications

MDPI

Article

Comparative Study of Metaheuristic Optimization of Convolutional Neural Networks Applied to Face Mask Classification

Patricia Melin, Daniela Sánchez, Martha Pulido and Oscar Castillo *

Tijuana Institute of Technology, TecNM, Calzada Tecnológico S/N, Fracc. Tomas Aquino, Tijuana CP 22379, BC, Mexico; pmelin@tectijuana.mx (P.M.); daniela.sanchez@tectijuana.edu.mx (D.S.); martha.pulido@tectijuana.edu.mx (M.P.)
* Correspondence: ocastillo@tectijuana.mx

Abstract: The preventive measures taken to curb the spread of COVID-19 have emphasized the importance of wearing face masks to prevent potential infection with serious diseases during daily activities or for medical professionals working in hospitals. Due to the mandatory use of face masks, various methods employing artificial intelligence and deep learning have emerged to detect whether individuals are wearing masks. In this paper, we utilized convolutional neural networks (CNNs) to classify the use of face masks into three categories: no mask, incorrect mask, and proper mask. Establishing the appropriate CNN architecture can be a demanding task. This study compares four swarm intelligent metaheuristics: particle swarm optimization (PSO), grey wolf optimizer (GWO), bat algorithm (BA), and whale optimization algorithm (WOA). The CNN architecture design involves determining the essential hyperparameters of the CNNs. The results indicate the effectiveness of the PSO and BA in achieving an accuracy of 100% when using 10% of the images for testing. Meanwhile, when 90% of the images were used for testing, the results were as follows: PSO 97.15%, WOA 97.14%, BA 97.23%, and GWO 97.18%. These statistically significant differences demonstrate that the BA allows better results than the other metaheuristics analyzed in this study.

Keywords: face mask classification; swarm intelligence metaheuristics; convolutional neural network; particle swarm optimization; whale optimization algorithm; bat algorithm; grey wolf optimizer

Citation: Melin, P.; Sánchez, D.; Pulido, M.; Castillo, O. Comparative Study of Metaheuristic Optimization of Convolutional Neural Networks Applied to Face Mask Classification. *Math. Comput. Appl.* **2023**, *28*, 107. https://doi.org/10.3390/mca28060107

Academic Editor: Oliver Schütze

Received: 18 August 2023
Revised: 27 October 2023
Accepted: 27 October 2023
Published: 1 November 2023

1. Introduction

The COVID-19 pandemic has shown the importance of using face masks, avoiding the spread of the virus, and preventing the infection of millions of people [1,2]. However, it is important to mention that various studies on its use were performed several years before the COVID-19 pandemic, where the importance and efficacy of its use to prevent other respiratory infections were demonstrated [3,4]. Two of the most widely used subsets of artificial intelligence related to face masks are deep learning (DL) and machine learning (ML). Different works on the detection of the facial mask using pre-trained models of convolutional neural networks can be found in [5–7], which allowed us to observe the potential of this technique in the detection and classification of facial masks [8–10]. In Ref. [11], the authors studied the architectures of different pre-trained models such as EfficientNet, InceptionV3, MobileNetV1, MobileNetV2, ResNet-101, ResNet-50, VGG16, and VGG19. Based on their study, they proposed a model for face mask detection based on MobileNetV2, applying data augmentation techniques to increase the number of images for the training phase. In Ref. [12], an application for mobile devices was developed to identify face masks using the Google Cloud ML API, while analyzing the progress of cloud technology and the benefits of machine learning. In Ref. [10], the authors developed five ML models for face mask classification. The developed models were Naïve Bayes (NB), Support Vector Machine (SVM), Decision Tree (DT), Random Forest (RF), and K-Nearest Neighbors (KNN).

The test of the models was performed using 1222 images, where the results demonstrated the effectiveness of the DT over the other models. The use of neural networks is related to metaheuristics, which are utilized to find the optimal architectures that improve the results depending on the application for which the network is used [13]. Metaheuristics are a great option for finding optimal parameters in applications in different areas. These algorithms have been classified according to their inspiration: based on evolutionary algorithms, physics-based algorithms, and algorithms based on swarm intelligence [14–16]. Nature-inspired algorithms are mainly inspired by collective behavior, where the main characteristics of a particular species are analyzed and represented in a computational way to be used in solving complex problems in the search for optimal solutions [17,18]. In recent works, comparisons have been made between metaheuristics to compare the performances applied to find CNN hyperparameters. Some of these metaheuristics are grey wolf optimizer (GWO), whale optimization algorithm (WOA), salp swarm algorithm (SSA), sine cosine algorithm (SCA), multiverse optimizer (MVO), particle swarm optimization (PSO), moth flame optimization (MFO), and bat algorithm (BA), to mention a few. The authors have concluded the advantages of combining convolutional neural networks and metaheuristics for the search of hyperparameters [19–21]. These techniques have been combined to solve applications related to pattern recognition [19,22,23], image classifications [18,24], and medical diagnosis [21,25,26], among other applications.

In this work, convolutional neural network hyperparameters are optimized by different nature-inspired algorithms [27–29]. The optimized hyperparameters are the number of convolutional layers, filters, fully connected layers, neurons, batch size, and epochs. The contribution of this work includes the optimal design of the convolutional neural network architectures to increase classification accuracy and its application to face mask classification: no mask, incorrect mask, and mask. Recent works applied to face mask classification based their model architectures on pre-trained models, which does not guarantee optimal architecture. As a novelty, this paper proposed optimizing CNN architectures instead of basing them on other architectures. The optimal hyperparameters are found using four metaheuristics used in recent works to make a statistical comparison and analysis, providing better accuracy for face mask classification.

This paper is presented as follows. In Section 2, the metaheuristics applied in this work are presented in a succinct manner. Section 3 shows the optimization proposed for the convolutional neural networks. The results obtained by each swarm intelligence metaheuristic are shown in Section 4. The statistical test results are shown in Section 5. The conclusions are presented in Section 6.

2. Background

The optimal design of architectures and models has allowed the realization of important practical applications. In Ref. [30], optimal convolutional neural network architectures was designed to identify various types of damage on reinforced concrete (RC) to avoid further structure deterioration. The results achieved show good accuracy of six types of damage. The design of convolutional neural network architectures using a particle swarm optimization algorithm was proposed and applied to sign language recognition using three study cases of sign language databases: the Mexican Sign Language alphabet, the American Sign Language MNIST, and the American Sign Language alphabet [31]. In Ref. [32], the authors proposed face detection and face classification by developing adaptive sailfish moth flame optimization (ASMFO) to the parameter optimization using a deep learning approach. In Ref. [19], the authors analyzed the importance of the CNN hyperparameters, such as filters, kernel, epoch, batch size, and pooling size of the convolutional neural networks applied to classify human movements. They compared seven metaheuristic algorithms: GWO, WOA, SSA, SCA, MVO, PSO, and MFO, concluding the advantages of the metaheuristics to optimize the hyperparameters of CNNs. The results led the authors to the conclusion that the implementation of GWO achieved higher accuracy than the other metaheuristics. In Ref. [20], the authors proposed a PSO to determine optimal hyperparameters of convolutional neural networks. They used the simplest CNN model as a base: LeNet. Their results achieved better results when the PSO designed the CNN architectures. The results achieved by their study were obtained using MNIST, Fashion-MNIST, and CIFAR-10 datasets.

In previous works [33,34], nature-inspired algorithms have optimized modular neural network architectures applied to human recognition using different biometric measures. In those works, comparisons using genetic algorithms (GAs) and swarm intelligence algorithms were performed, and significant evidence of the advantage of the swarm intelligence algorithms was proven. More recently, in [22], and based on the advantages offered by the swarm intelligence algorithms, the architecture of convolutional neural networks was optimized and applied to face recognition. In this work, algorithms such as particle swarm optimization and grey wolf optimizer offer advantages when designing convolutional neural network architectures. It is important to mention that the databases used for this work were small, with 400 and 165 images. In Ref. [35], the non-optimized design of convolutional neural network architectures applied to the facial mask classification was performed, and the best architecture was implemented in a real-time system using a Raspberry Pi 4 in combination with a camera to obtain the image in real time. The Raspberry Pi 4 sends a signal through its GPIO Board, and a result is provided by lighting an LED. If the mask is correctly used, the green LED is turned on. If the mask is incorrectly used, the yellow LED is turned on, and if a mask is not used, the red LED is turned on.

3. Intelligence Techniques

This section shows a description of the intelligence techniques utilized in this work.

3.1. Convolutional Neural Networks

Artificial neural networks (ANNs) are mainly based on the behavior of the human nervous system and its way of processing information. An artificial neural network is a type of distributive processor made up of simple processing units known as neurons, simulating two main aspects of the human brain: it acquires knowledge of its environment through a learning process and the use of synaptic weights to store the required knowledge [36,37]. Learning methods are categorized into supervised, semi-supervised, and unsupervised learning. Among the main properties that can be found in ANNs that make them one of the main techniques used in artificial intelligence, we can find their capacity for generalization, adaptation, learning, and parallelism [38,39]. Convolutional neural networks (CNNs) are an improvement of ANNs with some characteristics that make them powerful in applications where images are used. This type of network consists of other layers in addition to those already existing in conventional neural networks: the convolutional and the pooling layers. One of the advantages provided by this type of network is the extraction of features from the given images before proceeding to the learning phase, which makes it possible to reduce the amount of information that must be learned by the ANN [38]. In the convolutional layers (CLs), the inputs are multiplied by a filter with the size $m \times n$. Each layer contains a height, width, and depth. When talking about depth concerning the layer, it refers to the number of channels (primary colors) that contain the input images [40]. The most used grouping layers with the maximum, average, and minimum are responsible for grouping the feature map produced in the convolution layer, thus reducing the amount of information that will pass to the fully connected layers [23,41]. In Figure 1, a representation of a convolutional neural network is shown.

3.2. Nature-Inspired Algorithms

The nature-inspired algorithms used in this study are described below.

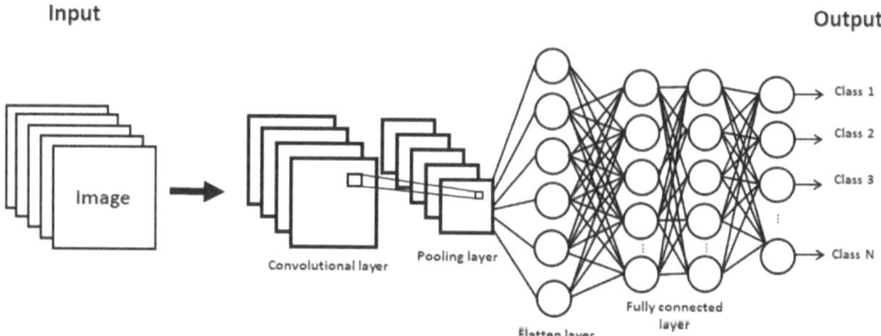

Figure 1. Representation of the architecture of a convolutional neural network.

3.2.1. Particle Swarm Optimization

In Ref. [42], the particle swarm optimization (PSO) based on the fish or bird social performance was proposed. A set of particles is known as a swarm, and each particle is a solution [43]. A particle defines their next position by Equation (1).

$$x_{id}(t+1) = x_{id}(t) + v_{id}(t+1) \tag{1}$$

where $x_{id}(t)$ indicates at time t, in the dimension d, the actual position of the particle i. A velocity $v_i(t+1)$ is designated to establish the next position. In Ref. [44], this algorithm was enhanced by adding the parameter: inertia weight (w). The particle velocity is defined by Equation (2).

$$v_{id}(t+1) = w \times v_{id}(t) + c_1 \times r_{1d}(t) \times [y_{id}(t) - x_{id}(t)] + c_2 \times r_{2d}(t) \times [\hat{y}_d(t) - x_{id}(t)] \tag{2}$$

where r_1 and r_2 are random values in [0, 1]. The best position of a particle i in dimension d is connoted by $y_{id}(t)$; the best position of the swarm in d dimension is denoted by $\hat{y}_d(t)$. c_1 and c_2 are the cognitive and social components.

w has a decreased value during the algorithm execution to allow exploitation and exploration. The linear decrease in the inertia weight applied in this work is given by Equation (3).

$$w_t = (w_s - w_e) \times \frac{(t_{max} - t)}{t_{max}} + w_e \tag{3}$$

where t_{max} denotes the maximum number of time steps, and w_s and w_e are the initial and final values of the inertia weight, respectively. The recommended values are $w_s = 0.9$ and $w_e = 0.4$ [45].

3.2.2. Grey Wolf Optimizer

In Ref. [46], the grey wolf optimizer (GWO) was proposed. This metaheuristic uses a dominant hierarchy applied by the wolves in hunting as inspiration. This dominant hierarchy is shown in Figure 2, where leaders known as alphas are at the top of the pyramid, and they make the main hunting and sleeping decisions. The betas are subaltern wolves that help the alpha wolves in making decisions. A delta wolf does not belong to any level already mentioned and can dominate only the lowest level. The delta wolves have different roles as scouts, sentinels, elders, hunters, and caretakers. The wolves in the lowest level are known as the omegas. They are always submitted by the wolves that are in the superior hierarchies [47,48].

The description of the principles used to define this algorithm and its mathematical representation is described below:

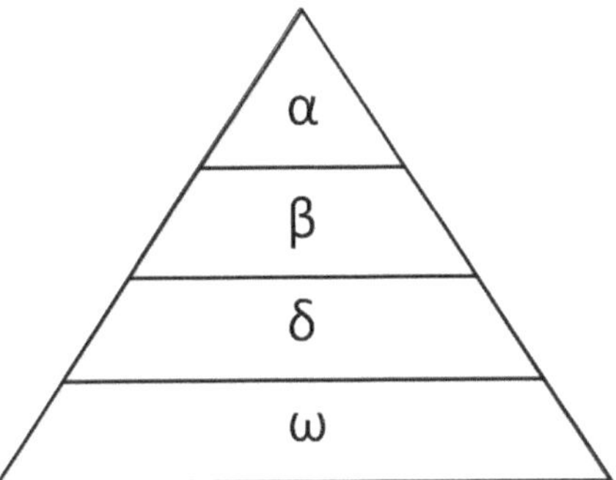

Figure 2. The dominance hierarchy of wolves.

- **Social hierarchy:** The three best solutions are alpha (α), beta (β), and delta (δ). The wolves belonging to the lowest level are the omegas (ω).
- **Encircling prey:** The process of prey encircling during hunting are represented by Equations (4) and (5).

$$\vec{D} = |\ \vec{C} \times \vec{X_p}\ (t) - \vec{X}\ (t)| \tag{4}$$

$$\vec{X}(t+1) = \vec{X_p}(t) - \vec{A} \times \vec{D} \tag{5}$$

where $\vec{X}$ denotes the agent position in the t iteration, and $\vec{X_p}$ represents the position of the prey. The coefficient vectors are $\vec{A}$ and $\vec{C}$. Equations (6) and (7) are used to determine their values.

$$\vec{A} = 2\vec{a} \times \vec{r_1} - \vec{a} \tag{6}$$

$$\vec{C} = 2 \times \vec{r_2} \tag{7}$$

where $\vec{r_1}$ and $\vec{r_2}$ represent vectors with random values in [0, 1]. During the algorithm execution, the vector $\vec{a}$ has linear decreasing values in [2, 0] given by Equation (8) [49].

$$\vec{a}(t) = 2 - \frac{2 \times t}{t_{max}} \tag{8}$$

where t denotes the current iteration, and t_{max} denotes the maximum number of iterations.

- **Hunting:** The first three levels in the dominant hierarchy know the prey position. With their positions, the wolves belonging to the lowest level (omega) can update their position using Equations (9)–(11).

$$\vec{D_\alpha} = \left|\ \vec{C_1} \times \vec{X_\alpha} - \vec{X}\right|, \quad \vec{D_\beta} = \left|\ \vec{C_2} \times \vec{X_\beta} - \vec{X}\right|, \quad \vec{D_\delta} = \left|\ \vec{C_3} \times \vec{X_\delta} - \vec{X}\right| \tag{9}$$

$$\vec{X_1} = \vec{X_\alpha} - \vec{A_1} \times \left(\vec{D_\alpha}\right), \quad \vec{X_2} = \vec{X_\beta} - \vec{A_2} \times \left(\vec{D_\beta}\right), \quad \vec{X_3} = \vec{X_\delta} - \vec{A_3} \times \left(\vec{D_\delta}\right) \tag{10}$$

$$\vec{X}(t+1) = \frac{\vec{X_1} + \vec{X_2} + \vec{X_3}}{3} \tag{11}$$

- **Attacking prey:** The process is also known as exploitation, where the current position of an agent and the prey allows it to establish the next position of the agent. This position is calculated using $\vec{a}$ and vector $\vec{A}$ with random values in an interval $[-2a, 2a]$.

- **Search for prey:** The process is also known as exploration, where vector $\vec{C}$ is used with values in $[0, 2]$ to provide diversity to the population and avoid local optimal.

3.2.3. Whale Optimization Algorithm

In Ref. [50], the whale optimization algorithm (WOA) was proposed. This algorithm uses as inspiration the hunting method applied by the whales. These marine mammals usually live in groups and are considered killers and predators [51]. One of the main characteristics shared with the grey wolf optimizer is the process of encircling prey, present also in WOA. The description of the processes used to define this algorithm based on humpback whales and its mathematical representation is described below:

- **Encircling prey:** The whales encircle the prey because they know its position. The whale closest to the prey becomes the best solution. Equations (3) and (4) allow the update of the position of the rest of the agents.

- **Bubble-net attacking method:** This process is also known as exploitation and is very similar to the one in the GWO, where the distance between the agent and the prey is determined. The process can be accomplished using two approaches:

    1. **Mechanism of shrinking encircling:** In Equation (5), the values of $\vec{a}$ decrease every iteration, and an interval $[-a, a]$ is used to generate random values for the vector $\vec{A}$.

    2. **Spiral updating position:** The helix-shaped movement of whales between the whale and prey position is mimicked by Equation (12).

$$\vec{X}(t+1) = \vec{D'} \times e^{bl} \times \cos(2\pi l) + \vec{X_p}(t) \tag{12}$$

where the distance between prey and whale is connoted by $\vec{D'}$, and b is a constant that represents the shape of the logarithmic spiral. A random value in an interval $[-1, 1]$ is represented by l.

- **Search for prey:** This process is also known as exploration, where the whales seek randomly based on the position of others. To force the exploration, the $\vec{A}$ vector has numbers less than -1 and greater than 1. The process is defined by Equations (13) and (14).

$$\vec{D} = \left| \vec{C} \times \vec{X_{rand}} - \vec{X} \right| \tag{13}$$

$$\vec{X}(t+1) = \vec{X_{rand}} - \vec{A} \times \vec{D} \tag{14}$$

where a random whale of the current iteration is represented by $\vec{X_{rand}}$. This random whale or the best solution found is utilized to help the other whales update their position.

3.2.4. Bat Algorithm

The bat algorithm (BA) [52] is based on their echolocation behavior due to the ability they have to identify their prey even in darkness. There are different types of bats depending on their size. Microbats have the characteristic of using a type of sonar known as echolocation, allowing them to detect prey and avoid obstacles. The bats make a loud sound pulse and listen for the echo that is reflected off of nearby objects. The rate of pulse is established in an interval $[0, 1]$ [53].

The author established some important rules to delimit the behavior and knowledge that the bats can have:

- Echolocation is used for all the bats to sense distance, and they know the difference between the prey and other kind of elements.
- To search for prey, each bat flies randomly in a position x_i with a velocity v_i. This task is performed by changing loudness A and wavelength λ. Depending on the closeness of its objective, the bat regulates the wavelength of its emitted pulses and regulate the rate of pulse emission $r \in [0,1]$.
- The loudness is assumed to be a large value positive number A to a minimum constant value A_{min}.

To define the update of position and velocities, the next equations are given by Equations (15)–(17).

$$f_i = f_{min} + (f_{max} + f_{min}) \times \beta \tag{15}$$

$$v_i(t) = v_i(t-1) + (x_i(t-1) - x_*) \times f_i \tag{16}$$

$$x_i(t) = v_i(t-1) + v_i(t) \tag{17}$$

where $x_i(t)$ and $v_i(t)$ represent the new position and velocity, respectively, at time step t. A vector with random values in [0, 1] is represented by β. The current global best solution is denoted by x_*. For the local search, the best solutions are used to select one of them and locally generate a new solution using a random walk given by Equation (18) [54].

$$x_{new} = x_{old} + \varepsilon \times A(t) \tag{18}$$

where ε represents a random value in an interval $[-1, 1]$, and the average loudness of all the bats at time step t is represented by $A(t)$.

4. Proposed Method

The proposed optimization is applied to face mask classification (no mask, incorrect mask, and mask). To perform this task, the method combines CNNs and optimization algorithms. The metaheuristics allow the optimal design of CNN architectures to be found. The optimization algorithm designs the CNN architectures, seeking their number of convolutional layers, filters, fully connected layers, neurons, batch size, and epochs. Each CL is followed by a max-pooling layer with a pool size of 3×3 to reduce image size. Figure 3 shows an example of the CNN architecture applied to face mask classification. As input to the convolutional neural network, images of people wearing (correctly and incorrectly) or not using face masks are used for the training phase of the convolutional neural network. The first layers of the convolutional neural network (convolutional and pooling) will extract features and reduce the image so that the fully connected layers learn the most relevant information. As output, when an image is simulated, a classification will be obtained (no mask, incorrect mask, and mask). A correct classification will depend on correct learning and the convolutional neural network architecture. For this reason, an optimization algorithm is an excellent option for designing the architecture because it allows a specific model to be applied to a particular application.

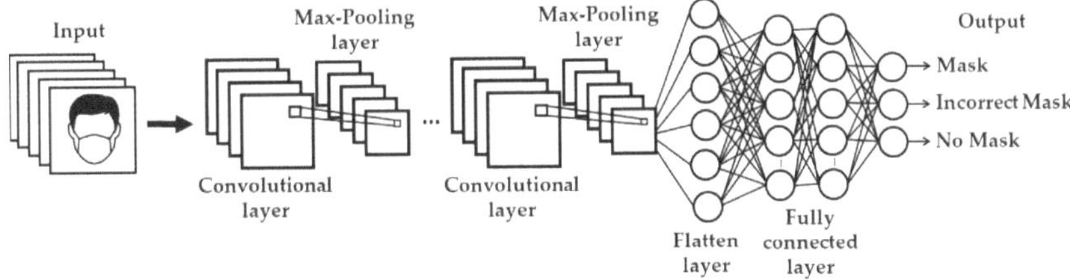

Figure 3. Illustration of the CNN architecture applied to face mask classification.

4.1. Description of the Optimization

The parameters used to execute any optimization algorithm have great importance because these depend on its performance. For each optimization algorithm, 10 solutions (particles, bats, or search agents) and 10 iterations are used. The configuration of the optimization algorithms used in this work is presented in more detail in Table 1. The parameters presented are based on previous works [22,33,55].

Table 1. Configuration of characteristic/tuning parameters of the optimization algorithms.

PSO		BAT		WOA and GWO	
Parameter	**Value**	**Parameter**	**Value**	**Parameter**	**Value**
Particles	10	Bats	10	Search Agents	10
Maximum Iterations (t_{max})	10	Maximum Iterations (t_{max})	10	Maximum Iterations (t_{max})	10
C_1	2	f_{min}	0	-	-
C_2	2	f_{max}	2	-	-
w_s	0.9	Loudness (A)	0.5	-	-
w_e	0.4	Pulse rate (r)	0.5	-	-

Each solution seeks to minimize the face mask classification error. In this work, the accuracy equation is used and given by Equation (19).

$$Accuraccy = \frac{TP + TN}{TP + FP + TN + FN} \tag{19}$$

where TP, TN, FP, and FN mean True Positive, True Negative, False Positive, and False Negative, respectively. The objective function used in this work is expressed by Equation (20).

$$f = 1 - \frac{TP + TN}{TP + FP + TN + FN} \tag{20}$$

The search space used for each solution (particle or agent) is determined by the minimum and maximum ranges shown in Table 2. These ranges are established based on previous works [18,22]. The convolutional neural networks use the Adaptive Moment Estimation (Adam) as a leaning algorithm and the rectified linear activation function (ReLU) as an activation function. The batch size is determined using a range from 1 up to 5, which means 8, 16, 32, 64, or 128.

Each particle or agent represents a solution, where each solution has 14 dimensions, which allow the creation of a CNN. In Figure 4, the dimensions of the solution are shown. The first four dimensions allow the determination of the number of convolutional layers, epoch, batch size, and the number of fully connected layers. Meanwhile, the rest of the dimensions allow us to determine the number of neurons and filters.

All the metaheuristics have, as a stopping criterion, 10 iterations or when the best solution has a cost equal to zero. The Keras Python package based on TensorFlow was used to implement the optimization algorithms and to build and train the CNN models.

Table 2. Definition of the search space to determine the solutions.

Hyperparameter		Minimum	Maximum
Convolutional layers (CLs)		1	5
Number of filters	CL 1	8	16
	CL 2	8	16
	CL 3	16	32
	CL 4	16	32
	CL 5	32	64
Fully connected layers (FCL)		1	5
Neurons		10	150
Epoch		5	50
Batch Size		1	5

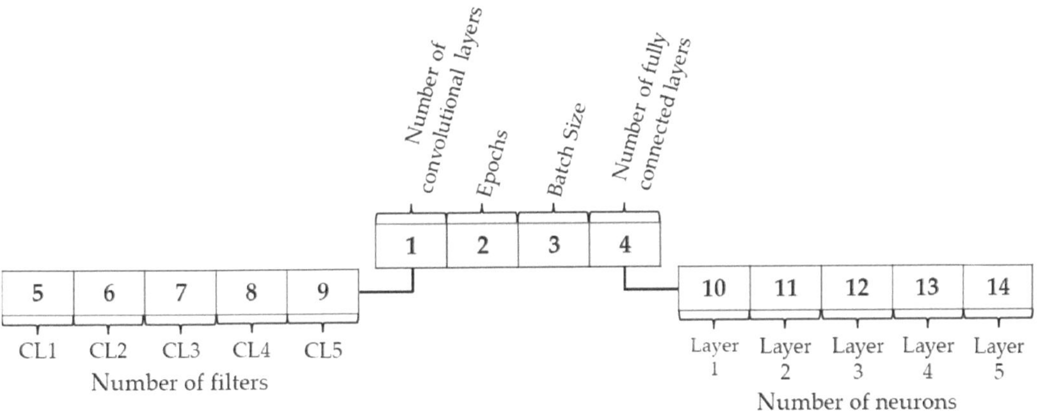

Figure 4. Dimensions of the solutions to design CNN architectures. CL indicates convolutional layer, and Layer indicates fully connected layer.

4.2. Database

To perform the face mask classification, the convolution neural networks are trained, validated, and tested using images of three classes (no mask, incorrect mask, and mask). The first two classes are obtained from the MaskedFace-Net dataset [56], and the no mask class is obtained from the Flickr-Faces-HQ Dataset (FFHQ) [57]. The MaskedFace-Net dataset consists of 137,016 images, and it is based on the Flickr-Faces-HQ (FFHQ) dataset. In this work, 3000 images were used, where each class contains 1000 images of the dataset. In Figure 5, a sample of the dataset is shown. The images used in this work were separated into training, validation, and testing. To help prevent bias in our models, when the images are split into sets, stratified sampling is utilized to guarantee a consistent distribution. Stratified sampling is a functionality provided by the Keras Python package.

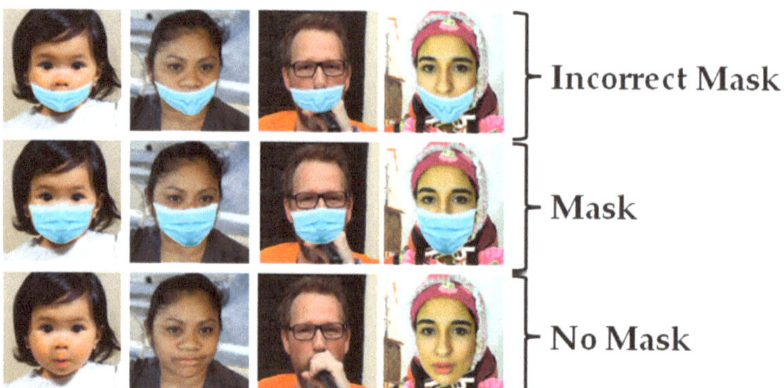

Figure 5. Examples of the database with 3 classes: incorrect, mask, and no mask.

4.3. Preprocessing

The original images have a resolution of 1024 × 1024 pixels. The region of interest (ROI) for this work is the face region, and it is automatically found using the Caffe model. The Caffe model was developed by the Berkeley Vision and Learning Center (BVLC). This model was trained to perform object detection and classification [58]. In Figure 6, an example of the face detection is shown.

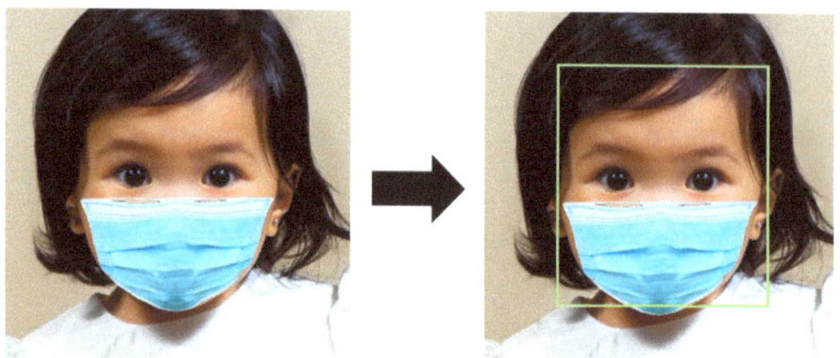

Figure 6. Application of the Caffe Model to detect the region of interest (face detection).

When the face region is detected, the image is resized to 100 × 100 pixels. Once the image is resized, an RGB subtraction technique is implemented to the ROI in order to help counteract slight variations [59]. In Figure 7, an example of the RGB subtraction technique is shown.

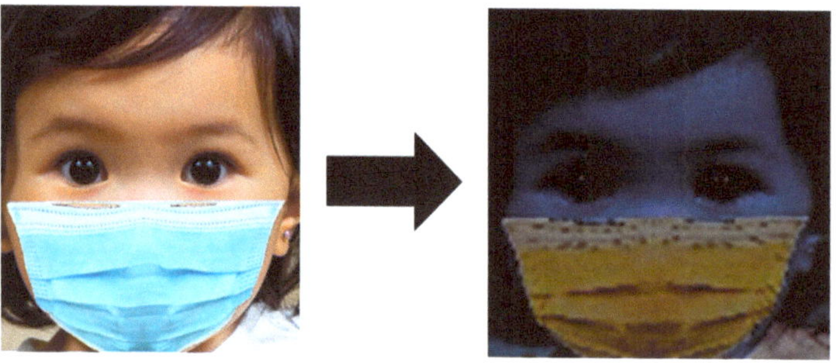

Figure 7. Example of the RGB subtraction applied to the ROI.

The proposed method is shown in Figure 8, which begins with the input images that go through preprocessing. The database is partitioned into three sets (training, validation, and testing), and the optimized CNN architecture is obtained with the metaheuristic.

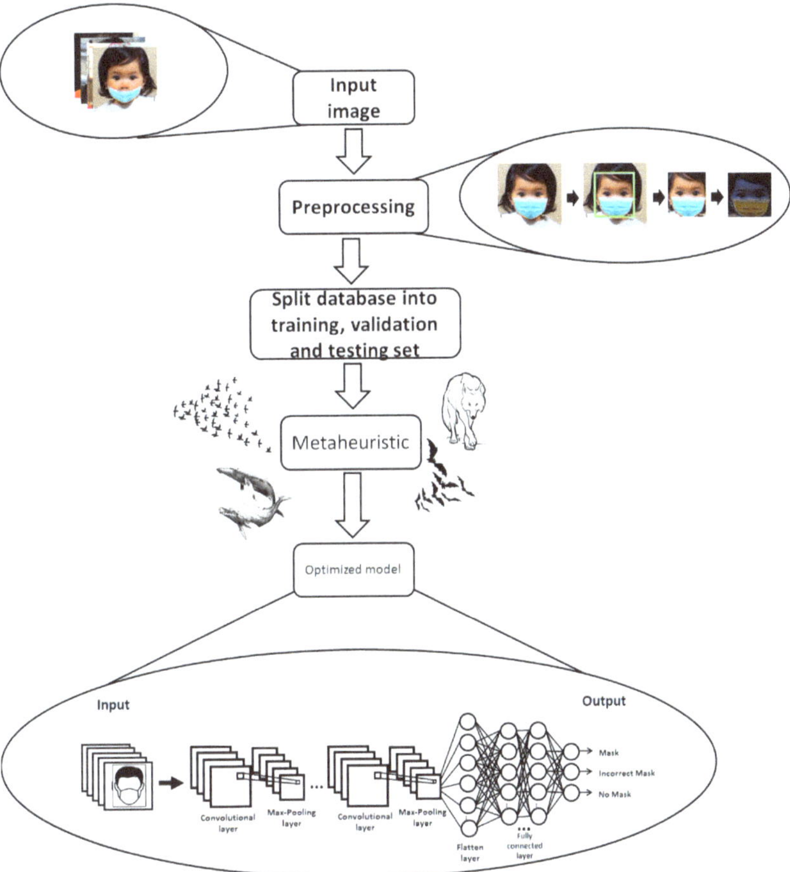

Figure 8. The flowchart of the proposed method begins with the input images up to the optimized CNN architecture using a metaheuristic.

5. Experimental Results

The database previously described is used to prove the proposed hyperparameters optimization. As previously mentioned, 3000 images were used to train, validate and test each convolutional neural network. In this work, 20 runs were performed using 10, 20, 30, 40, 50, 70, 80, and 90 percent of the images for testing, leaving the rest for training and validation. These experiments are performed with all the previously mentioned metaheuristics.

5.1. PSO Results

The best architectures achieved by the PSO with different percentages of images for the testing phase are summarized in Table 3. The best results are obtained with 10% and 20% of images for the testing phase, where an accuracy of 100% is achieved (marked with bold text in Table 3). We can define the best architecture as the one that uses less information for the training phase, which would be when 20% is used for the testing phase. This CNN model is structured as follows: four convolutional layers with 16, 16, 28, and 23 filters, with a size of 3×3. This architecture uses four FCLs with 150, 10, 117, and 19 neurons and a batch size of 8 with 12 epochs.

Table 3. The best accuracy results and architectures obtained by the PSO. CLs indicates the number of convolutional layers with their number of filters, and FCLs indicates the number of fully connected layers with their number of neurons.

% Images for Testing	CLs (Filters)	FCLs (Neurons)	Epoch	Batch Size	Error	Accuracy (%)
10	4 (12, 10, 17, 28)	3 (65, 40, 73)	12	32	0	100
20	4 (16, 16, 28, 23)	4 (150, 10, 117, 19)	12	8	0	100
30	3 (16, 11, 22)	3 (150, 10, 78)	20	8	0.0022	99.78
40	5 (8, 16, 16, 32, 64)	3 (10, 10, 10)	20	8	0.0017	99.83
50	4 (13, 12, 24, 32)	3 (99, 109, 54)	15	8	0.0033	99.67
60	4 (8, 16, 32, 32)	3 (150, 10, 150)	12	8	0.0056	99.44
70	4 (14, 16, 25, 23)	5 (104, 150, 10, 21, 50)	17	8	0.0067	99.33
80	4 (16, 8, 32, 18)	5 (150, 128, 10, 100, 10)	15	8	0.0125	98.75
90	1 (16)	5 (105, 150, 108, 100, 47)	19	8	0.0249	97.51

The results achieved by the PSO are shown in Table 4. The results illustrate how the accuracy (best and average) decreases as the percentage of images for the testing phase increases, and this occurs because the CNN is trained with less information.

Table 4. The best, average, and worst accuracy values obtained by the PSO.

Images (Testing) %	Best %	Average %	Worst %
10	-	100	-
20	100	99.66	99.17
30	99.78	99.59	99.22
40	99.83	99.51	99.25
50	99.67	99.49	99.20
60	99.44	99.17	98.72
70	99.33	98.72	98.00
80	98.75	98.08	97.54
90	97.51	97.15	96.14

5.2. WOA Results

In Table 5, the best architectures achieved by the WOA with different percentages of images for the testing phase are shown. The best results are also obtained with 10% and 20% of images for the testing phase, where an accuracy of 100% is achieved (marked with bold text in Table 5). The best architecture can be defined as the one that uses 20% for the testing phase. This CNN model is structured as five CLs with 16, 16, 32, 32, and 64 filters, with a size of 3×3 with five FCLs with 150, 88, 150, 100, and 50 neurons and a batch size of 32 with 20 epochs.

Table 5. The best accuracy results and architectures obtained by the WOA. CLs indicates the number of convolutional layers with their number of filters, and FCLs indicates the number of fully connected layers with their number of neurons.

% Images for Testing	CLs (Filters)	FCLs (Neurons)	Epoch	Batch Size	Error	Accuracy (%)
10	3 (9, 15, 21)	4 (77, 84, 83, 27)	19	8	0	100
20	5 (16, 16, 32, 32, 64)	5 (150, 88, 150, 100, 50)	20	32	0	100
30	4 (13, 16, 32, 27)	2 (150, 143)	20	8	0.0022	99.78
40	5 (16, 13, 32, 32, 46)	4 (150, 26, 136, 56)	20	8	0.0017	99.83
50	5 (16, 13, 32, 32, 64)	5 (150, 80, 150, 100, 50)	20	8	0.0033	99.67
60	5 (16, 12, 32, 32, 64)	5 (150, 137, 53, 55, 50)	20	16	0.0056	99.44
70	4 (16, 14, 30, 32)	4 (150, 150, 114, 26)	20	8	0.0067	99.33
80	5 (16, 9, 32, 32, 54)	3 (53, 150, 150)	20	8	0.0121	98.79
90	3 (14, 10, 23)	3 (11, 96, 102)	20	8	0.0223	97.77

The results achieved by the WOA is shown in Table 6. The results show how the accuracy (best and average) also decreases as the percentage of images for the testing phase increases, except the best result using 40% of the images in the testing phase, which is superior to the best value obtained using 30%.

Table 6. The best, average, and worst accuracy values obtained by the WOA.

Images (Testing) %	Best %	Average %	Worst %
10	100	99.92	99.33
20	100	99.76	99.50
30	99.78	99.53	99.11
40	99.83	99.46	99.17
50	99.67	99.48	99.27
60	99.44	98.94	98.27
70	99.33	98.76	98.14
80	98.79	97.94	97.24
90	97.77	97.14	96.51

5.3. BA Results

The best architectures achieved by the BA with different percentages of images for the testing phase are shown in Table 7. The best result is obtained with only 10% of images for the testing phase, where an accuracy of 100% is achieved (marked with bold text in Table 7). This CNN model is structured as follows: three CLs with 11, 10, and 28 filters, with a size of 3×3, three FCLs with 121, 61, and 63 neurons, and a batch size of 16 with 14 epochs.

This architecture uses less convolutional and fully connected layers than the previous ones, which also obtained 100% accuracy.

Table 7. The best accuracy results and architectures obtained by the BA. CLs indicates the number of convolutional layers with their number of filters, and FCLs indicates the number of fully connected layers with their number of neurons.

% Images for Testing	CLs (Filters)	FCLs (Neurons)	Epoch	Batch Size	Error	Accuracy (%)
10	3 (11, 10, 28)	3 (121, 61, 63)	14	16	0	100
20	3 (14, 15, 20)	3 (66, 69, 34)	15	8	0.0017	99.83
30	4 (14, 13, 17, 31)	4 (12, 43, 10, 75)	20	8	0.0022	99.78
40	4 (15, 8, 16, 16)	3 (150, 150, 10)	20	8	0.0033	99.67
50	4 (16, 15, 32, 24)	5 (150, 150, 150, 33, 28)	20	8	0.0027	99.73
60	4 (15, 16, 26, 32)	5 (35, 150, 50, 36, 10)	20	8	0.0050	99.50
70	5 (8, 8, 32, 32, 64)	5 (42, 150, 150, 100, 50)	20	8	0.0072	99.28
80	3 (16, 8, 32)	4 (50, 29, 150, 100)	20	8	0.0109	98.91
90	2 (12, 11)	4 (71, 75, 96, 25)	12	8	0.0245	97.55

Table 8 shows the results achieved by the BA. For this metaheuristic, the accuracy (best and average) also decreases as the percentage of images for the testing phase increases, except the best result using 50% of the images in the testing phase, which is superior to the best value obtained using 40%.

Table 8. The best, average, and worst accuracy values obtained by the BA.

Images (Testing) %	Best %	Average %	Worst %
10	-	100	-
20	99.83	99.72	99.50
30	99.78	99.54	99.22
40	99.67	99.47	99.00
50	99.73	99.53	99.33
60	99.50	99.23	99.05
70	99.28	98.89	98.33
80	98.91	98.16	97.70
90	97.55	97.23	96.84

5.4. GWO Results

Table 9 shows the best architectures achieved using the GWO with different percentages of images for the testing phase. The best results are obtained with 10% and 20% of images for the testing phase as PSO and WOA, where an accuracy of 100% is achieved (marked with bold text in Table 9). The best architecture can be defined as the one that uses 20% for the testing phase. This CNN model is structured in the following way: four convolutional layers with 10, 8, 23, and 25 filters, with a size of 3×3 with three FCLs with 42, 139, and 32 neuron and a batch size of 8 with 20 epochs.

Table 9. The best accuracy results and architectures obtained by the GWO. CLs indicates the number of convolutional layers with their number of filters, and FCLs indicates the number of fully connected layers with their number of neurons.

% Images for Testing	CLs (Filters)	FCLs (Neurons)	Epoch	Batch Size	Error	Accuracy (%)
10	4 (13, 8, 27, 24)	2 (122, 104)	17	32	0	100
20	4 (10, 8, 23, 25)	3 (42, 139, 32)	20	8	0	100
30	3 (9, 8, 22)	2 (38, 93)	10	8	0.0033	99.67
40	4 (16, 16, 16, 30)	4 (150, 67, 106, 10)	20	8	0.0025	99.75
50	5 (8, 9, 32, 19, 64)	3 (120, 81, 10)	20	8	0.0033	99.67
60	4 (9, 12, 16, 29)	5 (63, 10, 53, 15, 15)	20	8	0.0067	99.33
70	4 (8, 8, 26, 32)	3 (14, 102, 37)	16	8	0.0081	99.19
80	3 (16, 13)	1 (48)	15	8	0.0175	98.25
90	1 (15)	4 (107, 131, 117, 53)	11	8	0.0241	97.59

Table 10 shows the results obtained by the GWO. The accuracy decreases as with the other metaheuristics, but when 30% and 50% for the testing phase are used, the same result (the best value) is obtained. It is important to mention that when using 40%, the accuracy is better (the best value).

Table 10. The best, average, and worst accuracy values obtained by the GWO.

Images (Testing) %	Best %	Average %	Worst %
10	100	99.93	99.67
20	100	99.62	99.00
30	99.67	99.40	99.11
40	99.75	99.47	99.17
50	99.67	99.34	98.80
60	99.33	98.84	98.22
70	99.19	98.76	98.33
80	98.25	97.91	97.62
90	97.59	97.18	96.81

5.5. Comparison of Results

In Tables 3, 5, 7 and 9, the best architectures generated by each metaheuristic are presented, where it can be seen how the architectures can vary and still provide good results without using architectures as complex as those of the pre-trained models.

In Figure 9, the accuracy values (best, average, and worst) shown in Tables 4, 6, 8 and 10 are graphically shown. We can see that the PSO (Figure 9a) and BA (Figure 9c) always achieve an accuracy of 100% when 10% of the images are used for the testing phase (90% for training and validation). Meanwhile, the WOA and GWO only achieved the same value in some experiments using the same percentage of images. Using 50% of the percentage of images, we can see how the BA and PSO have very parallel values, which indicates that there is not much difference between their values (best, average, and worst), which could indicate greater stability between the results obtained in their experiments.

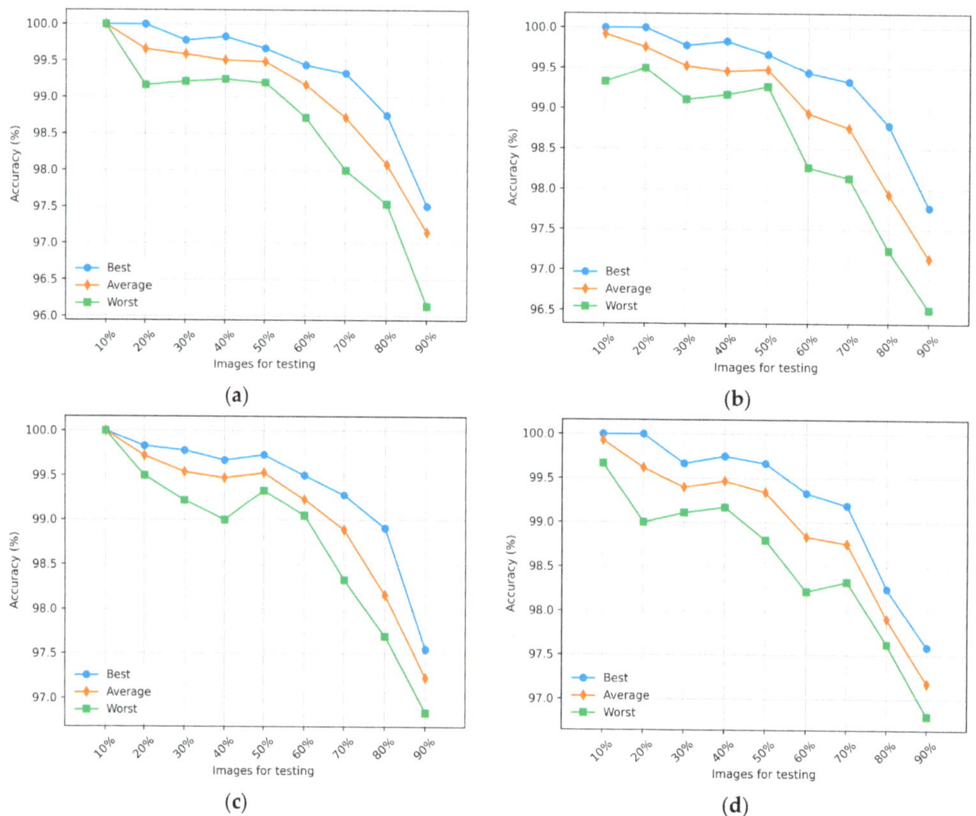

Figure 9. Accuracy values achieved by: (**a**) PSO; (**b**) WOA; (**c**) BA; (**d**) GWO using different percentages of images for the testing phase.

The average convergence during the learning phase of the 20 runs for each percentage of images (from 10 up to 90) obtained with each metaheuristic is depicted in Figure 10. It can be observed that when different percentages of images are used for the testing phase, the behavior of the PSO and BA is very similar. Even with only 10% of the images used for testing, both the PSO and BA achieve an error of 0 by iterations 6 and 4, respectively. Meanwhile, WOA and GWO exhibit similar behavior when 60%, 70%, 80%, and 90% of the images are used for the testing phase.

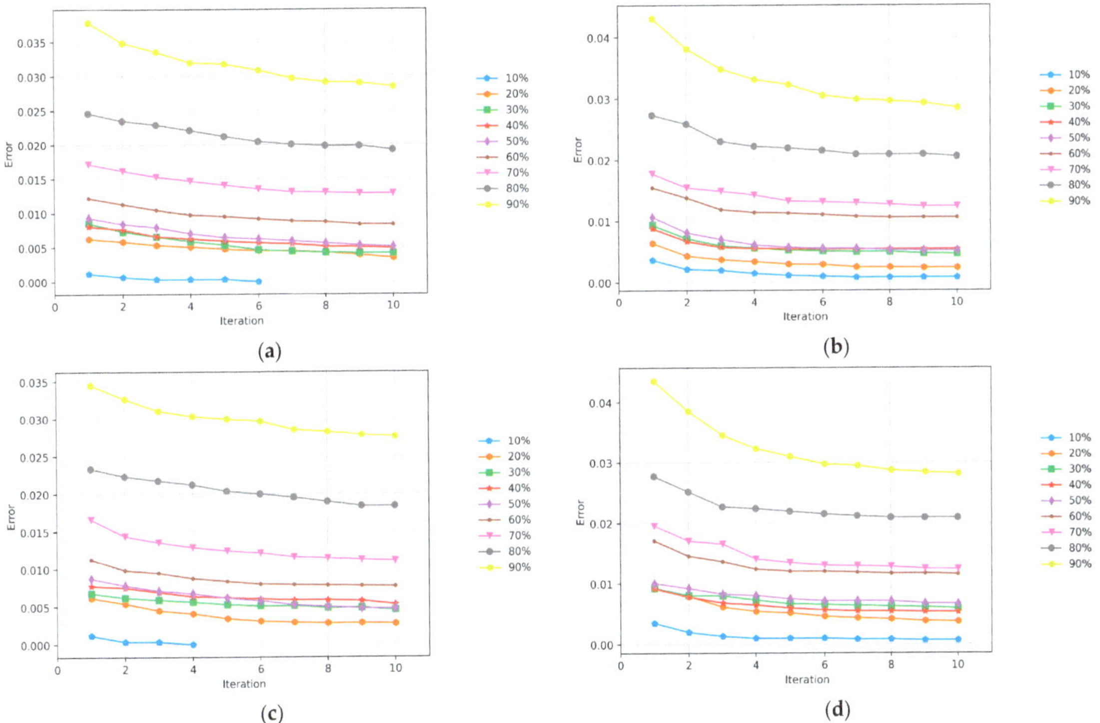

Figure 10. Convergence of accuracy error for: (**a**) PSO; (**b**) WOA; (**c**) BA; (**d**) GWO using different percentages of images for the testing phase.

The accuracy and loss curves with their respective validation of the best models are depicted in Figure 11. These models achieved an accuracy of 100%. The figure shows how the accuracy and loss have similar behavior to their validation.

Table 11 shows the averages (accuracy) obtained by each optimization algorithm. As results show, the average decreases when the percentage of images for testing increases, which means the CNN has less information to learn. Only two metaheuristics can achieve 100% accuracy: PSO and BA. Figure 12 shows graphically the accuracy achieved by the metaheuristics.

Table 11. Summary of accuracy results obtained by the metaheuristics.

Images (Testing) %	PSO %	WOA %	BA %	GWO %
10	100	99.92	100	99.93
20	99.66	99.76	99.72	99.62
30	99.59	99.53	99.54	99.40
40	99.51	99.46	99.47	99.47
50	99.49	99.48	99.53	99.34
60	99.17	98.94	99.23	98.84
70	98.72	98.76	98.89	98.76
80	98.08	97.94	98.16	97.91
90	97.15	97.14	97.23	97.18

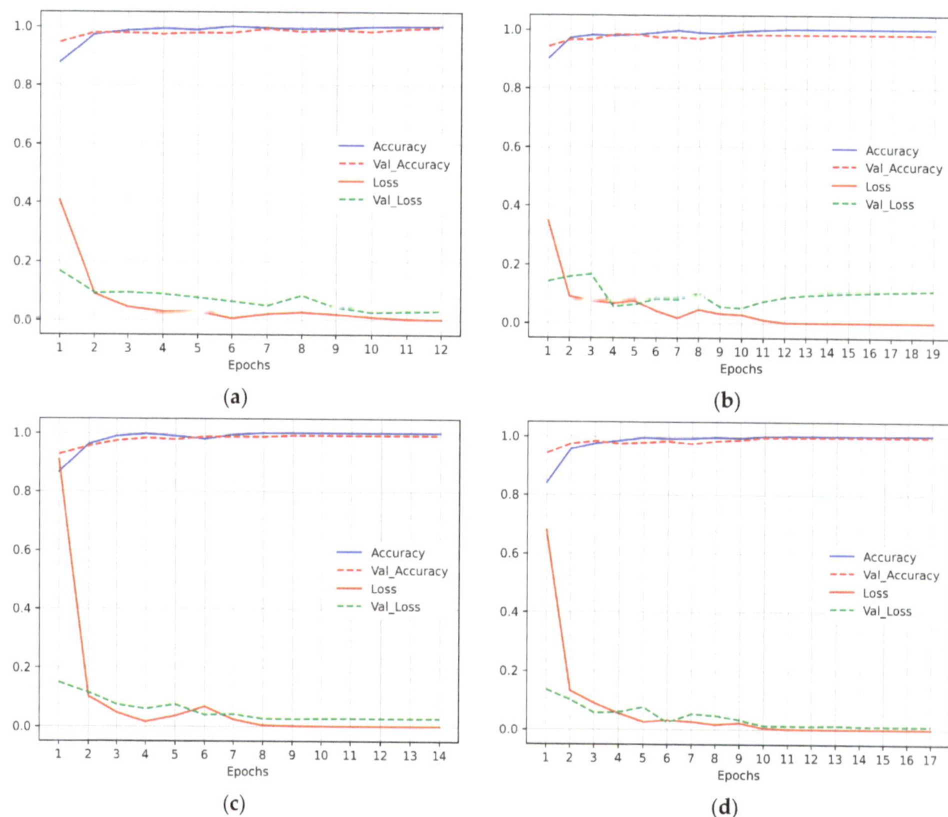

Figure 11. The accuracy and loss curves of the best models: (**a**) PSO; (**b**) WOA; (**c**) BA; (**d**) GWO.

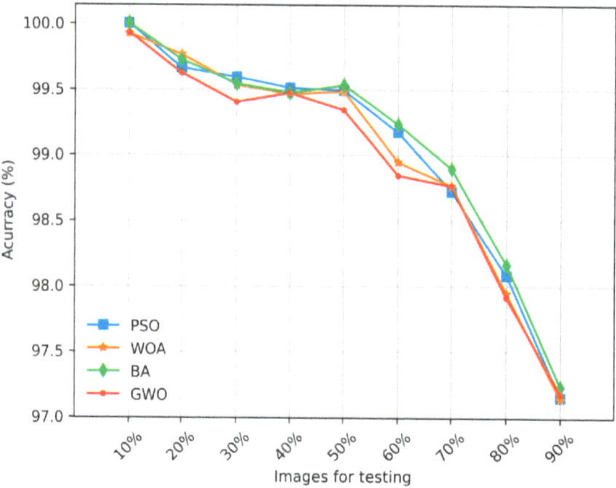

Figure 12. The average accuracy obtained by the metaheuristics.

The errors achieved by each metaheuristic are shown in Table 12. These results are utilized to perform statistical comparisons in the next section. These errors are graphically shown in Figure 13.

Table 12. Summary of error results obtained by the metaheuristics.

% Images (Testing)	PSO	WOA	BA	GWO
10	0	0.0008	0	0.0007
20	0.0034	0.0024	0.0028	0.0038
30	0.0041	0.0047	0.0046	0.006
40	0.0049	0.0054	0.0053	0.0053
50	0.0051	0.0052	0.0047	0.0066
60	0.0083	0.0106	0.0077	0.0116
70	0.0128	0.0124	0.0111	0.0124
80	0.0192	0.0206	0.0184	0.0209
90	0.0285	0.0286	0.0277	0.0282

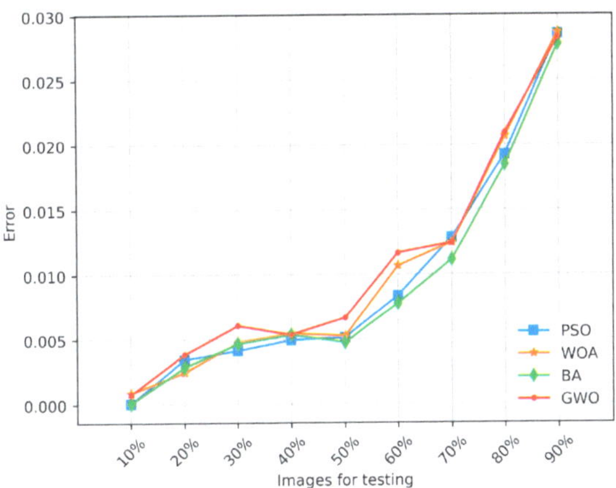

Figure 13. The accuracy error achieved by the metaheuristics.

Figures 12 and 13 graphically show the results obtained in this study. It can be seen that when a percentage between 10% and 50% is used for the testing phase, the PSO, WOA, and BA have similar good behavior. Meanwhile, when the percentage increases, it can be observed that the BA has a better accuracy, which means less error. In Table 13, the results achieved with the best average of accuracy are shown using other metrics (Recall, Precision, and F1 Score). The results show that the BA achieved better results in the other metrics, proving the effectiveness in metrics such as the F1 Score, where a combination of Recall and Precision is performed.

Table 13. Average results using Accuracy, Recall, Precision, and F1 Score.

Metric	PSO	WOA	BA	GWO
Accuracy	100	99.92	100	99.93
Recall	97.05	96.28	99.77	95.80
Precision	80.07	82.67	84.47	81.38
F1 Score	86.18	87.31	90.54	86.40

6. Statistical Comparison

This section shows statistical comparisons where the averages (errors) achieved by each optimization algorithm are used. In this work, the Wilcoxon signed-rank tests are utilized, where the value of α depends on the statistical significance. Table 14 shows the critical values with different statistical significance levels. A significance level of 0.10 is used in this work.

Table 14. Critical values for the Wilcoxon signed-rank test.

n	α		
	0.02	0.05	0.10
9	3	6	8

Table 15 shows the results of the statistical tests performed among all the metaheuristics. The null hypothesis assumes that means are equal, which contradicts the alternative hypothesis. The null hypothesis can be rejected if the column "W" value is equal to or smaller than the "W_0" based on the critical value with a 0.10 significance level. All possible comparisons were performed among the four metaheuristics studied in this work. The results exhibit a significant difference between the PSO and GWO. Meanwhile, the BA achieves significant differences against the other metaheuristics, allowing a better face mask classification.

Table 15. Summary of Wilcoxon test results.

Methods	Negative Sum (W−)	Positive Sum (W+)	Test Statistic (W)	Degrees of Freedom (m)	$W_0 = W\alpha,m$
BA PSO	41	3	3	9	8
BA WOA	41	3	3	9	8
BA GWO	44	0	0	9	8
PSO WOA	34	10	10	9	8
PSO GWO	39	5	5	9	8
WOA GWO	33	10	10	9	8

The results obtained with the method applying the bat algorithm allowed us to obtain better results, especially when less percentage is utilized for the training phase of the CNNs applied to face mask classification.

7. Conclusions

In this work, four swarm intelligence metaheuristics were applied to perform a comparison. A face mask database is used as a training, validation, and testing set to prove the proposed CNN design. This database has three classes: no mask, incorrect mask, and mask. The metaheuristics applied to CNN architecture design were PSO, WOA, BA, and GWO. These algorithms were implemented to CNN optimization applied to face mask multiclass classification, where hyperparameters of CNN were sought: the number of convolutional layers, filters, number of fully connected layers, neurons, batch size, and epoch. The results showed that some average convergences of the metaheuristics have a similar behavior when different percentages of images for the testing phase are utilized. The PSO and BA achieved an average of 100% accuracy when 10% of the images for the testing phase were used (leaving 90% for training and validation), but the BA converged faster than the PSO. The Wilcoxon signed-rank tests are utilized to compare results, and there is a statistical difference when the PSO and GWO are compared. However, when comparing the BA against PSO, WOA, and GWO, there is a statistical difference, which indicates that the BA allows for achieving better results than the other metaheuristics analyzed in this study when hyperparameters of convolutional neural networks are searched for the face mask classification. Results achieved in previous works and the results obtained in this work show that the performance of each optimization algorithm will depend on its application. In this work, only 3000 images were used, and different percentages of images were used for each phase to find optimal architectures with fewer images performing comparisons among swarm intelligence algorithms. The real implementation implies that optimized models have learned enough with the idea of not invading privacy and not having to train the models with specific persons. The optimized architectures could perform a correct face mask classification independently whether images of a person were used or not to train the model. The comparison performed in this work will allow us, as future work, to select those optimized architectures with a better percentage of accuracy and continue with the implementation in a real-time system. Although metaheuristics allow for optimal architectures with high accuracy, several limitations must be addressed in future works, such as the use of novel types of face masks not considered in this work, which would lead us to the need to evaluate their behavior. The dataset used to train and evaluate the architectures uses different face positions. However, it would be important to work with images with different kinds of illumination, especially for future work on implementing these architectures in real systems. Also, in future works, the comparison of these metaheuristics will be implemented by applying them to other intelligent techniques, such as fuzzy logic for parameter adjustment or fuzzy control.

Author Contributions: Methodology and validation, P.M.; Software, validation, writing, D.S.; Conceptualization, creation on main idea, writing—review and editing, O.C.; formal analysis, M.P. All authors have read and agreed to the published version of the manuscript.

Funding: This research received no external funding.

Data Availability Statement: Not applicable.

Acknowledgments: We would like to thank TecNM and Conacyt for their support during the realization of this research.

Conflicts of Interest: The authors declare no conflict of interest.

References

1. Eikenberry, S.; Mancuso, M.; Iboi, E.; Phan, T.; Eikenberry, K.; Kuang, Y.; Kostelich, E.; Gumel, A. To mask or not to mask: Modeling the potential for face mask use by the general public to curtail the COVID-19 pandemic. *Infect. Dis. Model.* **2020**, *5*, 293–308. [CrossRef] [PubMed]
2. Garcia Godoy, L.; Jones, A.; Anderson, T.; Fisher, C.; Seeley, K.; Beeson, E.; Zane, H.; Peterson, J.; Sullivan, P. Facial protection for healthcare workers during pandemics: A scoping review. *BMJ Glob. Health* **2020**, *5*, e002553. [CrossRef] [PubMed]
3. MacIntyre, C.; Cauchemez, S.; Dwyer, D.; Seale, H.; Cheung, P.; Browne, G.; Fasher, M.; Wood, J.; Gao, Z.; Booy, R.; et al. Face Mask Use and Control of Respiratory Virus Transmission in Households. *Emerg. Infect. Dis.* **2009**, *15*, 233–241. [CrossRef] [PubMed]
4. MacIntyre, C.; Chughtai, A.; Rahman, B.; Peng, Y.; Zhang, Y.; Seale, H.; Wang, X.; Wang, Q. The efficacy of medical masks and respirators against respiratory infection in healthcare workers. *Influenza Other Respir. Viruses* **2017**, *11*, 511–517. [CrossRef] [PubMed]
5. Pham-Hoang-Nam, A.; Le-Thi-Tuong, V.; Phung-Khanh, L.; Ly-Tu, N. Densely Populated Regions Face Masks Localization and Classification Using Deep Learning Models. In Proceedings of the Sixth International Conference on Research in Intelligent and Computing, Thủ Dầu Một, Vietnam, 3–4 June 2021.
6. Sethi, S.; Kathuria, M.; Kaushik, T. Face mask detection using deep learning: An approach to reduce risk of Coronavirus spread. *J. Biomed. Inform.* **2021**, *120*, 103848. [CrossRef]
7. Yu, J.; Zhang, W. Face Mask Wearing Detection Algorithm Based on Improved YOLO-v4. *Sensors* **2021**, *21*, 3263. [CrossRef]
8. Mar-Cupido, R.; Garcia, V.; Rivera, G.; Sánchez, J. Deep transfer learning for the recognition of types of face masks as a core measure to prevent the transmission of COVID-19. *Appl. Soft Comput.* **2022**, *125*, 109207. [CrossRef]
9. Umer, M.; Sadiq, S.; Alhebshi, R.; Alsubai, S.; Hejaili, A.; Eshmawi, A.; Nappi, M.; Ashraf, I. Face mask detection using deep convolutional neural network and multi-stage image processing. *Image Vis. Comput.* **2023**, *133*, 104657. [CrossRef]

10. Ramakrishnan, K.; Balakrishnan, V.; Wong, H.; Tay, S.; Soo, K.; Kiew, W. Face Mask Wearing Classification Using Machine Learning. *Eng. Proc.* **2023**, *41*, 13.
11. Habib, S.; Alsanea, M.; Aloraini, M.; Al-Rawashdeh, H.; Islam, M.; Khan, S. An Efficient and Effective Deep Learning-Based Model for Real-Time Face Mask Detection. *Sensors* **2022**, *22*, 2602. [CrossRef]
12. Wakchaure, A.; Kanawade, P.; Jawale, M.; William, P.; Pawar, A. Face Mask Detection in Realtime Environment using Machine Learning based Google Cloud. In Proceedings of the International Conference on Applied Artificial Intelligence and Computing, Salem, India, 9–11 May 2022.
13. Mirjalili, S. *Evolutionary Algorithms and Neural Networks: Theory and Applications*, 1st ed.; Springer: London, UK, 2019.
14. Du, K.; Swamy, M. *Search and Optimization by Metaheuristics: Techniques and Algorithms Inspired by Nature*, 1st ed.; Birkhäuser Cham: Berlín, Germany, 2018.
15. Hassanien, A.; Emary, E. *Swarm Intelligence: Principles, Advances and Applications*, 1st ed.; CRC Press: Boca Raton, FL, USA, 2015.
16. Iba, H. *AI and SWARM: Evolutionary Approach to Emergent Intelligence*, 1st ed.; CRC Press: Boca Raton, FL, USA, 2019.
17. Poma, Y.; Melin, P.; Gonzalez, C.; Martinez, G. Optimization of convolutional neural networks using the fuzzy gravitational search algorithm. *J. Autom. Mob. Robot. Intell. Syst.* **2020**, *14*, 109–120. [CrossRef]
18. Yang, X. *Nature-Inspired Computation and Swarm Intelligence: Algorithms, Theory and Applications*, 1st ed.; Academic Press: Cambridge, MA, USA, 2020.
19. Raziani, S.; Azimbagirad, M. Deep CNN hyperparameter optimization algorithms for sensor-based human activity recognition. *Neurosci. Inform.* **2022**, *2*, 100078. [CrossRef]
20. Yeh, W.; Lin, Y.; Liang, Y.; Lai, C.; Huang, C. Simplified swarm optimization for hyperparameters of convolutional. *Comput. Ind. Eng.* **2023**, *177*, 109076. [CrossRef]
21. Chawla, R.; Beram, S.; Murthy, C.; Thiruvenkadam, T.; Bhavani, N.; Saravanakumar, R.; Sathishkumar, P. Brain tumor recognition using an integrated bat algorithm with a convolutional neural network approach. *Meas. Sens.* **2022**, *24*, 100426. [CrossRef]
22. Melin, P.; Sánchez, D.; Castillo, O. Comparison of optimization algorithms based on swarm intelligence applied to convolutional neural networks for face recognition. *Int. J. Hybrid Intell. Syst.* **2022**, *18*, 161–171. [CrossRef]
23. Melin, P.; Sánchez, D.; Pulido, M.; Castillo, O. Convolutional Neural Network Design using a Particle Swarm Optimization for Face Recognition. In Proceedings of the International Conference on Hybrid Intelligent Systems, Online, 13–15 December 2021.
24. Fernandes Junior, F.; Yen, G. Particle swarm optimization of deep neural networks architectures for image classification. *Swarm Evol. Comput.* **2019**, *49*, 62–74. [CrossRef]
25. Bashkandi, A.; Sadoughi, K.; Aflaki, F.; Alkhazaleh, H.; Mohammadi, H.; Jimenez, G. Combination of political optimizer, particle swarm optimizer, and convolutional neural network for brain tumor detection. *Biomed. Signal Process. Control* **2023**, *81*, 104434. [CrossRef]
26. Murugan, R.; Goel, T.; Mirjalili, S.; Chakrabartty, D. WOANet: Whale optimized deep neural network for the classification of COVID-19 from radiography images. *Biocybern. Biomed. Eng.* **2021**, *41*, 1702–1708. [CrossRef]
27. Knypiński, Ł. Constrained optimization of line-start PM motor based on the gray wolf optimizer. *Maint. Eng.* **2021**, *23*, 1–10. [CrossRef]
28. Nazri, E.; Murairwa, S. Classification of heuristic techniques for performance comparisons. In Proceedings of the International Conference on Mathematics, Statistics, and Their Applications, Banda Aceh, Indonesia, 4–6 October 2016.
29. Kumar, A.; Bawa, S. A comparative review of meta-heuristic approaches to optimize the SLA violation costs for dynamic execution of cloud services. *Soft Comput.* **2020**, *24*, 3909–3922. [CrossRef]
30. Fan, C.; Chung, Y. Design and Optimization of CNN Architecture to Identify the Types of Damage Imagery. *Mathematics* **2022**, *10*, 3483. [CrossRef]
31. Fregoso, J.; Gonzalez, C.; Martinez, G. Optimization of Convolutional Neural Networks Architectures Using PSO for Sign Language Recognition. *Axioms* **2021**, *10*, 139. [CrossRef]
32. Shaban Naseri, R.; Kurnaz, A.; Farhan, H. Optimized face detector-based intelligent face mask detection model in IoT using deep learning approach. *Appl. Soft Comput.* **2023**, *134*, 109933. [CrossRef]
33. Sánchez, D.; Melin, P.; Castillo, O. A Grey Wolf Optimizer for Modular Granular Neural Networks for Human Recognition. *Comput. Intell. Neurosci.* **2017**, *2017*, 4180510. [CrossRef]
34. Sánchez, D.; Melin, P.; Castillo, O. Optimization of modular granular neural networks using a firefly algorithm for human recognition. *Eng. Appl. Artif. Intell.* **2017**, *64*, 172–186. [CrossRef]
35. Campos, A.; Melin, P.; Sánchez, D. Multiclass Mask Classification with a New Convolutional Neural Model and Its Real-Time Implementation. *Life* **2023**, *13*, 368. [CrossRef]
36. Haykin, S. *Neural Networks: A Comprehensive Foundation*, 1st ed.; Macmillan: London, UK, 1994.
37. Nunes Da Silva, I.; Hernane Spatti, D.; Flauzino, A.; Bartocci Liboni, L.; Dos Reis Alves, S. *Artificial Neural Networks: A Practical Course*, 1st ed.; Springer: London, UK, 2018.
38. Aggarwal, C. *Neural Networks and Deep Learning: A Textbook*, 1st ed.; Springer: London, UK, 2018.
39. Singh, M.; Singh, G. Two phase learning technique in modular neural network for pattern classification of handwritten Hindi alphabets. *Mach. Learn. Appl.* **2021**, *6*, 100174. [CrossRef]
40. Koonce, B. *Convolutional Neural Networks with Swift for Tensorflow: Image Recognition and Dataset Categorization*, 1st ed.; Apress: New York, NY, USA, 2021.

41. Ozturk, S. *Convolutional Neural Networks for Medical Image Processing Applications*, 1st ed.; CRC Press: Boca Raton, FL, USA, 2022.
42. Eberhart, R.; Kennedy, J. A New Optimizer using Particle Swarm. In Proceedings of the International Symposium on Micro Machine and Human Science, Nagoya, Japan, 4–6 October 1995.
43. Kennedy, J.; Eberhart, R. Particle Swarm Optimization. In Proceedings of the IEEE International Joint Conference on Neuronal Networks, Perth, WA, Australia, 27 November–1 December 1995.
44. Eberhart, R.; Shi, Y. Comparing Inertia Weights and Constriction Factors in Particle Swarm Optimization. In Proceedings of the IEEE Congress on Evolutionary Computation, La Jolla, CA, USA, 16–19 July 2000.
45. Xin, J.; Chen, G.; Hai, Y. A Particle Swarm Optimizer with Multi-stage Linearly-Decreasing Inertia Weight. In Proceedings of the International Joint Conference on Computational Sciences and Optimization, Sanya, China, 24–26 April 2009.
46. Mirjalili, S.; Mirjalili, S.; Lewis, A. Grey Wolf Optimizer. *Adv. Eng. Softw.* **2014**, *69*, 46–61. [CrossRef]
47. Mech, L. Alpha status, dominance, and division of labor in wolf packs. *Can. J. Zool.* **1999**, *77*, 1196–1203. [CrossRef]
48. Muro, C.; Escobedo, R.; Spector, L.; Coppinger, R. Wolf-pack (*Canis lupus*) hunting strategies emerge from simple rules in computational simulations. *Behav. Process.* **2011**, *88*, 192–197. [CrossRef]
49. Long, W.; Jiao, J.; Liang, X.; Tang, M. Inspired grey wolf optimizer for solving large-scale function optimization problems. *Appl. Math. Model.* **2018**, *60*, 112–126. [CrossRef]
50. Mirjalili, S.; Lewis, A. The Whale Optimization Algorithm. *Adv. Eng. Softw.* **2016**, *95*, 51–67. [CrossRef]
51. Watkins, W.; Schevill, W. Aerial Observation of Feeding Behavior in Four Baleen Whales: Eubalaena glacialis, Balaenoptera borealis, Megaptera novaeangliae, and Balaenoptera physalus. *J. Mammal.* **1979**, *60*, 155–163. [CrossRef]
52. Yang, X. A New Metaheuristic Bat-Inspired Algorithm. In *Studies in Computational Intelligence*, 1st ed.; González, J.R., Pelta, D.A., Cruz, C., Terrazas, G., Krasnogor, N., Eds.; Springer: London, UK, 2010; Volume 284, pp. 65–74.
53. Talbi, N. Design of Fuzzy Controller rule base using Bat Algorithm. *Energy Procedia* **2019**, *162*, 241–250. [CrossRef]
54. Yang, X. Review of meta-heuristics and generalised evolutionary walk algorithm. *Int. J. Bio-Inspired Comput.* **2011**, *3*, 77–84. [CrossRef]
55. Perez, J.; Valdez, F.; Castillo, O.; Melin, P.; Gonzalez, C.; Martinez, G. Interval type-2 fuzzy logic for dynamic parameter adaptation in the bat algorithm. *Soft Comput.* **2017**, *21*, 667–685. [CrossRef]
56. Cabani, A.; Hammoudi, K.; Benhabiles, H.; Melkemi, M. MaskedFace-Net—A dataset of correctly/incorrectly masked face images in the context of COVID-19. *Smart Health* **2020**, *19*, 100144. [CrossRef]
57. Karras, T.; Laine, S.; Aila, T. A Style-Based Generator Architecture for Generative Adversarial Networks. In Proceedings of the Conference on Computer Vision and Pattern Recognition, Long Beach, CA, USA, 15–20 June 2019.
58. Jia, Y.; Shelhamer, E.; Donahue, J.; Karayev, S.; Long, J.; Girshick, R.; Guadarrama, S.; Darrell, T. Caffe: Convolutional Architecture for Fast Feature Embedding. In Proceedings of the ACM Conference on Multimedia, Orlando, FL, USA, 3–7 November 2014.
59. Campos, A.; Melin, P.; Sánchez, D. Convolutional neural networks for face detection and face mask multiclass classification. In Proceedings of the International Conference on Hybrid Intelligent Systems, Online, 13–15 December 2022.

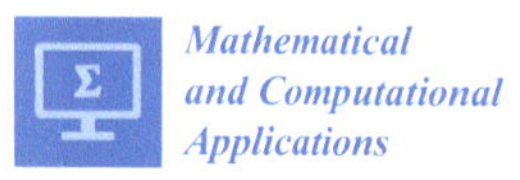

Article

Fokker–Planck Analysis of Superresolution Microscopy Images

Mario Annunziato [1,*] and Alfio Borzì [2]

1   Dipartimento di Fisica "E. R. Caianiello", Università degli Studi di Salerno, Via G. Paolo II 132, 84084 Fisciano, Italy

2   Institut für Mathematik, Universität Würzburg, Emil-Fischer-Strasse 30, 97074 Würzburg, Germany; alfio.borzi@mathematik.uni-wuerzburg.de

*   Correspondence: mannunzi@unisa.it or mannunzi@am-research.it

Abstract: A method for the analysis of super-resolution microscopy images is presented. This method is based on the analysis of stochastic trajectories of particles moving on the membrane of a cell with the assumption that this motion is determined by the properties of this membrane. Thus, the purpose of this method is to recover the structural properties of the membrane by solving an inverse problem governed by the Fokker–Planck equation related to the stochastic trajectories. Results of numerical experiments demonstrate the ability of the proposed method to reconstruct the potential of a cell membrane by using synthetic data similar those captured by super-resolution microscopy of luminescent activated proteins.

Keywords: super-resolution microscopy; Fokker–Planck equation; stochastic processes; numerical optimization

MSC: 49M05; 65K10; 93E03; 93E20; 92C37

Citation: Annunziato, M.; Borzì, A. Fokker–Planck Analysis of Superresolution Microscopy Images. *Math. Comput. Appl.* **2023**, *28*, 113. https://doi.org/10.3390/mca28060113

Academic Editor: Gianluigi Rozza

Received: 15 November 2023
Revised: 12 December 2023
Accepted: 12 December 2023
Published: 14 December 2023

1. Introduction

The pioneering works [1–3] mark the development of the revolutionary superresolution microscopy (SRM) that allows us to go beyond the Abbe limit for conventional light microscopy [4]. The SRM method consists of labeling the molecules moving on a biological support with fluorophores and then in sampling the microscopic images of the activated fluorescent molecules.

Observation of the frames of the sampled SRM microscopic images have suggested that the motion of the molecules could be modeled by a stochastic Langevin equation [5,6]. Clearly, the cell membrane at the microscope is a 3-dimensional object; however, it can be considered flat, and the observed motion is 2-dimensional since it results in the projection on the focal plane of the SRM microscope. It appears that an adequate model of the observed trajectories of 2-dimensional images is given by the following stochastic differential equation (SDE) [7]:

$$dX_t = b(X_t)\,dt + \sigma(X_t)\,dW_t \tag{1}$$
$$X_{t_0} = X_0, \tag{2}$$

where b represents the drift, σ is the dispersion coefficient, and $X_t \in \mathbb{R}^2$ denotes the position of the observed molecule at time t. In this framework, it is well-known that the drift and dispersion coefficients satisfy

$$\lim_{t \to s} \mathbb{E}\left[\frac{X_t - X_s}{t - s} \,\Big|\, X_s = z \right] = b(z), \qquad \lim_{t \to s} \mathbb{E}\left[\frac{|X_t - X_s|^2}{t - s} \,\Big|\, X_s = z \right] = \sigma^2(z),$$

where the expected values are computed with respect to the process having value z at $t = s$; the operator $\mathbb{E}[\cdot \mid X_s = z]$ denotes averaging with regard to the measure of the trajectories conditioned to be at z at time s.

The formulas above suggest that suitable approximations to b and σ can be obtained by tracking single molecules; see, e.g., [7,8]. However, this approach may suffer from the highly fluctuating values of the trajectories and the difficulty of discerning between different molecules that come closer to the resolution limit.

For this reason, already in [9] the authors have pursued an alternative strategy that allows us to build a robust methodology for the estimation of the drift based on the observation of an ensemble of trajectories. Our approach is built upon the assumption that this ensemble is driven by a velocity field (the drift), given by a potential velocity field $U(x)$, with $x \in \Omega \subset \mathbb{R}^2$, as follows:

$$b(x; U) = -\nabla U(x). \tag{3}$$

Moreover, one assumes a constant diffusion coefficient whose value is chosen consistently with estimates of laboratory measurement [10]. It is the purpose of our work to reconstruct the potential $U(x)$ by means of the observation of the motion X_t of the molecules modeled by Equation (1).

In agreement with our statistical approach based on ensembles, we focus on the evolution of the probability density function (PDF) of the positions of the molecules (not on the single trajectories) whose evolution is governed by the Fokker–Planck (FP) problem given by [6,11]:

$$\partial_t f(x,t) - \nabla \cdot (\nabla U(x)\, f(x,t)) - \frac{\sigma^2}{2} \Delta f(x,t) = 0, \quad (x,t) \in Q \tag{4}$$

$$F(f) \cdot \hat{n} = 0, \qquad (x,t) \in \Sigma, \tag{5}$$

$$f(x,0) = f_0(x), \qquad x \in \Omega, \tag{6}$$

where $Q = \Omega \times (0,T)$ and $\Sigma = \partial\Omega \times (0,T)$. In this formulation, $f(x,t)$ represents the PDF of a particle at $x \in \mathbb{R}^2$ at time t, $\nabla U(x)$ is the Cartesian gradient of the potential U, f_0 is the initial density, and Δ is the two-dimensional Laplace operator. Notice that we require zero-flux boundary conditions, where $F(f)(x,t)$ is the following flux of probability

$$F(f)(x,t) = \frac{\sigma^2}{2} \nabla f(x,t) - b(x; U)\, f(x,t). \tag{7}$$

We choose zero-flux boundary conditions since they reasonably model the situation where a similar number of particles enters and exits the domain; see, e.g., [6,12].

Our proposal is to construct an FP-based imaging modality that is based on the formulation of an inverse problem for U and the observation of a time sequence, in time interval $[0,T]$, of numerical PDFs (two-dimensional histograms), which are obtained from a uniform binning of SRM particles' positions. We denote this input data as $f_d(x,t)$ which is a piecewise constant function. In this setting, the initial condition is given by $f_0(x) = f_d(x,0)$.

This proposal is similar to that in our previous work [9]. However, in [9] the assumption of interacting particles was made that resulted in very involved and CPU time demanding calculations. It is the purpose of this work to demonstrate that accurate reconstruction results can be obtained assuming noninteracting particles, hence by using a linear FP model.

At the continuous level, our FP-based imaging tool is formulated as the following inverse problem:

$$\min J(f, U) := \frac{1}{2} \int_\Omega \int_0^T (f(x,t) - f_d(x,t))^2 \, dx \, dt$$

$$+ \frac{\zeta}{2} \int_\Omega (f(x,T) - f_d(x,T))^2 \, dx$$

$$+ \frac{\alpha}{2} \int_\Omega (|U(x)|^2 + |\nabla U(x)|^2) \, dx, \tag{8}$$

$$s.t. \quad \partial_t f(x,t) + \nabla \cdot [b(x,;U)f(x,t)] - \frac{\sigma^2}{2} \Delta f(x,t) = 0, \quad \text{in } Q$$

$$f(x,0) = f_0(x) \text{ in } \Omega, \qquad F(f) \cdot \hat{n} = 0 \text{ on } \Sigma,$$

with the given initial and boundary conditions for the FP equation, and $\alpha, \zeta > 0$.

In this problem, the objective functional J is defined as the weighted sum of a space–time best fit term $\int_\Omega \int_0^T (f(x,t) - f_d(x,t))^2 \, dx \, dt$, and at final time $\int_\Omega (f(x,T) - f_d(x,T))^2 \, dx$, and of a suitable 'energy' of the potential $\|U\|^2 = \int_\Omega (|U(x)|^2 + |\nabla U(x)|^2) \, dx$, which corresponds to the square of the $H^1(\Omega)$ norm of U. Notice that this formulation allows us to avoid any differentiation of the data and makes it possible to choose the binning size and, in general, the measurement setting, independently of any choice of parameters that are required in the numerical solution of the optimization problem.

Our second main concern in determining the potential U is to provide a measure of uncertainty, and thus of reliability, of its reconstruction. Statistically, this is achieved by many repetitions of the same experiment, that could not be feasible for (short) living cells. However, inspired by the so-called model predictive control (MPC) scheme [13] already used for optimal control problems [12,14], we propose a novel procedure to quantify the uncertainty of the estimation of U by using the data of a single experiment.

Our methodology is to consider a sequence of non-parametric inverse problems like (8) defined on time windows (t_k, t_{k+1}), $k = 0, \ldots, K-1$, that represent a uniform partition in K subintervals of the time interval $[0, T]$. Therefore, a statistical analysis can be performed on the set of the corresponding K solutions for U that are obtained in the subintervals.

For development and validation, we consider images of a cell's membrane structures (actin, cytoskeleton), as expression of potentials, that is pixel grey values where increased brightness stands for more repulsion, with which we generate our synthetic data. In particular, we use an image of actin from a cytoskeleton obtained with a Platinium-replica electron microscopy [15,16].

With this images taken as gray-level representation of potential functions, we perform Monte Carlo simulation of motion of particles to generate images of molecules at different time instants, thus constructing the datasets representing the output of measurements. This setting is illustrated in Figure 1, where the image of actin [17] and a plot of few trajectories of the corresponding stochastic motion of the particles in this potential are shown.

Once the synthetic measurement data is constructed, we perform a pre-processing step on this data to construct the numerical PDF required in our method and solve our inverse problem to find the estimated–measured potential U. The latter is compared with that one used in the MC simulation, by a measure of similarity based on the pixel cross-correlation between the two images.

In Section 2, we discuss a numerical methodology for solving our FP-based reconstruction method for the potential U. In Section 3, we provide all details of our experimental setting and introduce some analysis tools for determining the accuracy of the proposed reconstruction. In Section 4, we validate our reconstruction method, and use our uncertainty quantification procedure. In Section 5, we investigate the resolution of the proposed FP image reconstruction as an optical instrument. A section of conclusion and acknowledgements completes this work.

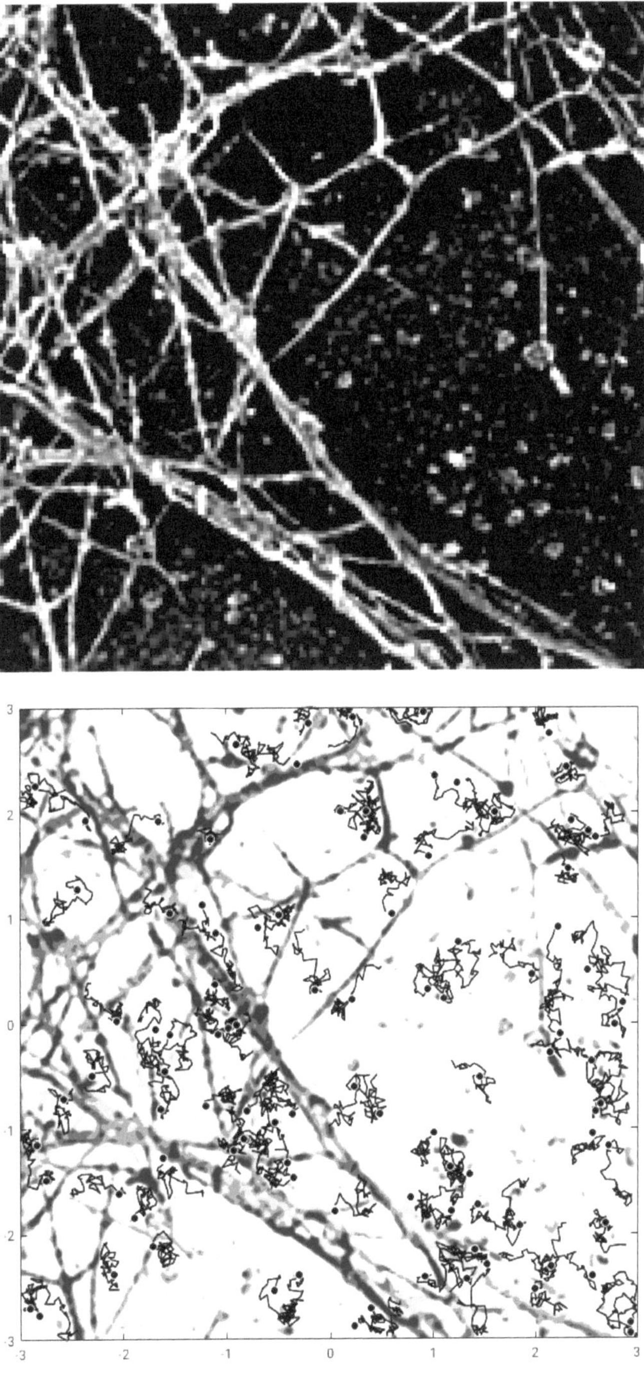

Figure 1. (**Up**) A picture of actin from a cytoskeleton as cell membrane potential (close up) (courtesy of Koch Institute [17]); (**Down**) a few simulated trajectories of particles (black dots) on the membrane (in reverse colors).

2. Numerical Methodology

Our aim is to reconstruct the potential U from the data consisting of a temporally sampled SRM images of the positions of particles subject to this potential; see Figure 2 (**up**) for a schematic snap-shot of this data. This image is subject to a pre-processing binning procedure in order to construct histograms by counting the number of particles in a regular square partition of Ω. The height of an histogram is proportional to the number of particles in a bin of the domain. This procedure for the image at time t defines the histogram function $f_d(\cdot, t)$; see Figure 2 (**down**).

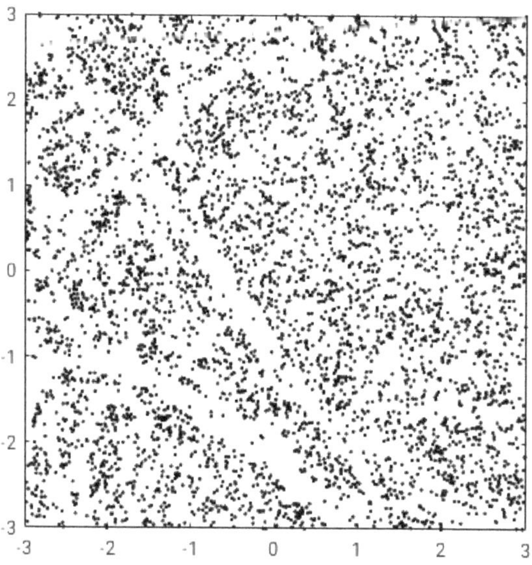

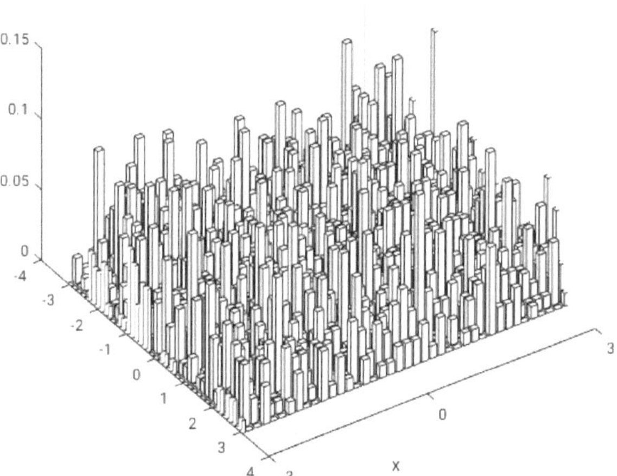

Figure 2. A frame of particles (**up**) and the corresponding histogram $f_d(x, t)$ on a mesh of 40×40 bins for a fixed time (**down**), from simulated data.

In order to illustrate our numerical framework, we introduce the potential-to-state map $U \mapsto f = S(U)$, that is, the map that associates to a given $U \in H^1(\Omega)$ the unique solution to our FP problem (4)–(6), with given initial condition f_0. For the analysis of well-posedness and regularity of the map S, we refer to [18].

Next, we remark that with the map S, we can define the reduced objective functional $\hat{J}(U) := J(S(U), U)$ and consider the equivalent formulation of (8) given by

$$\min_{U \in H^1(\Omega)} \hat{J}(U), \tag{9}$$

which has the structure of an unconstrained optimization problem. Thanks to the regularity of S and the quadratic structure of J, existence of an optimal U can be stated by well known techniques; see, e.g., [19].

Further, since S and J are Fréchet differentiable, it is possible to characterize an optimal U as the solution to the following first-order optimality condition

$$\nabla_U \hat{J}(U) = 0,$$

where $\nabla_U \hat{J}(U)$ denotes the so-called reduced gradient [20].

In the Lagrange framework, this condition results in the following optimality system:

$$\partial_t f(x,t) + \nabla \cdot [b(x; U) f(x,t)] - \frac{\sigma^2}{2} \Delta f(x,t) = 0,$$

$$f(x,0) = f_0(x) \text{ in } \Omega, \qquad F(f) \cdot \hat{n} = 0 \text{ on } \partial\Omega \times (0, T],$$

$$\partial_t p(x,t) + \frac{\sigma^2}{2} \Delta p(x,t) + \nabla p(x,t) \cdot b(x; U) = f(x,t) - f_d(x,t), \tag{10}$$

$$p(x,T) = -\xi \left(f(x,T) - f_d(x,T) \right) \text{ in } \Omega, \qquad \partial_{\hat{n}} p(x,t) = 0 \text{ on } \partial\Omega \times (0,T],$$

$$\alpha\, U(x) - \alpha\, \Delta U(x) - \int_0^T \nabla \cdot (f(x,t) \nabla p(x,t)) dt = 0 \text{ in } \Omega,$$

$$\partial_{\hat{n}} U = 0, \qquad \text{on } \partial\Omega,$$

where p denotes the adjoint variable, which is governed by a backward adjoint FP equation. Notice that the adjoint equation is a well-posed problem with $f_d(x,t) \in L^2$, which allows to use the irregular histograms as input data. The numerical solution obtained with a finite difference approximation is consistent with the interpretation of f_d as a piecewise constant function obtained by local averaging an L^2 function on subcells centered at grid points. Notice that the numerical grid is finer than that of the binning.

The last equation in (10) is the so-called the optimality condition equation, and the Neumann boundary condition $\partial_{\hat{n}} U = 0$ is our modeling choice. One can show that its left-hand side represents the L^2 gradient along the FP differential constraint with respect to U of the objective functional. We have

$$\nabla_U \hat{J}(U)(x) := \alpha\, U(x) - \alpha\, \Delta U(x) - \int_0^T \nabla \cdot (f(x,t) \nabla p(x,t)) dt. \tag{11}$$

Our approach for solving our FP optimization problem (8) is based on the nonlinear conjugate gradient (NCG) method; see, e.g., [20]. This is an iterative method that resembles the standard CG scheme and requires to estimate the reduced gradient $\nabla_U \hat{J}(U)$ at each iteration.

In order to illustrate the NCG method, we start with a discussion on the construction of the gradient. For a given U^n obtained after n iterations, we solve the FP equation and its adjoint, and use (11) to assemble the L^2 gradient. However, since the potential is sought in $H^1(\Omega)$, we need to obtain the H^1 gradient that satisfies the following relation

$$\left(\nabla_U \hat{J}(U)|_{H^1}, \delta U \right)_{H^1} = \left(\nabla_U \hat{J}(U)|_{L^2}, \delta U \right)_{L^2}, \tag{12}$$

where $(\cdot, \cdot)$ denotes the $L^2(\Omega)$ scalar product.

Now, using the definition of the H^1 inner product, we obtain

$$\int_\Omega [\nabla_u \hat{J}(U)|_{H^1} \cdot \delta U(x) + \nabla_x \nabla_u \hat{J}(U)|_{H^1} \cdot \nabla_x \delta U(x)]\, dx = \int_\Omega \nabla_u \hat{J}(U)|_{L^2}\, \delta U(x)\, dx, \quad (13)$$

which must hold for all the test functions $\delta U \in H^1(\Omega)$. Therefore we obtain

$$-\Delta[\nabla_u \hat{J}(U)|_{H^1}] + [\nabla_u \hat{J}(U)|_{H^1}] = \nabla_u \hat{J}(U)|_{L^2}, \quad (14)$$

with the boundary conditions $\frac{\partial}{\partial \hat{n}} \nabla_u \hat{J}(U)|_{H^1} = 0$ on $\partial\Omega$; see [9] for more details. In Algorithm 1 our procedure for computing the gradient is given.

Algorithm 1 Calculate H^1 gradient.

Require: control $U(x)$, $f_0(x)$, $f_d(x,t)$.
Ensure: reduced gradient $\nabla_u \hat{J}(U)|_{H^1}$
    Solve forward the FP equation with inputs: $f_0(x)$, $U(x)$
    Solve backward the adoint FP equation with inputs: $U(x)$, $f(x,t)$
    Assemble the L^2 gradient $\nabla_u \hat{J}(U)|_{L^2}$ using (11).
    Compute the H^1 gradient $\nabla_u \hat{J}(U)|_{H^1}$ solving (14).
    return $\nabla_u \hat{J}(U)|_{H^1}(x)$

In this algorithm, the FP problem and its optimization FP adjoint are approximated by the exponential Chang–Cooper scheme and the implicit BDF2 method; see [12].

Now, we can discuss the NCG method. The NCG iterative procedure is initialized with $U^0(x) = 0$. We denote the optimization directions with d^n. In the first update, we have $d^0 = -\nabla_u \hat{J}(U^0)|_{H^1}$ and perform the optimization step

$$U^1 = U^0 + \alpha_0\, d^0,$$

where α_0 is obtained by a backtracking linesearch procedure. After the first step, in the NCG method the descent direction is defined as a linear combination of the new gradient and the past direction as follows:

$$d^n = -\nabla_u \hat{J}(U^n)|_{H^1} + \beta_{n-1}\, d^{n-1},$$

where $\beta_{-1} = 0$, and $\beta_{n-1} = \|\nabla_u \hat{J}(U^n)|_{H^1}\|^2 / (d^{n-1} \cdot (\nabla_u \hat{J}(U^n)|_{H^1} - \nabla_u \hat{J}(U^{n-1})|_{H^1}))$, that is, the Dai–Yuan formula $b_{n-1} = \|r_n\|^2 / (-d_{n-1}(r_n - r_{n-1}))$, where here r_n stand for the deepest descent direction $r_n = -\nabla_u \hat{J}(U^n)|_{H^1}$ and d_{n-1} is the conjugate direction at the previous step; see, e.g., [20].

The tolerance *tol* and the maximum number of iterations $n_{\max}$ are used for termination criteria. Summarizing, in Algorithm 2 we present the NCG procedure.

Algorithm 2 Nonlinear conjugate gradient (NCG) method

Require: $U^0(x) \equiv 0$, $f_0(x)$, $f_d(x,t)$
Ensure: Optimal control $U(x)$ and corresponding state $f(x,t)$
    $n = 0$
    Assemble gradient $g^0 = \nabla_u \hat{J}(U^0)|_{H^1}$ using Algorithm 1; set $d^0 = -g^0$.
    while $\|g^n\|_{H^1} > tol$ **and** $n < n_{\max}$ **do**
        Use linesearch to determine α_n
        Update control: $U^{n+1} = U^n + \alpha_n\, d^n$
        Compute the gradient $g^{n+1} = \nabla_u \hat{J}(U^{n+1})|_{H^1}$ using Algorithm 1
        Calculate the new descent direction $d^{n+1} = -g^{n+1} + \beta_n\, d^n$
        Set $n = n + 1$
    end while
    return $U^n(x)$

For our numerical experiments these algorithms have been implemented with object oriented programming in C++, by using the numerical libraries Armadillo [21,22], Openblas [23], Lapack [24], SuperLU [25,26] and HDF5 [27].

3. Experimental Design and Analysis Tools

In a real application, the input data f_d is given by frames of a recorded sequence of a SRM experiment, and the desired output is the reconstructed potential denoted with U_r. In our case, we construct this data based on a sample potential U_s that determines the drift function in our stochastic model (1). Thus, we generate our frames of synthetic data first by time-integrating this SDE in the chosen interval $[0, T]$, and choosing the initial positions of the particles randomly uniformly distributed. Next, the positions of the particles at different times are collected in a sequence of 2-dimensional bins that result in the sequence of distributions $f_d(x, t_\ell)$, $\ell = 1, \dots, L$, where L is the length of the resulting time sequence of frames. With this preparation, we apply our algorithm to obtain U_r, which represents the proposed reconstruction of U_s. A comparison between these two potentials allows to validate the accuracy of our reconstruction method (see below).

We choose a domain $\Omega = [-3, 3] \times [-3, 3]$, and U_s corresponds to Figure 1 (up), where the values of U_s in Ω correspond to the gray scale pixel values of the picture mapped in $[0, 1]$. With this U_s, we perform a stochastic simulation of N_p particles for a time horizon T, and diffusion amplitude σ. The particles trajectories given by (1) with (3), are computed with the Euler-Maruyama scheme with a time step $\tau = 10^{-3}$, which results in a number of L frames. In this simulation, reflecting barriers for the stochastic motion are implemented. We remark that for the following calculations we are going to consider a relatively small value of density of particles; see [10,28].

Next, we perform a binning of the positions of the particles at each frame to construct f_d. Hence, we consider a uniform partition of Ω with non-overlapping squares; see Figure 2 for a plot of particles in Ω at a given time and the corresponding f_d. Notice that f_d is irregular; nevertheless, we do not perform smoothing of this data. The sequence of f_d values enter in our best fit functional in (8).

Once we have computed U_r with our optimization procedure, we aim at providing a quantification of its uncertainty. Thus, we compute the following normalized cross-correlation factor between the reconstructed potential U_r and the one used to generate the synthetic data U_s. We have

$$cc(U_r, U_s) = \frac{U_r \cdot U_s}{|U_r| \, |U_s|}. \tag{15}$$

In this formula, U_r and U_s are considered as vectors and $\cdot$ represents the scalar product. Therefore if $cc = 1$ we have that the two potentials match perfectly, whereas if its value is close to 0, the two potentials are dissimilar. Notice that cross-correlation is commonly used in medical imaging and biology; see, e.g., [29–31].

Clearly, one could consider many repetitions of the simulation of the motion of the particles with the same initial condition and make the final binning on the average of the resulting frames. This procedure would result in a less fluctuating $f_d(x, t_\ell)$ that allows a better reconstruction. However, this scenario seems difficult to realize in the real laboratory setting of a living cell. On the other hand, in SRM, imaging is able to visualize the motion of the particles on a cell membrane for a relatively long time ($T \gg 1$ in our setting), and our approach exploits this possibility considering a subdivision of the time interval in a number K of time windows, and solving our optimization problem in each of these windows almost independently. This approach allows us to improve the reconstruction U_r and makes it possible to quantify the uncertainty of the reconstruction.

Now, to illustrate our approach, consider a uniform partition of $[0, T]$ in time windows of size $\Delta t = T/K$ with K a positive integer. Let $t_k = k\Delta t$, $k = 0, 1, \dots, K$, denote the start- and end-points of the windows. At time t_0, we have the initial PDF f_0, and we solve our optimization problem (8) in the interval $[t_0, t_1]$. This means that the final time is t_1 and the terminal condition for the adjoined variable is given by $p(x, t_1) = -\xi \left(f(x, t_1) - f_d(x, t_1) \right)$.

The resulting potential is denoted with U_1. Thus, the solution obtained in this window also provides the PDF at $t = t_1$.

Clearly, we can repeat this procedure in the interval (t_1, t_2) with the computed PDF at $t = t_1$ as the initial condition and t_2 as final time, to compute U_2. This procedure is recursive and can be repeated for $k = 1, \ldots, K$, thus obtaining $U_k, k = 1, \ldots, K$.

Notice that small values of K in relation to L produce a rough estimate of the average potential and its standard deviation due to statistical fluctuations of the Monte Carlo experiments. On the other hand, for greater values of K, the number of frames for each window of our approach is reduced when L is kept fixed, thus resulting in a worsening of the reconstruction procedure.

For the purpose of our analysis, we apply a scaling of these potentials so that their point-wise values are in the interval $[0, 1]$. This scaling is performed as follows:

$$\hat{U} = \frac{U - \min(U)}{\max(U) - \min(U)}. \tag{16}$$

Thereafter, we the reconstructed potential by pixel-wise average of the U_k is given by

$$\langle U_r \rangle = \frac{1}{K} \sum_{k=1}^{K} \hat{U}_k. \tag{17}$$

Moreover, we can also compute the following pixel-wise standard deviation

$$sd(U_r) = \sqrt{\sum_{k=1}^{K} \frac{(\hat{U}_k - \langle U_r \rangle)^2}{K - 1}}. \tag{18}$$

Next, we provide conversion formulas for our parameters in order to accommodate data from real laboratory experiments. We introduce a unit of length u such that the side length l of our square domain Ω is given by $l = 6\,\mathrm{u}$, and the unit of the noise amplitude σ is given by $\sqrt{\mathrm{u}}/s$. In real biological experiments, the typical measure of the length $\tilde{l}$ of a cell membrane is given in μm. Further, the particle's diffusion constant $D = \sigma^2/2$ is given in $\mathrm{\mu m}^2/s$; hence, we have the correspondence $\sigma = 1/\tilde{l}\sqrt{2D}$ in unit $\sqrt{\mathrm{u}}/s$, whose value is used for MC simulations.

The depth of the potential $\tilde{U}$ is expressed in unit of $K_B\bar{T}$, where K_B is the Boltzmann constant and $\bar{T}$ the absolute temperature. In experimental papers, the Equation (3) is written with the diffusion constant D and $K_B\bar{T}$, i.e., $D\tilde{U}/(K_B\bar{T})$. As above, we obtain the relationship between the values of a potential U and the scaled $\tilde{U}$, as $\tilde{U} = U(\tilde{l}/l)^2/D$ in the unit of $K_B\bar{T}$.

As an illustration of the setting above, we see that in an experiment, the super-resolution of an acquired image frame can reach the value of 0.02 μm/pixel. With an image of 500×500 pixels, we have $\tilde{l} = 10$ μm. The average diffusion coefficient of particles (protein molecules) observed in SRM imaging is estimated with $D = 0.1\ \mathrm{\mu m}^2/s$ [10]. By super-resolution techniques, it is possible to activate a density of $0.5 \div 2/\mathrm{\mu m}^2$ visible particles, which in terms of image pixels corresponds to $0.5 \div 2$ particles in a square of 50 pixels of side. Each frame is usually sampled at time intervals of $\delta t = 30$ ms.

In order to set up a consistent MC simulation of a real experiment, by mapping an image of a square of side 10 μm on our domain Ω, we get from the above mentioned formula: $\sigma = 6/10\sqrt{0.2} \approx 0.268\,\mathrm{u}/\sqrt{s}$.

4. Numerical Validation

In this section, we discuss results of experiments in a setting that is close to real laboratory experiments involving SRM imaging. The results of these experiments demonstrate the ability of our methodology to reconstruct the potential from the simulations of the SRM measurements of the motion of particles on a cell's membrane.

We consider a potential that corresponds to a portion of cytoskeleton as depicted in Figure 3, with 200×200 pixels. We assume that the pixel is 50 nm, which corresponds to an area of 100 μm^2. In the figure, the white regions represent the structure of the cytoskeleton; the black ones are the 'valleys' where the proteins are supposed to be attracted.

For the MC simulations for generating the synthetic data, we choose $\sigma = 0.268 \, \mathrm{u}/\sqrt{s}$. This value of σ corresponds to a diffusion constant of $D \simeq 0.1$ μm^2/s. We consider $N_p = 1000$ particles, i.e., an average density of 10 particles per μm^2. In this case, we consider a sequence of $L = 3000$ frames and $T = 90$, obtained by the numerical integration of the stochastic differential equation with an integration step $\tau = 30 \times 10^{-3}$ s. The frames have $\delta t = 30$ ms, similar to a real experiment. The resulting (single run) particle trajectories are collected in a binning process based on a mesh Ω of 50×50 bins.

For our reconstruction method, we choose a numerical partition of Ω of 100×100 subdivisions, corresponding to a mesh size of 100 nm. The time integration step coincides with that of the frames. For the tracking functional, we set $\alpha = 10^{-4}$ and $\zeta = 1$. Further, in the FP setting, we have $\sigma = 0.7 \, \mathrm{u}/\sqrt{s}$. Notice that σ in the FP model is chosen to be larger than the one used in the MC simulations. This choice is dictated by numerical convenience and it appears that it does not affect the quality of the reconstruction. The calculations are performed according to the MPC procedure with $K = 5$ time windows.

With this setting, we obtain the reconstructed potential shown in Figure 3 (down). We see that the reconstruction is less sharp as we expected considering the much finer structure of the cytoskeleton and the small number of particles involved.

Further, in Figure 4, we depict the potentials obtained on each time window of the MPC procedure and the values of the corresponding cc. With these results, we have obtained the reconstructed potential U_r in Figure 3, which we re-plot in level-set format in Figure 5 for comparison. In Figure 5, we also depict the standard deviation that suggests that we have obtained a reliable reconstruction with small uncertainty.

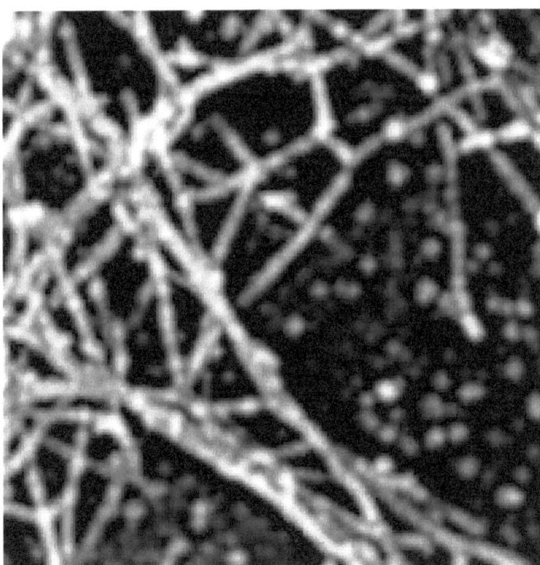

Figure 3. *Cont.*

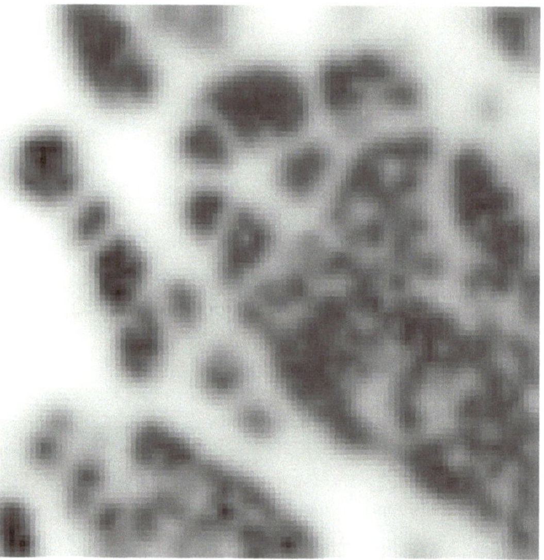

Figure 3. (**Up**): Portion of the cytoskeleton (Courtesy of [17]). (**Down**): reconstructed potential with the MPC scheme and $K = 5$.

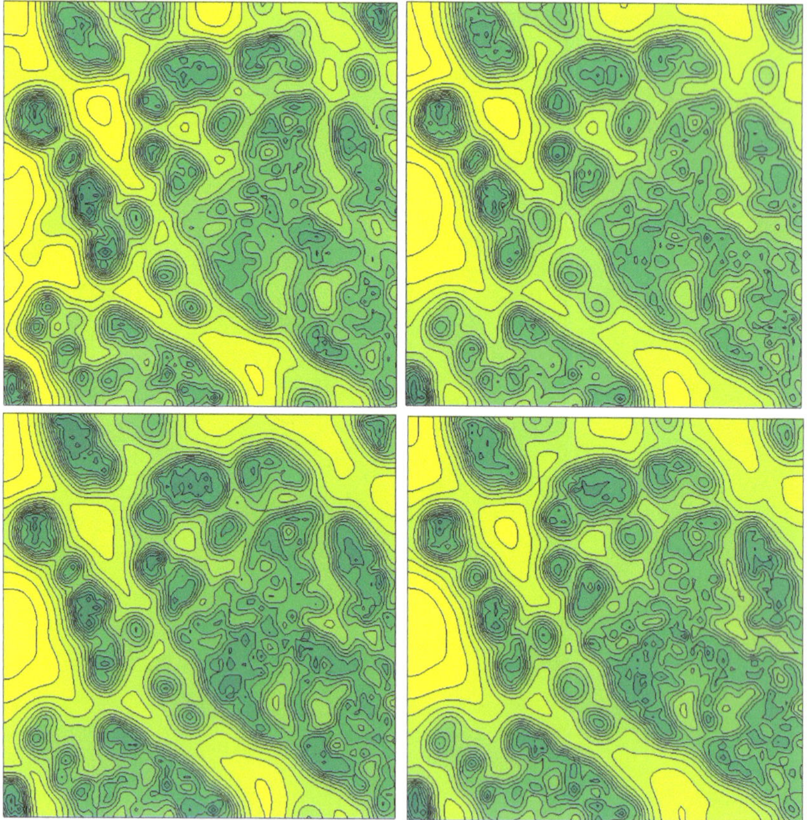

Figure 4. *Cont.*

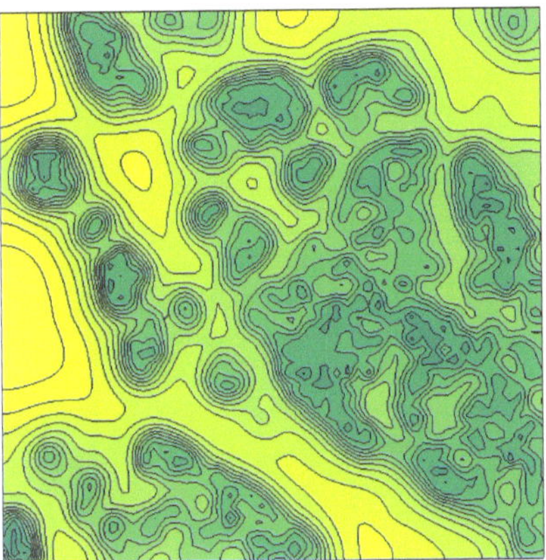

Figure 4. Sequence of the 5 (from **top-left** to **right-down**) calculated potentials obtained with the MPC procedure. Cross-correlation values: 0.82, 0.81, 0.82, 0.81, 0.82.

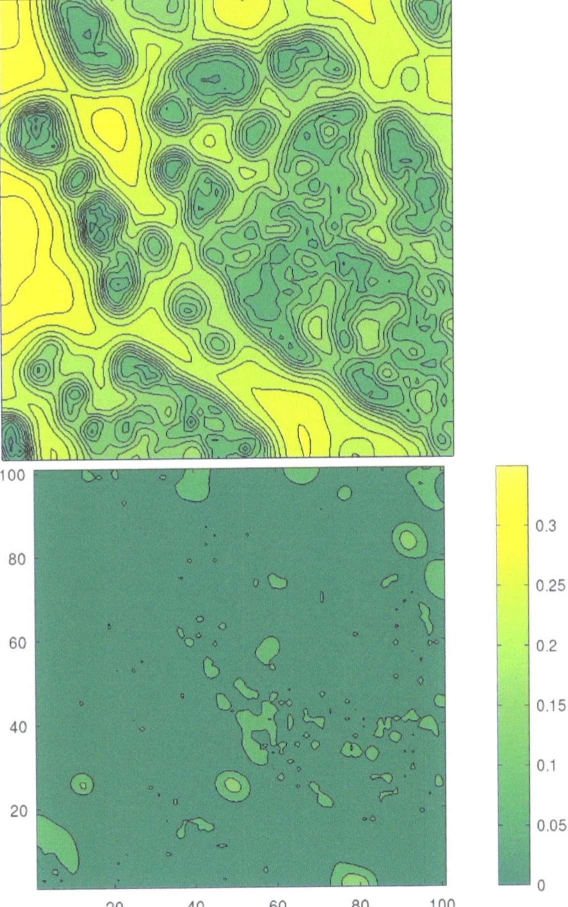

Figure 5. (**Up**): reconstructed mean potential. Its cross-correlation value with respect to the real image is 0.82. (**Down**): standard deviation of the reconstructed potential, in level set representation.

5. Resolution of FP-Based Image Reconstruction

In this section, we investigate the optical resolution of our reconstruction method, that is, try to determine a confidence value related to the scale at which our method can resolve variations of the potential. As a guideline, we remark that single molecule localization microscopy (SMLM) can distinguish distances of molecules of approximately 20 nm resolution. Therefore, we assume this resolution range of the fluorescently labeled particles images, and we attempt to quantify the smallest scale at which geometric features of the reconstructed potential U can be distinguished.

For our purpose, we consider the following 'target potential', appearing as an alternating sequence of black and white circles (likewise those in test targets used for the resolution measurement of optical instruments), to synthetically generate the motion data of particles. We have

$$U(x,y) = A\left(1 + \cos\left(\frac{2\pi}{dl}(x^2 + y^2)\right)\right), \qquad (x,y) \in \Omega, \tag{19}$$

where A denotes the semi-amplitude of the variation between the minimum and the maximum of the potential, l is the length of the side of the domain, d is the distance between two peaks of the potential as a fraction of l.

Now, we consider a single MC simulation of 500 particles with the setting: $\sigma = 0.5$, $T = 90$ and $L = 3000$ frames, integrated with the time step $\tau = 0.03$. In Figure 6, we show (left) the given potential with $A = 0.05$ and $d = 1/20$, with a gray-scale value representation conveniently adjusted for illustration pourpose. According to the above working hypothesis, we suppose that the pixel's width of the image is 20 nm. In Figure 6 (left), we depict U in a square of side of 500 pixels, corresponding to $\tilde{l} = 10$ μm. Hence, the distance between two peaks is $\lambda = 10/20 = 500$ nm. Further, the particle's density is 5 particles per μm^2, the diffusion coefficient $D \simeq 0.3472$ μm^2/s, and the potential depth, i.e., the difference between the maximum and the minimum, is $\tilde{U} = 0.8\, K_B T$. For the reconstruction process, we use a binning of 50×50, $\alpha = 10^{-4}, \xi = 1$. In the numerical setting, we use a grid of 100×100 points, and $K = 5$. Also in Figure 6 (right), we show the reconstructed potential $\langle U_r \rangle$ and notice its high accuracy that is also confirmed by the high value 0.82 of the cross correlation. Notice, that the quality of the reconstruction can be further improved by using post-processing techniques of images.

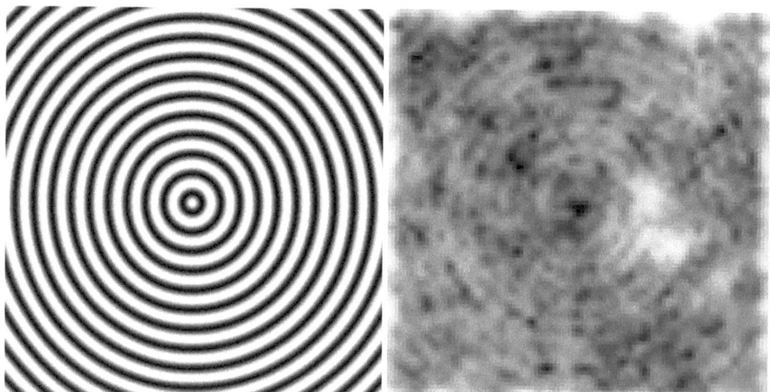

Figure 6. (Left): the potential (19) with $A = 0.05$ and $d = 1/20$ (the gray scale levels spans from $U = 0$ to 0.1). **(Right)**: result of the reconstruction with the gray levels expanded to the min/max of $\langle U_r \rangle$. The cross correlation between the two images is 0.82.

Now, with the aim to define a criteria to establish the resolution measure for the potential, we introduce a confidence level for the quality of the reconstructed potential by setting a threshold for the calculated cross-correlation. This approach has been adopted in [29] for the detection of cellular objects from images acquired from electron tomography.

For that purpose, the authors used the threshold value of 0.5, whereas in our case, we set a more strict threshold-*cc* level equal to 0.8. With this threshold, we can state that the test pattern depicted in Figure 6 (left) is satisfactorily reconstructed and determines that the resolution measure associated to our 'imaging instrument' is 500 nm. Notice that this value is affected by the value of the potential U and the diffusion D, and it can be further improved by changing the other parameters of the experiment, such as K or the time T of the motion sampling.

6. Conclusions

A novel method for the analysis of super-resolution microscopy images was presented and applied to the reconstruction of the structure of a cell membrane potential based on the observation of the motion of particles on the membrane.

The working principle of this method is the modeling with the linear Fokker–Planck equation of the ensemble of the stochastic trajectories of particles moving on the membrane of a cell, and the solution of an optimization problem governed by this equation, where the purpose of the optimization is to find a potential such that a least-squares best fit term of the computed and observed particles' density and a Tikhonov regularization term are minimized.

Results of numerical experiments were presented that demonstrated the ability of the proposed method to reconstruct the potential of a cell membrane by using the data of a super-resolution microscopy of luminescent activated proteins.

7. Brief Documentation of the Code

The program code has been written in C++ with the support of Armadillo, LAPACK, and others common public available linear algebra routines. The structure of the code follows the object oriented programming style, with some basic classes for the the definition of the mathematical model, numerical grids and solvers, and derived classes for the solution of the optimization problem. Notice that the code *is not computationally optimized*, that is, it has been written with the purpose to test the ability of the numerical scheme to accomplish the proposed reconstruction task. Further major improvements could be implemented, such as a better organization of the hierarchical class structure and a more efficient use of the pointers.

The package of the source codes is composed of 14 files (The code will be available in the repository of this journal as a supplemental material):

- `CALC_FIG3.cpp`, `CALC_FIG6.cpp` the main driver routines
- `Non_lin_conjug_grad.cpp/.hpp` class definitions for the nonlinear conjugate gradient method
- `Optim_problem.cpp/.hpp` class definition for the optimization problem
- `BDF_Chang_Cooper.cpp/.hpp` class definitions of the BDF2 solver associated with the Chang-Cooper method
- `Chang_Cooper.cpp/.hpp` class definitions of the Chang-Cooper numerical method
- `gradients.cpp/.hpp` support functions for the discrete gradients
- `model.cpp/.hpp` class definition for the Fokker–Planck model

For our tests, the following libraries were installed on the OS Linux Mint (20.3): Armadillo (version 12.6), Openblas (version 0.3.8), Lapack (version 3.9.0), SuperLU (version 5.2.1) and HDF5® (version 1.10.4). The `makefile` gives support for the compilation, provided that the user customizes the library paths according to his own OS configuration. The compilation process is invoked with the command `make CALC_FIG3` that produces the related executable file. The command `make clean` cleans all the object codes. At run time, the executable file loads the input data from the folder `input_data`, prints on the screen some computation information about the minimizing process, and finally saves the resulting reconstruction in data files and pictures inside automated created folders.

Here follows a brief description of the input/output data structure.

The main input data of the algorithm, namely f_d, is a 3-dimensional matrix t⬛ according to the syntax of Matlab®/Octave [32,33], has size as [Nx,Ny,Nt]=size(f⬛ where Nx,Ny are the number of histogram bins in each dimension on the domain space, a⬛ Nt is the number of time step samples of the particles trajectories, i.e., fd(:,:,n) represe⬛ the 2-dimensional matrices slice at the time step n with the values of the histograms of ⬛ particles positions. This matrix is saved in the HDF5® format [27] under the set data na⬛ value, as follows

```
name_h5 = [save_name '.h5'];
h5create(name_h5,'/value',size(fd));
h5write(name_h5,'/value',fd);
```

The library Armadillo provides the loading method.

The files CALC_FIG3.cpp and CALC_FIG6.cpp are the driver programs for calculati⬛ the figures of the papers. They differs on the input/output data name and some paramet⬛ value. Here follows the list of variables that can be customized by the user:

- base_data_dir: folder name containing the input data.
- data: file name of the input data, i.e., the trajectories, binned in histograms, of t⬛ particles. The HDF5 format is used.
- base_dir: is the parent folder name where the results of the computation will be sav⬛ This folder must be created in advance, otherwise the run-time is stopped by an err⬛ The program creates automatically the sub-folder and saves the results inside.
- save_name: is a root file name, used to create file names for saving the results of t⬛ reconstruction.
- nlcg_param_fname: file name for the file containing the parameters of the nonline⬛ conjugate gradient algorithm. See below the description.
- sig: intensity of the Gaussian noise σ in the FPE (4).
- alfa: weight α of the norm of the potential $U(x)$ in the objective functional (8).
- beta: weight of the norm of the gradient of the potential $\nabla U(x)$ in the objecti⬛ functional (8). Notice, here we assume beta= α.
- xi: weight ζ of the terminal condition in the objective functional (8).
- ax, bx, ay, by: are the boundaries of the 2-dimensional domain. It is not necessa⬛ to change the default values.
- TT: total time interval for the reconstruction. It is equal to the time of the Monte Car⬛ simulation or the time length of the sampled trajectories.
- Nx: number of grid points of the numerical domain. In this implementation it must ⬛ square.
- mux, muy, s0: parameters for the 2-dimensional Gauss PDF used as initial conditio⬛ $f_0(x)$ for the FPE (4). mux, muy are the (x,y) coordinates of the mean value, s0 is th⬛ standard deviation. If s0 $<$ 0 then the PDF is uniform.
- Nt_seq: number of windows of the Model Predictive Control. It corresponds to th⬛ number of the reconstructions computed along all available input time step data ⬛ the particles trajectories from the Monte Carlo simulation, as depicted in Figure 4.
- lam0: is the initial step length for a single cycle of the NLCG algorithm.
- kmax: maximum number of iterations for a single cycle of the NLCG algorithm.
- max_restart: maximum restart sequences of the NLCG algorithm.

Non_lin_conjug_grad.hpp/.cpp defines the class NLCG that implements the nonline⬛ conjugate gradient method. It contains some default values of parameters for the Armij⬛ condition during the linesearch algorithm (for details see [12,20]).

- BACK_TRACK 0.5 is the backtracking coefficient of the linesearch algorithm.
- MAX_LEVELS 10 is the maximum number of level search for the Armijo condition.
- GAMMA 0.1 is the coefficient for the Armijo sufficient decrease condition of the fun⬛ tional.
- EPS_CONV 1e-4 is the tolerance level for the search of the step-length

- `U_TOLERANCE 1e-4` is the tolerance level for the search of the step-length related to the control.
- `EPS_PDF 1e-8` is the minimum value of the PDF under which the control should be set vanishing (not used).

These values can be overwritten by those defined in an external file pointed out by the variable `nlcg_param_fname`. Such as an example, in the file `nlcg_params.dat` the following values are changed

- `BACK_TRACK 0.3`
- `GAMMA 0.2`

Outline of the classes usage.

The source codes `CALC_FIG3.cpp` and `CALC_FIG6.cpp` give the guidelines on how to use the classes for solving the optimization problem. After loading the input data and defining some parameters, the classes `Param` and `Model` must be instantiated. In particular, the class `Model` named `fokker_planck` that contains the mathematical model of the FPE. Afterward, it follows the definition of the Cauchy initial condition. Then it starts the iterations over the reconstruction windows related to the MPC. Each iteration solves an optimization problem, inside of that it instantiates a `Grid` class for the numerical grid and matrices for the data. The instance of the class `NLCG` provides the numerical optimizer for the reconstruction problem. The constructor of the class takes as input arguments the classes `Model` and `Grid`, and a reference to the output, i.e., the solution of the optimization stored in u. After loading the algorithm parameters, the optimization starts by invoking the public method `start_restart_sequence`. At the completion of the method, the solution is stored in u, and the final value of the PDF is set as the initial condition for the next temporal window.

Finally, notice that the classes can be easily reused, e.g., for solving the sole Fokker–Planck equation, or modified in order to implements others optimization algorithms.

Supplementary Materials: The following supporting information can be downloaded at: https://www.mdpi.com/article/10.3390/mca28060113/s1.

Author Contributions: All authors contributed equally to the formulation, analysis and the writing of the manuscript. The author M.A. made the main work in the development of the code. All authors have read and agreed to the published version of the manuscript.

Funding: This research received no external funding.

Data Availability Statement: The data presented in this study are available in Supplementary Material.

Conflicts of Interest: The authors declare no conflict of interest.

References

Betzig, E.; Patterson, G.H.; Sougrat, R.; Lindwasser, O.W.; Olenych, S.; Bonifacino, J.S.; Davidson, M.W.; Lippincott-Schwartz, J.; Hess, H.F. Imaging intracellular fluorescent proteins at nanometer resolution. *Science* **2006**, *313*, 1642–1645. [CrossRef]

Hofmann, M.; Eggeling, C.; Jakobs, S.; Hell, S.W. Breaking the diffraction barrier in fluorescence microscopy at low light intensities by using reversibly photoswitchable proteins. *Proc. Natl. Acad. Sci. USA* **2005**, *102*, 17565–17569. [CrossRef] [PubMed]

Moerner, W.E.; Fromm, D.P. Methods of single-molecule fluorescence spectroscopy and microscopy. *Rev. Sci. Instrum.* **2003**, *74*, 3597–3619. [CrossRef]

Abbe, E. Beiträge zur Theorie des Mikroskops und der mikroskopischen Wahrnehmung. *Arch. Mikrosk. Anat.* **1873**, *9*, 413–468. [CrossRef]

Saffman, P.G.; Delbrück, M. Brownian motion in biological membranes. *Proc. Natl. Acad. Sci. USA* **1975**, *72*, 3111–3113. [CrossRef] [PubMed]

Schuss, Z. *Theory and Applications of Stochastic Processes: An Analytical Approach*; Springer: Berlin/Heidelberg, Germany, 2009.

Holcman, D.; Hoze, N.; Schuss, Z. Analysis and interpretation of superresolution single-particle trajectories. *Biophys. J.* **2015**, *109*, 1761–1771. [CrossRef] [PubMed]

Dahan, M.E.B.M.M.; Masson, J.-B. InferenceMAP: Mapping of single-molecule dynamics with Bayesian inference. *Nat. Methods* **2015**, *12*, 594–595.

9. Annunziato, M.; Borzì, A. A Fokker–Planck approach to the reconstruction of a cell membrane potential. *Siam J. Sci. Comp* **2021**, *43*, B623–B649. [CrossRef]
10. Milo, R.; Phillips, R. *Cell Biology by the Numbers*; Garland Science; CRC Press: Boca Raton, FL, USA, 2015.
11. Risken, H. *The Fokker-Planck Equation: Methods of Solution and Applications*; Springer Science & Business Media: New York, N USA, 1996.
12. Annunziato, M.; Borzì, A. A Fokker-Planck control framework for multidimensional stochastic processes. *J. Comput. Appl. Ma* **2013**, *237*, 487–507. [CrossRef]
13. Grüne, L.; Pannek, J. *Nonlinear Model Predictive Control, Theory and Algorithms*; Communications and Control Engineerin Springer: Berlin/Heidelberg, Germany, 2011.
14. Annunziato, M.; Borzì, A. A Fokker–Planck control framework for stochastic systems. *EMS Surv. Math. Sci.* **2018**, *5*, 65– [CrossRef]
15. Dent, E.; Kwiatkowski, A.; Mebane, L.; Philippar, U.; Barzik, M.; Rubinson, D.A.; Gupton, S.; Van Veen, J.E.; Furman, C.; Zha J.; et al. Filopodia are required for cortical neurite initiation. *Nat. Cell Biol.* **2007**, *9*, 1347–1359. [CrossRef] [PubMed]
16. Lebrand, C.; Dent, E.W.; Strasser, G.A.; Lanier, L.M.; Krause, M.; Svitkina, T.M.; Borisy, G.G.; Gertler, F.B. Critical role Ena/VASP proteins for filopodia formation in neurons and in function downstream of netrin-1. *Neuron* **2004**, *42*, 37– [CrossRef] [PubMed]
17. Mebane, L. Portion of the Picture "Actin Cytoskeleton in a Neuronal Filopodia. Version #2", Gertler Lab at the Koch Institu MIT (2012). Available online: https://ki-images.mit.edu/2012/mebane-4 (accessed on 15 November 2023).
18. Körner, J.; Borzì, A. Second-order analysis of Fokker–Planck ensemble optimal control problems. *ESAIM COCV* **2022**, *28*, [CrossRef]
19. Lions, J.-L. *Optimal Control of Systems Governed by Partial Differential Equations*; Springer: Berlin/Heidelberg, Germany, 1971.
20. Borzì, A.; Schulz, V. *Computational Optimization of Systems Governed by Partial Differential Equations*; SIAM: Philadelphia, F USA, 2012.
21. Sanderson, C.; Curtin, R. Armadillo: A template-based C++ library for linear algebra. *J. Open Source Softw.* **2016**, *1*, 26. [CrossRe
22. Sanderson, C.; Curtin, R. A User-Friendly Hybrid Sparse Matrix Class in C++. *Lect. Notes Comput. Sci.* **2018**, *10931*, 422–430.
23. Openblas. Available online: https://github.com/OpenMathLib/OpenBLAS/wiki (accessed on 15 November 2023).
24. LAPACK Is a Software Package Provided by Univ. of Tennessee; Univ. of California, Berkeley; Univ. of Colorado Denver; ar NAG Ltd. Available online: www.netlib.org/lapack/ (accessed on 15 November 2023).
25. Demmel, J.W.; Eisenstat, S.C.; Gilbert, J.R.; Li, X.S.; Liu, J.W.H. A supernodal approach to sparse partial pivoting. *SIAM J. Mati Anal. Appl.* **1999**, *20*, 720–755. [CrossRef]
26. Li, X.S.; Demmel, J.W.; Gilbert, J.R.; Grigori, L.; Sao, P.; Shao, M.; Yamazaki, I. SuperLU Users' Guide. LBNL-44289 (199 Available online: https://portal.nersc.gov/project/sparse/superlu/ug.pdf (accessed on 15 November 2023).
27. The HDF Group. Available online: https://hdfgroup.github.io/hdf5/ (accessed on 15 November 2023).
28. Lukeš, T.; Glatzová, D.; Kvíčalová, Z.; Levet, F.; Benda, A.; Letschert, S.; Sauer, M.; Brdička, T.; Lasser, T.; Cebecauer, N Quantifying protein densities on cell membranes using super-resolution optical fluctuation imaging. *Nat. Commun.* **2017**, *8*, 173 [CrossRef] [PubMed]
29. Lebbink, M.N.; Geerts, W.J.; van der Krift, T.P.; Bouwhuis, M.; Hertzberger, L.O.; Verkleij, A.J.; Koster, A.J. Template matching a a tool for annotation of tomograms of stained biological structures. *J. Struct. Biol.* **2007**, *158*, 327–335. [CrossRef] [PubMed]
30. Rigort, A.; Günther, D.; Hegerl, R.; Baum, D.; Weber, B.; Prohaska, S.; Medalia, O.; Baumeister, W.; Hege, H.C. Automate segmentation of electron tomograms for a quantitative description of actin filament networks. *J. Struct. Biol.* **2012**, *177*, 135–14 [CrossRef] [PubMed]
31. Zhao, F.; Huang, Q.; Gao, W. Image matching by normalized cross-correlation. In Proceedings of the 2006 IEEE Internation. Conference on Acoustics Speech and Signal Processing Proceedings, Toulouse, France, 14–19 May 2006; p. II. [CrossRef]
32. MATLAB®-Mathworks®. Available online: https://www.mathworks.com/products/matlab.html (accessed on 15 Noven ber 2023).
33. GNU Octave. Available online: https://octave.org/ (accessed on 15 November 2023).

MDPI AG

Grosspeteranlage 5

4052 Basel

Switzerland

Tel.: +41 61 683 77 34

Mathematical and Computational Applications Editorial Office

E-mail: mca@mdpi.com

www.mdpi.com/journal/mca